THE STANDARD
PESTICIDE
USER'S GUIDE

THE STANDARD PESTICIDE USER'S GUIDE

Fourth Edition

Bert L. Bohmont, Ph.D.

Professor Emeritus
Pesticide Programs
College of Agricultural Sciences
Colorado State University

Prentice Hall
Upper Saddle River, New Jersey 07458

Library of Congress Cataloging-in-Publication Data
Bohmont, Bert L.
  The standard pesticide user's guide / Bert L. Bohmont. --Rev. and
enl.
    p. cm.
  Includes bibliographical references and index.
  ISBN 0-13-442443-3
  1. Pesticides--Handbooks, manuals, etc. I. Title.
SB951.B584 1996
632' .95--dc20   96-5695
                                                      CIP

Acquisitions Editor: Charles E. Stewart
Editorial/Production Supervision: Cindy Hass, Carlisle Publisher's Services
Cover Director: Jayne Conte
Manufacturing Buyer: Ilene Sanford
Managing Editor: Mary Carnis
Director of Production and Manufacturing: Bruce Johnson

 1997, 1990, 1983, 1981 by Prentice-Hall, Inc.
A Simon & Schuster Company
Upper Saddle River, New Jersey 07458

Printed in the United States of America

10 9 8 7 6 5 4

ISBN   0-13-442443-3

Prentice-Hall International (UK) Limited, *London*
Prentice-Hall of Australia Pty. Limited, *Sydney*
Prentice-Hall Canada Inc., *Toronto*
Prentice-Hall Hispanoamericana, S.A., *Mexico*
Prentice-Hall of India Private Limited, *New Delhi*
Prentice-Hall of Japan, Inc., *Tokyo*
Simon & Schuster Asia Ptc. Ltd., *Singapore*
Editora Prentice-Hall do Brasil, Ltda., *Rio de Janeiro*

Contents

3 PLANT DISEASE AGENTS   34

4 VERTEBRATE PESTS   50

5 WEEDS   73

6 PESTICIDE LAWS, LIABILITY, AND RECORDKEEPING   94

7 THE PESTICIDE LABEL   136

8 PESTICIDE SAFETY   185

9 PESTICIDES AND ENVIRONMENTAL CONSIDERATIONS   231

10 PESTICIDE FORMULATIONS AND ADJUVANTS   250

11 PESTICIDE APPLICATION EQUIPMENT   271

12 PESTICIDE EQUIPMENT CALIBRATION 324

13 PESTICIDE CALCULATIONS AND USEFUL FORMULAS 349

14 PESTICIDE TRANSPORTATION, STORAGE, DECONTAMINATION, AND DISPOSAL 371

15 INTEGRATED PEST MANAGEMENT 392

Contents

Preface

This book is a revision of *The Standard Pesticide User's Guide* last revised and copyrighted in 1990. Almost every chapter has had some additions and deletions made to bring information up-to-date. Major revisions were made in Chapter 6, to give updated information on new laws and requirements; in Chapter 7 to show new labels and labeling information; Chapter 11 incorporates new equipment and techniques; and Chapter 14 expands on the entire area of transportation, storage, decontamination, and disposal of pesticides. Chapter 15 titled "Integrated Pest Management" has been retained and is recommended reading to help pesticide users understand this important concept and to show the role of pesticides in an IPM program. Additional information has been added to the appendixes and telephone numbers and other pertinent information has been updated.

This publication is a valuable source of information to those who are new to pesticides or wish to have an understanding of the importance of pesticides in our society. Teachers should find it very useful in conveying to their students that pesticides are highly regulated chemicals that deserve respect in their uses and for their roles in food and fiber production.

Recognizing that pesticides are an essential tool in helping to control most pests, the information in this book is oriented toward those who apply pesticides. It should prove especially useful to commercial pesticide applicators, as well as employees of city, county, state, and federal agencies. Golf course superintendents, grounds maintenance supervisors, and tree nursery managers need the kinds of information contained herein. Pesticide dealers, salespeople, and consultants should also find it helpful in their work.

The science of pesticide use has become a highly specialized field. Federal laws require that individuals applying certain designated pesticides be able to substantiate or demonstrate that they have the knowledge and capabilities to use the materials safely and effectively. This book can help supply the necessary information for pesticide applicators to use pesticides in a responsible manner. Important information is presented throughout the 15 chapters as well as in the glossary and appendixes.

Information or suggestions for the use of specific pesticides for specific pest control problems is *not* included in this book. Specific control measures and recommendations have been intentionally omitted because they are subject to change and may soon become obsolete. Only current recommendations should be used, along with making sure that you are using the latest pesticide label. Pest control guides for the control of most pests are available through the State University Cooperative Extension Service in every state.

The information contained in this publication is supplied with the understanding that there is no intended endorsement of a specific product or practice, nor is discrimination intended toward any product or practice included in or omitted form this book.

ACKNOWLEDGMENTS

Information and illustrations for this book have been drawn from many sources, including publications from State Cooperative Extension Service Pesticide Programs across the United States. Some of these sources are listed in "Selected References" in Appendix A. Use of these materials is gratefully acknowledged.

I have made every effort to acknowledge all use of illustrations and materials, but may have missed an original source because of the extensive interchange of materials by Cooperative Extension Service writers, sometimes making the original source uncertain.

Special appreciation and acknowledgments are extended to the following people for their careful review of various chapters and for their constructive suggestions for improvement: Dennis M. Burchett, United Agri Services, Greeley, Colorado; Dr. Frank B. Peairs, Colorado State University; Tom Reed, TeeJet® Spray Products, Dillsburg, Pennsylvania; Robert G. Reeves, Loveland Industries, Loveland, Colorado; Dr. Larry D. Schulze, University of Nebraska; Dr. Howard F. Schwartz, Colorado State University; Edward Stearns, Environmental Protection Agency, Denver, Colorado; Dr. Philip H. Westra, Colorado State University.

Specimen labels are from the Agricultural Divisions of CIBA-GEIGY Corporation, E.I. duPont de Nemours and Company, FMC Corporation, and Sandoz Agro, Inc.s' 1995 label books. Use of the specimen labels for educational purposes is acknowledged and appreciated.

Some of the vertebrate pest illustrations are from "Furbearers of Colorado" published by the Colorado Division of Wildlife in cooperation, with the Colorado Department of Education. Weed illustrations in Chapter 5 are reductions from USDA Handbook Number 366, "Selected Weeds of the United States."

Bert L. Bohmont

Introduction to Pesticides

1

HISTORY

Historical records tell us that agricultural chemicals have been used since before the time of Christ. Ancient Egyptian records mention hemlock and aconite around 1200 B.C., and in 1000 B.C. sulfur was suggested by Homer for use on certain plants. The ancient Romans are known to have used burning sulfur to control insects. They were also known to have used salt to keep the weeds under control. The ninth-century Chinese used arsenic mixed with water to control insects. In the 1300s Marco Polo used mineral oil for treating mangy camels. Most of us have heard the phrase "the dose determines the poison." Actually, it was Paracelsus, a Swiss alchemist about 1500 A.D., who stated, "the right dose differentiates a poison from a remedy." Arsenic mixed with honey for ant bait (1669) appears to be the first use of an insect stomach poison. Not long after (1690), tobacco was used as a contact insecticide against pear insects, and in 1773 nicotine fumigation from heated tobacco proved to be effective on insect-infested plants. Early in the 1800s, pyrethrin and rotenone were discovered to be useful as insecticides for the control of many different insect species. Paris green, a mixture of copper and arsenic, was discovered in 1865 and was subsequently used to control the Colorado potato beetle. In 1882 a fungicide known as Bordeaux mixture, made from a mixture of lime and copper sulfate, was found to be useful as a fungicide for the control of downy mildew in grapes. Mercury dust was developed in 1890 as a seed treatment, and in 1915 liquid mercury was developed as a seed treatment to protect seeds from fungus diseases.

The first synthetic, organic insecticides and herbicides were discovered and produced in the early 1900s; this preceded the subsequent discovery and production of hundreds of synthetic, organic pesticides, starting in the 1940s. Chlorinated hydrocarbons

1

(cyclodienes), such as chlordane, aldrin, dieldrin, endrin, and heptachlor, came into commercial production in the 1940s. Organic phosphates (parathion, malathion, etc.) began to be commercially produced during the 1950s. In the late 1950s carbamates were developed and included insecticides, herbicides, and fungicides. During the 1960s there were trends toward specific and specialized pesticides that included systemic materials, as well as "prescription" types of pesticides. Many of the new families of pesticides being used in the 1990s are so biologically active that they are applied at rates of grams or ounces per acre. These include the pyrethroids, sulfonylureas, and imidazolinones. Most of the newer compounds offer greater safety to the user and the environment. Presently, there are approximately 750 active chemical ingredients being formulated into about 25,000 commercial preparations. Approximately 200 leading active ingredients, of the total 750 basic active chemicals, represent 90% of the agricultural uses in the United States.

CURRENT CIRCUMSTANCES

Pesticides are used by people as intentional applications to the environment for the purpose of improving environmental quality for humans, domesticated animals, and plants. Despite the fears and real problems they create, pesticides clearly are responsible for part of the physical well-being enjoyed by most people in the United States and the Western world. They also contribute significantly to the existing standards of living in other nations. In the United States, consumers spend less of their income on food (about 10%) than consumers in any other country spend. The chief reason for this is more efficient food production. Chemicals have made an important contribution in this area. In 1850 each U.S. farmer produced enough food and fiber for himself and three other persons. More than 100 years later (1960) he was able to produce enough food and fiber for himself and 24 other people, for himself and 45 other people in 1970, and for himself and 78 others in 1990. By the year 2000 each farmer will need to produce enough for more than 135 people, as world population is estimated to be more than 6 billion. There is great pressure on the farmers of the world to increase agricultural production in order to feed and clothe this extra population.

At the present time the world food supply does not satisfy the hunger of the total population. As much as one-half of the world's population is undernourished. The situation is worse in underdeveloped countries, where it is estimated that as much as three-fourths of the inhabitants are undernourished.

Agricultural Losses

In spite of pest control programs, U.S. agriculture still loses possibly 25% to 30% of its potential crop production to various pests. Without modern pest control, including the use of pesticides, this annual loss would probably double. If that happened, it is possible that (1) farm costs and prices would increase considerably, (2) the average consumer family would spend much more on food, (3) the number of people who work on farms would have to be increased, (4) farm exports would be reduced, and

(5) a vast increase in intensive cultivated acreage would be required. It has been estimated that 400 million fewer acres are required to grow food and fiber than might otherwise be required. This is said to be due to modern technology, which includes the use of pesticides.

Pest Competition

In most parts of the world today, pest control of some kind is essential because crops, livestock, and people live—as always—in a potentially hostile environment. Pests compete for our food supply, and they can be disease carriers as well as nuisances. Humans coexist with more than 1 million kinds of insects and other arthropods, many of which are pests. Insects transmit 15 major disease-causing organisms to humans. Cockroaches alone cause allergic reactions among 8% of the population. More than 25,000 people seek medical attention for fire ant stings annually. Fungi cause more than 1500 plant diseases, and there are more than 1000 species of harmful nematodes. Humans must also combat hundreds of weed species in order to grow the crops that are needed to feed our nation. Rodents and other vertebrate pests can also cause problems of major proportion. Many of these pest enemies of humans have caused damage for centuries.

Some good examples of specific increases in yields resulting from the use of pesticides in the United States are corn, 25%; potatoes, 35%; onions, 140%; cotton, 100%; alfalfa seed, 160%; and milk production, 15%.

Modern farm technology has created artificial environments that can worsen some pest problems and cause others. Large acreages, planted efficiently and economically with a single crop (monoculture), encourage certain insects and plant diseases. Advanced food production technology, therefore, actually increases the need for pest control. Pesticides are used not only to produce more food but also food that is virtually free of damage from insects, diseases, and weeds. In the United States, pesticides are often used because of public demand—supported by government regulations—for uncontaminated and unblemished food.

Environmental Concerns

In the past, pest problems have often been solved without fully appreciating the treatments and effects on other plants and animals or on the environment. Some of these effects have been unfortunate. Today, scientists almost unanimously agree that the first rule in pest control is to recognize the whole problem. The agricultural environment is a complex web of interactions involving (1) many kinds of pests, (2) relationships between pests and their natural enemies, and (3) relationships among all these and other factors, such as weather, soil, water, plant varieties, cultural practices, wildlife, and people.

Pesticides are designed simply to destroy pests. They are applied to an environment that includes pest, crops, people, and other living things, as well as air, soil, and water. It is generally accepted that pesticides specific to the pest to be controlled are very desirable, and some are available. However, these products can be expensive because of their limited range of applications.

Unquestionably, pesticides will continue to be of enormous benefit to humans. They have helped to produce food and protect health. Synthetic chemicals have been the front line of defense against destructive insects, weeds, plant diseases, and rodents. Through pest control, we have modified our environment to meet esthetic and recreational demands. In solving some environmental problems, however, pesticides have created others of undetermined magnitude. The unintended consequences of the long-term use of certain pesticides have been injury or death to some life forms. Much of the information on the effects of pesticides comes from the study of birds, fish, and the marine invertebrates, such as crabs, shrimps, and scallops. It is clear that different species respond in different ways to the same concentration of a pesticide. Reproduction is inhibited in some and not in others. Eggs of some birds become thin and break, while others do not.

Residues of some persistent pesticides apparently are *biologically concentrated.* This means that they may become more concentrated in organisms higher up in a food chain. When this happens in an aquatic environment, animals at the top of the chain (usually fish-eating birds) may consume enough to suffer reproductive failure or other serious damage. Research has shown that some pesticides decompose completely into harmless substances fairly soon after they are exposed to air, water, sunlight, high temperature, or bacteria. Many others also may do so, but scientific confirmation of that fact is not yet available. When residues remain in or on plants or in soil or water, they usually are in very small amounts (a few parts per million or less). However, even such small amounts of some pesticides or their breakdown products (which also may be harmful) sometimes persist for a long time.

Pesticides, like automobiles, can create environmental problems, but in today's world it is difficult to get along without them. Those concerned about pesticides and pest control face a dilemma. On the one hand, the modern techniques of food production and the control of disease-carrying insects require pesticides. On the other hand, many pesticides can be a hazard to living things other than pests, sometimes including people.

Human Concerns

No clear evidence exists on the long-term effects on humans from the accumulation of pesticides through the food chain, but the problem has been relatively unstudied. Limited studies with human volunteers have shown that persistent pesticides, at the normal levels found in human tissues at the present time, are not associated with any disease. However, further research is required before results are conclusive about present effects, and little information exists about the longer-term effects. Meanwhile, decisions must be made by extrapolating from the results of tests that have been done on experimental animals. Extrapolation is always risky, and the judgments concerning the chronic effects of pesticides on people are highly controversial.

Public concern about the possible dangers of pesticides is manifested in legal actions initiated by conservation groups. Pesticides, like virtually every chemical, may have physiological effects on other organisms living in the environment, including people. The majority of the established pesticides have no adverse effect on people, animals, or the environment in general as long as they are used only in the amounts sufficient to control pest organisms. Pest control is never a simple matter of applying a pesticide that removes only the pest species. For one thing, the pest population is seldom completely or permanently eliminated. Almost always there are at least a few survivors to re-create the problem later. Also, the pesticide often affects other living things besides the target species and may contaminate the environment.

There have been and continue to be unfortunate and generally inexcusable accidents where workers become grossly exposed due to improper and inadequate industrial hygiene or carelessness in handling and use. Children sometimes eat, touch, or inhale improperly stored pesticides. Consumers have been inadvertently poisoned by pesticides spilled carelessly in the transportation of pesticides in conjunction with food products. These cases are, however, no indictment of the pesticide itself or the methods employed to establish its efficacy and safety. They are solely due to the irresponsibility of the user.

Pesticides are rarely used in the form of a pure or technically pure compound, but rather are formulated to make them easy to apply. Formulations may be in the form of dust or granules that usually contain 5% to 10% of active ingredients, or wettable powders or emulsifiable concentrates that usually contain 40% to 80% active ingredients. It is important to remember that the formulations used as sprays are further diluted with water, oil, or other solvents to concentrations of usually only 1% or less before application. Therefore, the amount of active ingredient that is eventually released to the environment is generally quite small.

Few responsible people today fail to recognize the need for pesticides and the importance of striving to live with them. Several national scientific committees in recent years have stressed the need for pesticides now and in the foreseeable future. These same committees also recommended more responsible use and further investigation into long-term side effects on the environment. It is generally agreed among scientists that there is little, if any, chance that chemical pesticides can be abandoned until such time as alternative control measures are perfected.

TERMINOLOGY

Now that we understand the importance of pesticides and recognize the need for knowledge of these essential chemicals by the concerned citizen and user of pesticides, let's learn the terminology that is necessary to understand what pesticides are all about and to be able to communicate accurately.

The word *pesticide* is an all-inclusive term that includes a number of individual chemicals designed specifically for the control of certain pests. Figure 1–1 shows the wheel of pesticides, with 21 spokes denoting the generic terms for each specific purpose. All but six of these generic words end in *-cide*, which means "to kill" or "killer." Most of the specific pesticide names (insecticides, herbicides, fungicides) are specific to the pest for which they are intended to control. A few names, however, are not as readily recognized by the average person and need study to become familiar with them.

The term *pesticide,* as defined by federal and state laws, also includes chemical compounds known as growth regulators that stimulate or retard the growth of plants; those known as desiccants that speed the drying of plants and are used as an aid in

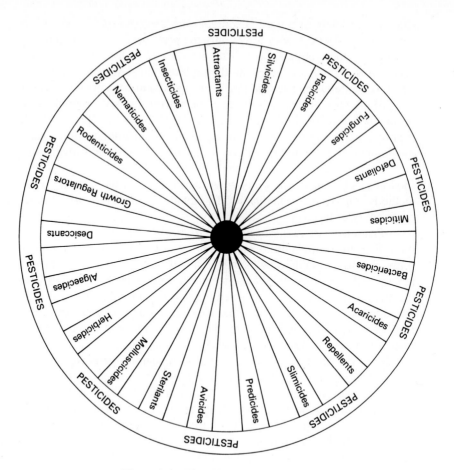

Figure 1–1 The all-inclusive pesticide wheel.

mechanical harvesting of cotton, soybeans, and other crops; and those chemicals known as defoliants that remove leaves and aid in the harvesting of potatoes and certain other crops.

The term *pesticide* also applies to compounds used for repelling, attracting, and sterilizing insects. The last two groups do not fit the original definition in that they are not "-cides." Rather, they are included only because they fit the legal definition of pesticides.

There are a few other *-cide* terms that you might encounter. The term *biocide* is sometimes used by environmental groups or others who are opposed to the use of pesticides. A true biocide would be one that kills a wide range of organisms and is toxic to both plants and animals. There are only a few pesticides that would fall in this class and, unless used carelessly or indiscriminately, they are used in specific situations with due care for humans and the environment. Therefore, the term *biocide* is incorrect when discussing materials used for specific purposes in controlling various pests.

PESTICIDES AS CLASSIFIED BY THEIR TARGET SPECIES

Acaricide	Mites, ticks
Algicide	Algae
Attractant	Insects, birds, other vertebrates
Avicide	Birds
Bactericide	Bacteria
Defoliant	Unwanted plant leaves
Desiccant	Unwanted plant tops
Fungicide	Fungi
Growth regulator	Insect and plant growth
Herbicide	Weeds
Insecticide	Insects
Miticide	Mites
Molluscicide	Snails, slugs
Nematicide	Nematodes
Piscicide	Fish
Predacide	Vertebrates
Repellents	Insects, birds, other vertebrates
Rodenticide	Rodents
Silvicide	Trees and woody vegetation
Slimicide	Slime molds
Sterilants	Insects, vertebrates

Other *-cide* terms include *adulticide, larvicide, aphicide,* and *ovicide.* These are all terms that refer to the use of insecticides. The use of insecticides that are more effective at certain stages of insect growth gives rise to the terms *adulticide* and *larvicide.* An insecticide that is more specific or best for aphid control might be referred to as an *aphicide.* There are a few insecticides that are best for destroying insect eggs and these are referred to as *ovicides.*

Some pesticides can be used for several purposes. For example, some insecticides can also be nematicides and at least one is an insecticide and also a bird repellent. Several restricted use pesticides will kill all life forms and, as mentioned earlier, need to be used with extreme care and caution.

WHO USES PESTICIDES?

During the early years of pesticide development, farmers were considered to be the primary users. As new chemicals were produced, however, new methods of formulation were developed, new application techniques were discovered, and new groups found uses for pesticides. The total U.S. annual pesticide consumption is estimated at 2.7 billion pounds of active ingredients. Of this amount, 1.6 billion pounds represents wood preservatives, disinfectants, and sulfur (a fungicide).

Today, pesticides are still a major part of agriculture's production tools, but they have also found uses by industry, state and federal governments, municipalities, commercial pesticide applicators, and the public as a whole, including homeowners and backyard gardeners. Purchases of pesticides by those who live in cities and towns is estimated to be approximately 20% of all pesticides sold in the United States. This includes household cleaners, bleaches, pet flea collars, lawn and garden products, rat

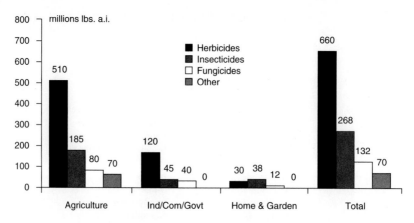

Figure 1–2 Amounts of major pesticides used.

poisons, and insect repellents. Industrial users, primarily water treatment plants, account for 20% of pesticide use. For example, chlorine, which is used to purify water, makes up 30% of the total volume of pesticides sold in California. About 10% of our pesticides are used by institutions and public areas. Hospitals and restaurants use pesticides to control germs and insects. Figure 1–2 illustrates the amounts of herbicides, insecticides, and fungicides used by the three main groups.

Pesticides are used on as many as 2 million farms, in 90 million households, and by 40,000 commercial pest control firms (a figure that includes structural, lawn care, and agricultural custom applicators). Together these users spent about $9.5 billion on all pesticides in 1995. While it is true that agriculture as an industry is still the largest single user of pesticides (about 900 million pounds), we can no longer automatically charge the agricultural segment with responsibility for all the problems these chemicals can create when used unnecessarily or irresponsibly. Of the 900 million pounds of pesticides used each year in the United States, the fungicide sulfur ranks first with 83 million pounds used. Sulfur is used by both organic and conventional growers to control crop diseases. One of the oldest pesticides, it is used at significantly higher rates than modern synthetic chemicals.

Approximately 2.7 billion pounds of pesticides are used in the United States annually. If this usage were equally distributed to all the people in the United States, it would amount to almost 10.4 pounds of pesticides for every man, woman, and child in this country! This may seem high, but if we consider the disinfectants (yes, disinfectants are pesticides) that are used in restaurants, hospitals, and elsewhere, we get a better appreciation for these chemicals and the amounts used.

PUBLIC PERCEPTION OF PESTICIDES

The general public, most of whom are far removed from daily food production, have a poor understanding of pesticides. This is partly due to lack of understanding of chemical uses and agricultural production practices, but it is also due to fear and misunderstanding brought about through publicity of accidents and misuses involving pesticides. The perception of widespread hazards associated with presumed long-term accumulations of pesticides in people, other organisms, or the environment often appears to stem from lack of knowledge of the processes of metabolism, elimination, and degradation that largely preclude such perceived problems.

A study done about 10 years ago vividly illustrates the misperception the general public has concerning the risk from pesticides compared to other risk situations. Three groups (college students, League of Women Voters, and business people) were asked to rate various risks on a scale from 1 to 30. Figure 1–3 illustrates the results. The left-hand column indicates the factual placement of representative risks based on mortality data. The right-hand column indicates where the different groups placed pesticides on the 1 to 30 scale. Pesticides actually fall in the number 28 slot, although all three groups mistakenly placed them much higher on the scale. Motor vehicles of all kinds kill approximately 50,000 people annually, 3000 die while swimming, and 1000 people die annually in bicycle accidents. Approximately 30 people die each year from pesticides, which includes the tragic accidental ingestions mostly by children.

Individuals using pesticides and those concerned about pesticide use must seek all the facts and become better informed about the benefits as well as the risks of using these technological tools.

ACRONYM DICTIONARY

We rarely read a government document (or many popular articles for that matter) that does not contain abbreviations or letters standing for the name of an agency, law or amendment, or practice or measurement. These are known as *acronyms* (words formed from the initial letters of other words). They are not always pronounced as a word, but may be stated simply as letters. For example, a Washingtonian may say, "I work for the EPA as an ALJ and I'm going to the AAPCO meeting this afternoon to discuss FIFRA and EUPs. Tomorrow I will meet with FDA and USDA representatives to discuss how to implement the ESA and SARA."

The following AD (acronym dictionary) should help you sort out the alphabet soup and have a better understanding of federalized jargon.

AAPCO	Association of American Pesticide Control Officials, Inc.
ADI	Acceptable daily intake (pesticide intake in consumer diet)
ALJ	Administrative law judge
ACPA	American Crop Protection Association
ANPRM	Advance notice of proposed rulemaking
APHIS	Animal and Plant Health Inspection Service
ARAR	Applicable or relevant and appropriate requirement
BAT	Best available technology
BDAT	Best demonstrated available technology
BMP	Best management practices
CAA	Clean Air Act
CEQ	Council on Environmental Quality
CERCLA	Comprehensive Environmental Response, Compensation and Liability Act ("Superfund")
CFSA	Consolidated Farm Service Agency
CHEMTREC	Chemical Transportation Emergency Center
CMA	Chemical Manufacturers Association

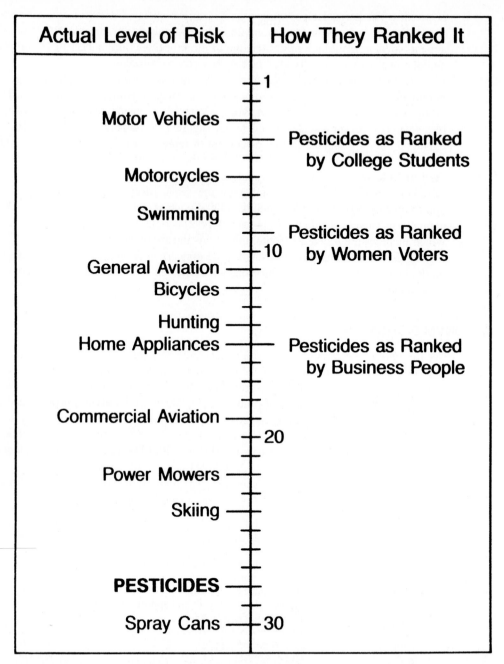

Actual Level of Risk	How They Ranked It
	1
Motor Vehicles	
	Pesticides as Ranked by College Students
Motorcycles	
Swimming	
	Pesticides as Ranked by Women Voters
General Aviation	10
Bicycles	
Hunting	
Home Appliances	Pesticides as Ranked by Business People
Commercial Aviation	
	20
Power Mowers	
Skiing	
PESTICIDES	
Spray Cans	30

Rating the pesticide risk. On the left are actual levels of risk of accidental death from various causes as established for insurance purposes. On the right are the risks from pesticides as ranked by college students, women voters, and business people. Reprinted Science of Food and Agriculture, January 1987.

Figure 1–3   Public perception of pesticides.

CPSC	Consumer Product Safety Commission
CRP	Conservation Reserve Program
CSREES	Cooperative State Research, Education, and Extension Service
CWA	Clean Water Act
DOL	Department of Labor
DOT	Department of Transportation
EPA	Environmental Protection Agency
OCM	Office of Compliance Monitoring (enforcement)
OPTS	Office of Pesticide and Toxic Substances
OPP	Office of Pesticide Programs
OSW	Office of Solid Waste
ODW	Office of Drinking Water
OES	Office of Endangered Species
EPCRA	Emergency Planning and Community Right to Know Act
ESA	Endangered Species Act
EUP	Experimental Use Permit
FDA	Food and Drug Administration
FFDCA	Federal Food, Drug, and Cosmetic Act
FIFRA	Federal Insecticide, Fungicide, and Rodenticide Act
FR	Federal Register
FSA	Farm Service Agency
FWS	Fish and Wildlife Service
GAO	General Accounting Office
GLP	Good Laboratory Practice
GRAS	Generally recognized as safe
GRGL	Groundwater residue guidance level
GWPS	Groundwater protection strategy
HCS	Hazard Communication Standard
HHS	Health and Human Services Department
IPM	Integrated pest management
LUST	Leaking underground storage tanks
MCL	Maximum contaminant levels
MCLG	Maximum contaminant level goals
MSDS	Material Safety Data Sheets
NAPIAP	National Agricultural Pesticide Impact Assessment Program
NAS	National Academy of Sciences
NASDA	National Association of State Departments of Agriculture
NEPA	National Environmental Protection Act
NIOSH	National Institute for Occupational Safety and Health
NIH	National Institutes of Health
NOEL	No-observed-effect level
NPDES	National Permit Discharge Elimination System
NPIRS	National Pesticide Information Retrieval System
NPRM	Notice of proposed rulemaking
NRCS	Natural Resources Conservation Service
OMB	Office of Management and Budget
OSHA	Occupational Safety and Health Administration
OTA	Office of Technology Assessment
PIMS	Pesticide Incident Monitoring System
PQA	Plant Quarantine Act
PRP	Potentially responsible party
P&TCN	Pesticide and Toxic Chemicals News
RCRA	Resource Conservation and Recovery Act

RMCL	Recommended maximum containment levels
SAB	Science Advisory Board (EPA)
SAP	Scientific Advisory Panel (FIFRA)
SARA	Superfund Amendments and Reauthorization Act
SDWA	Safe Drinking Water Act
SFIREG	State/FIFRA Issues Research and Evaluation Group
SWMU	Solid Waste Management Unit
TLV	Threshold limit value
TME	Test marketing exemption
TPQ	Threshold planning quantity
TSCA	Toxic Substances Control Act
USDA	United States Department of Agriculture
USDI	United States Department of the Interior
USDOT	United States Department of Transportation
USGS	United States Geological Survey
UST	Underground storage tanks
WPS	Worker Protection Standard

PUTTING NUMBERS IN PERSPECTIVE

We often see figures representing parts per million, parts per billion, or even parts per trillion being used to report measurements of pesticides. Some perspective on the pertinent analytical numbers is important. Before we go on to the next chapter, let's consider some examples using objects that are more familiar to us.

One part per million equals:
1 penny in $10,000
1 ounce of sand in 31¼ tons of cement
1 ounce of dye in 7530 gallons of water
1 square foot in 22.957 acres
1 second in 11.574 days
1 minute in 694.445 days

One part per billion equals:
1 penny in $10,000,000
1 square foot in 36.085 square miles
1 pinch of salt in 10 tons of potato chips
1 inch in a 16,000-mile trip
1 second in 11,575 days
1 bogey in 3,500,000 golf tournaments

One part per trillion equals:
1 penny in $10,000,000,000
1 square inch in 249.1 square miles
1 second in 317 centuries
1 postage stamp in an area the size of Dallas, Texas
1 second in 11,574,079 days

Putting these statistics in more familiar terminology may help the pesticide user and the concerned public have a better understanding about the numbers used in the measurement of chemicals in food, water, and the environment.

2
Insects

Success in a hostile environment is determined by our ability to adapt to or change our surroundings to our benefit. An area in which people do not exist has no pests. *Pest* is an artificial concept and is generally considered to include those organisms that come into conflict with people by competing with us for our crops and livestock, affecting our health or comfort, and destroying our property. Many species of insects are important pests that affect almost all our activities.

INSECTS AS PESTS

There are well over 1 million known species of insects in the world. Some are of public health importance, some cause wood and structural damage, and some are nuisance pests. A very small percentage of insects are considered to be economically important pests. However, this relatively small number of insect species causes an estimated annual loss of $4 billion in the United States. They damage crops by attacking the seed, cutting off young plants, chewing foliage, sucking sap, boring and tunneling in stems and branches, and transmitting diseases. After the crop is harvested and stored, it is subject to damage by stored-product insects.

What Is an Insect?

Insects are spineless; that is, they are invertebrates and belong to a group of organisms called Arthropoda. These animals are characterized by having a segmented exoskeleton and jointed appendages. Examples of arthropods other than insects are the spiders, ticks, and mites; the lobsters, crayfish, and crabs; and the centipedes and millipedes.

These are classes within the larger group, or Arthropoda. Therefore, insects are categorized as follows:

Kingdom: Animal
Phylum: Arthropoda
Class: Insecta or Hexopoda

The class *Insecta* is characterized as follows:

1. Segmented exoskeleton divided into three body regions: the head, thorax, and abdomen
2. Three pair of legs on the thorax
3. May have one or two pair of wings (some have no wings)
4. One pair of antennae

BASIC ANATOMY

It is necessary to know how an insect is "put together" to describe it accurately or to understand the way in which insects are identified. Proper identification is basic to controlling the pest.

The segmented exoskeleton of an insect is divided into three parts: the head, thorax, and abdomen (Figure 2–1). On the head are found the eyes (usually compound eyes), the antennae, and the mouthparts. The head is like a hollow capsule formed by the fusion of a number of segments into plates called *sclerites*. The characteristics of the mouthparts and antennae are used in insect identification.

The thorax is made up of three ringlike segments: the prothorax, mesothorax, and metathorax. There is a pair of legs on each thoracic segment. When wings are present, they arise from the mesothorax and metathorax. In some insects part of the prothorax extends back dorsally over the other thoracic segments and is called a *pronotum*.

The thoracic segments are joined by flexible membranes that permit movement. Located on each side of the mesothorax and metathorax in the membranes are openings, or *spiracles,* through which the insect respires. The characteristics of the legs, wings, and thorax are used a great deal in identification.

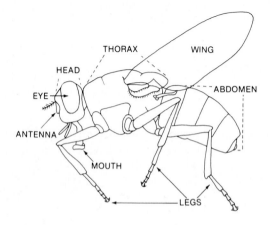

Figure 2–1 Generalized insect.

The abdomen consists of up to 11 ringlike segments. In most insects there are fewer than 11 distinguishable segments in the abdomen. Each segment is composed of two plates or sclerites: one above, or dorsal; and one below, or ventral. The plates and segments are joined by membranes that permit movement. There is usually one spiracle on each side of each of the first seven or eight abdominal segments. There are frequently appendages on the end of the abdomen.

If we put this all together, we have a generalized type of insect. The various modifications in this basic structure found in different insects are used in classification.

Digestive System

Insects have a highly developed digestive system. Starting with the mouth, the food passes into the pharynx through the esophagus into a crop. From the crop it moves to a stomach (called the *proventriculus),* then into a mid-gut, an ileum, a colon, and a rectum. The wastes are discharged from the anus. This is a generalized form and there are many highly specialized types of digestive tracts found in insects.

Circulatory System

Insects have an open blood system that functions to transport food and wastes. It is not involved in the movement of oxygen, as is the blood system of mammals. There is only one blood vessel, the dorsal aorta and heart, which is a tubelike structure extending through the thorax and abdomen. The blood is pumped forward and out of the front end of the tube. It then flows back through the body cavity and appendages and reenters the tube through slits from where it is again pumped forward.

Nervous System

There is a nerve cord on the ventral side of the body of insects instead of along the dorsal surface or "back," as in the case of vertebrate animals. This nervous system starts with a "brain" in the head and a series of ganglia throughout the length of the body connected by the ventral nerve cord. Nerve cells and the branching processes from them reach the muscles, glands, and organs.

Respiratory System

Insects have a unique system to obtain oxygen. It might be compared to a ventilation system using inlets and ducts. Along the sides of the thorax and abdomen are spiracles that are openings through which air is taken in and carbon dioxide is discharged. The spiracles connect to a complex system of tubes called *tracheae*. The tracheae divide and branch and finally end in tiny tubes called *tracheoles* that carry the oxygen directly to the tissues. They also pick up carbon dioxide from the tissues and carry it to the tracheae and then out the spiracles. The spiracles of some insects are equipped with valves that open or close.

Mouthparts

An understanding of the mouthparts and feeding habits of insects is important in selecting effective control measures and as an aid in identification. There are two major types of mouthparts: chewing and sucking.

Figure 2–2   Chewing mouthparts.   **Figure 2–3**   Sucking mouthparts.

Chewing mouthparts (Figure 2–2) are generally composed of a labrum (upper lip); a pair of cutting, crushing, or pinching mandibles; a pair of maxillae; a labium (lower lip); and a tonguelike hypopharynx. The mandibles and maxillae, or jaws, work sideways and are used to cut off and chew or grind solid food. A typical example is the type of mouthparts found in a grasshopper or cricket. In some forms of insects, mainly predators, the mandibles are long and sickle-shaped. In others, such as honey bees, the hypopharynx, or tongue, is greatly modified.

Insects with chewing mouthparts include the adult Orthoptera (grasshoppers and crickets), Odonata (dragonflies and damselflies), Neuroptera (lacewing flies), Coleoptera (beetles), Hymenoptera (bees, ants, wasps), and the larvae of these and other orders.

Sucking mouthparts (Figure 2–3) are those in which the parts we have described are highly modified into some form of organ for securing liquid food. They may be piercing-sucking as in the mosquitoes, true bugs, aphids, and stable flies; lapping or sponging, as in the house fly; rasping-sucking, as in the thrips; or tubelike, as in the moths and butterflies. Adults of the following orders have sucking mouthparts: Thysanoptera (thrips), Hemiptera (bugs), Homoptera (aphids, scale insects, leafhoppers), Diptera (flies and mosquitoes), Siphonaptera (fleas), Anoplura (sucking lice), and Lepidoptera (moths and butterflies).

Damage from chewing insects (Figure 2–4)

1. *Defoliators:* pests that chew portions of leaves or stems, stripping or chewing the foliage of plants (leaf beetles, flea beetles, caterpillars, cutworms, grasshoppers)
2. *Borers:* pests with chewing mouthparts that bore into seeds, stems, tubers, fruit trees, and others (corn borer, white grubs, granary weevil, coddling moth)
3. *Leaf miners:* pests that bore into and then tunnel in between epidermal layers (blotch leaf miners, serpentine miners)
4. *Root feeders:* pests that feed on and damage the roots and underground portions of the plant (seed and root maggots, corn rootworm, wireworms)

Damage from piercing and sucking insects (Figure 2–4)

1. *Distorted plant growth:* pests that cause leaves, fruit, or stems to wilt, curl, or become distorted (aphid injury, spruce gall, conelike growths, catfacing of peaches from lygus bugs)
2. *Stippling effect on leaves:* pests that may leave many small discolored spots on the leaves, which eventually turn yellow (spider mite injury)

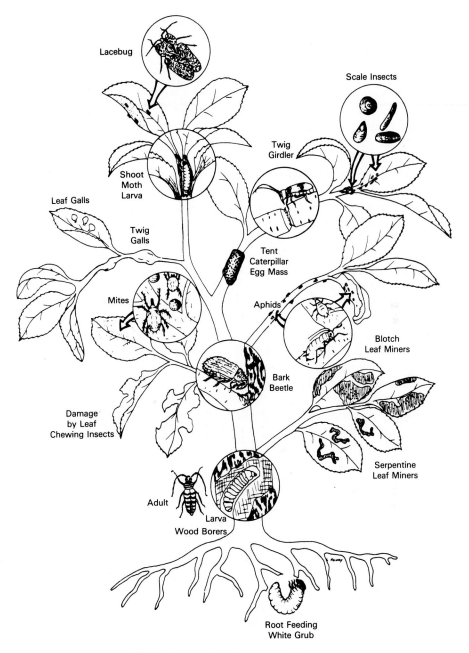

Figure 2–4 Ways in which insects affect plants.

3. *Burning on leaves:* pests that secrete toxic secretions in the host tissue, causing foliage to appear burned (leafhopper injury or "hopper burn," greenbug injury on sorghum and wheat)

DEVELOPMENT AND METAMORPHOSIS

Most insects change from the time they hatch from eggs until they are fully grown. This change in form is called *metamorphosis*. It may range from a rather gradual change, involving little more than an increase in size, to a very dramatic difference between the young and the old. There are several ways of characterizing the types of metamorphosis, but the most generally used method is to divide them into *gradual* (incomplete) and *complete*.

In gradual types of metamorphosis (Figure 2–5) the insects that hatch from the eggs are called *nymphs*. As they feed and grow, they shed their skins, or molt. In the winged species, wings first appear as padlike buds on the nymphs. Each stage between molts is referred to as an *instar*. There is no prolonged resting period before the adult stage is reached. The common orders of insects that have incomplete or simple metamorphosis are Odonata (dragonflies), Orthoptera (grasshoppers, crickets), Isoptera (termites), Mallophaga (chewing lice), Anoplura (sucking lice), Thysanoptera (thrips), Hemiptera (bugs), and Homoptera (aphids, leafhoppers). Incomplete metamorphosis is also referred to as *simple* metamorphosis as opposed to the more complicated complete metamorphosis.

Complete metamorphosis (Figure 2–6) involves a major change in form between the young and adult. In the winged forms, the wings develop internally instead of externally. The typical development involves the egg, larva, pupa, and adult. The larvae may go through a number of instars and molts as they grow. The pupae may take several forms; they may be exposed or contained in a capsulelike *puparium* or in a silken cocoon.

Some orders having this type of development are Neuroptera (lacewings), Coleoptera (beetles), Lepidoptera (moths and butterflies), Diptera (flies and mosquitoes), Siphonaptera (fleas), and Hymenoptera (bees, ants, and wasps).

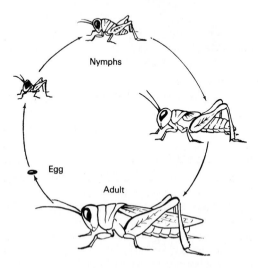

Nymphs

Egg

Adult

Figure 2–5 Incomplete metamorphosis.

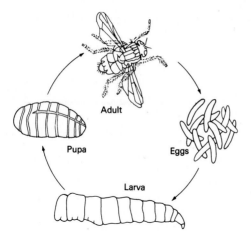

Figure 2–6 Complete metamorphosis.

CLASSIFICATION AND IDENTIFICATION

Proper identification of the pest is basic to its control. Various aids are available, including pictures, written descriptions, and keys. Keys are the best means of identifying organisms of any kind. There are pictorial, field, and couplet keys. The following is an example of an analytical couplet key to the major orders of insects. To use such a key the specimen is examined for the characteristics described after number 1. A choice must be made and this results in a referral to the next couplet, either 2 or 15 in this case. The process is continued until the name of the order is reached.

KEY TO SELECTED ORDERS OR ADULT INSECTS

1. Wings absent 2
  Wings present 15

2. Sedentary insects, legs usually absent; body scalelike or grublike and covered with a waxy secretion (females of many Coccidae)...HOMOPTERA
  Not sedentary; with distinct head and jointed legs...3

3. Mouthparts adapted for biting and chewing...4
  Mouthparts not adapted for biting and chewing; piercing-sucking, rasping-sucking, or siphoning type 9

4. Body covered with scales or mouthparts retracted within the head so that only their apices are visible..5
  Body not covered with scales; mouthparts fully exposed 6

5. Abdomen composed of not more than six segments; usually provided with a spring near the caudal end; first abdominal segment with a forked appendage (sucker) on ventral surface ....COLLEMBOLA
  Abdomen composed of 10 or 11 segments; abdomen may terminate in long caudal appendages or immovable forcepslike appendages, body may be covered with scales........ ....THYSANURA

6. Small louselike insects, usually markedly flattened, soft, or leathery in appearance...........
....MALLOPHAGA
Not louselike in form; not markedly flattened; exoskeleton well differentiated.............7

7. Abdomen sharply constricted at base..HYMENOPTERA
Abdomen not sharply constricted at base, broadly joined to the thorax............................8

8. Abdomen with cerci, body not antlike in form...ORTHOPTERA
Abdomen without cerci, body antlike in form..ISOPTERA

9. Mouthparts consisting of a proboscis coiled up beneath the head; body more or less covered with long hairs or scales..LEPIDOPTERA
Mouthparts not as described above; body not covered with long hairs or scales 10

10. Body strongly compressed; legs nearly always long and fitted for jumping ....
....SIPHONAPTERA
Body not strongly compressed; legs fitted for running...11

11. Mouthparts consisting of a jointed beak (the labium) with which are the piercing stylets
....12
Mouthparts consisting of an unjointed flesh or horny beak or the beak may be absent..13

12. Beak arising from the anterior end of the head .... HEMIPTERA
Beak arising from the posterior end of the head, apparently between the coxae of the front legs ....HOMOPTERA

13. Tarsi with the apical joints terminating in bladderlike enlargements; well-defined claws absent; mouthparts forming a triangular or cone-shaped, unjointed beak ....
....THYSANOPTERA
Tarsi not as described above; with well-developed claws 14

14. Antennae hidden in pits, not visible in dorsal view; tarsi with two claws...........DIPTERA
Antennae exposed, not hidden in pits, visible in dorsal view, tarsi with one claw..............
....ANOPLURA

15. Two wings present, hind legs represented by halters ....DIPTERA
Four wings present..16

16. Fore wings and hind wings similar in texture, usually membranous.................................17
Fore wings and hind wings dissimilar in texture; fore wings thickened, leathery, or horny; hind wings membranous 22

17. Wings entirely or for the most part covered with scales; mouthparts consist of a coiled tube beneath the head, formed for siphoning ....LEPIDOPTERA
Wings not covered with scales; mouthpart not as described above.................................18

18. Wings long and narrow with only one or two veins or none; last joint of tarsus bladderlike
....THYSANOPTERA
Wings not as described above; last joint of tarsus not bladderlike...................................19

19. Mouthparts enclosed in a jointed beak (labium) and fitted for piercing and sucking; beak arises from the rear of the head, apparently between the front coxae.........HOMOPTERA
Mouthparts not enclosed in a jointed beak; not arising from the rear of the head...........20

20. Wings with few longitudinal veins, not net-veined or they may be veinless ....
....HYMENOPTERA
Wings with many longitudinal veins and cross veins, appearing net-veined (12 or more cross veins)..21

21. Antennae short, setiform or setaceous..ODONATA

Antennae longer, not setiform or setaceous ....NEUROPTERA

22. Mouthparts adapted for chewing and biting...23

Mouthparts adapted for piercing and sucking 24

23. Front wings horny or leathery, lacking veins (elytra) ....COLEOPTERA

Front wings parchmentlike with a network of veins (tegmina) ....ORTHOPTERA

24. Fore wings thickened at base, membranous and generally overlapping at the tips (hemely-tra); beaklike mouthparts arise from the anterior end of the head.................HEMIPTERA

Front wings not thickened at base, uniform in texture, not overlapping at tips; beaklike mouthparts arise from the posterior end of the head apparently between the fore coxae ....

....HOMOPTERA

    Keys to larvae of certain insects are also helpful as the larvae are often the only destructive stage. Many bulletins and leaflets are available from cooperative extension offices. USDA publications are usually very good, and industry often has superlative written materials, pictures, videotapes, and films to aid in identification. Pictured in Figures 2–7 through 2–19 are insects that belong to the major orders.

    To make it easier to understand and remember the orders of insects, their common names are given along with the order to which they belong. Identifiable characteristics of the adult insects are also listed.

INSECT RELATIVES

Mites, ticks, and spiders (Figures 2–20, 2–21, and 2–22) are all related to the insects. The spiders are perhaps the best-known members of this class of arthropods, the Arachnida. Although they bear superficial similarities, the arachnids are unlike insects. They have eight jointed legs; two major body regions rather than three; no antennae; simple, small eyes; and they never have wings. Chemical control measures include the use of miticides and acaricides.

CENTIPEDES AND MILLIPEDES

Centipedes (Figure 2–23), often called hundred-legged worms, are predators that feed on small insects and spiders. They have 15 or more pair of legs. They belong to the genus *Chilopoda* and are not insects.

    Millipedes (Figure 2–24), genus *Diploda,* sometimes referred to as thousand-legged worms, are wormlike animals that have 30 or more pair of legs. They feed on decaying plant material, but occasionally damage roots of young plants.

SNAILS AND SLUGS

Snails and slugs (Figure 2–25) are included in this section because they frequent many of the same places as insects, and entomologists are most often called on for information

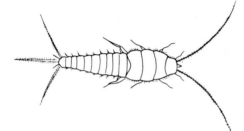

Figure 2–7 Order Thysanura (bristletails).
- No wings
- Usually two or three long tails
- Simple metamorphosis
- Often found in houses
- Feed on paper, cloth, and starches
- Are sometimes called "silverfish"

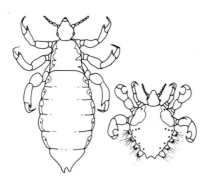

Figure 2–8 Order Anoplura (sucking lice).
- No wings
- Piercing-sucking mouthparts
- Simple metamorphosis
- Feed by sucking blood of animals and people

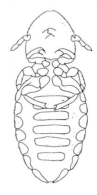

Figure 2–9 Order Mallophaga (chewing lice).
- No wings
- Chewing mouthparts
- Broad head
- Simple metamorphosis
- Feed on birds, where they are found on skin and feathers

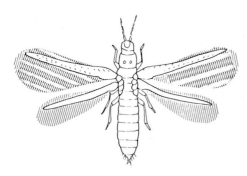

Figure 2–10 Order Thysanoptera (thrips).
- May have fringed wings, may lack wings
- Have a combination of sucking and chewing mouthparts
- Simple metamorphosis
- Usually feed in flowers or buds of plants
- Cause misshaped flowers, buds, fruits, and leaves

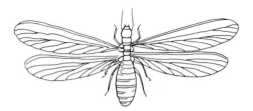

Figure 2–11 Order Isoptera (termites).
- Four wings of equal size, or no wings at all
- Chewing mouthparts
- Simple metamorphosis
- Feed on wood and wood products
- Live in complex societies

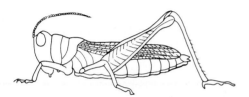

Figure 2–12 Order Orthoptera (grasshoppers, crickets, cockroaches, and katydids).
- Two pair of wings, or no wings at all
- Chewing mouthparts
- Simple metamorphosis
- Pests of agriculture and the home

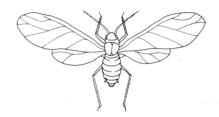

Figure 2–13 Order Homoptera (aphids, leafhoppers, and scale insects).
- May or may not have wings
- Piercing-sucking mouthparts
- Simple metamorphosis
- Suck juices from plants and carry plant diseases

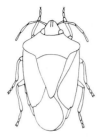

Figure 2–14 Order Hemiptera (true bugs).
- Two pair of wings, or no wings at all
- Top pair of wings partly leather, partly transparent
- Piercing-sucking mouthparts
- Simple metamorphosis
- Suck juices from plants and blood from animals and people

Figure 2–15 Order Lepidoptera (moths and butterflies).
- Two pair of wings, usually with scales
- Chewing mouthparts (larvae), sucking mouthparts (adults)
- Complete metamorphosis (larvae called "caterpillars" or "worms")
- Feed on many crops, damaging leaves, stems, tubers, and fruits

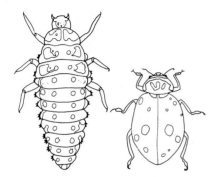

Figure 2–16 Order Coleoptera (beetles).
- Two pair of wings
- Top pair of wings hard and opaque, bottom pair transparent
- Chewing mouthparts
- Complete metamorphosis
- Larvae and adults feed on plants, in soil, in wood, and many other places
- Larvae and adults may be pests

Classification and Identification

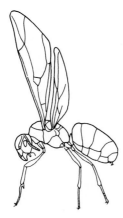

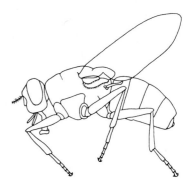

Figure 2–17 Order Hymenoptera (wasps, bees, ants, and sawflies).
- Two pair of wings or no wings
- Chewing mouthparts
- Adults usually have narrow waists
- Complete metamorphosis
- May have complex social organization

Figure 2–18 Order Diptera (flies and mosquitoes).
- One pair of wings
- Piercing-sucking mouthparts variously modified
- Complete metamorphosis (larvae called "maggots")
- May feed in rotting material, plant parts, or animal flesh
- May carry animal and human diseases

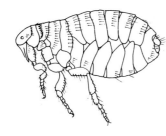

Figure 2–19 Order Siphonaptera (fleas).
- No wings
- Piercing-sucking mouthparts
- Complete metamorphosis
- Feed on blood
- May transmit animal diseases

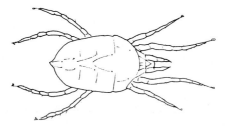

Figure 2–20 Order Acarina (mites).
- Usually very small
- Sucking mouthparts
- Damage plants through feeding; attack animals and people by irritating skin
- May transmit plant and animal diseases

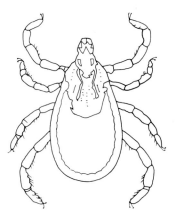

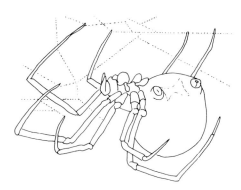

Figure 2–21 Order Acarina (ticks).
- Leathery soft body, without distinct head
- Sucking mouthparts
- Parasitic on animals and people
- May transmit animal and human diseases

Figure 2–22 Order Acarina (spiders).
- Fanglike sucking mouthparts
- Most are beneficial; a very few are poisonous

Centipede
($\frac{1}{2}$–$2\frac{1}{2}$ inches)

Figure 2–23 Centipede.

Millipede
($\frac{1}{2}$–$1\frac{1}{2}$ inches)

Figure 2–24 Millipede.

about them and for control recommendations. Snails and slugs are members of the large group of animals called *mollusks*. They often become pests around the home garden and in lawns, greenhouses, and ornamental plantings. These animals may damage plants by feeding on the foliage, by their presence in the plant after harvest, and by the slime trails they make as they move from place to place on their muscular "foot."

Snails are soft animals whose bodies are protected by a characteristic coiled shell. Because of the shell, snails are able to survive in fairly dry conditions. They generally reproduce by laying eggs. Control measures include the use of chemicals called *molluscicides*.

Snails and Slugs

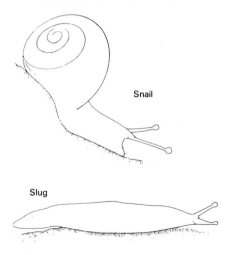

Figure 2–25 Slug and snail.

INSECT OUTBREAKS

Insect populations that are successful have two main factors that permit them to build up to large numbers: adaptability and reproductive capacity. Working against this force are a number of hazards or limiting factors in the environment. When the effect of limiting factors is low, outbreaks can result. People have had a direct influence on the environment, which may favor insects on the one hand or work against them on the other. For example, the development of large acreages of relatively few crops, a monoculture, has resulted in the provision of unlimited food supplies for certain insects.

The reproductive ability of most insects is difficult to imagine. Large numbers of eggs are usually produced and with a few exceptions the life cycles are relatively short. Even with insects that have one-year life cycles, hundredfold increases may occur in one season.

Insects have an amazing ability to adapt to changing conditions. This even includes factors that humans introduce into the environment—for example, the phenomenon of resistance to insecticides.

This all adds up to the fact that economically important numbers of certain insects can develop in a relatively short time and special techniques for their control must be developed. Examples of destructive, economically important insects are shown in Figure 2–26(a) and (b).

STEPS TO INSECT CONTROL

An orderly process of decision making must be used intelligently to effectively plan and carry out an insect control operation. The following steps are those that would be followed in handling a new or unfamiliar insect problem. For well-known insects, some of the steps would be passed over.

1. *Detection:* It is necessary to be watchful for those insects likely to be troublesome to a certain crop. It is important to detect infestations early before irreparable

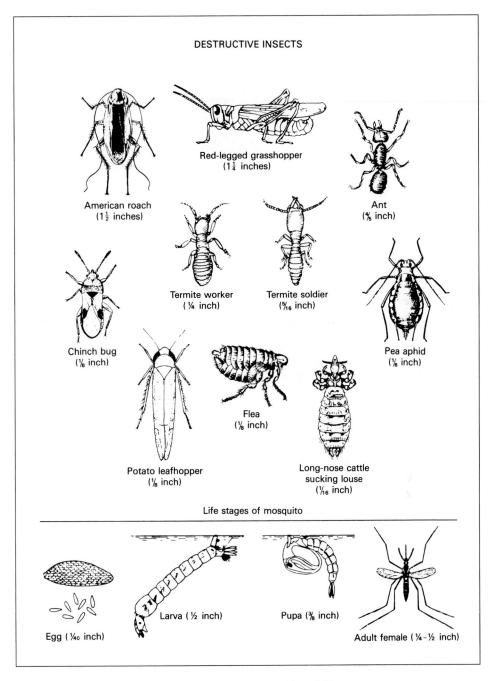

DESTRUCTIVE INSECTS

American roach
(1½ inches)

Red-legged grasshopper
(1¼ inches)

Ant
(⅘ inch)

Chinch bug
(⅛ inch)

Termite worker
(¼ inch)

Termite soldier
(5⁄16 inch)

Pea aphid
(⅙ inch)

Potato leafhopper
(⅛ inch)

Flea
(⅛ inch)

Long-nose cattle
sucking louse
(1⁄16 inch)

Life stages of mosquito

Egg (1⁄40 inch)

Larva (½ inch)

Pupa (⅜ inch)

Adult female (¼ - ½ inch)

Figure 2–26(a)   Destructive and harmful insects.

Steps to Insect Control

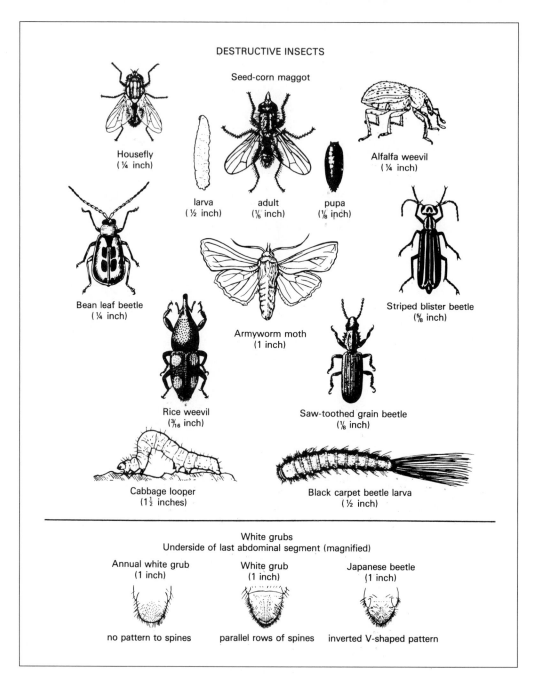

Figure 2–26(b) Destructive and harmful insects.

damage is done. In most cases this cannot be done through the windshield. It requires careful examination in the field. Areas threatened by new insects moving in from adjacent areas may require special detection methods.

2. *Identification:* A pest must be identified accurately before it can be controlled. Just because a certain practice worked on an insect that "looked something like it" is no reason it will be effective on this one. If local authorities are unable to identify the problem, assistance is available from the land grant universities or the U.S. Department of Agriculture.

3. *Biology and habits:* Knowledge of the seasonal cycles for the specific locality is important in order to pinpoint the most effective time of treatment. The principle is to determine the vulnerable stages in the insect's life cycle against which to direct the control effort to obtain the most effective, economical control.

4. *Economic significance:* Control efforts are not undertaken just for the sheer enjoyment of killing insects. There must be a return on the investment. For some insect pests, we have sufficient knowledge so that we can offer guidelines and survey methods for determining population levels required to make a given treatment pay. More studies are needed to obtain this information for more pests.

5. *Selection of methods:* After the pest problem has been identified, the life history and seasonal cycle understood, and the economic significance established, the proper method or combination of methods can be selected to do the most effective, practical, economical, and safe job of control.

6. *Application:* The control methods selected must be applied properly to be effective and safe. If the method involves the use of chemicals, the application must be done at the proper time, in the best place, at the right rate, and using suitable equipment. This means using the proper sprayer, duster, or other equipment for the formulation used and the type of coverage or delivery desired. This information is specified in the directions for use on pesticide labels and in written recommendations.

7. *Evaluation:* It is extremely important to check the field or otherwise evaluate the results of the control operation. This can be done in several ways, such as insect counts before and after treatment, comparative damage ratings, and yields. In most cases, it is difficult if not impossible to do an adequate evaluation without untreated checks to use as a basis for comparison. The results obtained should be recorded for future reference.

8. *Recording:* Records provide the basis for gaining from past experiences. It is especially important to have all chemical applications recorded.

METHODS OF INSECT CONTROL

1. Cultural
   a. Crop rotations
   b. Tillage methods
   c. Resistant or tolerant varieties

2. Mechanical
   a. Screens, traps
   b. Light and sound

3. Biological
   a. Parasites
   b. Predators
   c. Diseases
   d. Male sterile technique
   e. Temperature
   f. Moisture

4. Legal
   a. Inspections and quarantines
   b. Plant pest laws

5. Chemicals
   a. Kill
   b. Repel
   c. Attract
   d. Sterilize

6. Integrated control: combination of all possible compatible techniques to manage insect populations at subeconomic levels, usually involving the use of artificial or natural biological methods along with selective chemicals.

The real challenge in pest control today lies in making decisions for each situation and fitting the available techniques into a program that results in the maintenance of subeconomic levels of pest species and causes the least amount of insult to the environment. Often, troublesome insects can be held in check by natural predators or through the introduction of useful predator insects. Some examples of beneficial insects are shown in Figure 2–27.

USE OF CHEMICALS IN INSECT CONTROL

The development of effective, economical pesticides has had a profound effect on our continual battle with insects. In many cases, chemicals have been incorporated as tools in well-planned insect control programs without serious hazards to humans or to the environment. However, there has been a tendency for some people to regard these tools as cure-alls and to believe that for every insect problem there is a chemical that will magically solve it. This has resulted at times in some unwise, uneconomical, or hazardous applications of chemicals.

Certain insects have developed resistance to certain insecticides. This has become a problem in the last few years for people trying to design effective management programs.

It should be the responsibility of every user to learn as much as possible about the insect to be controlled, the chemicals to be used, and the potential hazards involved. Pesticides should be applied in such a way that pesticide resistance management is considered, hazards are minimized or avoided, and an effective, economical job of insect control results.

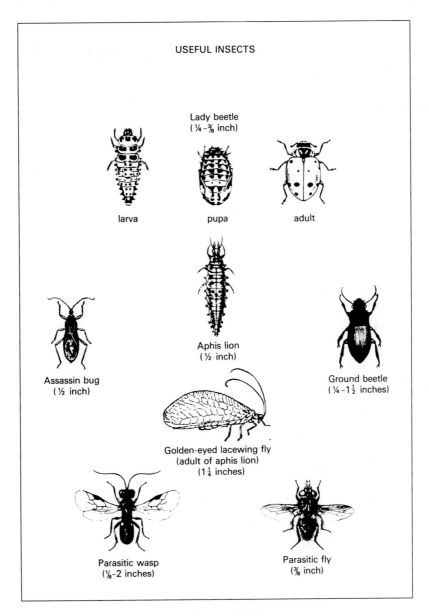

Figure 2–27 Beneficial insects.

INSECTICIDE CLASSIFICATION

Insecticides may be classified in various ways: by chemical composition, by formulation, or by the way they kill insects. Before synthetic organic insecticides were developed, the classification was usually made on the basis of whether the chemical killed as a stomach poison or as a contact poison. Because many modern insecticides act in both ways, this method of classification is outmoded. Therefore, most classifications in the literature are now by chemical composition. The following does not list all the insecticides, but is given as an example of chemical classification.

Chemical Groups

1. Inorganics
   a. Lead arsenate*
   b. Calcium arsenate
   c. Paris green
   d. Sodium fluoaluminate or cryolite
   e. Mercurous chloride or calomel
   f. Boric acid
   g. Sulfur
2. Organics
   a. Botanicals
      i. Pyrethrum
      ii. Rotenone
      iii. Ryania
      iv. Sabadilla
      v. Nicotine
   b. Organochlorines (chlorinated hydrocarbons)**
      i. Aldrin
      ii. Chlordane
      iii. Toxaphene
      iv. Lindane
      v. Endrin
      vi. Methoxychlor
      vii. DDT
      viii. Heptachlor
   c. Organophosphates
      i. Malathion
      ii. Parathion
      iii. Diazinon
      iv. Phorate (Thimet®)
      v. Demeton (Systox®)
      vi. Azinphosmethyl (Guthion®)
      vii. Chlorpyrifos (Dursban®)
   d. Carbamates
      i. Carbaryl (Sevin®)
      ii. Carbofuran (Furadan®)
      iii. Methomyl (Lannate®) (Nurdrin®)
      iv. Aldicarb (Temik®)
      v. Propoxur (Baygon®)

*Arsenicals and other heavy metals have been phased out because of their persistence and toxicity.
**Most uses for this group have been canceled because of danger to humans and the environment.

e. Formamidines
   i. Chlordimeform (Galecron®)
   ii. Amitraz (Baam®)

f. Organotins
   i. Cyhexatin (Plictran®)
   ii. Fenbutatin (Vendex®)

g. Biologicals
   i. *Bacillus thuringiensis* spores
   ii. Milky white disease spores

h. Pyrethroids: pyrethrin-like chemicals of synthetic origin
   i. Fenvalerate (Pydrin®)
   ii. Permethrin (Pounce®) (Ambush®) (Mavrik®) (Spur®) (Baythroid®)

i. Miscellaneous
   i. Dinitros
   ii. Petroleum oils
   iii. Soaps
   iv. Insect growth regulators (IGRs)
   v. Fumigants

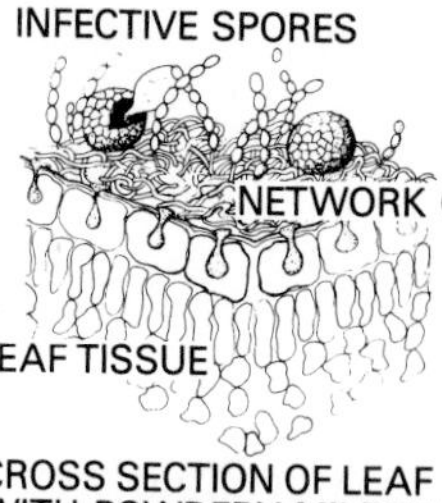

INFECTIVE SPORES

NETWORK OF FUNGUS

LEAF TISSUE

CROSS SECTION OF LEAF
WITH POWDERY MILDEW

3

Plant Disease Agents

WAYS IN WHICH DISEASES AFFECT PLANTS

The study of plant diseases is a challenging biological science. It pits the energy of a few against more than 50,000 destructive plant diseases. Because humans depend on plants for food and other products, plant disease control is essential to feeding the people of the world. By studying microorganisms, host plants, and their relationship to each other, plant pathologists aim at controlling diseases and thereby enhancing the quantity and quality of everyone's food.

The ancient Romans worshipped Robigus, a god named for the rust disease of grain. Indeed, people throughout history have feared the fantastic potential of plant diseases for destroying food and other essential crops. The Irish potato famine of 1843 to 1845, which was caused by a fungus, killed a million people and encouraged a mass migration of Irish citizens to America. Within just 25 years, the chestnut blight fungus from Asia essentially destroyed the magnificent stands of American chestnuts that once extended from Maine to Georgia. Likewise, millions of elm trees have been destroyed by the Dutch elm disease fungus (Figure 3–1) that moved across the United States from the East and into the Rocky Mountains. In 1970 the corn blight fungus ravished cornfields throughout our country and threatened the entire corn industry.

Plant diseases are caused by microscopic agents known as fungi, bacteria, nematodes, mycoplasmas, viruses, and viroids. These agents and nutrient deficiencies produce symptoms we readily recognize: rotted fruits, blighted and cankered trees, spotted flowers, shrunken grain, and yellowed, dying plants of all kinds (Figure 3–2).

No plant is immune to disease. Houseplants are as vulnerable as farm crops. Disease loss occurs during the shipping and storage of fruits, vegetables, flowers, and

34

ADULT EUROPEAN ELM
BARK BEETLE SPREADS
FUNGUS SPORES
INTO FEEDING SCARS
ON ELM TWIGS

FUNGUS SPREADS
THROUGH WATER
CONDUCTING SYSTEM
AND KILLS TREE

EMERGING BEETLE
PICKS UP FUNGAL SPORES

BARK BEETLE EGGS
LAID IN DEAD
OR DYING ELM WOOD

Figure 3–1 Disease cycle of Dutch elm disease.

cereals. Plant disease agents are everywhere. The earth and the air serve them as both conveyance and habitat. They can move around the world on a crumb of soil. They can move in water and on the wind or hitchhike on our vehicles.

Despite modern agricultural practices, plant diseases cause annual losses of billions of dollars. The everthreatening world food shortage, coupled with overpopulation in many areas of the world, emphasizes the need for more efficient crop production, one of plant pathology's prime challenges.

BIOLOGICAL ABNORMALITIES

A wide variety of disorders affect plants. These disorders include disease, injury, and malformations. Disease may be pathological, caused by a living microorganism or virus; or physiological, caused by the continued deficiency of a nutrient, such as lack of available iron in the soil.

Injury is similar to disease in some respects, but is caused by the momentary or transient impact of some agent, for example, injury from machinery or tools. Insects may cause injury; likewise, there may be environmental stress, such as hail, wind, or drought. Malformations are usually related to autogenetic factors.

THE PLANT DISEASE CONCEPT

Plant pathology is a well-established science based on a number of principles and concepts.

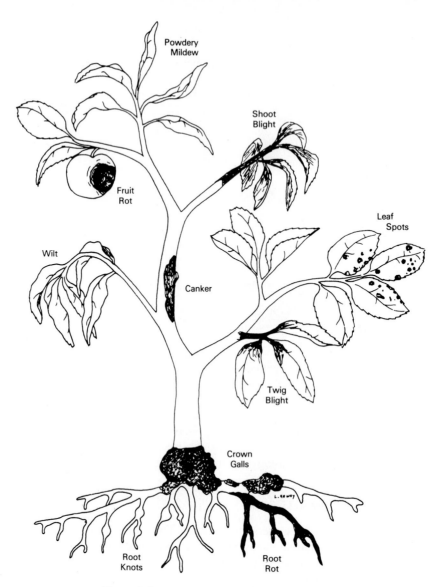

Figure 3–2 Ways in which diseases affect plants.

1. Disease consists of deleterious physiological activity caused by continued irritations from an external primary agent (or agents), resulting in disturbed cellular activity and expressed in characteristic pathological changes called *symptoms*.

2. Any particular disease is caused by one or more primary irritants or causal factors and several secondary factors of different degrees of importance.

3. The causal factor (or factors) of a disease has its own history and life cycle, which may vary in individual instances depending on its environment, and is of fundamental importance in the initiation and course of the disease.

4. Disease development passes through three essential stages: inoculation, incubation, and infection. To be meaningful, these stages must be correlated with the activities of the pathogen and interaction with the plant environment.

5. The damage resulting from any disease depends on the time of initiation, the degree of disturbance, the part or portion of plant affected, the previous development of the plant, and the extent of secondary involvement.

6. Control of any disease depends on one or more of the fundamental principles of control: exclusion, eradication, protection, and resistance. These principles must also be based on the concept and knowledge of the disease and therefore related to the stages of disease development and the factors that influence these stages.

PATHOLOGICAL DISEASE

The age-old phenomenon of parasitism occurred along with the evolution of plants from time immemorial. It is the inherent nature of certain microbes to be parasitic. As a result of parasitic activities, a complex and detrimental process of disease occurs. Causes of disease include fungi, bacteria, viruses, mycoplasmas, nematodes, and a few more developed parasites such as mistletoe.

Fungi are plants that lack the green coloring (chlorophyll) found in seedproducing plants and therefore cannot manufacture their own food. There are more than 100,000 different species of fungi of many types and sizes. Not all are harmful, and many are beneficial. Most are microscopic, but some, such as the mushrooms, are quite large. Most fungi reproduce by spores, which vary greatly in size and shape. Some fungi produce more than one kind of spore, and a few fungi have no known spore stage.

Bacteria (Figure 3–3) are very small, one-celled plants that reproduce by simple fission. They divide into two equal halves, each of which becomes a fully developed bacterium. This type of reproduction may lead to rapid buildup of population under ideal conditions. For example, if a bacterium can divide every 30 minutes (a generation time not especially short for some bacteria), in 24 hours a single cell could produce 281,474,956,710,656 offspring.

FIRE BLIGHT
CANKER

Figure 3–3 Fire blight, showing symptom development following spread of bacteria from blossom to limb.

Viruses are so small that they cannot be seen with the ordinary microscope and are generally detected and studied by their effects on selected indicator plants. Many viruses that cause plant disease are transmitted from one plant to another by insects, usually aphids or leafhoppers. Viruses also cause very serious problems in plants that are propagated by bulbs, roots, and cuttings because the virus is easily carried along in the propagating material. Some viruses are easily transmitted mechanically by rubbing leaves of healthy plants with juice from diseased plants. A few viruses are transmitted in pollen. Big vein, a lettuce virus, is transmitted by a soilborne fungus, and a few viruses are transmitted by nematodes.

Mycoplasmas are a relatively new group of cellular microorganisms, some of which were previously considered viruses, particularly in the yellow virus group.

Nematodes (Figure 3–4) are small eel-shaped worms that reproduce by eggs. The number of eggs produced by one female nematode, and the number of generations in a season, depends largely on soil temperature. Therefore, nematodes are usually more of a problem in warmer areas of the country. Most nematodes feed on the roots and lower stems of plants, but a few attack the leaves and flowers. Many crops, such as strawberries, root crops, bulbs, ornamentals, mint, alfalfa, and potatoes, are damaged by nematodes. Stunted or distorted plants may result from invasion by these pests. Symptoms vary from swellings, thickenings, galls, and distortions on aboveground parts to root conditions such as short stubby roots, lesions (dead spots),

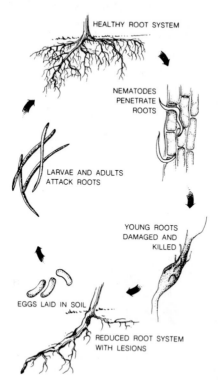

HEALTHY ROOT SYSTEM

NEMATODES PENETRATE ROOTS

LARVAE AND ADULTS ATTACK ROOTS

YOUNG ROOTS DAMAGED AND KILLED

EGGS LAID IN SOIL

REDUCED ROOT SYSTEM WITH LESIONS

Figure 3–4 Disease cycle of the lesion nematode.

Plant Disease Agents   Chap. 3

swellings, galls, and general breakdown. Heavy infestations by the root knot nematode may result in small roots resembling strings of beads. Infested potato tubers may have a bumpy or pebbly, uneven surface, even though the skin is not broken. Most plant parasitic nematodes range in size from $\frac{1}{50}$ inch to $\frac{1}{25}$ inch in length. These organisms may occur almost anywhere life exists.

All plant parasitic nematodes possess a spear or stylet, usually hollow, by which they puncture plant cells and feed on the cell's contents. Nematodes may develop and feed within plant tissue (endoparasites) or outside (ectoparasites). A complete life cycle involves the egg, four larval stages, and the adult. The larvae usually resemble the adults, except in size. The females of some, such as root knot and cyst nematodes, become fixed in the plant tissue, and the body becomes swollen and rounded. The root knot nematode deposits its eggs in a mass outside the body, whereas the cyst nematode retains part of its eggs within the body where they resist unfavorable environmental conditions and may survive for many years.

HOW PLANT DISEASES DEVELOP

Development of any parasitic disease is critically dependent on the life cycle of the pathogen. The life cycles of all disease organisms are greatly influenced by environmental conditions affecting both the host and the pathogen. Temperature and moisture are probably the most important factors that affect the severity of plant diseases. They not only influence the activities of the disease organism but also affect the ease with which a plant becomes diseased and the way the disease develops.

The life cycle (or life history) of a pathogen begins with the arrival of some portion (fungus spore, nematode egg, bacterial cell, virus particle) at a part of the plant where infection can occur. This step is called *inoculation.* If environmental conditions are favorable this infectious material (called *inoculum*) will begin to develop. Spores will germinate and bacterial cells will begin to multiply. This stage is called *incubation.* If the inoculum is on a plant part that it can enter, such as natural openings or wounds, or if it can penetrate the plant's surface directly, the pathogen will enter the host plant to begin the stage called *infection.* This is the real beginning of disease, but the plant is not yet diseased. Only when the plant responds to the invasion of the pathogen in some way (e.g., cells die or multiply abnormally) has disease developed.

The plant generally expresses its response to invasion by a pathogen and development of disease in the form of symptoms. These are outward expressions of the plant disease and consist of three general types.

1. *Overdevelopment of tissues:* galls (crown gall, club root, western gall rust); witches' broom (dwarf mistletoe, big vein of lettuce, alfalfa virus); swellings (white pine blister rust); leaf curls (peach leaf curl, leaf blister)

2. *Underdevelopment of tissues:* stunting (many virus diseases); chlorosis or lack of chlorophyll (iron deficiency); incomplete development of organs

3. *Necrosis (death) of tissue:* blights (fire blight); leaf spots, wilting (verticillium wilt); decays (soft rot); cankers (anthracnose, bacterial canker)

For a disease to develop, three factors must be present:

1. Susceptible host plant (the suscept)
2. Disease-producing agent (the pathogen)
3. Environment favorable to disease development

Eliminate any of these three factors and disease cannot develop. Because each of these components is variable, the interrelationships are variable and complex. Any of the three factors can influence one of the other two independently of the third, yet ultimately have an impact on disease development.

To understand the interrelationships of the three factors, look at the diagrams in Figure 3–5. There are three circles, each representing one of the three components. These circles are free to move in any direction. Only when the three overlap will disease develop. If all conditions are ideal for disease development, the circles will be almost one on top of the other. This represents a very severe disease situation or *epidemic*. If conditions are average or moderate, only a small portion of each circle will overlap the others and the disease situation will be mild. This would be an endemic

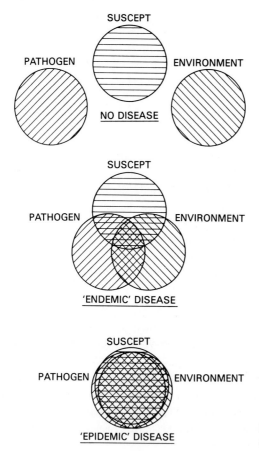

Figure 3–5 Ingredients for disease development.

disease. If one or more of the ingredients is not present under otherwise favorable circumstances, the circles cannot overlap and no disease will occur.

HOW DISEASES ARE IDENTIFIED

Symptoms by themselves may not allow an accurate diagnosis of a plant disease because several distinctly different causal agents may produce the same symptoms. However, symptoms used with other evidence, plus experience, can often produce a satisfactory diagnosis. The name of a disease is often derived from the symptoms produced.

Signs are more reliable indicators of the cause of a disease, but their detection and identification require more specialized training and techniques than does the observation of symptoms. Signs are structures produced by the causal organism. They may be fungus spores or other structures, nematodes or their eggs, bacterial ooze, or similar material. Usually signs are too small to be seen by the unaided eye, and lenses or microscopes of varying types must be used.

The extremely large number of plant diseases (about 50,000 in the United States) makes it impossible for any one person to be familiar with very many of them. However, certain facts help in the identification of plant diseases. In a diagnosis of a suspected plant disease there is a logical and convenient series of steps to follow.

1. The first piece of information to obtain is the identity of the plant affected. If possible, this should include the scientific name as well as the common name, because the same common name may have been used for several distinctive plant species. Even the variety of the diseased plant should be determined whenever possible since great variation in susceptibility to plant diseases may exist between different cultivars.

2. Careful examination of the diseased area, whether a bench in a greenhouse or a large field, is a logical second step in determining the cause of a plant problem. Note how the diseased plants are distributed over the affected area. Are they uniformly distributed or localized in certain spots? Definite patterns of distribution, such as the edges or the center of a greenhouse bench or along roadways, fences, corners, or low spots in a field, frequently indicate climatic soil factors or toxic chemicals but do not necessarily exclude parasitic causes.

   Is only one type of plant affected or are many unrelated plants involved? If more than one kind of plant is affected, disease organisms are probably not involved. Rather, one should suspect climate, insects, or chemicals.

   How did the disease develop in the affected area? Did it appear overnight? If so, suspect a climatic factor such as fog or the application of a toxic chemical. However, if the condition started at one point and spread slowly in extent and severity in a single species, a disease is probably responsible.

   Obtain a record of environmental conditions preceding the appearance of the disease. Check for short periods of below freezing temperature or prolonged drought periods. Did hail and/or lightning occur? Air pollutants might have been involved. If so, information on air inversions and a prevailing wind will be helpful in explaining the pattern of damage.

Has there been any treatment that could have resulted in plant injury? Soil sterilant herbicides can burn foliage on trees some distance away because extensive tree root systems have probably intercepted the downward movement of the herbicide. Excessive amounts of fertilizers can cause similar damage to plants.

3. The typical appearance of a diseased plant should be described. The symptom picture should not be based entirely on the early stages of the disease nor on a plant that has deteriorated to a point that secondary organisms have obscured the primary cause of the problem. Ideally, the symptom picture should be a composite based on the progression of symptoms from earliest to latest.

   The typical diseased plant should always be compared with a healthy or normal plant since normal plant parts are sometimes mistakenly assumed to be evidence of a disease. Examples are the brown spore-producing bodies on the lower surface of fern leaves. These are the normal propagative organs of ferns. Also in this category are the small, brown, clublike tips that develop on arborvitae foliage in early spring. These are the male flowers, not deformed shoots. Small galls on the roots of legumes such as beans and peas are most likely nitrogen-fixing nodules essential to normal development and are not indicative of nematode galls.

   Premature dropping of needles by conifers frequently causes alarm. Conifers normally retain their needles for 3 to 6 years and lose the oldest gradually during each growing season. This normal needle drop is not noticed. However, prolonged drought or other factors may cause the tree as a whole to take on a yellow color for a short period of time and may accelerate needle loss. If the factors involved are not understood, this often causes alarm. The needles that drop or turn yellow are usually the oldest needles on the tree, and the dropping is a defense mechanism that results in reduced water loss from the foliage as a whole.

   The portion or parts of a plant affected should be noted. Are they the roots, leaves, stems, flowers, fruit, or is the entire plant involved? The root system, which may constitute half of the plant, is often overlooked because it is underground and not visible.

4. What are the primary symptoms of the disorder under study? Symptoms are expressions of the affected plant that indicate something abnormal and are grouped into three general classes:
   • Underdevelopment of tissues or organs
   • Overdevelopment of tissues or organs
   • Necrosis or death of plant parts

   Diagnosis of a disease up to this point requires no special equipment other than a hand lens and knife.

5. Signs of disease are structures produced by the causal agent of that disease. Since signs are more specific than symptoms, they are much more useful in the accurate diagnosis of a disease. For example, distortion of leaves and shoots usually points to damage from a hormone-type herbicide such as 2,4-D, but similar symptoms are sometimes produced by frost, viruses, fungi, or insects (eriophyid mites).

   Discovery and identification of signs of disease may require special equipment and knowledge, but many signs can be found with only a hand lens and a knife. The presence of spore-producing bodies on bark cankers, mycelial mats of the root rot fungus and *Armillaria mellea,* nematodes in alfalfa roots, bacterial ooze from fireblighted pears, masses of rust spores, and the gray and white mycelia of powdery mildew are all signs of different diseases.

6. Sometimes neither symptoms nor signs are specific or characteristic enough to pin down the cause of a disease. Then additional specialized techniques are required to *isolate* and *identify* the causal agent in the laboratory.

Isolation of bacteria and fungi usually requires that small pieces of diseased tissue be placed on a nutrient medium. The organisms growing out of the tissue are then isolated into a culture. However, many disease-producing organisms, especially obligate parasites, cannot grow on artificial media, but can grow only on living tissue. Obligate parasites require still other special methods for their isolation. Obligate parasites include rust, powdery mildew, and viruses.

Invasion of diseased tissue by saprophytic organisms often makes isolation of the primary cause of disease difficult if not impossible. Because of their ability to grow rapidly and utilize the artificial substrate more efficiently than can many disease organisms, these saprophytes may rapidly overgrow and crowd out the primary pathogen on synthetic media. Experienced plant pathologists overcome this difficulty by isolating from the margins of the diseased tissue and by employing selective media especially developed to favor growth of the specific pathogen.

Assuming that isolation has been successful and you are reasonably certain you have isolated the primary cause of the disease, there is still a problem of identifying the organism. There are 20,000 to 40,000 species of bacteria and fungi, as well as many viruses, nematodes, and even higher plants that cause disease. The characteristics upon which the identification is based are often complex and not easy to determine. Frequently, only a specialist who deals with a small group of organisms can correctly identify the disease organisms in question. For example, identification of a parasitic nematode is most difficult for anyone except a trained nematologist.

Viruses are especially difficult to identify because their submicroscopic size makes them too small to be seen except by very special and costly equipment. Their identification usually is based on the reactions (symptoms) of selected hosts called *indicator plants* and on their physical and chemical properties. The determination of all characters requires very special knowledge and techniques. Recently, considerable progress has been made in serological identification of viruses, for example potato viruses.

Our primary interest is usually to control a given disease. Only after positively identifying the disease can the best control measures be employed. Therefore, when all the previously mentioned information has been successfully collected, we should consult the available literature to determine what is already known about the disease. This information may be found in books, technical journals, commodity newsletters, experiment station records, or correspondence with other experts on plant diseases. County extension agents and specialists in plant pathology should be consulted.

HOW PLANT DISEASES ARE CONTROLLED

The ultimate concern about a plant disease is to reduce or eliminate the economic or esthetic loss it causes. This is called the *control* of a disease. Plant disease control involves one or more of four basic principles:

1. *Exclusion* involves measures to prevent a disease organism from becoming introduced into and established in an area where it does not occur. Plant quarantines are one means of exclusion.
2. *Eradication* is the elimination of a pathogen from an area, usually when it has limited or restricted distribution.
3. *Protection* consists of the placement of a protective barrier—usually a chemical—between the plant and the pathogen.
4. *Resistance* involves the use of plants that are not susceptible to a disease. Immunity is the ultimate degree of resistance and is usually not obtained in genetic programs aimed at developing resistance in a given plant. The level of resistance may vary considerably depending on a large number of factors, such as the age of the host plant, aggressiveness of the pathogen, and relative favorability of the environment. Very often, a plant variety or selection that is resistant to disease lacks desirable qualities wanted for commercial purposes.

Many diseases can be controlled by cultural practices alone that include exclusion, eradication, and resistance. Such practices include:

* Selection of resistant or tolerant varieties of plants.
* Proper establishment of plants.
* Rotating planting locations.
* Maintenance operations, such as raking and destroying fallen leaves, thinning, pruning, and regulating fertilizer and water.

Chemical control of plant diseases involves the principle of protection, whereby a protective chemical barrier is placed between the plant and the pathogen. Chemicals used to control plant diseases are generally called *fungicides,* but correctly these include only those chemicals that kill or retard the development of fungus pathogens. Those that control bacteria are *bactericides* and those controlling nematodes are called *nematicides.*

In general, fungi are much more difficult to control with chemicals than are insects or weeds. A fungi pathogen is a plant that is living in or on another plant, its host. Thus the goal is to kill one plant without injuring a second plant. Finding chemicals with such selectivity is a time-consuming task, which is not always successful. In addition, most of the fungal pathogens have asexual repeating cycles only a few days long; thus the crop plant may be subjected to as many as 10 to 25 generations of a pathogen during the growing season. This requires repeated application of the fungicide to control the pathogens. In addition, many plant pathogens are either below ground or within the plant's interior tissues and hence are not reached by most common fungicides.

In general, application of fungicides differs markedly from the application of herbicides or insecticides. An herbicide placed anywhere on the plant foliage will often kill the plant. Likewise, an insecticide placed at random on the plant is likely to kill the insect because most insects tend to move around from leaf to leaf or branch to branch. However, with a fungicide, only that portion of a plant covered by the fungicide is protected. Thus it is essential that a fungicide give complete coverage of all tissues. A dust may be quite suitable as an insecticide, but because most dusts are deposited only on the upper surfaces of the leaf, they are relatively ineffective as fungicides. Sprays should be applied to both upper and lower surfaces for maximum protection.

Protection of plants with chemicals involves killing the pathogen (or limiting its growth) before it invades the suscept; this may be accomplished by two different means:

1. Application of the fungicide to the pathogen in place. Such a fungicide is termed a *contact fungicide.*
2. Application of the fungicide before the pathogen is in place. Such a fungicide is termed a *residual fungicide.*

The use of contact fungicides to eliminate potential sources of inoculum or to eradicate inoculum after it has been deposited on the target site is risky and hence rare. In most instances, residual protectant fungicides are used. A good residual protectant fungicide is one that:

- *Remains active for a relatively long period of time.* Obviously, if a fungicide must be applied every day to maintain the necessary level of protection around the clock, it would be next to impossible to apply and too expensive if it were possible.
- *Has good adhesive properties.* Since the dissemination of most pathogens is favored by rainy weather, the fungicides must resist the erosive action of water.
- *Has good spreading properties.* Since it is necessary to completely protect leaf and stem surfaces, the fungicide should spread evenly over the surface of the leaf. This is usually accomplished by adding a wetting agent to reduce surface tension. If too much wetting agent is added to a formulation, runoff may occur, which results in lower concentrations of fungicide than required. Wetting agents also facilitate penetration of fungicides into leaves or stems, and if too much is used, phytotoxicity may occur.
- *Is stable against photodeactivation.*
- *Is toxic to plant pathogenic microorganisms, but nontoxic to the plant and non-target organisms.*
- Is active against a wide range of pathogenic microorganisms. Actually, most of the presently available fungicides are rather specific. Thus a particular fungicide must be selected for control of a particular disease. Frequently, spray programs are devised involving the use of more than one fungicide, either as combination sprays or in alternate applications. Formulations of two or more fungicides in a single package are marketed for use in the home and garden, but commercial producers must prepare their own combination sprays as tank mixes.

- *Is compatible with other pesticides, such as insecticides and other fungicides.*
- *Is relatively easy to apply and does not present an undue hazard to the applicator or to the environment.*
- *Is noncorrosive to the application equipment.*

TYPES OF PROTECTANT FUNGICIDES

Fungicides can be classified either as organic or inorganic; they can be further broken down by the heavy metal element(s) involved in inorganic fungicides or by the basic class of chemical compounds in organic fungicides.

Inorganic Fungicides

Some inorganic pesticides date back to our early history. The use of these compounds has continued until relatively recently. Many of the heavy metal compounds have now been banned for use because of their persistence in or on the treated plants themselves, leading to pesticide poisoning of people and animals, or because of the buildup of toxic residues in the soil.

Sulfur. Probably the oldest pesticides known are various sulfur compounds. These were used either as dusts or mixed with fats and used as ointments to treat various diseases of people. It was probably sulfur dust that was used by the early Greeks and Romans to control certain insects and diseases on crops. Other formulations have been developed since those times. Elemental sulfur kills some insects, bacteria, and fungi by direct contact. Sulfur is still used to control powdery mildew because the mycelium of these pathogens is superficial on the leaf surface. Sulfur acts by interfering with electron transport in the cytochrome system.

Copper. The use of copper goes back to the early 1800s when it was discovered that copper inhibited germination of the spores of covered smut of wheat. In 1887 Millardet discovered Bordeaux mixture as a control for downy mildew of grapes. It was the first successful and widely used fungicide. It is still chemically undefined and no one is really sure of the active ingredients. It is formed by a mixture of copper sulfate and hydrated lime. In general, copper compounds are relatively insoluble in water and are not easily washed from the leaves by rain. A very small portion of the copper goes into solution. This is absorbed by living cells and then more copper goes into solution to replace that which has been removed. The living cells continue to absorb the toxic ions and eventually accumulate enough so that they are poisoned. Because of this factor, many copper compounds are fairly phytotoxic.

Mercury, Cadmium, and Other Heavy Metals. Some of the most toxic fungicides are those containing inorganic mercury, cadmium, or certain other heavy metals. However, these compounds are also very toxic to other forms of life, particularly to warm-blooded animals. Thus mercury and other heavy metal residues are not permitted in food or feeds, and these compounds have been banned from use.

Organic Fungicides

These are synthesized compounds and can be divided into two groups: nonsystemic and systemic.

Nonsystemic

Dithiocarbamates. A whole group of compounds belong to the dithiocarbamates, most of which were developed in the 1930s and 1940s and are some of the oldest organic fungicides known. These include Polyram®, Manzate®, Dithane®, and Fermate®.

Dicarboximides. The first of this group of fungicides, captan, appeared in 1949. Two others, folpet and captafol, appeared in the early 1960s. They are widely used as protective sprays or dusts for fruits, vegetables, ornamentals, and turf; and as seed treatments. They probably act by an inhibition of synthesis of proteins and enzymes containing sulfhydryl groups.

Dinitrophenols. Dinocap (Mildane®) was developed in the late 1930s not only as an acaricide, but also for control of powdery mildews. Because it acts in a vapor phase, it is effective against the powdery mildews, whose spores are often able to germinate in the absence of water. It acts by uncoupling oxidative phosphorylation, which upsets the energy systems within the cells.

Substituted Aromatics. This group contains several different fungicides that basically have a simple benzene ring with various attached radicals. Pentachlorophenol has been used by itself or as a salt of sodium, copper, or zinc. Pentachlorophenol is a restricted use pesticide, and its use is no longer permitted around desirable plants because it is phytotoxic. Pentachloronitrobenzene (PCNB) was also introduced in the 1930s as a seed treatment and as a foliage fungicide in some cases. It has also been used as a soil treatment to control damping-off fungi. Hexachlorobenzene was introduced in the mid-1960s for use in controlling diseases of turf grasses and cotton seedlings.

Quinones. Dichlone is a quinone that has been used on various food and vegetable crops as a foliar fungicide and also in control of blue-green algae in ponds. It apparently acts by affecting the sulfhydryl groups in enzymes, inhibiting them and affecting oxidative phosphorylation.

Aliphatic Nitrogenous Compounds. Dodine is an example of this class of fungicide, introduced in the mid-1950s for control of various fungal leaf spot diseases. It is quite specific. The mode of action is unknown in total, but it acts by affecting membrane permeability, allowing a leakage of metabolites from affected cells. It also inhibits synthesis of RNA.

Antibiotics. Many different antibiotics are used to control bacterial diseases of humans and other animals. Antibiotics are metabolites produced by microorganisms, particularly the actinomycetes. Three compounds have been used to control bacterial diseases on fruits and vegetables. Streptomycin is commonly used to control fire blight of apples and pears. The mode of action is unknown, but it presumably interferes with protein synthesis and with the synthesis of organic acids. Another compound is tetracycline, which has been used as a chemotherapeutic against mycoplasma diseases. A third compound is cycloheximide, (Actidione®), which has been used to control various foliage diseases. Label registration was recently discontinued on these last two compounds.

Organotins. Triphenyltin hydroxide (Du-Ter®) is an example of this group of fungicides. They were first introduced in the mid-1960s. The trisubstituted tins act to block oxidative phosphorylation. Most uses have been discontinued.

Systemic

Oxathiins. Two of these compounds, carboxin (Vitavax®) and oxycarboxin (Plantvax®), were introduced in 1966 as systemic fungicides. They are selectively toxic to some of the smut and rust fungi and to *Rhizoctonia*. They apparently act by inhibition of succinic dehydrogenase, a respiratory enzyme important in the mitochondrial systems.

Benzimidazoles. These compounds were first introduced in the late 1960s and have been used as systemic fungicides against many different types of diseases. One of the most widely used of these is benomyl, which is used to control many fruit, vegetable, ornamental, and turf diseases. Two others are thiabendazole (Mertect®) and thiophenate (Tops MZ®). These compounds have been used as foliar fungicides, seed treatments, soil drenches, and dips for fruit or roots. They apparently act by interfering with the synthesis of DNA, affecting spore germination and growth.

Pyrimidines. These fungicides were introduced in the late 1960s. They are particularly active on mildews. Examples are dimethirimol (Milcurb®) and buprimate (Nimrod®).

Acylalanines. These are effective against soilborne diseases as well as downy mildew and certain foliar diseases. Metalaxyl (Ridomil®) is in this category.

Organophosphates. Their mode of action is unclear, but they are believed to interfere with phospholipid synthesis in the invading microorganism. An example is fosetyl-aluminum (Aliette®).

Triazoles. One of the newer and more promising groups of fungicides. They are broad-spectrum systemic fungicides that offer both protective and curative effects. Examples are triadimefon (Bayleton®), propiconazole (Tilt®), triadimenol (Baytan®), and bitertanol (Baycor®).

Piperazines. This class of systemic fungicide is represented by triforine (Funginex®), which is active against scabs, rusts, and mildews. It was introduced in the 1970s and apparently interferes with the synthesis of sterols.

Imides. These are related to the dicarboximides, although they are significantly different structurally, and they are systemic. They are effective against a number of important diseases. Iprodione (Rovral®) and vinclozolin (Ronilan®) are examples of this group.

Carbamates. This newer category includes propamocarb hydrochloride (Banol® and Prevex®), which is used as a turf fungicide against pythium blight.

Systemic fungicides are absorbed by the plant through the leaf, stem, or root surface and translocated varying distances within the plant. This distance may be as small as from one leaf surface to the other or as far as from the roots to the shoot apex.

Advantages of systemic fungicides

- The plant can be continuously protected throughout the growing season without repeated applications of fungicides.
- The systemic may be taken up by the roots and transported to newly formed tissues.
- The systemic is not subjected to weathering, as are fungicides applied to the foliage.
- Unsightly residues on flowers and foliage can be avoided.
- The systemic may provide a means of controlling and eradicating vascular wilt diseases, as well as other internal disorders of plants.
- Since the toxicant is in the plant, there is minimal toxic effect to people working in the greenhouse or with the crop during the growing season.

Disadvantages of systemic fungicides

- Resistance to many of these compounds is becoming quite common. This is because many of the more selective organic fungicides have only a single mode of action, thus disrupting usually just one process in metabolism.
- Most systemic fungicides are actually fungistatic, not fungicidal. Thus an organism frequently can recover as the chemical dissipates. Also, although the growth of an organism may be inhibited, the organism may continue to reproduce.
- The systemic fungicide exerts a selection pressure, and any resistant genotypes will be selectively propagated, soon establishing a resistant or tolerant population.

It is best to employ several techniques to control, delay, or suppress plant disease pathogens. The integration of several techniques will help to delay or avoid fungicide resistance, breakdown of resistance in resistant crop varieties, and escapes due to carelessness.

4

Vertebrate Pests

Humans hold a unique and tenuous position in the complex world of existence. We are perhaps the most formidable creatures on earth. At the same time, we are one of the most vulnerable. Humans can modify and adapt to a myriad of climatic conditions and food niches. When people pick a place for themselves in each new system, it is at a very great expense to the creatures already present. In a system that produces a finite amount of usable, digestible energy, creatures that take their cut survive. Humans, in their quest for survival, have done well in this power struggle because they have reasoned methods and tools for eliminating or suppressing competition. During our 2 million year existence, we have written or drawn our thoughts for about 10,000 years. From the beginning of recorded thought, our contests with wildlife for food, shelter, and protection have held a place of honor in song, legend, and ritual. Those contests remain with us today. If you doubt it, read the national and state agricultural presses' regular features on the coyote, prairie dog, and starling problems.

In a period of affluence and food abundance, the competitive feeling with wildlife is suppressed by elements of society. Most people feel there is an abundance for all. Moreover, people can tolerate greater competition with wildlife than was the case 35 or 350 years ago. The result has been governmental management regulations to reduce the extent and effectiveness of wildlife pest control. This has led to severe financial losses on the part of segments of the agricultural industry, particularly livestock producers. It has also led to severe dilemmas among wildlife management agencies trying to satisfy widely divergent views and needs.

The current theme of wildlife pest control is to center control on managing out the opportunity for a wildlife species to cause a problem, chiefly by nonlethal means. The minimum action that can be taken is usually promoted over higher-impact action. Prophylactic reduction of wildlife populations to lower the probability of depredations has been largely discarded because the cost-effectiveness of such programs has been discounted. The effects of prophylactic population suppression have been considered too severe for environmental harmony.

Wildlife generally encompasses fish, amphibians, reptiles, birds, and mammals. Wildlife and humans share backbones, sexual reproduction, and red blood and have to hustle food by eating plants, lower forms of life, or each other. Wildlife is generally larger than invertebrate forms of life and has a more visible impact on the ecosystems in which they live, even though their overall effects are generally less than those by invertebrate life.

SOME VERTEBRATE PEST ANIMALS

Small Animals

Bats (Figure 4–1). Bats are unique in the animal kingdom in that they are the only true flying mammals. A high degree of structuralization makes it possible for bats to fly. Bats' wings are thin membranes of skin stretched from fore to hind legs and from hind legs to tail. This skin is barely or thinly furred.

Most bats in the United States are members of the evening bat family. By choice, none are active during the brighter hours of the day; they prefer late afternoon, evening, and early morning for their feeding flights. When not in flight, they rest in the dark seclusion of natural places such as caves, hollow trees, and rock crevasses. They may also occupy vacant buildings, church steeples, attics, spaces between walls, and belfries. Bats can enter places of refuge through very small openings; some observers say a small bat can squeeze through a crack no wider than ⅜ inch thick.

Most bats eat insects that are captured in flight. During the winter when food is in short supply, bats must either hibernate or migrate to warmer climates. Because of

Figure 4–1 Bat about to catch an insect on the fly.

their insectivorous habits, bats are beneficial. They occupy a very special niche in the animal kingdom because they can fly. They do not compete with other mammals for food and shelter. For these reasons, even though bats are unprotected by law, they should not be needlessly destroyed.

Bats frequently congregate in significant numbers at favored roosting sites. If these sites are in buildings, the accumulation of droppings and the odor of bat urine are objectionable. Their squeaks and the noise they make as they enter or leave the roost may also prove bothersome to the buildings' occupants. The incidence of bats transmitting disease to man is not high, but if such a situation occurred, it would suggest a need for control. As a precaution against exposure to disease, do not handle live bats. Bat bites are dangerous. In case of bat bites, prompt medical attention should be obtained and, if possible, the bat should be captured without damaging the head. The head should be placed in a jar or a plastic bag, kept under refrigeration, and submitted to health authorities for rabies tests.

Chipmunks, deer mice, ground squirrels, moles, pocket gophers, and shrews (Figure 4–2). The small mammals in this group are known to consume a large number of seeds during an evening's feeding. Deer mice, shrews, ground squirrels, and chipmunks may consume large numbers of pine or fir seeds. Such seedeating activities are not readily apparent to the casual observer. These rodents can cause serious problems in attempts at reforestation by means of reseeding.

Pocket gophers develop extensive burrow systems among roots for feeding purposes. The pocket gopher can be a problem in alfalfa fields, pastures, and lawns. They may seriously damage a crop of alfalfa by feeding on the leaves, stems, and roots. A 20% to 50% reduction in yield has been caused in some areas. Plants may be killed when their roots are cut or when they are covered by dirt from the mound. Gopher mounds in hayfields cause breakdowns and wearing of sickle mowers. The scattered, flat, fan-shaped mounds created by the gophers may also provide good seed beds for noxious weeds.

Moles and shrews in their natural environment cause little damage. They are seldom noticed until their tunneling activities become apparent in lawns, gardens, golf courses, pastures, or other grass and turf areas. These are the rare times that moles and shrews may require control. Otherwise, these mammals are beneficial because they feed on insects (both mature and larvae), snails, spiders, small vertebrates, earthworms, and only a small amount of vegetation.

Marmots (woodchucks, rockchucks) (Figure 4–3). These mammals can cause damage to many crops. Damage to legumes and truck gardens is often severe. The majority of the damage is caused by the animals consuming the plants. Earthen

Figure 4–2   Short-tailed shrew.

Figure 4–3 Marmot.   **Figure 4–4** Porcupine.   **Figure 4–5** Prairie dog.

mounds from their burrows and the burrows themselves may damage haying and other farm equipment. The burrows may also cause a loss of irrigation water when dug along water-conveying ditches.

Marmot damage is usually evidenced by areas where plant production has been terminated or reduced by the grazing of these rodents. Supplemental signs include droppings, burrows, and trails leading to and from the damaged area to dens or loafing areas. Their burrows and mounds may also be hazardous to horses and riders.

Porcupines (Figure 4–4). This nocturnal creature is most active at night, usually spending the day asleep in a cave or perched in a tree. Porcupines eat succulent plants of many species and are especially fond of garden and truck crops during the summer. In the fall, they may eat fruit in orchards. In the winter, the porcupine moves to the forested areas and feeds largely on the bark of certain evergreens, such as white, ponderosa, and piñon pine and hemlock. Basal girdling may occur on seedlings, the thinner barked Douglas fir, and young, pole-sized trees. Damage can be particularly severe during winter months when porcupines are concentrated around den areas. In addition to damaging crops and forests, porcupines cause considerable damage to structures such as camps and summer homes that are unoccupied during the winter.

Porcupines are extremely fond of salt and are attracted to anything that may contain salt or salt residue. They have been known to gnaw on saddles, harnesses, belts, and salt placed for livestock.

Porcupine damage may be identified by broad incisor marks on the exposed sapwood. Also, oblong droppings about 1 inch in length may be found under freshly damaged trees. Roost trees are recognized by the large deposits of droppings beneath them. Sometimes porcupines girdle on tops of pines, producing a characteristically bushy crown.

Prairie dogs (Figure 4–5). Prairie dogs can cause damage to rangeland by feeding on and cutting vegetation within their "towns." Studies have shown that prairie dog grazing changes the composition of the vegetation toward species that are more tolerant of their grazing, but not all these changes are detrimental. One possible benefit is that prairie dog grazing appears to increase the nitrogen content of grasses. In general, prairie dogs cause a reduction in grasses such as blue grama and an increase in buffalo grass. In areas of tallgrass and mixed-grass prairies, overgrazing often leads to infestations of prairie dogs. Prairie dog towns are easily visible, particularly in early spring when the green grass makes the light-colored mounds easier to spot.

Rabbits (Figure 4–6).   Rabbits are responsible for damage to young trees, truck crops, grainfields, gardens, and ornamentals. They can also strip the bark from established orchard trees. All rabbits, including both cottontails and jackrabbits, produce similar clipping injuries to tree seedlings. Repeated clippings can suppress the height of or deform seedlings, seed grains, and other crops. Rabbits may also girdle trees by gnawing off the bark around the base.

Jackrabbits may cause damage to cereal grains, alfalfa, and even range plants under high population densities. Jackrabbits often cut distinctive paths through fields or can be observed doing damage.

Figure 4–6   Rabbits.

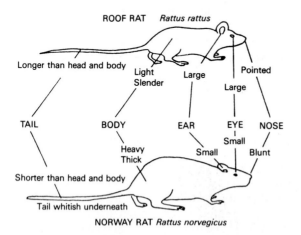

Figure 4–7   Rat identification.

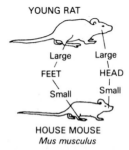

Figure 4–8   Differences between rats and mice.

Rats and mice (Figures 4–7 and 4–8). Rats feed on garbage, meat, fish, cereal, grain, and fruits. They require about an ounce of food and an ounce of water daily. The normal roaming range is about 100 to 150 feet. Commensal (domestic) rodents will move greater distances when their shelter is destroyed or disturbed.

Human health problems associated with rats involve direct contamination of human food from excrement and from the rodent filth. Indirectly, rat parasites such as fleas can transmit disease to humans.

The rat and its cousin, the mouse, have been important pests to humans since ancient times, especially as disease carriers. Rats and mice are carriers of a number of serious viral and bacterial diseases including plague, murine typhus, leptospirosis, and salmonellosis. They also cause enormous destruction and loss of food and property. Control is not easy due to their ability to adapt to changes and their capacity to reproduce. These rodents consume, contaminate, and cause extensive damage to food and agricultural crops. For every $2 worth of food they eat, they cause $20 worth of damage.

Rat behavior is influenced by thirst, hunger, sex, maternal instinct, and curiosity. Rats cannot go without water for more than 48 hours or without food for more than 4 days. Thirsty or hungry rats become desperate and are therefore easier to control because they are less wary. Judicious use of traps, poisons, and other control measures thus becomes doubly effective. Properly utilized environmental and physical controls along with rodenticides will prevent rapid population buildups and reinfestations of treated areas.

Carnivores and Other Animals

Bears (Figure 4–9). Bears are powerful animals with omnivorous feeding habits. They occasionally can create problems by killing domestic livestock. Torn and mutilated carcasses, often with many broken bones, characterize the victims of bear attacks. Often, the udders of female victims are removed and eaten. While feeding, bears will often peel back the hide similar to an animal that has been skinned. They move the carcass to a more secluded spot to eat if the kill was made in the open.

Bears may inflict severe damage to apiaries in attempts to obtain the honey contained in them. Bears can also cause damage to trees. A feature of bear girdling is an array of long, vertical grooves on exposed sapwood and large strips of bark at the base of the tree. Barking of trees by bears differs markedly from that caused by rodents, which eat the bark and leave horizontal or diagonal tooth marks.

Figure 4–9   Black bear.

Some Vertebrate Pest Animals

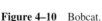

Figure 4–10   Bobcat.

Figure 4–11   Coyote.

Bobcats and lynx (Figure 4–10).   These related animals occasionally create problems by killing domestic livestock, especially sheep and lambs. Bobcats usually attack sheep by leaping on their backs and biting at the top of the neck or the throat. If the carcass is skinned, hemorrhaging and numerous small punctures through the skin will be noted on the shoulders and possibly the hips. These wounds are inflicted by the claws and are diagnostic of a bobcat kill. Animals fed upon by bobcats usually have the skin peeled back neatly around the area that was fed upon.

Coyotes and domestic dogs (Figure 4–11).   Coyotes more commonly take lambs rather than mature sheep and characteristically kill with a bite in the throat. The throat wound, if visible, usually consists of pairs of punctures by the coyote's large canine teeth. Death is usually by suffocation, caused by damage or compression of the trachea. Blood on the throat wool is usually indicative of predation, but external bleeding is not always apparent. In that case, the hide should be skinned from the neck, throat, and head of the carcass. If the animal has been killed by a coyote, this will reveal subcutaneous hemorrhage (bruises and clots), fang holes in the hide, and tissue damage. The hemorrhage occurs only if the animal was bitten while still alive. Although coyotes typically attack sheep in the neck, it is not a hard-and-fast rule. In late winter, when the winter wool is long and thick on the neck, coyotes may attack sheep at the more exposed hindquarters.

Domestic dogs can be a serious problem to livestock, particularly in areas near towns or cities. Dogs can occasionally cause severe losses and injuries to penned sheep or turkeys without touching an animal. Harassment from outside the pen or simply running the enclosed animals can cause them to stampede and pile up.

It is sometimes difficult to distinguish between a coyote kill and a dog kill. In general, a dog will mutilate an animal much more severely than will a coyote. Dogs typically attack the hindquarters and front shoulders, but little flesh is actually consumed. The ears of mature sheep are often badly torn by attacking dogs. Sheep-killing dogs often work in pairs or larger groups and can inflict a considerable amount of damage in a short period of time. A few dogs are very efficient killers and will kill sheep by attacking the throat or neck. These kills are indistinguishable from kills made by coyotes without searching for further evidence. Dogs often kill or injure large numbers of animals in a single episode, whereas coyotes normally will kill only one or two in a night. Any time more than three sheep are killed in a single episode, it would be wise to investigate the possibility of dogs.

Killing behavior and feeding patterns are useful in making a first general determination of cause of death. Often, however, it is necessary to look for more evidence before you can be certain what did the killing, particularly in situations where either dogs or coyotes could be involved. Additional evidence such as tracks, droppings, hair on fences or vegetation, or past experience are useful in making a more positive determination.

Foxes (Figure 4–12).   Gray, red, and kit foxes are sometimes a problem with small domestic animals such as lambs, pigs, and turkeys. Foxes will feed upon larger animals as carrion, but their largest prey are young lambs and pigs, adult jackrabbits, and turkeys. Mice and other rabbits are their major food.

Foxes may carry their prey some distance from the kill location, often to a den. The remains may be cached or partly buried in a hole scratched in the soil. Foxes often urinate on uneaten remains and that odor is an identifying sign. A fox can climb over a fairly high chickenwire fence while removing poultry.

The breast and legs of birds killed by foxes are eaten first and the other appendages are scattered about. Foxes often kill more young birds than they can eat when they find a nest. The young birds are left where killed, with tooth marks under the wing on each side of the body, and the head is often missing.

Eggs are usually opened enough to be licked out, and the shells are left beside the nest and are rarely removed to the den. Fox dens are noted for containing the remains of the prey, particularly the wings of birds.

Mountain lions (Figure 4–13).   This large animal, also called *cougar* or *puma,* preys on deer, elk, and domestic stock. They particularly like colts, lambs, kids, and occasionally grown horses and cattle.

Mountain lions have relatively short, powerful jaws and kill with bites inflicted from above, often severing the vertebral column and breaking the victim's neck. They may also kill by biting through the skull. They usually feed first upon the front quarters and neck region of their prey. The stomach generally is untouched. The large leg bones may be crushed and ribs broken. They frequently drag their prey from the kill site to a more remote area, and have a tendency to cover a partially eaten carcass with leaves or loose soil.

Figure 4–12   Red fox.

Figure 4–13   Mountain lion.

Some Vertebrate Pest Animals

Figure 4–14   Raccoon.

Figure 4–15   Striped skunk.

Raccoons (Figure 4–14).   There is little evidence that raccoons prey heavily on mammals. They may, however, occasionally kill poultry. They also prey on birds and their eggs. The heads of adult birds are usually bitten off and left some distance from the body. The breast and crop may be torn and chewed, the entrails are sometimes eaten, and there may be bits of flesh near water. In a poultry house, the heads of many birds may be taken in one night. One or more of the eggs may be removed from poultry or game birds' nests and may be eaten away from the nest. The shells are heavily cracked, with the line of fracture being along the long axis of the egg. There is often some disturbance of nest materials. In addition to eggs, raccoons sometimes dig a small hole in watermelon and rake out the contents.

Skunks (Figure 4–15).   Skunks, although often accused of killing poultry and small game species, kill few adult birds compared to the number of nests they rob. Most rabbits, chickens, and pheasants are eaten as carrion, even though they are dragged to the skunk's den. When skunks do kill poultry, they generally will kill only one or two birds and maul them considerably. Spotted skunks are quite effective in controlling rats and mice in grain and corn buildings. They kill these rodents by biting and chewing the head and foreparts, and the carcasses are not eaten.

Striped and spotted skunks are notorious egg robbers and the signs of their work are quite obvious. Eggs are usually opened up at one end and the edges are crushed as the skunk punches its nose into the hole to lick out the contents. The eggs may appear as hatched, except for the edges. They are sometimes removed from the nest, though rarely more than 3 feet. Canine tooth marks on the egg will be at least ½ inch apart. When in a more advanced stage of incubation, the eggs are likely to be chewed into small pieces.

Skunks may sometimes damage lawns or golf courses by digging numerous small holes while looking for beetle larvae and other insects.

Large Hoofed Animals

Deer, elk, and moose (Figure 4–16).   These large animals can have adverse effects on forest regeneration and maintenance of habitat for other wildlife. They may also compete with livestock on rangelands and cause damage to crops, orchards, and hayfields. Damage to fences by elk and moose can result in secondary damage by permitting livestock access to other areas.

Figure 4–16 Bull moose.

Winter wheat or hayfields can be badly damaged by trampling and bedding of large herds, especially elk. Deer and elk will persistently raid haystacks and feedyards in the wintertime.

Tracks, droppings, and the location of the crop, orchard, or forest trees being damaged will provide evidence to make a determination of the species causing the damage. Damage to trees is generally in the form of nipped branches and broken smaller trees. Larger trees are sometimes ridden down to allow the animals to feed on the smaller branches by straddling the tree and pushing it over with their chests.

If damage is severe enough and the guilty species has been determined, the next step is to decide if the amount of damage justifies the expense and effort to attempt control.

Birds

Crows, ravens, magpies, and gulls (Figure 4–17). Crows, ravens, magpies, and gulls are well-known robbers of other birds' nests. Crows usually remove the egg from the nest before breaking a hole in it. The raven and magpie break a hole up to an inch in diameter. The raven leaves a clean edge along the break (never crushed), whereas the magpie often leaves dented, broken edges. Ravens eat young birds, as do gulls.

In certain areas, crows, ravens, and magpies may occasionally kill newborn lambs by pecking at their eyes, navel, and anal regions. At times, they may also damage cattle by picking on fresh brands or sores where they may cause infection of the area, loss of weight, or even severe wounds.

Ducks and geese (Figure 4–18). These migratory birds are capable of causing serious crop damage as they travel up and down their migration routes. They can cause damage to most grain crops and sometimes to melons, lettuce, corn, and soybeans.

These large birds can sometimes cause compaction to fields when they congregate in large numbers. Most damage is due to sprout pulling in the spring and eating of grain when it begins to ripen. The birds not only eat large quantities, but cause large amounts to shatter on the ground where harvest is impossible.

Research has revealed that grazing of waterfowl on dormant winter wheat during February and March does not reduce the yield. In fact, moderate grazing tends to increase the yield.

Figure 4–17 Crow.

Figure 4–18 Ducks and geese.

Eagles and hawks (Figure 4–19).   Eagles may occasionally kill small lambs in the western United States. The carcass will show puncture wounds from talons on the shoulders or lower back. These wounds should be looked for by skinning the carcass before making a positive determination of an eagle kill. Feeding takes place on the uppermost chest area. The meat between the ribs is characteristically picked clean by eagles. Eagles will readily feed on carrion, particularly when other food supplies are scarce. For this reason, a positive determination of an eagle kill requires the finding of talon wounds.

Hawks are sometimes accused of causing small domestic animal kills, but these are very rare. More often, hawks are seen feeding on small animals, such as lambs, that have died of natural causes. Remember that eagles, hawks, falcons, and owls—because of their value in eating insects and rodents—are protected by law.

Blackbirds, cowbirds, English or house sparrows, grackles, pigeons, and starlings (Figure 4–20).   Many of our conflicts with birds, whether crop damage or nuisance problems, really represent the inevitable consequences of attempting to coexist with wildlife in an environment modified to serve only our interests.

Blackbirds, such as the red-winged blackbird and Brewers blackbird, are often seen together with the common grackle and the brown-headed cowbird as well as starlings because these birds all have similar habits and are often found in mixed flocks. Although these birds do feed in ripening grainfields and are a fairly large problem in sunflowers and grain sorghum at times, only 5% to 10% of their total annual diet may consist of grain. They feed extensively on insects, including grubs; caterpillars such as armyworms, cutworms, and corn earworms; as well as beetles and other insects.

The most serious problem associated with these birds is their gregariousness and formation into rather large flocks that causes competition with songbirds and other more desirable species. Urban roosts pose a number of problems to residents of cities, suburbs, or towns. Starlings are the most common problem, although other species may be present. The filth caused by their droppings and nests is highly undesirable. Large roosting concentrations of birds can lead to a potential public health problem.

Figure 4–19 Bald eagle.   **Figure 4–20** Starling.

The droppings form a medium for the growth of bacteria and fungi. In addition, birds may act directly as carriers or vectors for some diseases. Histoplasmosis is a respiratory disease in humans caused by inhaling spores from the fungus *Histoplasma capsulatum.* Birds do not spread the disease directly; the spores are spread by the wind and the disease is contracted by inhaling them. But the birds' droppings enrich the soil and promote growth of the fungus.

Salmonellosis, a form of food poisoning, is a common disease. It is an acute gastroenteritis produced by members of the salmonella group of bacteria pathogenic to people and other animals. The organism can be spread in many ways, one being through food contaminated with bird feces or with salmonella organisms carried on the feet of birds.

Pigeons, similar to those now living in a semiwild state in towns and cities, have been closely associated with people since before recorded history. Pigeons utilize structures such as barns, city buildings, bridges, and overpasses almost exclusively for their roosting and nesting sites. Roosting on awnings, ledges, and the like often causes problems because of their droppings. Excessive numbers of pigeons can cause property damage and may constitute a health hazard.

The pigeon is the wild bird species most commonly associated with the transmission of ornithosis (psittacosis) to humans. Ornithosis is caused by a viruslike organism and is usually an insidious disease with primarily pneumonic involvement, but it can be a rapidly fatal infection. Birds have become adapted to the disease and show no symptoms, but act as healthy carriers, shedding the organism in their feces, which later become airborne in dust. The disease may also be contracted from parakeets or farm poultry.

Reptiles

Snakes. Snakes are secretive and usually prefer to move away when disturbed. During their active period, snakes feed on a variety of animal life, such as frogs, toads, salamanders, insects, worms, small rodents, and birds.

Some Vertebrate Pest Animals

The majority of snakes are harmless. However, poisonous snakes do occur in the United States, and it is important to know the difference between a harmless snake and a poisonous one. Poisonous snakes, such as copperheads, rattlesnakes, and cottonmouth moccasins, belong to the pit viper family. The name comes from the pit or opening in the side of the head between the eye and the nostril (Figure 4–21). The pit is absent in nonpoisonous snakes. The poisonous snakes mentioned have a vertically elliptical eye pupil, and nonpoisonous species have a round eye pupil.

Another distinguishing characteristic is the pattern of scales on the underside of the tail. Poisonous snakes have one row of scales, whereas harmless species have two rows of scales on the underside of the tail (Figure 4–22). Because observation of these distinguishing characteristics requires rather close examination and handling of the snake, they have limited usefulness as field identification markings, but they are useful in correctly identifying a dead snake suspected of being poisonous.

The incidence of snake bites is very low in the United States, yet sensible precautions should be observed. Do not handle snakes, alive or dead, without being thoroughly familiar with harmless or poisonous ones. There is no reason for concern about the occasional nonpoisonous snake found in the field, forest, or home garden. Even an occasional poisonous snake should not cause panic. However, when snakes invade homes or become common in urban areas or when poisonous species are frequently observed, reductional measures may be required.

The most effective and lasting method to get rid of snakes is to make the area unattractive to them. Remove cover or other shelter such as board piles, debris, or trash piles, all of which are good protective cover. Keep vegetation mowed around such areas and be sure that there are no rodents available as they are important prey of many snakes. All cracks and crevices that snakes may use to enter houses and other buildings should be plugged and snake-proofed. There are no pesticides registered to repel or kill snakes.

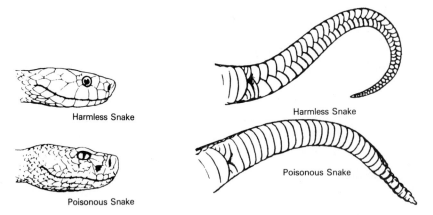

Harmless Snake

Poisonous Snake

Harmless Snake

Poisonous Snake

Figure 4–21 Snake "pit" identification characteristics.

Figure 4–22 Tail identification characteristics of snakes.

Other Animals

Other animals that can, on certain occasions, cause problems include armadillos, badgers, weasels, mink, opossums, beavers, muskrats, nutria, snapping turtles, and a few other bird species. Tracks, droppings, tooth marks, dens, feathers, burrows, and trails cut through vegetation by small rodents will help determine if the damage was caused by wildlife and, if so, by what species.

GENERAL WILDLIFE DAMAGE CONTROL MEASURES

1. All wildlife species have both positive and negative social and economic values. When a wildlife species or congregation causes serious economic losses or becomes a public health threat or an unbearable nuisance, control of the problem is the legitimate right or obligation of the individual or agency in the position of responsibility.

2. Wildlife damage control techniques should be applied in proportion to the problem.

3. Damage control techniques include the following categories, listed in general order of public acceptance:
   a. Tolerance of losses
   b. Mechanical exclusion or protection
   c. Repellent devices and sounds
   d. Live trapping and transfer
   e. Habitat manipulation: flooding, clearing, burning
   f. Biological control: disease introduction, encouraging predators
   g. Mechanical lethal control
   h. Chemical control

4. Control methods should be selected on the basis of cost-effectiveness and after careful weighing of available alternatives.

5. The most feasible control methods are those that result in the minimum effect on the problem species, maximum reduction of losses, and the least negative impact on related natural systems.

6. Control methods should be as efficient, safe, economical, selective, and humane as possible under the circumstances presented.

7. Control methods should be selected in conformity with local, state, and federal law.

8. In situations where endangered species may be affected, special care must be taken to ensure that provisions of the Endangered Species Protection Act are complied with.

9. Control processes should be applied in a low-key, responsible manner, without vindictiveness or public display.

DAMAGE CONTROL TECHNIQUES

The second law of thermodynamics states: For every action in the universe there is an equal and opposite reaction. This concept applies equally to wildlife damage control.

Tolerance of Losses

Human tolerance of wildlife damage is generally in direct proportion to the intensity and cost of the damage, and the directness of the impact on the human who suffers as a result of that damage. Damage by wildlife is easy to tolerate when it happens to a neighbor.

Some people are willing to tolerate very serious depredations; others will tolerate very little depending on their backgrounds and personal beliefs.

Considerations

- Toleration of the damage allows the pest to continue to exist and perhaps multiply to further compound damage.
- The pest may infest neighboring situations and become a hardship for others because of the lack of control.
- The pest may reach carrying capacity and the population may decline to a more suitable tolerance level without action.
- The pest may overshoot the carrying capacity, contract an epizootic, and die back to tolerable levels.
- Pest species may produce an effect on other creatures that reduces overall damage. The pest may pay its way by its action on mice and rats, insects, or weeds.
- The pest may provide recreation, food, or income values outweighing its negative values.
- If it costs more to control a species than that species can be expected to destroy, the logical approach is to put up with the damage.

Mechanical Exclusion or Protection

The most effective method of wildlife damage control is to fence out or exclude depreciating species. The construction or modification of buildings, fences, and machinery to exclude wildlife is often the cheapest method in the long run.

Considerations

- Some crops and livestock production require large areas, making exclusion fencing prohibitively expensive and impractical.
- Fencing may pen the depreciating species in and compound the damage.
- Fencing may severely affect nontarget species; for example, fencing to exclude coyotes may result in a serious disruption of deer, antelope, and elk migration lanes. In the case of antelope, it can result in a large reduction of carrying capacities by fencing out water and necessary plant groups and space.
- Fencing may create added mortality as a result of accidents and restricted escape routes from predators.
- Fencing an open range and forested areas is difficult and costly to maintain. Wildlife often figure out means of defeating the fence.
- Temporary netting and screening work well for bird and bat control but again may be too costly in time and labor for practical use and do not add to the esthetics of the building.

- Exclusion from food or protection often ultimately results in lower carrying capacity for the pest species involved. Exclusion may result in hardship or death of the pests.

Repellent Devices and Sounds

Repellents generally work on the principle of scaring or warding off pests without consequence to the health of pests.

Sound. Zon guns, firecrackers, firearms, music, distress calls, and ultrasonic frequencies are examples of sound repellents, which work through the sense of hearing and animal communication. They are generally irritating to the animals, or as a result of past experience associating danger with the noise, the pest is frightened off or cannot tolerate the sound and leaves.

Advantage

- The pest is not killed and can go about its beneficial role with little impact on nontarget critters and the surrounding environment.

Disadvantages

- The offending pests may be driven over to the neighbors to create a problem there.
- The pest may get used to the noise and continue the damage.
- Cost and time involvement may be high with questionable results.
- When the pest becomes accustomed to the noise associated with food, the noise may attract the pest.

Visual repellents. Scarecrows, silhouette images, and similar devices have been used for centuries to scare off pests. Lighting or flashing lights may be effective for nocturnal species, especially when accompanied with noise.

Considerations

- The effectiveness of repellents depends on the pest involved, the duration of time over which damage occurs, the intensity of the pests' past experience, and the delivery method of the repellent stimulus.
- Repellents all vary considerably in effectiveness.
- The longer they are used, the less effective they are.
- Several repellent devices delivered on alternate days, separately and then in concert, are more effective than one method used continuously.
- Repellents in many cases simply transfer problems from one place to another, defer losses to a later date, or condition the pest to tolerate higher noise and light levels. As one sheep producer put it, "Sure, I put up high density light in my feedlot and I also put on hard rock music from dark 'til dawn. I had the most contented coyotes in the state. They picked out the choicest lamb by the light and sat down to eat it to music."

Live Trapping and Transfer

Live trapping and transfer is a preferred method for endangered species and valuable game animals, and in areas of delicate public relations. It involves catching the pests, removing them to a sufficient distance to ensure against their return, and releasing them alive.

Advantage. Valuable animals are placed out of a deprecative situation without destruction of the animal. Black bear, mountain lion, eagles, and other protected species are often handled this way. It is popular with the public.

Disadvantages. It is expensive in time and materials cost. It often transports the problem to a new area. Migration back to the complaint site is common with some species. Survival of transported animals is often very low as a result of intraspecific territorial strife that results with the animals already at the introduction area. Diseases such as rabies may also be spread to new areas. The net effect of the transplant is often death of the animal at a tremendous cost in time and money.

Habitat Manipulation

Habitat manipulation is a widely accepted method of control. It entails destruction of food, shelter, protection, or other habitat requirements of the pest species. Burning of fence rows and irrigation ditch banks, cleanup of old wood piles and machinery lots, and destruction of trees are effective in removing nest and roosting areas of pest birds and mammals. Destruction of winter food sources by cultivation and burning is also used successfully to reduce depredations of some wildlife species. Flooding areas to drown out rodents (gophers, ground squirrels) is an example of habitat manipulation.

Advantages. Once these methods are established and the initial costs are absorbed it is fairly easy to maintain depredation-free situations for many pest species. It is acceptable to the ignorant protectionist since wildlife is not directly killed. It is often extremely effective in a negative way.

Disadvantages. It sterilizes the area for both pest and beneficial species. The end result is very low levels of wildlife for recreational and esthetic enjoyment. *Of all methods listed, this one is perhaps the most damaging to wildlife and should be recommended only as a last resort.*

Biological Control

There is a story about an old range rider who used to pick up black plague stricken prairie dogs, put them in a gunnysack, and release them into healthy prairie dog colonies. The black plague reduced the prairie dogs very effectively. The rider somehow survived it. This is an example of biological control.

Biological control has been highly touted as the way of the future. It, however, has some serious problems. Environmentalists have applauded this as a general alternative to chemical control.

Biological control agents can be categorized into infectious disease agents, parasites, or predators. Myxomatosis virus disease was introduced into Australia to control rabbits. It worked. However, each species of pest requires different disease organisms to affect it. Infectious diseases with the capacity to greatly reduce some pests have the capacity to reduce humans as well. Bubonic plague, rabies, and tularemia might be developed for use on mammal pests, but the public health implications are too negative to allow it. Generally, parasitic species are host-specific and do not offer the potential for application and effective treatment in pest control situations under the most optimistic assessments.

Predators are commonly given general credit for controlling rodent populations. The literature of wildlife does not support those claims; the reverse is more accurate. Depending on the pest involved, sometimes predators can be of assistance (e.g., bull snake and Norway rats), but very few creditable examples are available. The livestock producer losing lambs or calves to coyotes is not generally impressed with the amount of grass a coyote saves for his cows by eating mice and rabbits. The space-territorial requirements of predators are usually much larger than the area over which these predators could feasibly control or eradicate rodent pest populations.

Although biological control is a popular idea, it has not been shown to be of much practical help in solving wildlife pest management problems. A regime of excellent range and wildlife management may contribute to lower livestock losses to large predators by encouraging a surplus of rabbits, mice, and other alternate prey. It also encourages resistant, alert, and healthy livestock to better ward off predator attack. Many losses of livestock are random events, impossible to predict or protect against. They may occur again the next day, or not for another 5 years.

Occasionally, an animal species is affected by some sort of disease that helps to reduce the total population. Unfortunately, diseases frequently occur in species whose numbers are small (this is probably why they are small), and often the species is not one that is of concern as far as damage or other problems. An example of this is lung disease in bighorn mountain sheep and sometimes elk.

Mechanical Lethal Control

Mechanical lethal control methods (traps, snares, and shooting from airplanes and snowmobiles or with the aid of calls and spotlights) have come under considerable attack recently by preservationist groups. Each tool has limitations in effectiveness and to some people the tools violate the "fair chase" hunting ethic. Each tool has a necessary place in control. Contrary to popular belief, the direct killing of a depredating individual animal or animals has the least environmental impact of all the methods discussed to this point except tolerance and perhaps live trapping. In the overall management of pest species, the killing of a few individuals that are a problem does little more than duplicate the natural mortality forces working on that species. There is no destruction of habitat, no transfer and continuance of problems, or other adverse effects. Widely used mechanical lethal control means include the following:

Steel leghold traps, snares, killer traps. Used properly, these are effective, humane, and efficient. They have no secondary effects and used properly are very selective. Traps are inexpensive and can be applied and withdrawn at will to meet temporary problems.

Shooting: Calling, aerial shooting, shooting from a vehicle, catching with hounds. These methods are selective, of moderate efficiency, and of varying expense—from very inexpensive as in calling to very expensive as in shooting from helicopters. These methods have been often condemned by some groups as being unfair, cruel, and inhumane.

Advantages. These methods are generally very selective to individual pests or pest species causing depredations. They are of varying expense but response time is quick and termination of losses may be immediate.

Disadvantages. These methods are susceptible to weather conditions and are based on human involvement. Control with air or land vehicles is dangerous and expensive. Some pests have habits that do not leave them vulnerable to these types of control methods. Cost in human effort is often high. In situations where pest density levels require reduction, mechanical methods have not been shown to be adequate to achieve population reduction.

Chemical Control

Chemical control of wildlife can be broken into seven major categories:

1. Lethal toxicants applied orally (strychnine, warfarin, 1080)
2. Lethal toxicant contact poisons (endrin)
3. Fumigants (gasoline, phostoxin, chloropicrin, sulfur dioxide)
4. Chemosterilants (diethylstilbestrol)
5. Aversive conditioning agents (lithium chloride)
6. Saponificants (soaps)
7. Repellents (mothballs, bone tar oil, lion dung, Thiram, and cinnamonaldehyde)

In recent years, chemical use on vertebrate pests has diminished considerably. Options on chemicals and application alternatives are extremely narrow. According to the EPA, no chemicals are registered for use on amphibians and reptiles. Several of the commonly recommended repellent and fumigant chemicals are not formally registered for such use; for example, mothballs cannot be legally recommended for repelling skunks but can be for bats. Aluminum phosphide is registered for control of prairie dogs and other rodents. Several of the old faithful toxicants, including strychnine and 1080, have had the registrations canceled for use on predatory wildlife, leaving only sodium cyanide used in the M-44 device available for controlling canine damage. Appeals have been made by livestock interests for continued use of 1080 and strychnine, but have not met with favor. There are no orally delivered toxicants registered for noncanine predators. Check local agencies and recommendations for up-to-date registrations for wildlife control.

Use of chemical toxicants to control wildlife is not popular and has been the subject of heated debate, public pressure, and a great deal of inaccurate description and legend. Like any tool people use, chemicals can be and have been abused. Judiciously and carefully used, chemicals are safe, selective, economical, and often the most feasible of alternatives.

Lethal toxicants applied orally (strychnine, warfarin, and 1080). These chemicals are ingested into the digestive tract through baits and lead to the death of the animal. They have a wide range of effects, residual half-lives, modes of action, and effectiveness. Each chemical carries unique qualities of taste, smell, acceptance, physiological reaction, and delivery modes.

Toxicity of each chemical varies with the physiology of each species; for example, warfarin is deadly to rats and mice, but is relatively harmless to chickens and predators. Strychnine is lethal to anything that eats it in sufficient dosages. Strychnine in fresh dead animals can cause secondary poisoning to scavengers. It does, however, break down quickly in the soil or dead animals, rendering it harmless. The baiting system used, placement of baits, dosage levels used, and time of placement all affect the selectivity of the toxicant to the target animal.

Advantages

- A great degree of selectivity to target animals can be attained.
- A range of chemicals can be used for the conditions presented.
- Often very effective in terms of time and cost investment.
- Often the most humane method available.
- Offer a great deal of flexibility in terms of bait systems, placement, and exposure time for handling the wide variety of terrain and weather conditions encountered in wildlife damage control.
- Usually quite efficient in solving problems quickly and economically.

Disadvantages

- Risk to humans, pets, and livestock varies with each chemical and the skill of the user.
- Emotionally and politically a hot issue.
- Some risk of primary and secondary poisoning to nontarget species with some chemicals and delivery systems.

Lethal toxicant: Contact poisons (endrin). These chemicals are placed such that the chemical is absorbed through the skin by contact. Much of what was covered in the previous section applies here. Very few chemicals are registered for this application method; most of these are for birds and bats. Advantages and disadvantages are the same as for orally ingested toxicants.

Fumigants (aluminum phosphide, sulfur dioxide, calcium cyanide). These chemicals release toxic gases that are inhaled and result in death of the pest. They are widely used and are effective for some building and burrow dwelling wildlife.

Advantages

- Extremely effective and usually economical. When used correctly they are quite selective.
- Little risk of secondary hazard to people or nontarget species if properly applied.

- Usually results in immediate termination of damage.
- Results in quick, humane death of the pest.

Disadvantages

- Requires trained people to handle them. Fumigants are hazardous and highly toxic to people unless proper precautions are taken.
- Not publicly popular.

Chemosterilants (diethylstilbestrol). Chemosterilants have been researched quite extensively during the past 20 years. Ornitrol® is registered for use on pigeons. There is speculation as to the possibilities for others. There are two basic problems with regard to development of chemosterilants for carnivores:

1. How can chemosterilants be delivered to 2.5 million coyotes over the western United States with enough efficiency to make a difference? Some coyotes will not eat dead bait materials. If they did, 1080 would have controlled them with more efficiency.
2. No chemical is available that has all the needed characteristics (that is, tasteless, odorless, residual effect for 2+ months, 100% sterility for the life of the animal, no effect on nontarget birds and animals, and biodegradable, among other requirements).

Frankly, science is not very close to releasing a practical chemosterilant for most wildlife pests.

Aversive conditioning agents (lithium chloride, Bitrex®, cinnamonaldehyde). Basically, these are chemicals that when ingested or tasted make the pest sick or give off a bad taste. The pest associates its illness with the ingestion of the bait and supposedly will not eat that material again. Lithium chloride was given a lot of publicity recently as showing great promise for use in coyote damage control, but subsequent research has failed to show an efficient delivery system or successful field application. There are many problems involved with the idea.

Advantages

- Quite specific to the pest animal.
- Nonlethal to the pest animal.

Disadvantages

- A delivery system must be developed to get the material to the coyote and the coyote must eat it.
- The coyote must make the association between illness, the bait, and the live animal that needs protection (the sheep).
- The coyote has to have a memory so it will avoid eating the sheep again.
- The chemical has to fulfill all other requirements listed under the discussion of chemosterilants except sterility, plus others to get EPA registration for that use.

Aversive conditioning chemicals are a long way from being a field-ready tool.

Saponificants (soaps). These are used primarily in controlling large concentrations of bird pests (starlings and blackbirds). The chemicals are sprayed on in water solution during cold weather. The soap breaks down the feather oils and wets the bird so that its body heat dissipates. The bird dies of hypothermia.

Advantages

- It can be quite selective, effective, and economically feasible.
- Very little secondary or negative side effect.

Disadvantages

- The effectiveness of this method varies with weather conditions (temperature), tree cover, and densities of the roosting birds.
- It requires planning and high-pressure spraying equipment or aerial spraying equipment.
- Some states and communities require permits before this method is used.

Repellents. Examples of repellents include thiram (very effective against deer and rabbit browsing), bone tar oil, and mothballs. These chemicals by their action scare off or irritate the pest, resulting in less economic loss.

Advantages

- The pest is not affected, so it can continue its positive roles.
- Often inexpensive to apply and effective, depending on the pest.

Disadvantages

- Materials may adversely affect domestic livestock and handlers.
- May just delay the problems or drive them over to the neighbors.
- Odor may become acceptable or even an attractant to the pest after prolonged use.
- Most repellent chemicals are not readily available.

WHAT DOES THE LAW SAY?

Laws affecting wildlife damage control are extensive, specific, and generally allow for quick relief for most serious problems. Persons and agencies dealing with wildlife damage control are advised to read local ordinances and state and federal laws carefully. Legal control measures for some species are very limited, especially with regard to endangered species.

Wildlife is considered publicly held property, the use of which is regulated by the federal and state governments. The Migratory Bird Treaty protects all migratory birds, with authority for regulation of hunting seasons and management given generally to the USDI, U.S. Fish and Wildlife Service, and state game and fish agencies. Certain pest species are listed as exceptions (English sparrow, European starling, and domestic pigeon). State laws generally are applied to these. Blackbirds, magpies, crows, and great horned owls can, under certain conditions, be destroyed without federal permit by mechanical and chemical means. Eagles are protected under the Migratory Bird

Treaty and the Bald Eagle Act. The Endangered Species Protection Act protects threatened wildlife. The Lacey Act prohibits interstate shipment of wildlife taken in violation of state laws. These laws and treaties were enacted and are enforced from federal and state levels.

Very few chemicals are registered for use on wildlife. The few labels that are cleared are specific to a narrow range of target species and circumstances. Chemicals (lethal or repellent in nature) cannot be legally used on species other than those listed on the label. The 1972 amendment of FIFRA and EPA's restrictive interpretations of FIFRA have resulted in severe restrictions on lethal and nonlethal chemical control of wildlife. Registration requirements are extensive and expensive. Since chemical use on vertebrates is periodic and minor in terms of amounts of chemicals and monetary return to the chemical industry, it is not feasible for the industry to register many effective vertebrate control products. This leaves the responsibility for registration of wildlife control chemicals to state or federal agencies who have shown little enthusiasm for the task.

Because of the biopolitics and public pressures involved, many historically effective chemicals like strychnine and 1080 are not appreciated by the federal agencies, which are the only groups with the money and capability for collecting the data required for reregistration of the products.

SUMMARY

Wildlife damage control is presently in an adjustment period from one of wide flexibility and personal freedom to one of narrowed alternatives and heavy state and federal agency regulation. Control methods and tools should be carefully selected after consultation with responsible agencies. Anticipate problems before they occur and manage around them. Most problems with wildlife can be simply and inexpensively handled with the proper information, a little imagination, and good sense.

CANADA THISTLE

5
Weeds

People are more responsible for the spread of weeds than any other single factor. Most of the plants that are commonly thought of as weeds have existed in parts of the world for many years, but were relatively insignificant before people started growing plants for food. Thus weeds have evolved along with crops and are objectionable because of their ability to compete with the plants we try to grow for food or fiber. The original habitat of many weeds is unknown. Many weeds are closely related to cultivated crops or to ornamental plants, and it is this close relationship that makes some weeds more difficult to control than others.

A weed can be defined in many ways. Some of the more common definitions are: a plant species growing where it is not desired (Figure 5–1), a plant out of place, or a plant that is more detrimental than beneficial. Kentucky bluegrass that spreads from a lawn into a flower bed is very much a weed; likewise, volunteer corn in a sugar beet or bean field is as much a weed as lambsquarters or redroot pigweed. Thus a plant is a weed only in terms of human or, even more specific, individual definition. A plant that is a weed to one person may be a desirable plant to another. Any plant can be a weed in a given circumstance.

CHARACTERISTICS OF WEEDS

Plants that are commonly referred to as weeds have certain characteristics that give them the ability to spread and exist where most cultivated plants would soon die out. Some of these characteristics are as follows:

- Most weeds produce an abundance of seed.
- Many have unique ways of dispersing and spreading their seed.

Figure 5–1 Any plant can be considered a weed if it grows where you don't want it to grow.

- Seeds of many weeds can remain dormant in soil for long periods of time.
- Most weedlike plants have the ability to grow under adverse conditions.
- Weeds can usually compete for soil moisture, nutrients, and sunlight better than crop plants.

Modern agricultural practices favor invasion by weeds. Plant communities are complex, and under any particular set of environmental conditions (climate, temperature, rainfall, soil), there is a natural progression to a climax vegetation. Humans change this natural progression by growing crops in pure stands or as individual plants. Under natural conditions, single plant species will usually not monopolize an area because any single species will have too short a growth cycle to cover the ground for the entire growing season, or cannot use all the available sunshine because of leaf area development, or will not use all the available water or nutrients because of its root system. In nature, other plant species (weeds) move in to use these wasted resources. Our ability to manage vegetation to meet our needs for food, livestock feed, and fiber is one of the most important factors to our survival. To produce crops efficiently or to grow plants for ornamental purposes, it is necessary to minimize the competitive effect of weeds. This must be accomplished selectively. A weed control method or combination of methods ought to minimize the yield-reducing effect of weeds, while, in turn, assisting the crop.

WAYS IN WHICH WEEDS INJURE PLANTS OR AFFECT LAND USE

Weeds affect our lives in many ways. Not only do they cost us money, but they can cause untold misery and grief to hay fever sufferers, as well as being poisonous to humans, livestock, and wildlife. Weeds may affect us in the following ways:

- Reduce yields: competition for moisture, light, and nutrients
- Reduce crop quality: weed seeds, dockage, weeds in hay, straw, and the like
- Increase production costs: additional tillage of farm crops and cultivation of nursery crops

- Increase labor and equipment costs: machinery wear and tear, for example
- Act as insect and disease carriers or hosts: wheat stem rust, corn borer, pine needle rusts, and numerous viruses
- Poisonous or irritating to animals or people: cocklebur seedlings, poison ivy, hay fever, and the like
- Increase upkeep of home lawns and gardens
- Create problems and increase costs in recreation areas such as golf courses, parks, and fishing and boating areas
- Increase upkeep and maintenance along highways, railroads, and irrigation ditches
- Reduce land values, especially the presence of perennial weeds
- Limit cropping system choice (Some crops will not compete effectively against heavy weed growth.)
- Increase reforestation costs because of a slower rate of growth due to weed competition

WEED CLASSIFICATION

Green plants are basic for life and are indispensable in our environment. They are a complex life form that utilizes energy from the sun—combined with minerals, water, and carbon dioxide—to provide food for people and wildlife, to beautify the landscape, and to reduce soil erosion. Plants can be classified in several ways; one simple classification is according to life cycles. Weeds may be classified into four major groups: summer annuals, winter annuals, biennials, and perennials.

Summer annuals (Figures 5–2, 5–3, and 5–4) live 1 year or less. They grow every spring or summer from seed; they produce seed, mature, and die in one growing season. Seed from most summer annual weeds germinates during a 2-month period in the spring. The seed produced then lies dormant until the next spring. There are usually a few seeds of every species that germinate during the summer or fall, but seldom in numbers equal to the spring flush. Of the millions of weed seeds lying beneath every square foot of land, probably not more than 50% will germinate in any growing season. Summer annual weeds are the greatest problem in spring when annual crops are planted. The new seedlings of field crops are competing directly with weeds because their seeds germinate most readily in disturbed soil. Examples of summer annuals are Russian thistle, redroot pigweed, lambsquarters, puncture vine, wild oats, and crabgrass.

Winter annuals (Figures 5–5, 5–6, and 5–7) germinate in the fall or early winter and overwinter in a vegetative form (without flowering). In the spring, they flower, mature, and set seed. They die in late spring or early summer. Seed from most winter annual weeds is dormant in the spring and germinates in the late summer or fall. Some species such as common chickweed can germinate under snow cover. These weeds start growth at the first sign of spring, and many species bloom and produce ripe seed by mid-May or June. This means that some winter annuals can reseed themselves before late crops are planted. Examples of winter annuals are tansy mustard, blue mustard, downy bromegrass, chickweed, and shepherd's purse.

Figure 5–2 Large crabgrass: *Digitaria sanguinalis* (L.) Scop.

Figure 5–3 Lambsquarters: *Chenopodium album* L.

Figure 5–4 Redroot pigweed: *Amaranthus retroflexus* L.

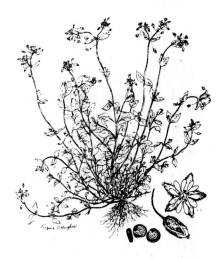

Figure 5–5 Common chickweed: *Stellaria media* (L.) Cyrillo.

Figure 5–6 Shepherd's purse: *Capsella bursa-pastoris* (L.) Medic.

Figure 5–7 Downy brome: *Bromus tectorum* L.

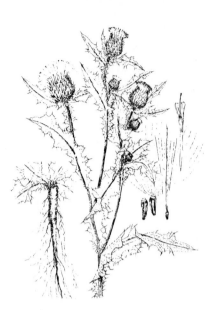

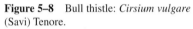

Figure 5–8 Bull thistle: *Cirsium vulgare* (Savi) Tenore.

Figure 5–9 Common mullein: *Verbascum thapsus* L.

Biennial plants (Figures 5–8 and 5–9) have a similar life cycle to annuals because they die after flowering and setting seed, but they require 2 years to complete the sequence. Growth during the first year is usually vegetative (no flowering activity) and low-growing, frequently as a rosette. Flowering and seed production occur during the second year. Biennial weeds, though relatively few, may become established from seeds anytime during the growing season. Because true biennials never produce flowers or seeds the first year, but instead form a rosette of leaves that usually lie close to the ground, such weeds are well adapted to lawns, pastures, hayfields, and orchards where grazing or mowing removes very few of the rosette leaves. Consequently, growth and energy storage in the roots or crowns continue until a killing frost. This large amount of stored energy supports spring growth ahead of other pasture and forage species. If biennial weeds are not controlled before this time, there are no forage species that will crowd them out. Examples are bull thistle, wild carrot, common mullein, and musk thistle.

Perennial weeds (Figures 5–10, 5–11, 5–12, and 5–13) become established by seed or by vegetative parts, such as rootstocks or rhizomes. Once established they live for more than 2 years. Perennials are usually herbaceous (top growth usually winter kills), woody brush, or trees. Since perennial weeds live indefinitely, their persistence and spread is not as dependent on seeds as are the other three groups. Seed is the primary method of introducing these weeds into new areas, but perennial weeds are often spread during soil preparation and cultivation. Most perennial weeds that spread by rhizomes or rootstocks will spread in circular patches if left undisturbed. In crop fields, patches of perennial weeds such as Canada thistle spread in oblong patches in the direction the field is worked.

Figure 5–10 Yellow nutsedge: *Cyperus esculentus* L.

Figure 5–11 Dandelion: *Taraxacum officinale* Weber.

Figure 5–12 Quack grass: *Agropyron repens* (L.) Beauv.

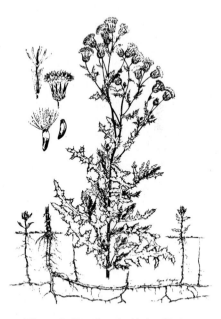

Figure 5–13 Canada thistle: *Cirsium arvense* (L.) Scop.

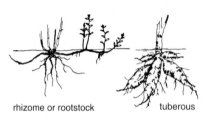

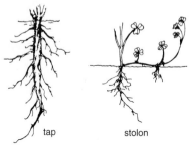

rhizome or rootstock   tuberous

tap   stolon

Figure 5–14 Types of underground structures.

Perennial weeds grown from seed are no more difficult to control than any other kind of weed coming from seed, but once established, perennial weeds are the most difficult to control. They compete with any crop, especially row crops where competition between rows is not too great. Perennial weeds also get an early start in the spring and compete with perennial crops, such as alfalfa and other forage species. Examples are field bindweed, Canada thistle, Russian knapweed, johnsongrass, yellow nutsedge, dandelion, quack grass, and many tree and brush species.

Some perennials have underground structures in addition to true roots (Figure 5–14). Such structures are *rhizomes,* a modified underground stem (johnsongrass, Bermuda grass); *bulbs* (wild onion, death camas); *tubers* (nutsedge); *creeping roots* (field bindweed, Canada thistle); and *taproots* (dandelion and plantain). These underground structures serve as food storage organs and produce new shoots. A few perennials have aboveground roots called *stolons.* Perennial plants are the most difficult to control because of their ability to reproduce in several ways.

WEED IDENTIFICATION

Correct identification of weeds is essential for an effective weed control program. It may not be necessary to identify plants down to the individual species, but it is necessary that plants be properly identified as to major weed groups. Many plant species can create weed problems and most of them are flowering plants (the higher forms of life that produce seed).

Some herbicides are effective on some groups of weeds, while other herbicides perform best on other plant groups. Very often, poor results with herbicides can be attributed to using the wrong herbicide for control of a weed. For example, alachlor (Lasso®) does not generally kill mustard species or lambsquarters, atrazine (AAtrex®) will not usually control puncture vine or Russian thistle, and trifluralin (Treflan®) does not usually control nightshade. Often a crabgrass herbicide is applied on turf that is infested with perennial weedy grasses. Crabgrass herbicides control annual grasses but have little effect on perennial species. Therefore, it is important that an applicator of herbicides know the weed species to be controlled and use the herbicide that is most effective on that weed.

Weeds can be divided into two broad groups depending on their leaf characteristics.

1. *Narrowleafed weeds.* These include grasses, sedges (bullrush, nutsedge), rushes, cattails, and several other plants that are less often weed problems, such as iris. These plants can be identified by the characteristic parallel veins in the leaves.

2. *Broadleafed weeds.* This is the larger of the two groups and includes all of the broadleafed weeds, as well as most trees and many other brush species. In this group, the leaf veins are netlike, not all parallel. This group includes mustard, dock, pigweed, purslane, field bindweed, and many others.

The decision to use a preplant or preemergence herbicide has to be made before weeds germinate. Therefore, it is important to know the history of the weed species in a field (what weeds were there the year before) in order to make the right decision as to what herbicide to use.

Seedling identification is necessary to properly select postemergence herbicides in many cases. Wrong identification can lead to less desirable results. Numerous publications will help an applicator identify weeds. These are available from state universities or the U.S. Department of Agriculture. Weeds can be identified by simply looking at pictures. A more specific method uses keys that systematically describe different characteristics of plants and require decisions between two possibilities. However, the use of plant identification keys requires knowledge of plant characteristics and experience in distinguishing between plant parts that are not always obviously different.

The following is an example of a small part of an analytical couplet key to the major genera of one plant family:

KEY TO SELECTED GENERA

1. Fruit a group of achenes; flowers not spurred..2
   Fruit a group of follicles; flowers spurred..4

2. Petals none..3
   Petals present ....RANUNCULUS

3. Sepals usually 4; involucre none..CLEMANTIS
   Sepals usually 5; involucre present...ANEMONE

4. Flowers regular; spurs 5 ....AQUILEGIA
   Flowers irregular; spur 1 ....DELPHINIUM

To use such a key, the specimen is examined for the characteristics described after a choice is made and results in a referral to the next applicable couplet.

It must be remembered that there are approximately 130 different plant families, ranging from *Aceraceae* to *Zygophyllaceae,* and each family may have hundreds of subfamilies, genera, and species.

WEED CONTROL METHODS

Methods that can be used for weed control include (1) prevention, (2) crop competition and rotation, (3) burning, (4) mechanical, (5) biological, and (6) chemical. Although chemical control is often thought of first, it is not a panacea for all weed problems. Alternative control methods should be identified and included. Good farming or other good land use practices should always be foremost in any weed control program.

1. *Prevention* is the most practical method of dealing with weeds. If weeds are not allowed to infest an area and seed is not produced, there is much less chance of weeds becoming established. Once weeds have become established, they are difficult and costly to control and may persist for many years if the seeds can lie dormant in soil. Field bindweed seed can remain dormant and alive in soil for 40 years or more.

   Preventive measures should be adopted where practical and should be the first step in any effective weed management program. These should include the following:
   - Always use clean seed.
   - Do not feed grains or hay containing weed seeds without destroying their viability.
   - Do not spread manure unless the viability of the weed seeds has been destroyed.
   - Livestock should not be moved directly from infested to clean areas.
   - Make sure harvesting equipment is clean before moving from infested areas.
   - Avoid the use of soil from infested areas.
   - Inspect nursery stock for presence of weed seeds or other weed parts.
   - Keep irrigation ditches, fencerows, roadsides, and other noncropped areas free from weeds.
   - Prevent the production and spread of weed seeds by wind when possible.
   - Use weed screens for trashy irrigation water.

   Obviously, the implementation of an effective weed prevention program requires alertness and perseverance.

2. *Crop competition and rotation* is probably the cheapest and easiest method of managing weeds; it is based on the law of nature—"survival of the fittest." A crop will survive and flourish if it can compete more efficiently for sunlight, water, and nutrients than the unwanted plants. It is important to understand growth habits of crops in relation to weeds. Early weed competition is usually more detrimental to a crop than later competition when the crop is well established. Crop rotations can be a means of controlling weeds; certain weeds are more common in some crops than in others, and some crops are more competitive with certain weeds than others.

3. *Burning* has been used for many years to control unwanted vegetation. Selective burning is used to kill weeds in row crops. Controlled burning is a valuable tool in conversion of brushlands to productive grazing lands. Fire is an effective tool for removal of vegetation from ditch banks, roadsides, fencelines, and other waste areas. Intense heat can sear green vegetation that often

grows abundantly in irrigation ditches. Burning can kill small weed seeds on the soil surface.

4. *Mechanical control* includes cultivation, mowing, hoeing, hand pulling, root plowing, chaining, and tree grubbing. All these methods involve the use of tools to physically cut off, cover, or remove undesirable plants from soil. Cultivation or tillage is the most common method of weed control. It is effective on small annual weeds, but when the plants are larger, tillage effectiveness may be reduced. Tillage can also be used to disturb perennial root systems. However, repeated tillage operations are usually required to effectively control perennial weeds.

5. *Biological control* includes the use of insects and diseases to control weeds, but not harm desirable crops. Biological controls will usually not eradicate weeds because the host plant is necessary as a food source for the predator of the weed species. Examples of biological controls are a moth borer from Argentina used to destroy prickly pear cactus in Australia and the *Chrysolina* beetle used to control St. Johnswort.

6. *Chemical controls or herbicides* are used for killing or inhibiting plant growth. Selective herbicides date back to the turn of the century, but the first breakthrough with selective herbicides occurred with the introduction of 2,4-D in the early 1940s. Since then, the development of organic chemicals for weed control has expanded. There is now a large number of herbicides on the market. With the different formulated combinations, the total number available is over 200. Herbicides are now applied to more agricultural land in the United States than insecticides and fungicides combined.

CLASSIFICATION OF HERBICIDES

There are several ways to classify herbicides. One is based on chemical structures and effect on plants, and another is the separation of herbicides into families. The most practical classification for field use is based on how herbicides are used.

Classification of Herbicide by Action

The following classification is based on whether the herbicide is applied to the foliage or the soil, the pathway by which the herbicide enters the weed, and whether the herbicide is selective or nonselective.

Foliage, contact, nonselective (Figure 5–15). An herbicide applied to the weed's foliage, in the absence of a crop or directed underneath a crop, kills the foliage the herbicide contacts with little or no translocation to underground or protected parts of the weed. Because biennial and perennial weeds normally have dormant and protected buds with stored energy in the crown or root systems that produce new growth, these weeds will recover after treatment. Small annual weeds are completely and permanently controlled. For example, paraquat applied before a no-tillage corn planting or used as a directed spray underneath fruit trees will control seedling weeds. Because contact herbicides normally do not leave any residue beyond the initial contact, a residual herbicide is commonly mixed with the contact herbicide to control the regrowth of the biennial and perennial weeds plus weeds coming from seed.

Figure 5–15 Foliage, contact, nonselective.

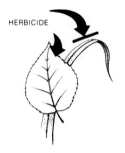

Figure 5–16 Foliage, contact, selective.

Figure 5–17 Foliage, translocated, nonselective.

Foliage, contact, selective (Figure 5–16). These herbicides kill weeds by a contact-burning effect on the foliage. Due to differences in the waxy covering on the leaves, certain crops and weeds are not injured. These herbicides are effective only on weeds in the seedling stage, and biennial and perennial weeds with dormant buds near the soil level will regrow after the foliage is killed. Generally, these herbicides are not translocated, although some may have a short residual effect in the soil.

Foliage, translocated, nonselective (Figure 5–17). These herbicides are applied to foliage of weeds and are absorbed and translocated throughout the plant. Because they are nonselective, they cannot be applied when a crop is present, but they can be used before planting or after harvesting. For example, Amitrol-T®, which can be used on noncrop land only, is nonselective and will kill many plant species. Amitrol-T® has only a short residual life, so dormant weeds or weeds germinating from seed after herbicide application will not be controlled. A residual herbicide, such as atrazine or simazine, is commonly mixed with this herbicide to give continued weed control for the rest of the year.

Foliage, translocated, selective (Figure 5–18). This group includes some of the oldest and most widely used herbicides. These herbicides are applied to the foliage and are absorbed primarily through the foliage and translocated throughout the plant. They are also selective, so weeds may be treated while the crop is present with little or no injury to the crop. The best example of this type of herbicide is 2,4-D, which has many uses. It takes dandelions out of lawns or broadleafed weeds out of oats, wheat, barley, and corn. Dicamba is used for tough-to-control weeds such as this-

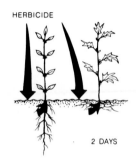

Figure 5–18 Foliage, translocated, selective.

Figure 5–19 Soil, short residual, nonselective.

Figure 5–20 Soil, short residual, selective.

tle, common milkweed, horse nettle, and hedge or field bindweed in corn, grass pasture, small grains, and rangelands.

Soil, short residual, nonselective (Figure 5–19). Only a few herbicides belong to this group. The most common, methyl bromide, is a gas at normal air temperature and dissipates into the air if the area to be treated is not covered. A sheet of plastic or a gastight cover must be used. The gas is released under the cover. All organisms, including weed seed, are killed by methyl bromide. The cover must be kept on the treated area for about 24 hours. Another 24 hours is required to aerate the soil after the cover is removed. After this 48-hour period, anything can be planted without injury, making methyl bromide one of the shortest residual herbicides available.

Soil, short residual, selective (Figure 5–20). Any herbicide that is applied to the soil prior to planting a crop or immediately after planting and has a residue of less than 1 year belongs to this group. This includes almost all those herbicides used for weed control in vegetable and field crops, as well as for other specialized crops, such as flowers and ornamentals. Because the crop is present at the time the herbicide is applied or it is planted soon after application, the herbicide must be selective and should not cause injury to the crop. These herbicides are often referred to as preplant incorporated or preemergence-type herbicides. For example, atrazine on corn, linuron on potatoes, or EPTC on ornamentals have residues that last from 6 to 8 weeks to all summer. Because these herbicides are applied to the soil, they are primarily root

Figure 5–21 Soil, long residual, nonselective.

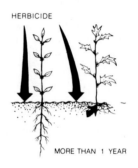

Figure 5–22 Soil, long residual, selective.

absorbed, although some may be absorbed through the foliage if any is present at the time of application.

Soil, long residual, nonselective (Figure 5–21). Herbicides that are used to control all vegetation for as long as possible are desirable for use near parking lots, buildings, oil storage areas, factory supply storage areas, and the like. For example, bromacil and simazine, at high rates of 10 to 25 pounds per acre, last for 3 to 5 years. Bromacil will kill brush and trees present in a treated area. Simazine will not kill deep-rooted trees since it does not leach as readily as bromacil.

Soil, long residual, selective (Figure 5–22). These herbicides can be used only in deep-rooted crops, such as fruit trees, nut trees, cane fruits, and grapes. They may be applied to the foliage of the weeds, although most of the herbicide is eventually absorbed through the root system. These herbicides have a low solubility in water and do not leach readily; therefore, they seldom get down to the root systems of the deeper-rooted crop. They are residual and will give weed control for more than 1 year. For example, simazine applied at rates of about 5 pounds per acre will control shallow-rooted weeds without injuring deeper-rooted crops. At high rates, selectivity is lost and it becomes a nonselective herbicide.

Classification of herbicides by chemical families

The following list does not include all herbicides, but is given as an example of chemical classification:

1. Inorganic compounds
    a. AMS
    b. Boron
    c. Copper sulfate
    d. Sodium chlorate
2. Organic compounds
    a. Acetanilides
        i. Metolachlor (Dual®)
        ii. Butachlor (Machette®)

b. Arsenicals (organic)

c. Amino acids

   i. Glyphosate (Roundup®)

d. Benzoic acids

   i. Dicamba (Banvel®)

e. Benzonitriles

   i. Dichlobenil (Casoron®)

   ii. Bromoxynil (Brominal®)

f. Carbamates

   i. Chlorpropham (Chloro-IPC®)

   ii. Barban (Carbyne®)

g. Carbanilates

   i. Phenmedipham (Betanal®)

h. Dinitroanilines

   i. Benefin (Balan®)

   ii. Trifluralin (Treflan®)

i. Diphenyl ethers

   i. Acifluorfen (Blazer®)

   ii. Diclofop-methyl (Hoelon®)

j. Dipyridyls

   i. Diquat dibromide (Reglone®)

   ii. Paraquat (Gramoxone®)

k. Halogenated aliphatic acids

   i. Dalapon (Dowpon®)

l. Imidazolinones

   i. Imazapyr (Arsenal®)

   ii. Imazamethabenz-methyl (Assert®)

   iii. Imazaquin (Scepter®)

m. Phenols

   i. Dinoseb (Sinox®)

n. Phenoxy compounds

   i. Acetic
   2,4-D
   MCPA
   2,4,5-T

   ii. Butyric
   2,4-DB
   MCPB

   iii. Propionic
   2(2,4-DP)
   2(MCPP)
   Silvex (2,4,5-TP)

o. Phenyl ethers

   i. Nitrofen (TOK®)

   ii. Bifenox (Modown®)

p. Phthalic acids
  i. DCPA (Dacthal®)
  ii. Endothal (Aquathol®)
  iii. Napthalam (Alanap®)
q. 2-Popp acids
  i. Oxadiazon (Ronstar®)
r. Pyridines
  i. Picloram (Tordon®)
s. Pyridazinones
  i. Pyrazon (Pyramin®)
t. Substituted amides
  i. Alachlor (Lasso®)
  ii. Bensulide (Betasan®)
  iii. Propachlor (Ramrod®)
u. Sulfonylureas
  i. Chlorsulfuron (Glean®)
  ii. Sulfometuron methyl (Oust®)
v. Thiocarbamates
  i. Butylate (Sutan®)
  ii. Diallate (Avadex®)
  iii. EPTC (Eptam®)
  iv. Pebulate (Tillam®)
  v. Triallate (Fargo®)
  vi. Vernolate (Vernam®)
w. *S*-Triazines (symmetrical)
  i. Atrazine (AAtrex®)
  ii. Cyanazine (Bladex®)
  iii. Cyprazine (Outfox®)
  iv. Prometon (Pramitol®)
  v. Propazine (Milogard®)
  vi. Simazine (Princep®)
x. *as*-Triazine (asymmetrical)
  i. Metribuzin (Sencor®, Lexone®)
y. Triazoles
  i. Amitrole (Weedazol®, Amino-triazole®)
  ii. Methazole (Probe®)
z. Uracils
  i. Bromacil (Hyvar®)
  ii. Terbacil (Sinbar®)
aa. Ureas
  i. Diuron (Karmex®)
  ii. Fenuron (Dybar®)
  iii. Linuron (Lorox®)
  iv. Monuron (Telvar®)
  v. Siduron (Tupersan®)

A Word about Premix Herbicide Combinations

The proliferation of newly introduced herbicide premixes, mostly combinations of existing products, with trade names different from either contributing product, poses great potential for weed control but also causes a great deal of confusion. Advertisements present the products as "new" when in fact they are usually just established herbicides that are given a new name when they are combined.

These new combinations offer ease of handling, elimination of compatibility problems, and sometimes a cost savings, especially when generics are used. But the maze of premixes may create problems for chemical dealers and their customers. One limitation is the ratios in the premixes. They do not always fit the specific conditions of a field. The user may need to add additional amounts of one or the other product.

The tendency toward "prescription" herbicides is good, but the user must have made an accurate identification of the weed problems or the mixture may not be correct for the situation. Additional efforts in weed identification and education will be necessary on the part of extension weed specialists, manufacturers, and dealers.

Table 5–1 lists some of the premix combinations that have been formulated and marketed.

Basic Rules for Herbicide Tank Mixes

Tank mixes are different from premix combinations. When tank mixing, several herbicides are mixed together on the spot and it is hoped that they will be compatible, whereas compatibility problems have already been worked out for premixes. Therefore, it is important to follow some basic rules that will help ensure tank mixing success:

1. Read labels of every product being mixed. Look for and observe tank mixing recommendations. The most restrictive labeling of any product used applies to the mixture.

2. Follow the WALE mixing sequence. Fill the spray tank ¼ full of water and add wettable powders and water-dispersible granules (W) first. Next, agitate thoroughly and continue agitation (A). Add liquids, flowable liquids, and suspension concentrates (L). Finally, fill the tank to the desired level with water and add emulsifiable concentrates (E).

3. Premix the materials to ensure compatibility. For example, make a slurry of water-dispersible granules and wettable powders before adding them to the tank. Likewise, make a one-to-one mixture of water and emulsifiable concentrates before adding them to the tank.

4. Consider using a compatibility agent. This step is called for on some product labels. For example, the label for Bicep 6L®, in covering tank mixes with AAtrex®, Princep®, or Dual®, points out, "Always use Unite® (or an equivalent compatibility agent) at 2 pints per 100 gallons. . . ." Unite® and E-Z mix® are two widely used compatibility agents. Loveland's LI700®, a nonionic surfactant, enhances compatibility of many products, but check labels before adding any surfactant.

5. Run a compatibility test. Many herbicide labels outline procedures for conducting the jar test. Test product mixes in each of two jars—one with compatibility

TABLE 5–1 HERBICIDE PREMIX COMBINATIONS

Product name	Company	Active ingredients	Formulation trade name/equivalent
Access	DowElanco	picloram + triclopyr	Tordon + Garlon
Arrosolo	ZENECA	molinate + propanil	Ordram + Stam
Betamix	AgrEvo	desmedipham + phenmedipham	Betanex + Betanal
Bicep	Ciba	metolachlor + atrazine	Dual + Atrazine
Broadstrike Plus	DowElanco	clopyralid + flumetsulam	Stinger + Broadstrike
Broadstrike + Dual	DowElanco	flumetsulam + metolachlor	Broadstrike + Dual
Broadstrike + Treflan	DowElanco	flumetsulam + trifluralin	Broadstrike + Treflan
Bronate	Rhone-Poulenc	bromoxynil + MCPA	Buctril + Rhonox
Bronco	Monsanto	alachlor + glyphosate	Lasso + Roundup
Buckle	Monsanto	triallate + trifluralin	Far-Go + Treflan
Buctril/Atrazine 3S	Rhone-Poulenc	bromosynil + atrazine	Buctril + Atrazine
Bullet	Monsanto	alachlor + atrazine	Lasso + Atrazine
Canopy 75DF	DuPont	metribuzin + chlorimuron	Lexone + Classic
Cheyenne FM	AgrEvo	fenoxaprop-P-ethyl + MCPA	Bugle + MCPA
Cheyenne X-TRA	AgrEvo	thifensulfuron + tribenuron	Pinnacle + Express
Commence 5.25E	FMC	trifluralin + clomazone	Treflan + Command
Concert	DuPont	chlorimuron + thifensulfuron	Classic + Pinnacle
Conclude B	BASF	acifluorfen + bentazon	Blazer + Basagran
Confront	DowElanco	clopyralid + triclopyr	Stinger + Garlon
Contour	American Cyanamid	atrazine + imazethapyr	Atrazine + Pursuit
Croak	Drexel	fluometuron + MSMA	Cotoran + MSMA
Crossbow	DowElanco	triclopyr + 2,4-D	Garlon + 2,4-D
Curtail	DowElanco	clopyralid + 2,4-D	Stinger + 2,4-D
Dakota	AgrEvo	fenoxaprop-P-ethyl + MCPA	Bugle + Rhonox
Envert 171	Rhone-Poulenc	2,4-D + dichlorprop	2,4-D + Weedone 2,4-DP
Extrazine II 4L	DuPont	cyanazine + atrazine	Bladex 4L + Atrazine 4L
Extrazine II DF	DuPont	cyanazine + atrazine	Bladex 90DF + Atrazine 90DF
Fallow Master	Monsanto	glyphosate + dicamba	Roundup + Banvel
Finesse	DuPont	chlorsulfuron + metsulfuron	Glean + Ally
Freedom	Monsanto	alachlor + trifluralin	Lasso + Treflan
Gemini 60DF	DuPont	linuron + chlorimuron	Lorox + Classic

Product	Company	Active ingredients	Brand combination
Ginstar	AgrEvo	diuron + thidiazuron	Karmex + Dropp
Guardsman	Sandoz	atrazine + dimethenamid	Atrazine + Frontier
Harmony Extra	DuPont	thifensulfuron + tribenuron	Pinnacle + Express
Horizon 2000	AgrEvo	fenoxaprop-ethyl + fluazifop-P	Horizon + Fusilade 2000
Krovar	DuPont	bromacil + diuron	Hyvar + Karmex
Laddok 3.33L	BASF	bentazon + atrazine	Basagran + Atrazine
Landmaster BW	Monsanto	glyphosate + 2,4-D	Roundup + 2,4-D
Lariat	Monsanto	alachlor + atrazine	Lasso + Atrazine
Lorox Plus 60DF	DuPont	linuron + chlorimuron	Lorox + Classic
Marksman 3.2L	Sandoz	dicamba + atrazine	Banvel + Atrazine
New Lorox Plus	DuPont	chlorimuron + linuron	Classic + Lorox
Passport	American Cyanamid	imazethapyr + trifluralin	Pursuit + Treflan
2 Plus 2	ISK Biosciences	2,4-D + mecoprop	2,4-D + Mecomec
Preview 75DF	DuPont	metribuzin + chlorimuron	Lexone + Classic
Prompt	BASF	atrazine + bentazon	Atrazine + Basagran
Pursuit Plus	American Cyanamid	imazethapyr + pendimethalin	Pursuit + Prowl
Ramrod/Atrazine 4F	Monsanto	propachlor + atrazine	Ramrod + Atrazine
Salute 4EC	Bayer	trifluralin + metribuzin	Treflan + Sencor
Scepter O.T.	American Cyanamid	acifluorfen + imazaquin	Blazer + Scepter
Squadron 2.33L	American Cyanamid	pendimethalin + imazaquin	Prowl + Scepter
Surpass 100	ZENECA	acetochlor + atrazine	Surpass + Atrazine
Synchrony	DuPont	chlorimuron + thifensulfuron	Classic + Pinnacle
Tiller	AgrEvo	2,4-D + fenoxaprop-P + MCPA	2,4-D + Bugle + Rhonox
Topsite	American Cyanamid	diuron + imazapyr	Karmex + Arsenal
Tordon RTU	DowElanco	2,4-D + picloram	2,4-D + Tordon
Tornado	ZENECA	fluazifop-P + fomesafen	Fusilade 2000 + Reflex
Tri-Scept	American Cyanamid	imazaquin + trifluralin	Scepter + Treflan
Trimec	PBI/Gordon	dicamba + 2,4-D+ mecoprop	Banvel + 2,4-D + Mecoprop
Turbo 8EC	Bayer	metolachlor + metribuzin	Dual + Sencor
Weedmaster	Sandoz	dicamba + 2,4-D	Banvel + 2,4-D
Weedone 170	Rhone-Poulenc	2,4-D + dichlorprop	2,4-D + Weedone 2,4-DP
Weedone CB	Rhone-Poulenc	2,4-D + dichlorprop	2,4-D + Weedone 2,4-DP

agent, and one without. Let the jars stand for 30 minutes. If settling occurs, shake the jars and observe how well the products stay in combination. This is especially important when mixing herbicides with liquid fertilizers.

6. Consider how the tank mix partners will be affected by mixing. Your goal is for herbicides to work together, each performing its own function, or for their effectiveness to be enhanced by mixing (which occurs in some cases).

What you want to avoid is herbicide "antagonism." That's when the combined effect of two or more tank-mixed herbicides is less than the effect they would have if applied separately.

For example, certain postemergence herbicides, such as Cobra®, Blazer®, and Reflex®, tend to "burn" surfaces of grass leaves. This makes it tough for grass herbicides in the tank mix, such as Assure®, Fusilade®, Option®, or Poast®, to enter plants. There are times when sequential treatments, rather than tank mixes, should be used.

HARVEST AIDS AND GROWTH REGULATORS

Harvest aids and plant growth regulators are not herbicides per se, but they are included in this section because they are used to control plant growth and they do act on plants much the same as do many herbicides.

Harvest Aids

Materials generally referred to as harvest aid chemicals fall into two classes: (1) defoliants that induce the plant to drop its leaves, but do not kill the plant, and (2) desiccants that kill plant foliage. These classifications often overlap, depending on the amount of chemical applied. Before the use of chemicals to defoliate plants, some crops (for example, potatoes) were defoliated with machines, such as beaters with rubber flails or chains. Contact chemical vine killers have replaced machines and are favored by growers because they provide the means of artificially hastening the maturity of the crop.

Desiccants applied to alfalfa grown for seed purposes are common in some areas and have replaced mowing and windrowing, which were the common practice for harvesting alfalfa seed. Desiccation for preharvest drying of dry beans has been used in some years when the weather is not right for good drying and a bean field can remain green up until frost time.

Some of the commonly used defoliants or desiccants registered by the EPA are paraquat, diquat, endothal, pentachlorophenol, sodium borate, and sodium chlorate.

Plant Growth Regulators

Plant growth regulators are used to regulate or modify growth. These chemicals are used to thin apples, control the height of turfgrass, control the height of some floral potted plants, promote dense growth of ornamentals, and stimulate rooting. They are used in minute amounts to change, speed up, stop, retard, or in some way influence vegetative or reproductive plant growth.

A plant is made up of many cells with specialized functions. Plant growth regulators can change or regulate cell development. Maleic hydrazide is a plant growth regulator. When it is sprayed on some plants, it is translocated and restricts development of new growth by preventing further cell division. The older cells, affected by the chemical, may continue to grow. In other words, cell division—but not cell maturity—is prevented.

Potted plants such as chrysanthemums, Easter lilies, and poinsettias may grow too tall. Plant height can be regulated with growth regulators that are applied at the proper stage of growth. Plant growth regulators are used to control vegetative growth and promote earlier flowering and development in some species. Some materials are used to retard apical dominance, which results in additional branching.

Plant growth regulators are used on apples and peaches. They are used to increase fruit color and may result in earlier and more uniform ripening of fruit. Plant growth regulators can also be used to thin the fruit, widen branch angles, produce more flower buds, prevent fruit drop, increase fruit firmness, reduce fruit cracking, reduce storage problems, and encourage more uniform fruit bearing.

Rooting hormones are used to increase root development and speed up rooting of certain plants. Gibberellic acid is a growth regulator that is used to initiate uniform sprouting of seed potatoes and increase the size of sweet cherries.

6

Pesticide Laws, Liability, and Recordkeeping

As the use of chemicals to control pests has increased in the United States, the scope of federal and state laws and regulations has also increased to protect the user of pesticides, the consumer of the treated products, and the environment from pesticide pollution. In 1910 the federal government passed the Federal Insecticide Act. It covered only insecticides and fungicides and was intended primarily to protect farmers against adulterated or misbranded products. Prior to that time, there was no protection for the buyer. For 37 years, the continual development of new and more effective chemicals to control pests was an exciting story for farmers and the consuming public. Synthetic organic compounds, including chlorinated hydrocarbons, organic phosphates, carbamates, and phenoxy herbicides were added to the early arsenal of arsenic, copper, and sulfur compounds. At the end of this period, the U.S. Department of Agriculture (USDA) laws were broadened to regulate the labeling and interstate distribution of insecticides, herbicides, fungicides, nematicides, germicides, plant growth regulators, defoliants, desiccants, and rodenticides for use by farmers, homeowners, and industry.

The Federal Insecticide, Fungicide, and Rodenticide Act (FIFRA) of 1947 added a new concept. It placed the burden of proof of acceptability of a product on the manufacturer prior to its being marketed. The act was oriented to protect the user, the consumer, and the public from pesticides, some of which are highly toxic and all of which are subject to limitations in application. In 1948 the Food and Drug Administration (FDA) began establishing safe levels of residue tolerances in foods. In no case were tolerances established that exceeded a safety factor of 100 to 1. In addition to this safety factor, tolerances were never approved for levels higher than necessary to accommodate registered uses. The Pesticide Chemicals Amendment (the Miller Pesticide Amendment) amended the Federal Food, Drug, and Cosmetic Act in 1954 and formalized the tolerance setting procedures of the FDA. As a matter of policy the USDA registered only pesticide uses that would result in no residue or residue levels declared safe by the FDA. The pesticide industry was required to submit to the USDA data proving the efficacy of the chemical to control the pest and to submit to the FDA proof of the safety of any measurable residue in the food produced. In 1958 a Food Additives Amendment to the Federal Food, Drug, and Cosmetic Act was passed, which prescribed regulations for the safe use of food additives. This amendment included the Delaney Clause, which prohibits any residue of carcinogenic (cancer producing) chemicals. The Federal Environmental Pesticide Control Act of 1972 completely revised FIFRA, which had been the basic authority for federal pesticide regulations since 1947.

FEDERAL AGENCIES REGULATING PESTICIDES

Environmental Protection Agency*

The Environmental Protection Agency (EPA), created in 1970, is now responsible for the pesticide regulatory functions previously delegated to the departments of Agriculture; Health, Education, and Welfare; and the Interior. These responsibilities include the registration of pesticides as required under FIFRA; the setting of tolerances as required by the Miller Amendment to the Federal Food, Drug, and Cosmetic Act; and many of the pesticide-related research and monitoring programs previously conducted by the three departments.

FIFRA.   The new FIFRA, as revised in 1972 and amended in 1975, 1978, 1980, 1984, and 1988, regulates the use of pesticides to protect people and the environment and extends federal pesticide regulation to all pesticides, including those distributed or used within a single state.

There are eight basic provisions:

1. All pesticides must be registered by the EPA.
2. For a product to be registered, the manufacturer is required to provide scientific evidence that the product, when used as directed, (a) will effectively control the pests listed on the label, (b) will not injure humans, crops, livestock, wildlife, or damage the environment, and (c) will not result in illegal residues in food or feed.
3. All pesticides will be classified into general use or restricted use categories.
4. Restricted use pesticides must be applied by a certified applicator.
5. Pesticide-producing establishments must be registered and inspected by the EPA.
6. Use of any pesticide inconsistent with the label is prohibited.
7. States may register pesticides on a limited basis for local needs.
8. Violations can result in heavy fines and/or imprisonment.

All persons using and applying pesticides should have a general understanding of the laws pertaining to pesticide use and application. The following is a section-by-section synopsis of the provisions of the Federal Environmental Pesticide Control Act

*Refer to Appendix D for addresses of EPA offices.

of 1972, including the 1975, 1978, 1980, 1984, and 1988 amendments. A copy of the complete act may be obtained from a regional office of the EPA.

Section 1.   Short title and table of contents.

Section 2.   Definitions.

Section 3.   Registration of pesticides: with certain exceptions, any pesticide in U.S. trade must be registered with the administrator.

All information except trade secrets and privileged information in support of registration must be made available to the public within 30 days of the registration.

A pesticide that meets the requirements of the act shall be registered and the fact that it is not essential cannot be a criterion for denial of registration.

Registered pesticides will be classified for either general use or restricted use. A pesticide would be restricted because of its potential for harm to either human health or the environment.

The administrator may require that the packaging and labeling of a pesticide for its restricted uses be clearly distinguishable from its general uses.

A restricted use pesticide can only be applied by a certified pesticide applicator who is certified by the state according to standards set by the administrator. The applicator may be either a commercial or private certified applicator or under the supervision of a certified applicator.

Section 4.   Reregistration of registered pesticides: requires that all pesticides registered before 1984 must meet the same standards as the pesticides registered after the 1984 amendment to FIFRA. Reregistration is to take place in five phases under a sequence of deadlines. Pesticide registrants are responsible for supplying the complete test databases necessary for the EPA to make pesticide reregistration decisions.

Section 5.   Experimental use permits: the administrator may issue permits and set the terms for the experimental use of a pesticide in order to gather data for registration, and may establish a temporary pesticide residue tolerance level for that purpose. The administrator may authorize a state to issue experimental use permits under an approved plan.

Section 6.   Administrative review, suspension: if the administrator has reason to believe a registered pesticide does not comply with the act or that it, when "used in accordance with widespread and commonly recognized practice, generally causes unreasonable adverse effects on the environment," the administrator may move to cancel the registration or change the classification of the pesticide.

Suspensions, hearings, emergency orders, judicial reviews, and conditional registrations are also provided for.

Section 7.   Registration of establishments: pesticide-producing establishments are required to be registered with the administrator.

Producers are required to submit upon initial registration, and annually thereafter, information on the amount of pesticides produced, distributed, and sold.

Section 8.   Books and records: allows for EPA officer inspection provided a written reason is presented.

Section 9.   Inspection of establishments: the administrator's agent may enter any establishment or other place where pesticides or devices are held for distribution or sale, inspect pesticides or devices, and take samples. The agent must state a sufficient reason for the action taken.

Section 10.   Protection of trade secrets and other information: prohibits disclosure of information to the public or to foreign producers by the administrator and federal employees.

Section 11.   Use of restricted use pesticides; applicators: the administrator sets minimum standards for the certification of applicators. The standards are to be separate for private and commercial applicators.

Certification will be accomplished by state programs whose plans are approved by the administrator. The administrator may withdraw approval of a state plan if the program is not maintained in accordance with it. In any state for which a state plan has not been approved by the administrator, the administrator shall conduct a program for certification of pesticide applicators.

The administrator cannot require private applicators to take written examinations. Instructional materials concerning integrated pest management techniques are to be made available to interested individuals.

Section 12.   Unlawful acts.

Section 13.   Stop sale, use, removal, and seizure: when it appears a pesticide is in violation of the act or its registration has been suspended or canceled by a final order, the administrator may issue a "stop sale, use, or removal" order to any person.

Pesticides in violation of the act may be seized, as may misbranded devices and pesticides that have an unreasonable adverse effect on the environment even when used in accordance with requirements and as directed on the labeling.

Section 14.   Penalties: any registrant, commercial applicator, wholesaler, dealer, retailer, or other distributor is liable to a $5000 civil penalty for an act violation. Private applicators and other persons are liable to a $1000 civil penalty on their second and subsequent offenses. The opportunity for a hearing is required prior to assessment of a civil penalty.

Any registrant, commercial applicator, wholesaler, dealer, retailer, or other distributor is liable to a penalty of $25,000 or 1 year in prison or both upon conviction of a misdemeanor. Private applicators and other persons are liable to a penalty of $1000 or 30 days in prison or both. Persons who reveal formula information are liable to a penalty of $10,000 or 3 years in prison, or both.

Section 15.   Indemnities: persons owning a pesticide whose registration is suspended and later canceled are eligible for payment by the administrator of an indemnity for the pesticide owned. Manufacturers who had knowledge prior to the issuance of a suspension notice that their products did not meet registration requirements and who continued to produce the pesticide without notifying the administrator of the deficiency would not be eligible.

In lieu of indemnification the administrator may authorize, under certain limited conditions, the use or other disposal of a pesticide.

Section 16. Administrative procedure; judicial review.

Section 17. Imports and exports: the administrator is required to notify foreign governments through the State Department whenever the registration or the cancellation of suspension of the registration of a pesticide becomes effective or ceases to be effective.

The administrator is authorized to examine imported pesticides or devices and have those in violation of the act refused entry, seized, and destroyed.

Section 18. Exemption of federal and state agencies: the administrator may exempt any federal or state agency from the act if it is determined that emergency conditions exist that require such exemptions.

Section 19. Storage, disposal, transportation and recall: the administrator shall establish procedures and regulations for storage and disposal of pesticides and containers.

The administrator may require the registrant or applicant for registration of a pesticide to submit data or information regarding methods for safe storage and disposal of excess quantities of the pesticide. The labeling must contain requirements and procedures for the transportation, storage, and disposal of the pesticide, pesticide container or rinsate containing the pesticide.

The administrator may require either voluntary or mandatory recalls of pesticides that have been suspended and/or canceled under section 6.

Section 20. Research and monitoring: authority for pesticide research and monitoring is provided, including research contracts and grants. A National Monitoring Plan is required to be formulated and carried out by the administrator.

Section 21. Solicitation of comments; notice of public hearings: the administrator is required to solicit the views of the Secretary of Agriculture before publishing regulations under the act.

The administrator is authorized to solicit the views of all interested persons concerning any action under the act and to seek advice from qualified persons.

Timely notice of any public hearing to be held shall be published in the *Federal Register.*

Section 22. Delegation and cooperation: the administrator shall cooperate with the USDA, other federal agencies, and any appropriate state agency in carrying out the provisions of this act.

Section 23. State cooperation, aid, and training: the administrator may delegate to any state or Native American tribe the authority to cooperate in enforcement of the act, train state personnel in cooperative enforcement, and assist states to implement cooperative enforcement with grants.

The administrator may assist states and Native American tribes in developing and administering applicator certification programs under cooperative agreements and under contracts.

The administrator shall, in cooperation with the Secretary of Agriculture, utilize the Cooperative State Extension Service to inform and educate pesticide users about accepted uses and other regulations under this act.

Section 24. Authority of states: a state may regulate the sale or use of any federally registered device in the state but only if and to the extent the regulation does not permit any sale or use prohibited by this act. However, different state packaging or labeling requirements are prohibited.

A state may provide registration for additional uses of federally registered pesticides formulated for distribution and use within that state to meet special local needs in accordance with the purposes of this act.

Section 25. Authority of the administrator: the administrator is authorized to prescribe regulations to carry out the provisions of this act, after consulting with the Secretary of Agriculture and notifying the appropriate committees in the Senate and House of Representatives.

The administrator is authorized to declare that which is a pest; determine highly toxic substances in pesticides; specify classes of devices subject to certain misbranding provisions; and, under section 7, prescribe pesticide discoloration requirements and determine suitable pesticide names.

Peer review is required of all major scientific studies. Congressional and judicial review is required of all rules and regulations promulgated by the EPA under the act.

Section 26. State primary enforcement responsibility: a state shall have primary enforcement responsibility for pesticide use violations, providing the state has adopted adequate pesticide use laws and regulations and has adequate procedures for their enforcement.

Section 27. Failure by the state to ensure enforcement of state pesticide use regulations: authorizes the administrator to rescind a state's primary enforcement responsibility if it is determined that the state has been inadequate in its program.

Section 28. Identification of pests; cooperation with Department of Agriculture's program: the administrator, in coordination with the Secretary of Agriculture, shall identify those pests that must be brought under control.

Section 29. Annual report.

Section 30. Severability.

Section 31. Authorization for appropriations.

FIFRA authorizes the EPA to regulate *pesticides,* a term that, under the act, includes almost anything intended to be used to kill, repel, or control any nonhuman form of life—from bacteria and fungi to prairie dogs and weeds. Under the law, if a product is represented as having pesticidal properties, the product is a pesticide under section 2 of FIFRA.

In addition to insecticides, herbicides, fungicides, and rodenticides, products to control algae, bacteria, viruses, nematodes, and other life forms are considered to be pesticides. Also included are herbicide products used to stop plant growth, desiccate plants, or defoliate crop plants.

Disinfectants used to kill bacteria in hospitals and chemicals used to retard growth of unwanted bacteria, fungi, or other living organisms in cloth, wood, leather, and other materials are also among the products regulated by the EPA under the act.

Along with chemical pesticides, the EPA controls plant growth regulators and biological products, such as pheromones (sex attractants), juvenile growth hormones (substances that keep insects from maturing and reproducing), and even bacteria and parasites if they are used as pesticides. In 1984 the EPA included in this category new organisms derived through recombinant-DNA technology for use as pesticides.

Mechanical devices (traps for mice and other animals) are also covered by FIFRA, but are not subject to the full registration requirements imposed on chemical products. Only microorganisms or drugs that affect people or animals are excluded from regulation under FIFRA, because these are regulated by the FDA. These would include such items as vaccine products and preparations for killing lice on humans or parasites in animals.

As intended by Congress, FIFRA has four main thrusts:

1. To evaluate the risk's posed by pesticides by requiring stringent screening and testing of each pesticide and eventual registration with the EPA before being offered for sale.
2. To classify pesticides for specific uses and to certify pesticide applicators and thus control exposure to humans and the environment.
3. To suspend, cancel, or restrict pesticides that pose a risk to the environment.
4. To enforce these requirements through inspections, labeling notices, and regulation by state agencies.

SARA Community Right-to-Know Act. The Superfund Amendments and Reauthorization Act (SARA), Title III, the Community Right-to-Know Act, requires that anyone handling and storing hazardous chemicals must report to state and local emergency groups, including local fire departments. An inventory of hazardous chemicals being stored must be reported to these agencies. This law is under EPA jurisdiction.

There are four major reporting requirements under Title III: emergency planning notification, emergency release notification, community right-to-know, and toxic chemical release forms. Each reporting provision has different requirements for the chemicals and facilities covered.

Emergency Planning Notification. Because the law was designed generally to identify all facilities that have any of the listed 360 extremely hazardous substances present in excess of its threshold planning quantity (TPQ), farms were not exempted from this provision. The TPQ is based on the amount of any one of the hazardous substances that could, upon release, present human health hazards that warrant emergency planning. The TPQ emergency planning trigger is based on these public health concerns rather than the type of facility where the chemicals might be located. The type of facility and degree of hazard presented at any particular site, however, are relevant factors for consideration by the local emergency planning committees.

Although many facilities with chemicals in such quantities may not present a significant hazard to their communities due to their rural location or short holding times, other facilities may well present a potentially significant hazard if the chemicals are located in a suburban, high-population area or near a school, hospital, or nursing home. Although the substances may only be stored or used periodically, there is always the possibility of accidents that could present a hazard to the community. In the event of a fire or other emergency on the farm, local responders should know what chemicals they might encounter in order to take appropriate precautionary measures.

If one or more of the 360 substances is present in excess of its TPQ, the facility must notify its state emergency response commission, which in turn will notify the appropriate local emergency planning committee. The local fire department should also be made aware of the hazardous chemicals storage.

Emergency Release Notification. Releases (spills) of any of the extremely hazardous pesticides in excess of its reportable quantity (RQ) must be reported to the state emergency response commission and local emergency planning committees. Reportable quantities are the amounts of the substances that, if released (spilled), must be reported.

The release (other than spills or other accidents) of a pesticide registered under FIFRA when used generally in accordance with its intended purpose (during routine agricultural applications according to approved product label instructions) is exempted from this reporting.

Community Right-to-Know. Community right-to-know reporting pertains to the material safety data (MSD) sheets for hazardous chemicals from facilities (mainly manufacturers and importers) that are required to report and make the MSD sheets available to wholesale and retail outlets.

Toxic Chemical Release Forms. The chemical release reporting is limited to facilities that are manufacturing or importing hazardous chemicals. A complete listing of hazardous pesticides with their TPQs is available from the county offices of the USDA Natural Resources Conservation Service. Remember, too, that the EPA may from time to time add additional chemicals to the list. For assistance in meeting these requirements, the EPA's chemical emergency preparedness program hotline is (800) 535-0202.

Agricultural chemicals in groundwater: Pesticide strategy. In October 1991, the EPA published the final version of *Agricultural Chemicals and Groundwater: Pesticide Strategy.* The strategy addresses three major issue areas:

1. The agency's goal in addressing the groundwater contamination concern
2. The management approach for preventing groundwater contamination
3. The agency's policy for responding to contamination that has already occurred

A summary of the three key issues follows:

1. *EPA's environmental goal in addressing the pesticides in groundwater concern.* EPA's strategy and goal is to manage pesticides to protect the groundwater resource. Specific attention will be given to preventing unacceptable contamination of current and potential drinking water supplies. The agency will use maximum contaminant levels (MCLs)—the enforceable drinking water standards under the Safe Drinking Water Act—as reference points for helping to determine unacceptable levels of pesticides in underground sources of drinking water. When an MCL is not yet available for a particular pesticide, the EPA will develop interim protection criteria for use as reference points for pesticide management decisions.

2. *EPA's prevention approach.* For prevention efforts, EPA strategy envisions national registration of pesticides with state-directed monitoring and pesticide use restrictions based on local groundwater concerns. The EPA will establish groundwater protection measures that will be uniformly applicable across the country, such as restricting the use of certain pesticides to trained and certified applicators. Still other EPA-directed management measures will be applicable only at sites with certain conditions (for example, shallow water tables or location within wellhead protection areas). The strategy describes the agency's preferred management approach as being one that is directed by individual states.

3. *EPA policy for response to contamination that has already occurred.* The agency's strategy for responding to pesticide contamination of groundwater emphasizes federal-state coordination and statutory enforcement activities. Under its management plan, a state should have the capability to take necessary action to prevent further contamination when unacceptable levels are reported. A state should also take the lead in addressing immediate public health concerns. The proposed strategy calls for the EPA and the states to place greater emphasis on coordinating FIFRA, the Safe Drinking Water Act (SDWA), and Comprehensive Environmental Response, Compensation And Liability Act CERCLA enforcement activities to identify parties responsible for groundwater contamination as a result of the misuse of pesticides, including illegal disposal or leaks and spills.

A complex set of federal, state, and local laws authorizes a wide variety of activities and programs that can be used to protect groundwater from pesticide contamination. Many of these laws were written before pesticide contamination of groundwater became a concern, and subsequently have been broadly interpreted and expanded to cover this issue. Sometimes these laws provide conflicting direction regarding how pesticides in groundwater should be managed.

At the federal level, the primary agencies that have jurisdiction over pesticides in groundwater are the U.S. Department of Interior (DOI), the USDA, and the EPA.

The U.S. Geological Survey (USGS) within the DOI has a principal role for gathering hydrogeologic information on and assessing the quality of the nation's aquifers. Through a number of federal-state property programs, the USGS and cooperative states compile information for planning, developing, and managing the nation's water resources.

The second key player at the federal level is the USDA. Through its Cooperative State Research, Education, and Extension Service and the Natural Resources Conservation Service, the USDA provides technical assistance to individual landowners and a range of incentives that can affect the way landowners choose to manage their land and water resources. Ultimately, landowners make choices regarding pesticide use and agricultural practices at their specific sites. Agencies (such as those found in the USDA) that advise landowners play a significant role in ensuring that landowners make environmentally sound decisions.

The third critical player at the federal level is the EPA. This agency has the lead responsibility for regulating pesticide use in the United States and for protecting the quality of the nation's groundwater. Within the EPA, several environmental statutes and programs address one or more discrete sources of groundwater contamination (Figure 6–1).

FIFRA authorizes the agency to regulate the marketing and use of pesticides in the United States. Before allowing a pesticide on the market, the Office of Pesticide Programs weighs the risks of a pesticide against its benefits. Manufacturers must submit extensive data to the EPA, including environmental fate data, which will indicate the leaching potential of a pesticide.

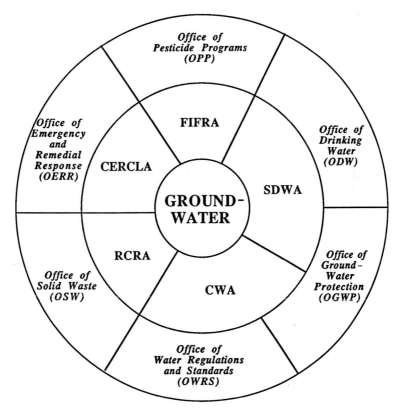

Figure 6–1 EPA offices working to protect groundwater.

The SDWA is designed to ensure that public water systems provide water meeting minimum standards for protection of public health. As required by the act, the EPA establishes drinking water standards (maximum contaminant levels) and water supply monitoring requirements for public water supplies. Under recent amendments to the act, the agency has been authorized to provide resources for states to establish wellhead protection areas (WHPAs) for public drinking water wells. Other recent amendments restrict underground injection of hazardous waste and establish a sole-source aquifer demonstration program.

The Clean Water Act (CWA) has the basic mission to restore and maintain the chemical, physical, and biological integrity of the nation's waters. The EPA provides grants to states for development and implementation of state groundwater protection strategies. Under the CWA's nonpoint source authorities, the EPA also provides financial assistance to states for nonpoint source monitoring and assessments, planning, program development, and demonstration projects.

The Resource Conservation and Recovery Act (RCRA) regulates disposal of waste, including pesticides, that may create a hazard. Pesticide-containing wastes that are considered hazardous wastes under RCRA are subject to extensive regulatory requirements governing storage, transportation, treatment, and disposal.

The Comprehensive Environmental Response, Compensation, and Liability Act (CERCLA) establishes a trust fund (Superfund) to finance government responses to releases or threats of releases of hazardous substances. However, if groundwater contamination results from normal application of pesticides, the law does not allow the agency to recover costs from pesticide applicators or private users.

State-level Involvement. States have traditionally been responsible for implementing and enforcing federal policies and standards. With the assistance of FIFRA and CWA grants, practically all states are now developing and implementing groundwater protection strategies addressing various sources of contamination, including pesticides. Although many states have at least some activities addressing pesticide contamination, only a few have begun efforts that could be defined as prevention programs. Several states have established or are developing regulatory programs to require safety measures on chemigation systems and requirements for pesticide mixing, loading, and storage areas.

A few states are preparing for or have passed legislation aimed at protecting groundwater from contamination. In addition, many states have initiated groundwater monitoring programs and have identified areas where pesticide contamination of groundwater is a problem.

State efforts to address the concern regarding pesticides in groundwater have some of the same coordination problems that the federal government has. Often the agency responsible for pesticide control is housed in the state agricultural department, although primary responsibility for water quality and waste disposal programs is located in a state environmental protection agency. Responsibility for ensuring safe drinking water often rests in still another agency, such as a public health department. This fragmentation of responsibilities among various agencies makes it difficult to establish an effective federal-state relationship.

Local-level Involvement. The United States has more than 91,000 units of local government, which by and large operate under state control. The local governments are the first point of contact for controlling contamination of groundwater. Through their land use and zoning powers, local governments can also closely influence groundwater use and quality.

In summary, protection of groundwater from pesticide contamination will require a coordinated approach among a number of agencies at each level of government. The EPA feels that development of a national EPA strategy is one of the first steps needed in moving toward an integrated and successful management approach. Pesticide applicators will have to be aware of the laws pertaining to groundwater contamination. The pesticide label should be carefully checked for restrictions concerning groundwater. Be sure to consult state and local agencies if there is a question regarding the use of a certain pesticide.

Worker Protection Standard for Agricultural Pesticides. The Worker Protection Standard (WPS) for Agricultural Pesticides is another rule administered by the EPA. Fully implemented on January 1, 1995, it covers nurseries, forests, and greenhouses, as well as farms. It requires certain posting of fields, training, notification of workers, and the like. A summary of the requirements is as follows:

Scope. Owners, lessees, and operators of the mentioned types of establishments and their contractors, workers, and supervisors have duties and are subject to enforcement action for violations. Family farms (no employees) are exempt from many requirements. However, all users of pesticides must conform to product-specific requirements on the labeling.

Training. Handlers and workers (mixers, loaders, applicators, flaggers, disposers, cleaners) must be trained in a manner they can understand. This portion of the rules was amended to require that, beginning January 1, 1996, agricultural employers must ensure that untrained workers receive basic pesticide safety information before they enter a treated area. No more than 5 days after their initial employment has commenced, all untrained agricultural workers must receive the complete WPS pesticide safety training. Workers and handlers must be retrained after a period of 5 years.

Notification. The act takes a four-tiered approach, including oral warnings, treated area posting, central notice board, and information on request. The following is a summary of requirements:

- Information must be posted at a central location. It must include (1) an EPA-approved safety poster, (2) the name, address, and telephone number of the nearest emergency medical facility, (3) facts regarding pesticide applications, including the product name, EPA registration number, active ingredients, location and description of treated areas, time and date of application, and restricted-entry intervals (REIs). All information must be legible, current, and in an easily accessible location.
- Pesticide safety training must be provided. Unless the workers possess a valid EPA-approved training card, handlers and workers must be trained before

they begin work and at least once every 5 years, in a manner they can understand. **Note:** Written and audiovisual training materials are available from the EPA as well as most cooperative extension offices in most states. Several mail order catalog companies sell almost everything you need. These include Gempler's, Inc., (800) 382-8473; Graingers, (800) 521-5585; Lab Safety Supply, (800) 356-0783; and Direct Safety Company, (800) 528-7405.

- Decontamination sites must be established. Sites must be within ¼ mile of all workers and handlers, supplying ample, safe, cool water for routine and emergency whole-body washing. Eye flushing equipment must be immediately available for all employees on the job as well as at the decontamination site. A clean change of clothing must be available for handlers at the decontamination site. The same supplies should be available at mixing and loading sites and when personal protective equipment (PPE) is washed. Decontamination sites must be protected from applications in the field and mixing and loading sites.

- Employer information must be exchanged. A commercial chemical handler must make sure the farm operator knows details of applications, such as location, time and date, product name, EPA registration number, REI, and other safety requirements. In turn, farm operators must inform commercial applicators about recent pesticide applications on the farm and any restrictions on entering those areas.

- Emergency assistance must be provided. If a chemical injury is suspected, transportation must be available to an appropriate facility and medical personnel must be provided with emergency guidelines printed on the product label and other relevant information about the injury.

Additional Duties for Employers of Handlers

1. **Application restrictions and monitoring.** Do not allow handlers to apply a pesticide so that it contacts, directly or through drift, anyone other than trained and PPE-equipped handlers. Make sight or voice contact at least every 2 hours with anyone handling pesticides labeled with a skull and crossbones. Additional rules apply for greenhouse fumigation procedures.

   Under a final EPA rule amendment effective May 1, 1996, certified or licensed crop advisors and persons under their direct supervision while performing crop advisor tasks are exempt from WPS provisions *except* for pesticide safety training. The EPA considers crop advisors to be handlers and as such, they may enter treated areas during application and the REI without time limitations, if provided with the required personal protective equipment specified on the product labeling and other protection provided for handlers.

2. **Specific instructions for handlers.** Inform handlers how to use equipment properly and how to follow safety procedures outlined on the label. Keep labels readily available throughout the procedure.

3. **Equipment safety.** Inspect pesticide handling equipment before each use, repairing and replacing as needed. Only trained and equipped handlers can clean, repair, or adjust equipment containing pesticides or residues.

4. **Personal protective equipment.** Employers have many duties regarding the cleaning use, and maintenance of PPE, including respirators, coveralls, and other gear.

Additional Duties for Employers of Workers

1. **Restrictions during applications.** Only trained and equipped handlers can be in areas being treated with pesticides. Nursery workers must be at least 100 feet away from nursery sites being treated. There are additional requirements for greenhouse situations.

2. **Restricted-entry intervals.** During any REI, do not allow workers to enter a treated area or have contact with anything treated with the pesticide. REIs are established based on the acute toxicity of the active ingredients in the pesticide. The most toxic of the applicable toxicity categories that are obtainable for each active ingredient is the one used to determine the REI. If the product contains only one active ingredient and it is in toxicity category I, the REI is 48 hours. If, in addition, the active ingredient is an organophosphorus ester that inhibits cholinesterase and may be applied outdoors in an area where the average annual rainfall for the application site is less than 25 inches per year, the REI is 72 hours.

   If a pesticide product contains only one active ingredient and it is in toxicity category II, the REI is 24 hours. If a pesticide product contains only active ingredients that are in toxicity category III or IV, the REI is 12 hours. In April, 1995, the EPA established reduced REIs for low risk pesticides with active ingredients in toxicity categories III and IV. The pesticides with reduced REIs will have an REI of 4 hours. If a pesticide contains more than one active ingredient, the REI, including any associated statement concerning use in arid areas, is designated for the most toxic active ingredient with the longest REI. Check the label carefully to determine the REI for the pesticide you intend to use.

   An exception, granted by the EPA in April, 1995, allows workers the flexibility during an REI to perform *irrigation tasks* that could not have been foreseen and which, if delayed, would cause significant economic loss. There are a number of conditions that must be met in order to allow early entry for the irrigation activities. Be sure to check WPS regulations in order to comply with the stated conditions.

   Another 1995 exception grants workers the flexibility during an REI to perform *limited contact tasks* that could not have been foreseen and which, if delayed, would cause significant economic loss. As with the exception for irrigation tasks, the exception for limited contact tasks includes significant provisions to limit pesticide exposure and risk to employees performing the tasks. Again, be sure to check WPS regulations in order to comply with all of the stated conditions.

3. **Oral notice about applications.** Orally warn workers and also post treated areas if the product label requires it. Otherwise, either warn workers or post entrances to treated areas. Tell workers which method is in effect. Post all greenhouse applications.

4. **Post warning signs.** Post legible 14″ × 16″ WPS-design approved signs at all entrances to treated areas just before application; keep posted during the REI; remove before workers enter and within 3 days after the end of the REI.

5. Oral warnings. Before each application, tell workers who are on the property (in a manner they can understand) the location and description of treated areas, the REI, and not to enter during the REI. Workers who enter the property after application starts must also be warned at the start of their work period.

Endangered Species Act. The U.S. Fish and Wildlife Service of the DOI holds the responsibility for protecting endangered and threatened species in the United States under the Endangered Species Act. The EPA is responsible for protection of endangered species from pesticides. Under the act, if an endangered species is determined to exist on public lands marked for some action that might be disruptive to that species, a biological assessment of the impact of the planned action must be conducted. Thus a pesticide application by a public agency would have to await the results of the study on the impact of applying that particular pesticide at the intended rate. All pesticide applicators, whether private or public, should make every attempt to avoid any impact on threatened or endangered species.

The Endangered Species Act was passed by Congress in 1973, but it has only been in recent years that the EPA has undertaken to uniformly enforce the act. In 1987 the EPA attempted to promulgate a pesticide labeling regulation that would require manufacturing companies to put endangered species information on the pesticide labels and to list the counties in each state where the pesticide was prohibited for use. The EPA had undertaken preliminary work on county maps showing areas within the county that contained endangered species and where that particular pesticide could not be used. Public opposition to the scheme caused the national Congress to order the EPA to rethink their plans and to delay implementation of the Endangered Species Regulations indefinitely.

Congress gave additional attention to the program in the 1988 Endangered Species Act Amendments, in which the EPA, in cooperation with the USDA and DOI, was directed to set up a program for educating and fully informing the agricultural community about the pesticide labeling program. The amendments also require the EPA, USDA, and DOI to conduct a study of the following: (1) reasonable and prudent means to implement an endangered species conservation program that would allow continued production of food and fiber commodities, (2) the best methods for mapping restricted use areas, (3) alternatives to prohibitions of pesticides, (4) methods to improve coordination among agencies, and (5) means to implement the program to promote conservation and minimize impacts on pesticide users.

Be sure to be aware of the Endangered Species Act and regulations whenever they are finalized. Contact the local cooperative extension office or the Fish and Wildlife Service for information. See also Chapter 9, Environmental Considerations.

EPA policy decisions. Many questions have been asked of the EPA over the years as to what constitutes "under the direct supervision of a certified applicator." The following is the EPA's policy [FIFRA Section 2(e)(4): FIFRA Compliance Program Policy No. 2/3 Under the Direct Supervision of a Certified Applicator in EPA-Administered Programs].

Issues. What specific operational relationship and oversight constitute work "under the direct supervision of a certified applicator" in EPA-administered programs?

Policy. This statement of compliance program policy sets forth the standard that the agency will employ in determining whether the use of a restricted use pesticide (RUP) was performed under the direct supervision of a certified applicator.

The certified applicator is responsible for ensuring that those working under his or her direct supervision are qualified to handle pesticides in general and are competent in the use of specific RUPs.

The certified applicator must maintain oversight of the pesticide use operation from onset to completion, being at all times aware of hazards presented by the situation. Furthermore, the certified applicator in supervising the work of a noncertified applicator must:

- Determine the level of experience and knowledge of the noncertified applicator in the use of a pesticide.
- Provide verifiable, detailed guidance on how to conduct each individual pesticide use performed under his or her direct supervision.
- Accompany the noncertified applicator to at least one site that would be typical of each type of pesticide use that the noncertified applicator performs.
- Be accessible to provide further instruction in the use of the RUP.
- Be able to be physically on the site within a reasonable period of time, should the need arise, where the pesticide use or storage is taking place.
- In the case of a certified commercial applicator, maintain for at least 2 years records regarding the types, amounts, uses, dates, and places the RUP was used as required by 40 CFR 171.7(b)(1)(iii)(E).

Responsibilities. FIFRA Section 2(e)(4) defines the phrase "under the direct supervision of a certified applicator" and 40 CFR 171.6 establishes "Standards for Supervision of Noncertified Applicators by Certified Private and Commercial Applicators." These provisions place on the supervising certified applicator authority and responsibility for ensuring that any noncertified applicators working under his or her direct supervision will use an RUP safely and effectively.

To effectively discharge these responsibilities, the certified applicator must ensure that each of the following steps has been taken:

1. **Experience and knowledge of the noncertified applicator.** The certified applicator must review the noncertified applicator's experience and knowledge of both general and restricted use pesticides. This review will be used by the certified applicator to determine the extent and type of training and oversight the noncertified applicator will require. The certified applicator must ensure that the noncertified applicator is familiar with and understands pesticide labeling instructions, especially those instructions relating to the prevention of hazards to people and the environment. Furthermore, the certified applicator must ensure that the noncertified applicator is able to recognize poisoning symptoms and knows the appropriate procedures to follow in the event of a poisoning incident.

2. **Instructions to noncertified applicators.** The certified applicator must provide the noncertified applicator with certifiable, detailed guidance regarding the pesticide use to be performed (the agency suggests written instructions as an effective

method of providing verifiable, detailed guidance to the noncertified applicator) and must address such matters as the types and amount of pesticides to be used, the maintenance and calibration of the equipment to be used, and the presence and nature of any unique risk of environmental or health exposure. If a noncertified applicator will be using several different RUPs at different types of sites, then the certified applicator must accompany the noncertified applicator to at least one site for each different type of pesticide use. The noncertified applicator must also be informed of the pesticide labeling instructions, especially those instructions relating to the prevention of hazards to people and the environment.

3. **Assurance that the noncertified applicator is competent.** The certified applicator must ensure that the noncertified applicator is competent in the use of an RUP and understands the instructions provided. At a minimum, the instructions must be given at a level and in a language understood by the noncertified applicator. In addition to the required verifiable, detailed instructions, the certified applicator must ask questions of the noncertified applicator to ensure his or her understanding of the instructions. When the noncertified applicator has not previously used the particular RUP, or if there is any doubt about the noncertified applicator's understanding of the instructions, the certified applicator must, in addition, provide on-the-job training and observe the performance of the noncertified applicator before leaving the site.

4. **Accessibility of certified applicator.** Section 171.6 states "the availability of the certified applicator must be directly related to the hazard of the situation." The agency has interpreted this statement to require the physical presence of the certified applicator on site when the use of an RUP poses a potentially serious hazard to people or the environment. In addition, the physical presence of the certified applicator is required on site when it is specified on the label.

Both 40 CFR 171.6 and FIFRA Section 2(e)(4) require that in cases where the physical presence of the certified applicator is not required, the certified applicator "is available if and when needed." The agency interprets this provision to require, at a minimum, the availability by telephone or radio of the certified applicator immediately before, during, and after the noncertified applicator's use of an RUP. The certified applicator must also have the capability to be physically on site, within a reasonable period of time, should the need arise. The potential or real consequences of a delay in arriving on site will be taken into consideration when determining what is "a reasonable period of time."

5. **Records.** Commercial applicators are required by 40 CFR 171.7(b)(1)(iii)(E) to maintain for 2 years records regarding the types, amounts, uses, dates, and places of application of RUPs. The requirement of 40 CFR 171.7(b)(1)(iii)(E) applies to both commercial applicators applying RUPs and commercial applicators supervising the application of RUPs.

6. **Classification of pesticide dealers as commercial applicators.** Any dealer who sells RUPs to persons who are not certified and who then supervises the uses of those products must be certified as a commercial, not private, applicator. If the dealer has received payment from the sale of the RUP and the product will

be used on land owned by the purchaser or by some third party, the dealer cannot be deemed under FIFRA Section 2(e)(2) or 40 CFR 171.2(a)(20) to be a private applicator.

7. **Liability.** The certified applicator is responsible for ensuring the safe and effective use of any pesticide used under his or her direct supervision. Accordingly, the certified applicator is subject to enforcement liability for violations that are committed by the noncertified applicator who is operating "under his or her direct supervision." The certified applicator cannot use as a defense against such liability the ignorance or negligence of the applicator who is operating under his or her direct supervision, as it is his or her responsibility to ensure that the RUP application was performed "under the direct supervision of a certified applicator." *Note:* Some states do not allow noncertified applicators to apply RUPs.

Restricted use pesticides. Certain pesticides have been classified as "restricted use" by the EPA. Some of these pesticides have been classified by regulations; that is, they have been reviewed by the EPA and determined to fit the requirements for RUPs. Approximately one-third of the pesticides have been classified as restricted at the time of their registration. This list will grow as the EPA completes its review of all previous registrations and whenever a newly registered material requires it.

The list of restricted use pesticides can be found in Appendix C. No endorsement is intended of those trade names shown, nor is discrimination intended against any trade names not shown.

The pesticide label will tell whether or not a pesticide is classified as an RUP. The statement on the label will look like this:

RESTRICTED USE PESTICIDE

FOR RETAIL SALE TO AND USE ONLY
BY CERTIFIED APPLICATORS OR PER-
SONS UNDER THEIR DIRECT SUPERVI-
SION AND ONLY FOR THOSE USES
COVERED BY THE CERTIFIED APPLI-
CATOR'S CERTIFICATION.

An RUP can only be applied by a certified applicator or under his or her direct supervision. Direct supervision means you, as a certified applicator, have given the uncertified applicator verifiable and specific directions to apply the pesticide. It also means that you have informed him or her of all directions for its proper use and of the safety precautions to follow during application and that you are available in case you are needed. In some situations the label may require the actual physical presence of a certified applicator. *Remember:* If something goes wrong, the certified applicator is responsible.

Some states have their own restricted use list in addition to the EPA list. Some states also have additional requirements that must be satisfied before a permit can be issued for the use of certain pesticides. A pesticide applicator contemplating the use of RUPs must not only be certified to use those pesticides, but must be aware of any state laws that may be in effect regarding the use and application of certain pesticides.

Occupational Safety and Health Administration

On September 23, 1987, Congress put into law the Hazard Communications Standard and OSHA was charged with regulation of the standard. Briefly, the purpose of the standard is to provide employers and employees with information regarding hazardous chemicals. The basic manufacturer or importer is required to perform a hazard determination on each product or chemical it sells. Products or chemicals considered hazardous under this specified standard are required to have material safety data (MSD) sheets.

The standard requires that all shipments of hazardous products or chemicals be accompanied by an MSD sheet. The sheet must also be included with all subsequent downstream sales of the product. No specific format is required for the MSD sheets nor is there a rigid training schedule to which employers must adhere. This provides flexibility but also puts responsibility directly on the manufacturing industry.

United States Department of Agriculture

The consumer and marketing services of the meat and poultry inspection program monitor the quality of meat and poultry products. Sampling is scheduled on a national basis through the use of a computer. Samples are identified as to source, and questionable samples are followed up with inspections at the product's origin, where appropriate action is taken to prevent the marketing of contaminated meat.

The USDA, through its Agricultural Marketing Service Pesticide Records Branch, is responsible for Recordkeeping Requirements for Certified Private Applicators of Federally Restricted Use Pesticides. These requirements were included in section 1491 of the Food, Agriculture, Conservation, and Trade (FACT) Act of 1990, commonly referred to as the 1990 Farm Bill. Final regulations went into effect in May, 1993 and amendments were added effective August, 1995. Refer to the recordkeeping section later in this chapter.

Food and Drug Administration

The FDA still retains the responsibility to monitor food for humans and feed for animals. Any products violating pesticide residue tolerances are subject to seizure by the FDA. The Federal Food, Drug, and Cosmetic Act delegates this authority.

Federal Aviation Administration

The FAA under the Federal Aviation Regulation, part 137, Agricultural Aircraft Operations, January 1, 1966, regulates the dispensing of pesticides by aircraft. Under these regulations, as amended, it is a violation for an aerial applicator to apply any pesticide except according to federally registered use.

Federal Department of Transportation

The hazardous nature of pesticides requires strict interstate regulation for their transport. The federal Department of Transportation (DOT) is the major agency involved in enforcing hazardous material laws. Individuals directly involved in transporting hazardous pesticides should know the following classifications of hazardous materials:

- Class A poisons: extremely dangerous poisons, so poisonous that a small amount of gas, vapor, or liquid mixed with air would be dangerous to life.
- Class B poisons: less dangerous poisons, including liquids, solids, pastes, and semisolids that are known to be toxic to people and afford a health hazard during transportation. *Most pesticides fall in this group.*
- Class C poisons: tear gas and irritating substances. Materials that upon contact with fire or exposure to air give off dangerous fumes. Does not include any material included under class A.

Each hazardous material transported must be packaged in the manufacturer's original container. Each container must meet the DOT standards. Each vehicle carrying a hazardous material must have the proper placard (sign). The manufacturer is responsible for placing the proper warning signs on each box or container containing class A or B poisons.

DOT regulations prohibit carriers from hauling class A and B poisons or other hazardous materials in the same vehicle with food products. DOT regulations require carriers of hazardous materials to immediately notify DOT after each accident when a person is killed, receives injuries requiring hospitalization, or when property damage exceeds $50,000.

Refer to Chapter 14 for further information on transporting and placarding hazardous materials.

STATE AGENCIES REGULATING PESTICIDES

Various state agencies, including the Department of Agriculture, State Health Department, State Department of Environmental Protection (or similar designation), Labor Department, State Highway Patrol, and others have one or more laws pertaining to pesticide use and application. These include laws regulating pesticide registration and sale, commercial applicators, and structural pest control applicators. Many of these agencies now regulate and enforce EPA laws (as described earlier in this chapter) under cooperative agreements.

Individuals planning to sell, transport, or apply pesticides should obtain copies of the applicable laws from the appropriate agencies in their state. Check Appendix D for the addresses of the various state regulatory agencies.

CANADIAN PEST CONTROL PRODUCTS REGULATIONS

All persons using and applying pesticides in Canada should have a general understanding of the laws pertaining to pesticide use and application. The following is a section-by-section synopsis of the major provisions of the Pest Control Products Regulations established in 1972 and amended in 1973 and 1977. The complete act is published in Part II of the *Canada Gazette* and should be consulted for all purposes of interpreting and applying the regulations. A copy may be obtained from Agriculture Canada or from the regulatory offices in provinces. Refer to Appendix D for the appropriate offices and addresses.

The reader should be aware that the Canadian Government has proposed extensive changes to the present law. These changes were still being considered at the time of publication of this book and may soon be implemented. The Canadian Government proposes to implement the changes by establishing a Pest Management Regulatory Agency within the Department of Health. A mandate is proposed to protect human health, safety, and the environment by minimizing risks associated with pesticides, while enabling access to pest management tools, namely, pest control products and pest management strategies.

Regulations Made Pursuant to the Pest Control Products Act

A Short Title

1. These regulations may be cited as the Pest Control Products Regulations.

Interpretation

2. Definitions of act, active ingredient, applicant, assessed or evaluated, certificate of registration, device, director, display panel, district director, metric unit, Plant Products Division, registrant, residues, and seed.

Exemption of Certain Control Products

3. The following control products are exempt from the act:
   a. A control product that is subject to the Food and Drugs Act and is only used for:
      i. Control of viruses, bacteria, or other microorganisms on or in humans or domestic animals,
      ii. Control of arthropods on or in humans, livestock, or domestic animals if the control product is to be administered directly and not by topical application,
      iii. Control of microorganisms on articles that are intended to come directly into contact with humans or animals for the purpose of preventing or treating disease when associated with medical care,
      iv. Control of microorganisms in premises in which food is manufactured, prepared or kept, or
      v. Preservation of foods for humans during cooking or processing; and
   b. A control product that is a device other than a device of a type and kind listed in Schedule I.

Application

4. These regulations do not apply to a control product, other than a live organism, that is imported into Canada for the importer's own use, if the total quantity of the control product being imported does not exceed 500 grams by mass and 500 millimeters by volume and does not have a monetary value exceeding $10.

Exemption from Registration

5. A control product is exempt from registration if

    a. It is a control product, other than a live organism, or other than 2,4-D, also known as 2,4-dichlorophenoxy acetic acid, that is used only in the manufacture of a registered control product and conforms to the relevant specifications of that registered control product, set forth in the register of control products;

    b. It is for use by a person for research purposes

        i. On premises owned or operated by that person, or

        ii. On any other premises not owned or operated by that person, if such use has been approved by the director; or

    c. It is a control product that

        i. Is a substance or a thing the primary purpose of which is not for controlling, preventing, destroying, mitigating, repelling, or attracting any pest, but is represented as having such properties or contains an active ingredient possessing such properties, and

        ii. Is of a type and kind listed in Schedule II and meets the conditions relevant to that substance or thing set forth in that schedule (SOR/81-87, 5. 1).

Registration of Control Products Required

6. Subject to Section 5, every control product imported into or sold in Canada shall be registered in accordance with these regulations.

Application for Registration

7. An application for a certificate of registration shall contain the name and address of the applicant or agent; name and address of the product manufacturer; brand name, if any; active ingredient; size, type, and specification of the package; guarantee statements; and other relevant information, including safety and need for the product. The application must be signed by the applicant or agent.

8. An applicant or registrant who is not resident in Canada shall appoint an agent who is a permanent resident in Canada to whom any notice or correspondence under the act and these regulations may be sent.

9. **a.** In addition to the information required by Section 7, an applicant shall provide the minister with such further or other information as will allow the minister to determine the safety, merit, and value of the control product.

    b. Without limiting the generality of subsection (a), where a control product

        i. Is a device that has not been previously assessed or evaluated for the purposes of the act and these regulations, or contains an ingredient that has not been so assessed or evaluated, the applicant shall provide the minister with the results of scientific investigations respecting effectiveness; safety to persons occupationally exposed to it when it is manufactured, stored, displayed, distributed or used; safety to the host plant, animal, or article in relation to which it is used; effects of the control product on nontarget organisms relative to the intended use;

degree of persistence, retention, and movement of the control product and its residue; suitable methods of analysis for detecting the active ingredient and measuring the specifications of the control product; suitable methods of analysis for detecting significant amounts of the control product in food, feed, and the environment; suitable methods for the detoxification or neutralization of the control product in soil, water, air, or on articles; suitable methods for the disposal of the control product and its empty packages; stability of the control product under storage and display conditions; and the compatibility of the control product with other products; or

 ii. Is intended for use on living plants or animals or products derived therefrom which plants, animals, or products are for human consumption, the applicant shall provide the minister with the results of scientific investigation respecting effects of the control product or its residues when administered to test animals for the purposes of assessing any risk to humans or animals; and the effects of storing and processing food or feed, in relation to which the control product was used, on the dissipation or degradation of the control products and any of its residues.

10. Every application for a certification of registration shall be accompanied by five copies of the proposed label for the control product or reasonable facsimiles thereof.

11. When requested to do so by the minister, the applicant shall provide a sample of the control product, a sample of the technical grade of its active ingredient, and a sample of the laboratory standard of its active ingredient.

Fees on Applications

12. Fees in various amounts are required for certain products or devices.

Registration

13. Requirements for registration include specifications of each control product, label for each product, registration number assigned to each product, and such other information as the minister deems necessary.

Duration and Renewal of Registration

14. Duration of certificates of registration, expiration, and renewal requirements.
 a(i). Registration of devices remains valid as long as the control product complies with the conditions of registration.
 (ii). Products that are not devices are registered for periods up to 5 years.
 b. Provides for renewal periods not exceeding 5 years.

15. Notwithstanding Section 14, where a control product was, immediately preceding 1978, registered under these regulations, the registration of the control product expires on December 31, 1980.

16. Where the registrant intends to discontinue the sale of a control product, he shall so inform the minister and the registration of that control product shall, on such terms and conditions, if any, as the minister may specify, be continued to allow any stocks of a control product to be substantially exhausted through sales.

Temporary Registration

17. The minister may, upon such terms and conditions, if any, as he may specify, register a control product for a period not exceeding 1 year where the applicant agrees to endeavor to produce additional scientific or technical information in relation to the use for which the control product is to be sold; or the control product is to be sold only for the emergency control of infestations that are seriously detrimental to public health, domestic animals, natural resources, or other things.

Refusal to Register

18. The minister may refuse to register a control product if, in his opinion, the application or label does not comply with the act and these regulations; the information provided is insufficient to enable the control product to be assessed or evaluated; the merit or value of the control product has not been established in accordance with its label directions; the use of the control product would lead to an unacceptable risk of harm to things on or in relation to which the control product is intended to be used, or public health, plants, animals, or the environment; or the control product is not required to be registered.

Cancellation and Suspension of Registration

19. During the period of registration of a control product, the registrant shall, when requested to do so by the minister, satisfy the minister that the availability of the control product will not lead to an unacceptable risk of harm to things on or in relation to which the control product is intended to be used; or public health, plants, animals, or the environment.

20. The minister may, on such terms and conditions, if any, as he may specify, cancel or suspend the registration of a control product when, based on current information available to him, the safety of the control product or its merit or value for its intended purposes is no longer acceptable to him.

21. Requires the minister to provide information to the applicant regarding his reasons for cancellation or suspension.

22. Provides for continued dealer sales providing the dealer had the product for sale prior to the minister's suspension order.

23. Provides for hearing for those applicants affected under Section 20.

24. Provides for the appointment of a review board for hearings as provided in Section 23.

25. Sets forth the conditions and requirements of the board hearing and requirements by the board and the minister.

Records

26. Every registrant shall make a record of all quantities of a control product stored, manufactured, or sold by him and the record shall be maintained for 5 years from the time it is made; and be made available to the director at his request at such time and in such manner as the director may require.

Labeling

27. No label shall be used on a control product unless it has been approved by the minister and, unless the minister otherwise directs, every label shall show the information required by Sections 27–37. This section provides for labeling of "RESTRICTED" and "DOMESTIC" pesticides as well as the requirement for "READ THE LABEL BEFORE USING"; "GUARANTEE"; "REGISTRATION NUMBER"; "PEST CONTROL PRODUCT ACT"; "FIRST AID INSTRUCTIONS"; "TOXICOLOGICAL INFORMATION"; and "NOTICE TO USER."

28. The label for a control product that is a device of a type and kind listed in Schedule I shall contain the information referred to in Section 27.

29. The display panel shall consist of one principal display panel and at least one secondary display panel.

30. Additional requirements to labeling set forth in Section 27.

31. Further clarification of requirements for registration under Section 27.

32. Clarification of requirements in Section 31.

33. Notwithstanding Sections 29, 30, and 31, the minister may, for reasons satisfactory to him, approve the inclusion of the information required by those sections elsewhere than on the display panel.

34. Additional requirements regarding the product class designation "RESTRICTED."

35. When the information required to be shown on the label is, pursuant to Section 32, not included in the display panel, the display panel shall contain the words in capital letters "READ ATTACHED BROCHURE (OR LEAFLET) BEFORE USING" prominently displayed thereon.

36. Subject to the approval of the minister, additional information relating to the control product and any graphic design or symbol may be shown on the label if it does not unreasonably detract or obscure the information required to be shown on the label.

37. Pertains to a statement regarding limitation of warranty.

38. Requirements for bulk container information.

39. The information on every label shall be printed in either the English or the French language, or both.

Units of Measurement on Labels

40. All units of measurement shown on a label shall be expressed only in metric units in accordance with the Weights and Measures Act. This section provides for additional information regarding net quantities and the metric units required in the declaration of net quantities.

41. All information shown on a label shall be printed in a manner that is conspicuous, legible, and indelible.

Denaturation

42. Where the physical properties of a control product are such that the presence of the control product may not be recognized when it is used and is likely to expose a person or domestic animal to severe health risks, the control product shall be denatured by means of color, odor, or such other means as the minister may approve to provide a signal or a warning as to its presence.

Storage and Display

43. A control product shall be stored and displayed in accordance with any condition set forth on the label, and a control product bearing the POISON symbol superimposed on a DANGER symbol shall be stored and displayed apart from food for humans or feed for animals in a separate room; or separated by a physical barrier so as to avert the contamination of the food or feed.

Distribution

44. A control product shall be distributed in a manner that is consistent with any special conditions specified by the minister and, when required by the minister, the conditions shall be shown on the label, and on the shipping bill respecting the control product or on a statement accompanying the shipment.

Prohibitions Respecting Use

45. a. No person shall use a control product in a manner that is inconsistent with the directions or limitations respecting its use shown on the label.

b. No person shall use a control product imported for the importer's own use in a manner that is inconsistent with the conditions set forth on the importer's declaration respecting the control product.

c. No person shall use a control product that is exempt from registration under paragraph 5a for any purpose other than the manufacture of a registered control product.

Packaging

46. a. The package for every control product shall be sufficiently durable and be designed and constructed so that it will contain the control product safely under practical conditions of storage, display, and distribution.

b. Every package shall be designed and constructed to permit the withdrawal of any or all of the contents in a manner that is safe to the user; and the closing of the package in a manner that will contain the control product satisfactorily under practical conditions.

c. Every package shall be constructed so as to minimize the degradation or change of its contents resulting from interaction or from the effects of radiation or other means.

d. When the package is essential to the safe and effective use of the control product, it shall be designed and constructed to meet specifications acceptable to the minister on registration of a control product.

Standards

47. Every control product shall conform to the specifications and bear the label contained in the register of control products.

48. Every control product shall have the chemical and physical composition and uniformity of mix necessary for it to be effective for the purposes for which it is intended.

General Prohibitions

49. A control product shall not contain an active ingredient unless it is present in an amount sufficient to add materially to the effectiveness, merit, or value of the control product.

50. A label shall not contain any information respecting any organism or causative agent of a disease of humans mentioned in Schedule A to the Food and Drugs Act.

51. Unless otherwise authorized by the minister a label shall not contain any information respecting any organism or causative agent of a disease of domestic animals that is required to be reported under the Animal Contagious Diseases Act.

Sampling

52. Provides for sampling of control products by an inspector in a manner approved by the director.

Detention

53. A control product seized pursuant to Section 9 of the act may be detained by an inspector at any place by attaching a detention tag to at least one package of the control product in the lot that has been seized. This section stipulates further provisions for detention of products.

54. Where a control product has been seized and detained by an inspector, the registrant shall be entitled to a hearing if he so requests, and Sections 23 and 24 apply in respect of the hearing.

Import

55. A control product may be imported into Canada if it is accompanied by a declaration, in a form specified by the minister, which form shall be signed by the importer and shall state the name and address of the person who is shipping the control product; the name and brand, if any, of the control product;

common or chemical names; total amounts being imported; name and address of the importer; and purpose of the importation of the control product using the words "for resale" or "for manufacturing purposes" or "for research purposes."

56. Where the collector of customs at a port of entry is not satisfied that an importer's declaration is complete and in order, he shall hold the control product at the port of entry or place the control product in bond and forthwith advise a district director.

57. The collector of customs at a port of entry shall forward one copy of every importer's declaration to a district director.

SCHEDULE I

1. Garment bags, cabinets, or chests that are manufactured, represented, or sold as a means to protect clothing or fabrics from pests;

2. Apparatuses that are manufactured, represented, or sold as a means to attract or destroy flying insects, or to attract and destroy flying insects;

3. Devices that are manufactured, represented, or sold as a means to repel pests by causing physical discomfort by means of sound or touch;

4. Devices for attachment to garden watering hoses that are manufactured, represented, or sold as a means to dispense or apply a control product;

5. Devices that are manufactured, represented, or sold as a means of providing the automatic or unattended application of a control product;

6. Devices that are sold for use with chemical products containing cyanide as a means to control animal pests.

SCHEDULE II

1. Feed for animals;

2. Fertilizer that is subject to the Fertilizers Act if the control product contained therein is registered under these regulations;

3. Seed that has been treated with a control product registered for the purpose of treating seed if
   a. The seed is sold and shipped in bulk and the shipping documents bear information setting forth the common name or chemical name of the active ingredient of the control product used to treat the seed; and
   b. Where the seed is packaged, the package bears a label with the words "This seed is treated with," followed by the name of the control product including the common name or chemical name of its active ingredient together with the appropriate precautionary symbols and signal words selected from Schedule III and such other statements as are required by these regulations and are applicable to the control product used to treat the seed.

SCHEDULE III

PRECAUTIONARY SYMBOLS AND SIGNAL WORDS

Signal word: Degree of hazard	Symbol	Signal word: Nature of primary hazard	Symbol
1. Danger		1. Poison	
2. Warning		2. Corrosive	
3. Caution		3. Flammable	
		4. Explosive	

Canadian Regulations in the Provinces

Each province in Canada has its own regulations regarding pesticide use and application. Individuals contemplating pesticide application in any province should contact the proper authorities regarding local laws. Appendix D contains the addresses of the Canadian provincial regulatory offices.

LIABILITY

Responsible pesticide applicators rarely have legal claims brought against them. However, most pesticides are considered hazardous material and even the most careful applicators may have claims for damages brought against them. The usual claims are for nonperformance when the grower feels that a pesticide application did not do the job that was expected. More often, the claims are for injury to crops that were treated or to crops in nearby fields. It is important for all licensed pesticide applicators to be informed of the most common legal claims that can be made against them. (The information presented here is intended only as a guide and is not represented as complete or all-inclusive. The author assumes no responsibility for interpretation of these suggestions or for wrongful acts as a result of the use of them.)

Drift

Drifting pesticides are a major cause of environmental contamination and damage to nontarget areas. In general, the courts have held that the applicator and the grower who hired him or her are jointly liable in drift cases. The grower is responsible when he or she hires or contracts for a "particularly dangerous operation" such as the application of pesticides. However, don't depend on the grower to share costs. The grower may file a countersuit against you (the applicator) claiming that you agreed not to cause drift damage. The manufacturer of the pesticide may be held liable in drift cases in certain instances. If the label does not clearly warn about the possibility of drift, the manufacturer may have to share the liability.

Crop Injury

Claims of injury to the crop that was treated or claims that the pesticide did not perform as expected involve the dealer, the manufacturer, and the applicator. The courts must decide which of the three recommended or guaranteed the product for that specific use on that crop. The party in error must accept the blame and pay damages. Applicators must make sure that all the pesticides they use are recommended on the label for that purpose.

If the crop injury was not great or total, the grower must show how much damage was from the pesticide and how much was from other conditions such as weather, disease, and so on. This breakdown is not required in cases where there is total injury.

Personal Injury

The application of pesticides is considered an especially dangerous or in legal terms an *ultrahazardous* activity. As a result, the pesticide applicator is liable for any injury to a person from a pesticide. Usually, the injured person can recover damages without proving negligence on the part of the applicator. The injured party must only prove he or she was free of any negligence and did not assume the risk of pesticide exposure. Pest control operators are in a somewhat different category. The liability in most cases involving personal injury or death depends on proving the pest control operator is negligent.

Wrong Field

If the pesticide is applied on a field, crop, or area other than the one it was intended for, serious problems can result. If the owner did not want the area treated, the applicator may be charged with trespass. In the event that the damage involves residue or overtolerance, the applicator may be liable for damages involving the entire crop. Some pesticides are highly persistent and last for long periods of time, while others are not persistent and last only a few days. Naturally, the persistent pesticides are more likely to cause long-lasting residues. Some examples of insecticide persistence in the environment include Sevin®, 7 to 10 days; diazinon, 2 to 4 weeks; Dursban®, 10 to 14 days; parathion, 3 to 7 days; chlordane, 2 to 4 weeks; and malathion, 5 to 10 days. The pesticide applicator must do everything he or she can to eliminate residue problems. Defense is very difficult. It is important to double-check on addresses, field locations, and all landmarks before treating an area. Applying pesticides to the wrong field or area can be costly.

Honey Bees

Honey bees are very important to the farmer and to the beekeeper who makes a living raising and caring for honey bees. Bees are insects, and unfortunately they are very susceptible to many pesticides. If the bees in hives are killed as the result of drift from nearby fields, the applicator is usually held legally responsible and often must pay damages. However, if the bees contacted the pesticides while in the sprayed fields, the applicator is usually not liable. The courts have ruled that the bee is trespassing and that the land doesn't need to be safe to uninvited animals. Pesticide applicators should know where the bee hives are located in their area, and warn the beekeeper before spraying (see Chapter 9).

Attractive Nuisance

The rulings on "attractive nuisance" usually involve cases when children are attracted to ground equipment or aircraft and injure themselves. The owner and/or applicator is held liable for leaving the "nuisance" where a child could be "attracted" to it. Therefore, be very careful. Do not leave ground equipment with exposed drive belts, drive wheels, gears, or any moving parts alone in areas where children can get to them. Aircraft should never be parked where curious children can find them. Empty containers and aerosol cans are also attractive and dangerous to children. Be sure to store or dispose of them properly (see Chapter 14).

Noise

Recently, claims have been brought against applicators for noise damage. Owners of specialty operations such as mink raising, poultry farms, and occasionally cattle ranches may claim injury to their animals from fright caused by noise of aircraft and ground equipment operating above or near their place of operation. They must prove direct loss of property as a result of noise from machinery operated carelessly or negligently. In some cases, the ranch owner may claim that the applicator made an unlawful flight over his property without permission. This is especially important in aerial applications when pull-ups over nearby property are necessary. Successful defense is possible when the applicator can show that the noise was not the cause of injury or that no injury occurred.

Cross Contamination

There are three main ways that cross contamination may occur.

1. The manufacturer may make a mistake in labeling or formulating a product. In these cases the pesticide container has not only the pesticide named on the label,

but another pesticide also. The materials in the container may then damage crops being treated.

2. The applicator may err in mixing or filling the spray tank or may not have removed from the spray tank all the pesticide left over from a previous application.

3. Open containers of herbicides such as 2,4-D can vaporize and penetrate other pesticides stored nearby. When the other pesticides are applied, the 2,4-D contamination can seriously injure the crops.

The applicator must know which container of pesticide was used on the crop so that laboratory tests can be made. The lab test can help determine whether the contamination occurred during mixing and filling or was the fault of the manufacturer. In cases involving herbicide contamination, it is difficult to prove whether it is a result of vaporization during storage or a manufacturer formulation error. The courts must decide who is to blame.

All applicators should keep a careful record of the location and the date that a particular pesticide was applied. They should also record the lot number of the pesticide in case of cross contamination (see the section on recordkeeping, later in this chapter).

INSURANCE

To protect your business and yourself as an applicator, you should have insurance for possible pesticide mishaps. State laws may require commercial applicators to carry minimum amounts of insurance. There are many different types of insurance plans, ranging from bodily injury, property damage, and restricted chemical liability, to comprehensive chemical liability. The plan you choose should fit your needs and your business. If you are in a relatively low-risk position from a legal standpoint, then minimum coverage may be enough. On the other hand, aerial applicators will probably need more insurance. Be sure to explore the costs, benefits, and drawbacks of insurance before you buy. You need to know exactly what you are covered for. An insurance agent who specializes in pesticide insurance is the best person to advise you on your individual insurance needs.

What to Do When You Are Involved

No matter how careful you are, accidents will happen. You must be prepared for any emergency. If you are sued or become involved in any legal problem, act carefully and promptly. Always be friendly and helpful. *Never* admit liability. Be careful who you give information to about your spray operation. Offer to look into the matter at once.

- Examine your records to make sure that you were actually operating in the area at the time of the alleged injury.
- Make sure that all your records are up-to-date, particularly as to the identity of the equipment used, temperature, wind direction and velocity, and all other pertinent information.
- Proceed to the scene immediately and make notes of all essential information when you get there.

- Record the presence of any adverse condition that you observed at the time of your investigation, particularly insect infestations, disease, plant stress, late planting, carryover effect from other materials or herbicides, and general condition of the crop.
- Photograph any adverse conditions with color film at a sufficiently close focal length so that the symptoms can be examined by an expert.
- Collect samples of crop injury, including roots, so that the materials can be examined by an independent expert.
- If the container from which the product was removed that was used on the job is still available, it is wise to keep it for laboratory examination. If the container is not available, check your records to see if you at least have the lot number for that particular pesticide.
- Notify your insurance company immediately.
- If you do not have insurance for the loss involved, request permission to have an expert examine the crop or the property in order that you may have the benefit of his or her opinion.
- If a chemical company is likely to be involved because of a particular pesticide used, notify them immediately. The company will probably want to send experts to the site also.
- Obtain the names and addresses of all witnesses who might testify as to the nature of the operation and conditions of the crop before and after treatment. In the event that the crop is perennial, such as fruit trees or wind breaks, USDA aerial photographs may be available for your examination to ascertain the condition of the trees or other perennial plantings in years prior to the year of alleged injury.

RECORDKEEPING

The federal recordkeeping rule mandated by the 1990 Farm Bill took effect in 1993 and requires private pesticide applicators to maintain records for all federally restricted use pesticides. (Commercial applicators were already required to keep records under state and federal regulations.) A restricted use pesticide (RUP) is one available for purchase and use only by certified pesticide applicators (private or commercial), or persons under their direct supervision. An RUP must be applied in accordance with the applicator's certification. These requirements have been placed on this group of pesticides because of their potential hazards to humans and/or the environment.

Applicators are required to record details of RUP applications, including but not limited to the name of the pesticide; EPA registration number of the chemical; total amount applied; size of the area treated; the crop, commodity, stored product, or site to which the product was applied; location of the application (legal property description); month, day, and year of the application; and the certified applicator's name and certification number.

Amendments were added to the regulations in 1995 and include reducing the time to record an RUP application from 30 days to 14 days; changing the way the locations for spot applications are recorded; changing the definition of a medical emergency; changing the definition of a licensed health care professional; clarifying access to and release of record information obtained for purposes of medical treatment; and clarifying provisions for penalties.

Private pesticide applicators need to obtain a copy of the regulations and become familiar with them.

The Need for Records

Records of pesticide usage are important to help protect yourself and your investments. In many cases the pesticide usage history of a piece of property is as important as the cropping history of the land. Often the type of crop to be planted is determined by the chemicals applied previously. This is especially important with certain herbicides. Good records can sometimes be the difference between winning or losing a lawsuit when you have been accused of using pesticides wrongly. Courts have looked favorably upon those who routinely keep records. This establishes a "system of records" that shows that you have been regularly keeping records and have not suddenly tried to conjure up some sort of records.

The record will help you to:

- Comply with federal and state laws.
- Improve pest control practices and efficiency.
- Avoid pesticide misuse.
- Compare applications made with results obtained.
- Purchase only amounts of pesticides needed.
- Reduce inventory carryover.
- Establish proper use in case of residue questions.
- Establish where the error was, if any was made.
- Establish proof of use of recommended procedures in case of lawsuits.
- Plan cropping procedures for next year.
- Plan pesticide needs for next year.

How to Keep Records

A number of different types of forms have been devised for keeping pesticide application records. Some forms are for use in agricultural field applications and others are for use in agricultural field applications and others are for use in keeping records of pesticide applications in vegetable and fruit crops. Still other forms are needed for keeping records of pesticide applications on livestock, poultry, and buildings.

Carry a pocket notebook with you and write down information as it happens; don't trust your memory. This information can be transferred to a permanent record that can be kept in your home or office. Meister Publishing Company offers a recordkeeping system called MARKS™. Their address is 37733 Euclid Ave., Willoughby, OH 44094. A number of other companies also offer recordkeeping schemes or computer software.

What Information Should Be Kept?

- Crop, animal, or building treated
- Crop variety or animal species treated
- Pest(s) treated
- Location and number of acres or animals treated
- Time of day, date, and year of application
- Type of equipment used
- Pesticide used, including the name and percent of active ingredient, type of formulation, trade name, manufacturer, lot number, and EPA registration number
- Amount used per acre or per 100 gallons of water
- Amount of active ingredient applied per acre or per animal
- Stage of crop or animal development
- Pest situation; that is, severity of infestation and presence of beneficial species
- Weather conditions: temperature, wind, rainfall
- Harvest date
- Results of application

Miscellaneous Information

- Record information including damage to crop from spray application or other damages caused by storms or other situations. Information such as estimated yield losses due to storm damage would be valuable.
- A sketch of fields treated, showing locations of ditches and surrounding crops could also be valuable.
- Note any mechanical or calibration problems.
- Ask your customer to sign or initial your written record.

Example Forms

The examples of recordkeeping forms on the following pages (Figures 6–2 through 6–8) are suggested for your consideration. You may wish to use commercially developed forms or devise your own recordkeeping forms and include other information in addition to those items already mentioned.

GROWER'S PERMANENT RECORD OF CHEMICAL PESTICIDE APPLICATIONS

INCLUDE ALL TREATMENTS: SOIL, DORMANT, PRE-PLANT, GROWING SEASON AND POST-HARVEST.

CROP		GROWER'S NAME & FIELD IDENTIFICATION	TOTAL ACRES IN FIELD OR ORCHARD
VARIETY			

IDENTITY OF TREATED AREA	DATES OF TREATMENT START	DATES OF TREATMENT FINISH	NUMBER ACRES TREATED	AIR OR GRND	- COPY THIS INFORMATION FROM THE PESTICIDE LABEL - PESTICIDE COMMON NAME OR TRADE NAME	LIQUID WET.PWDR. DUST GRANULES	ACTIVE INGREDIENTS, % OR LBS. PER GALLON	SPRAYS ONLY: AMOUNT OF MATERIAL PER 100 GALS.	OR PER ACRE	GALLONS OF MIXED SPRAY, LBS. DUST OR LBS. GRANULES PER ACRE	PEST TO BE CONTROL'D

Figure 6-2 Typical all-purpose record form.

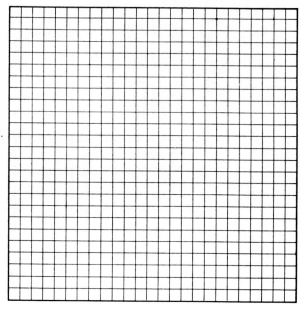

CAUTION . . .
BE CAREFUL WITH PESTICIDES
READ THE LABEL CAREFULLY AND COMPLETELY.
FOLLOW THE LABEL INSTRUCTIONS.
STORE PESTICIDES IN A SAFE PLACE.

Figure 6–3   Grid for sketching field and/or treated area.

CHEMICAL APPLICATION RECORD

Farm: _____   Fertilization: _____

Field: _____   Acres: _____

Soil Type: _____   Date Harvested: _____

Crop: _____ Variety _____   Yield: _____

Crop Last Year: _____   Notes: _____

RECORD EACH APPLICATION OF EACH CHEMICAL TO THIS FIELD	APPLICATION NO. ( )	APPLICATION NO. ( )	APPLICATION NO. ( )	APPLICATION NO. ( )
DATE (DAY, MONTH, YEAR)				
ACRES TREATED				
CHEMICAL USED				
FORMULATION OF CHEMICAL				
TOTAL AMOUNT APPLIED				
STAGE OF CROP GROWTH				
PURPOSE OF APPLICATION (NAME OF WEED, INSECT, DISEASE OR OTHER REASON)				
STAGE OF DEVELOPMENT OF WEEDS, INSECTS, DISEASE				
METHOD OF APPLICATION				
SOIL CONDITIONS (WET, DRY, CLODDY)				
TEMPERATURE				
WIND (DIRECTION AND SPEED)				
CLOUD COVER				
EFFECTIVENESS OF TREATMENT				
COMMENTS				

Figure 6–4 Typical farm record form.

CROP & VARIETY				Grower Field Foreman			Block or Area		Acres		Planted	
Seed Source & Lot No.											Date	Year

CROP TREATMENT AND DEVELOPMENT HISTORY: Include all pre-plant, plant bed, soil, seed, growing, and post-harvest treatments, with date and hour.

Dates of Treatment		Acres Treated	Method-Spray, Dust, Granule, Etc.	Type of Equipment & Who Applied Aircraft, Sprayer, Dusters, Etc.	Pesticide-Name & Amount Active Ingredient	Amount Used Per 100 Gal. Water	Amount Applied Per Acre	Lbs. Active Per Acre	Stage of Development of Crop	Remarks-Weather, Rainfall, Pest Infestation
Start	Finish									
EXAMPLE: 7-16	7-16-90	25	Spray	400 gallons - Boom Sprayer	Parathion 4E	1 Pint	1½ pint/ 150 gal. water	.75	Tomatoes-first bloom	Leafminers & worms, Bad-Dry, warm, clear, no wind-No disease

ESTIMATED DATE OF FIRST HARVEST＿＿＿＿＿＿＿＿＿   ACTUAL HARVEST DATES: Start＿＿＿＿＿＿＿   Finish ＿＿＿＿＿＿＿

Figure 6–5 Vegetable crop record form.

COMMERCIAL PESTICIDE APPLICATION RECORD

Patron Name _____

Date and Time
of Application _____

Field Location _____

Acres Tested _____

Chemical(s) Applied	Formulation of Chemical	Rate Per Acre		Total Added To Sprayer
_____	_____	_____		_____
_____	_____	_____		_____

Carrier
(water, liq. fert. etc.)

Rate Per Acre _____

Total Added To Sprayer _____

_____

_____

_____

Wind Direction _____

Temperature _____

Wind Velocity _____

Soil Texture

Soil Moisture _____

Calm
5-10 mph
Over 10 mph

Level, fine texture
Rough, cloddy
Plant debris on surface

Very wet
Wet
Moist
Dry

Method of Application: Preplant   Pre-emergence   Post-emergence
_____

Crop _____   Stage of Crop Growth _____

Stage of Weed or Insect Development _____

Comments: _____

Field Map (sketch to show field & direction of application)

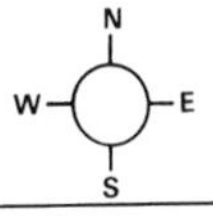

_____

_____

_____

_____

_____

_____

_____

_____

Ticket No. _____

Applied by _____

Accepted by

Figure 6–6 Commercial pesticide applicator record form.

Address ....

Producer's Name ....

Year ....

CHEMICAL USE RECORD

Dairy Animals, Livestock, Poultry, and Animal Housing

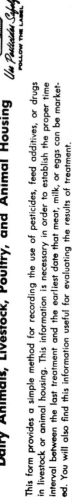

This form provides a simple method for recording the use of pesticides, feed additives, or drugs in livestock or animal housing. This information is necessary in order to establish the proper time interval between the last treatment and the earliest date that meat, milk, or eggs can be marketed. You will also find this information useful for evaluating the results of treatment.

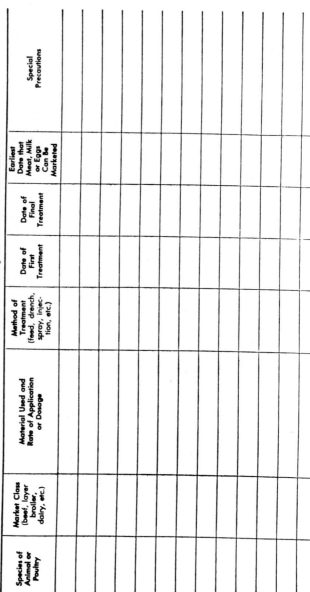

Species of Animal or Poultry	Market Class (beef, layer broiler, dairy, etc.)	Material Used and Rate of Application or Dosage	Method of Treatment (feed, drench, spray, injection, etc.)	Date of First Treatment	Date of Final Treatment	Earliest Date that Meat, Milk or Eggs Can Be Marketed	Special Precautions

Figure 6–7 Record form for animal treatments.

READ THE LABEL | LIVESTOCK TREATMENT RECORD | STORE CHEMICALS IN A SAFE PLACE

USE ONLY AS DIRECTED | DISPOSE OF EMPTY CONTAINERS PROPERLY

Data	Number of Animals Treated	Breed and Average Weight	Pests Controlled	Pesticide Used (Kind & Amount)	Method of Treatment	Weather Condition	Other Operations Performed (Dehorning, Castrating, Weaning, etc.)

Figure 6–8 Alternate record form for livestock pesticide treatments.

7

The Pesticide Label

The label is the key to the information about any pesticide you are planning to use. The label on the bottle, can, or package of a pesticide tells you the most important facts you must know about that particular pesticide for safe and effective use. The information on the label gives not only the directions on how to mix and apply the pesticide, but also offers guidelines for safe handling, storage, and protection of the environment.

Information on pesticide labels has been called the most expensive literature in the world. The research and development that leads to the wording on a label frequently costs millions of dollars. The combined knowledge of laboratory and field scientists—including chemists, toxicologists, pharmacologists, pathologists, entomologists, weed scientists, and others in industry, universities, and government—is used to develop the information found on the label. To appreciate the value of the information on the label, consider the time, effort, and money spent in gathering and documenting the label information.

Chemical companies all over the world continually make new compounds and then test them in the laboratory and greenhouse for possible pesticide use. For each material that finally meets the standards of a potential pesticide, thousands of other compounds are tested and discarded for various reasons. When a promising pesticide is discovered, its potential use must be evaluated. If the company believes it has a worthwhile product and there is a possibility for reasonable sales volume, wide-scale testing, and label registration procedures are begun. From this point of development until the pesticide reaches the market, it requires from 10 to 15 years with costs reaching as high as $60 million.

Many kinds of carefully controlled tests must be made to determine the effectiveness and safety of each pesticide under a wide range of environmental conditions. Basic data requirements for a new food-use pesticide are many and varied (Figure 7–1). The following facts can help us better appreciate the effort that goes into the pesticide label and the value of the information on the label to the user.

136

Basic Data Requirements for a New Food-Use Pesticide

At the present time, data from the following tests must be submitted
to EPA by a manufacturer prior to registration:

Chemistry:

list of ingredients
description of manufacturing process
discussion of formation of impurities
physico-chemical properties
residue studies
metabolic studies
analytical methods
results of analytical procedures

Environmental Fate:

hydrolysis
leaching
terrestrial dissipation
photodegradation
soil metabolism
rotational crop study

Toxicology:

acute oral
acute dermal
acute respiratory
eye irritation
chronic toxicity
subchronic oral toxicity
reproduction and fertility
metabolism
mutagenicity
birth defects
carcinogenicity

Ecological Effects:

aquatic, acute toxicity
avian, dietary & acute oral

Figure 7–1 Basic data requirements for a new food-use pesticide.

- *Toxicological tests.* These tests determine the possible hazards of the new pesticide to humans, animal life, and the environment. The pesticide is fed by mouth and applied to the skin of test rats, rabbits, and other animals. Tests are conducted to determine if it has gases or vapors that would harm the skin or the eyes. Extended feeding of test animals is also carried out to determine if the pesticide will cause cancer or affect the offspring of the test animal.
- *Degradation studies.* These studies are made to determine how long it takes for the compound to break down into harmless materials under various conditions.
- *Soil movement tests.* Tests are performed to determine how the pesticide moves in the soil, how long it remains in the soil where it might be absorbed by crops planted later in the same field, or how it may migrate into groundwater.
- *Residue tests.* These tests determine how much of the pesticide or its breakdown products remain on the crops. Similar tests determine residues (if any) in the meat of cattle, swine, and poultry and also in milk and eggs. From these data, the number of days from the last pesticide application until harvest or slaughter is determined. A tolerance is set for the pesticide on each crop or commodity and is established by the Environmental Protection Agency (EPA) as the amount that can safely remain on a food or feed commodity without hazard to the consumer. All food or feed commodities must not be above the pesticide residue tolerance limits when ready for market or for farm feeding. If a residue is over the tolerance level, the commodity can be condemned and the producer penalized.
- *Wildlife tests.* Tests are performed to determine the immediate and long-range effects on wildlife of field application of pesticides. The possibilities of residue building up in wildlife are checked on mammals, fish, and birds.
- *Field performance tests.* The producing company must prove that the new pesticide has practical value as a crop-protection tool. Performance data must be collected for each pest and from each crop or animal species on which the material is to be used. Data on crop varieties, soil types, application methods and rates, and number of applications required are also needed. This information must prove that the pests are controlled, crops or animals are not injured, yield and/or quality has been improved, and the pesticide definitely provides a worthwhile benefit.
- *Label review.* After all the tests described have been completed, the chemical company is ready to take its data to the EPA's Office of Pesticide Programs in Washington, D.C.

The company petitions for a legal tolerance and seeks registration of labeled uses for the pesticide on as many pests, crops, and animals as they have test data to support their claim. The request for label registration must now pass three reviews: (1) a review to determine the effectiveness of the pesticide for the purpose or purposes claimed, (2) a review by the tolerance section where the environmental, degradation, residue, and toxicological data are studied (this section sets the tolerance levels for the pesticide, and they are published in the *Federal Register*), and (3) a final review by the Office of Pesticide Programs to determine if all claims, restrictions, and directions for use on the label are supported by data. Figure 7–2 summarizes the steps the company must go through from product discovery to label approval.

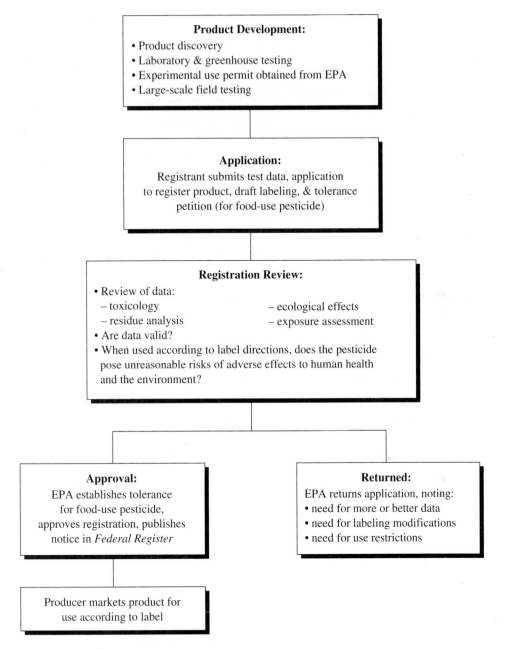

Figure 7–2 Product development from discovery to label approval.

WHAT IS ON A PESTICIDE LABEL?

Every pesticide you buy has a label that gives you instructions on how to use the product. The manufacturer may also provide additional forms of labeling. *Labeling* is all information that is received from the manufacturer about the product. Labeling includes not only the label on the product container, but also includes any supplemental information accompanying the product. This may include such things as brochures, leaflets, and information handed out by the dealer.

The *label* is the information printed on or attached to the pesticide container.

- To the manufacturer, the label is a "permit to sell."
- To the federal or state government, the label is a way to control the distribution, storage, sale, use, and disposal of the product.
- To the buyer or user, the label is a source of how to use the product correctly and legally.
- To physicians, the label is a source of information on proper treatment for poison cases.

The label, by law, must include all the necessary information and instructions for the effective and safe use of the pesticide. Some labels are easy to understand, while others are complicated.

Parts of the Label

Trade name (brand name), common name, and chemical name. The trade name is the producer's or formulator's proprietary name of the pesticide. Most companies register each brand name as a trademark and will not allow any other company to use that name. The brand name is usually the largest print on the front panel of the label. Some companies use the same basic trade name with only minor variations to designate entirely different pesticide chemicals. Applicators must be careful in choosing pesticides and not make their selection by brand name alone. Refer to Appendix E for a cross-reference of trade names and common names.

Use classification. All pesticides are evaluated, and if additional care over and above simple label compliance is required for their safe use, they receive a restricted use classification. Pesticides classified for restricted use will bear the following statement on the label:

RESTRICTED USE PESTICIDE

For retail sale and use only by certified applicators or persons under their direct supervision and only for those uses covered by the certified applicator's certification.

Type of pesticide. The type of pesticide usually is listed on the front panel of the pesticide label. This short statement indicates in general terms what the product will control. For example:

- Insecticide-ovicide for control of certain insects on cotton.
- Herbicide for nonselective weed control on industrial sites in noncrop areas.
- Fungicide for the control of certain diseases in broccoli, cabbage, cauliflower, cucurbit vegetables, onions, potatoes, and tomatoes.

Ingredients. The ingredient statement tells you what is in the container. Ingredients are usually listed as active and inert (inactive). The active ingredient is the chemical that does the job that the product is intended to do. The ingredient statement must list the official chemical names and/or common names for the active ingredients. Inert ingredients need not be named, but the label must show what percent of the total content they comprise. Inert ingredients may be the wetting agent, the solvent, the carrier, or the filler to give the product its most desirable quality for application. The *common name* is the generic name accepted for the active ingredient of the pesticide product regardless of the brand name. Only common names that are officially accepted by the EPA may be used in the ingredient statement on the pesticide label. The *chemical name* of the pesticide is the one that is long and difficult to pronounce unless you are a chemist. It identifies the chemical components and structure of the pesticide. For example, Benlate® is a brand name, and the common name for this fungicide is benomyl. The chemical name of the active ingredient in Benlate® is [methyl 1-(butylcarbamoyl) 2-benzimidazolecarbamate]. Always check the common or chemical names of pesticides when you make a purchase to be certain of getting the right active ingredient.

Contents. The bottom of the front panel of the pesticide label tells you how much is in the container. This is expressed as ounces, pounds, grams, or kilograms for dry formulations; and as gallons, quarts, pints, milliliters, or liters for liquids. Also listed is the amount of active ingredient per unit of product. This information is usually found in or near the ingredient statement.

Name and address of manufacturer. The law requires the manufacturer or distributor of a product to put the name and address of the company on a label. This is so you will know who made or sold the product in case you wish to contact them with specific questions about the product.

EPA registration number and establishment number. The EPA registration number signifies that the product has been registered with the Environmental Protection Agency. This means that the pesticide has passed the agency's review of the data and is ready for use according to the label's directions. The label may also contain an SLN number along with a state abbreviation. SLN means "special local need," and the state abbreviation means that the product is registered for use in that particular state.

The establishment number is a special number assigned to the plant that manufactured the pesticide for the company. Some companies manufacture pesticides in more than one manufacturing plant, and the establishment number identifies the place of manufacture. In case something goes wrong, the facility that made the product can be traced.

Signal words and symbols. Every label must contain a signal word giving you a clue as to how dangerous the product is to humans. Knowing the product's hazard helps you choose the proper protective measures for yourself, your workers, and other persons (or animals) that may be exposed. The signal word will be found in large letters on the front panel of the pesticide label. It immediately follows the statement "Keep out of reach of children," which must appear on every pesticide label.

- **CAUTION** This word signals you that the product is slightly toxic. An ounce to more than a pint taken by mouth could kill the average adult. Any product that is slightly toxic orally, dermally, or through inhalation or that causes slight eye or skin irritation must be labeled **CAUTION.**
- **WARNING** This word signals you that the product is moderately toxic. A teaspoonful to a tablespoonful by mouth could kill the average sized adult. Any product that is moderately toxic orally, dermally, or through inhalation or that causes moderate eye and skin irritation will be labeled **WARNING.**
- **DANGER-POISON** This word combination signals you that the pesticide is highly toxic. A taste to a teaspoonful taken by mouth could kill an average sized adult. Any product that is highly toxic orally, dermally, or through inhalation will be labeled **DANGER-POISON** printed in red and accompanied by the skull and crossbones symbol. The word **DANGER** alone (without the word **POISON** and the skull and crossbones) may be found on a few labels where the pesticide is corrosive or causes severe eye and skin burning, but is not highly toxic orally or through inhalation.

Precautionary statements. The statements that immediately follow the signal word, either on the front or side of the pesticide label, indicate which route or routes of entry (mouth, skin, lungs, eyes) you must particularly protect. Many pesticide products are hazardous by more than one route, so study these statements carefully. **DANGER-POISON** followed by "May be fatal if swallowed or inhaled" gives you a far different warning than "**DANGER: CORROSIVE**—causes eye damage and severe skin burns."

All pesticide labels contain additional statements to help you decide the proper precautions to take to protect yourself, your workers, and other persons (or domestic animals) that may be exposed. Sometimes these statements are listed under the heading "Hazards to Humans and Domestic Animals." The statements usually follow immediately after the route-of-entry statements. They recommend specific action you should take to prevent poisoning accidents. These statements are directly related to the toxicity of the pesticide product (signal word) and the route or routes of entry that must be protected. These statements contain such items as "Do not breathe vapors or spray mist" or "Avoid contact with skin or clothing" or "Do not get in eyes." There is a distinct difference between "May be harmful if swallowed" and "Harmful if swallowed," as well as the difference between "Do not get in eyes" and "Avoid getting in eyes." The variation and the strength of the statement depends on the signal word for the product. These specific-action statements help you prevent pesticide poisoning by taking the necessary precautions and wearing the correct protective clothing and equipment.

Protective clothing and equipment statements. The Worker Protection Standards require that each pesticide label give information on the personal protective equipment (PPE) that applicators and other handlers must wear. A good way to double-check for the correct type of clothing and equipment is to use the signal word, the route-of-entry statement, and the specific-action statement, along with the basic guidelines listed in Chapters 6 and 8.

Some pesticide labels fully describe appropriate protective clothing and equipment. A few list the kind of respirator that should be worn when handling and applying the product. Others require the use of a respirator but do not specify the type or model to be used. You should follow all advice for protective clothing or equipment that appears on the label. However, the lack of any statement with the mention of only one piece of equipment does not rule out the need for additional protection. Even though the label may not specifically require them, you should wear a long-sleeved shirt, long-legged trousers, and gloves whenever you are handling pesticides. You should consider wearing rubberized or waterproof clothing if you will be in prolonged contact with the pesticide or may be wet by an overhead spray application.

Other precautionary statements. Labels often list other precautions to take while handling the product. These are self-explanatory:

- Do not contaminate food or feed.
- Remove and wash contaminated clothing before reuse.
- Wash thoroughly after handling and before eating or smoking. Wear clean clothing daily.
- Not for use or storage in or around a house.
- Do not allow children or domestic animals in treated area.

These statements represent actions that a competent applicator should always follow. The absence of any or all of them from the label does not indicate that they need not be performed.

Statement of practical treatment. These statements will tell you the first aid treatments recommended in case of poisoning. Typical statements include:

- In case of contact with the skin, wash immediately with plenty of soap and water.
- In case of contact with eyes, flush with water for 15 minutes and get medical attention.
- In case of inhalation exposure, move from contaminated area and give artificial respiration if necessary.
- If swallowed, drink large quantities of milk, egg white, or water. Do not induce vomiting.
- If swallowed, induce vomiting.

All **DANGER-POISON** labels and some **WARNING** and **CAUTION** labels will contain a note to physicians describing the appropriate medical procedures for poisoning emergencies and may identify an antidote.

Environmental hazards. Pesticides may be harmful to the environment. Watch for special warning statements on the label concerning such hazards. If a particular pesticide is especially hazardous to wildlife, this fact will be stated on the label. For example:

- This product is highly toxic to bees.
- This product is toxic to fish.
- This product is toxic to birds and other wildlife.

These statements alert you to the special hazards that the use of the products may pose. They should help you choose the safest product for a particular job and remind you to take extra precautions. Other general environmental statements appear on nearly every pesticide label. They are reminders of common sense actions to follow to avoid contaminating the environment. The absence of any or all of these statements does not indicate that you do not have to take adequate precautions. Examples of general environmental statements include the following:

- Do not apply when runoff is likely to occur.
- Do not apply when weather conditions favor drift from treated areas.
- Do not contaminate water by cleaning of equipment or disposal of waste.
- Keep out of any body of water.
- Do not allow drift on desirable plants or trees.
- Do not apply when bees are likely to be in the area.

Physical or chemical hazards. This section of the label will tell you of any special fire, explosion, and chemical hazards the product may pose. For example:

- Flammable: Do not use, pour, spill, or store near heat or open flame. Do not cut or weld container.
- Corrosive: Store only in a corrosion-resistant tank.

Hazard statements. Hazards to humans and domestic animals, environmental hazards, and physical-chemical hazards are not always located in the same place on all pesticide labels. You should search the label for statements that will help you apply the pesticide more safely and knowledgeably.

Directions for use. The instructions on how to use the pesticide are an important part of the label. This is the best way to find out the correct manner in which to apply the product. The instructions will tell you the following:

- The pests that the manufacturer claims the product will control. You may legally apply a pesticide against a pest not specified on the labeling if the application is to a crop, animal, and site that the labeling approves.
- The crop, animal, or site the product is intended to protect.
- In what form the product should be applied.
- The proper equipment to be used. If application is permitted through an irrigation system, such as center pivot sprinkler, specific instructions will be given.

- How much to use.
- Mixing directions.
- Compatibility with other often-used products.
- Phytotoxicity and other possible injury or staining problems.
- Where the material should be applied.
- When the material should be applied.

Some pesticide labels with the signal words **DANGER-POISON** or **WARNING** (especially cholinesterase-inhibiting pesticides) contain a *restricted entry statement.* The restricted entry statement includes the amount of time that must elapse after a pesticide has been applied and before it is safe to enter the treated area without wearing full protective clothing and equipment. Refer back to Chapter 6 for information on restricted entry intervals (REIs) in connection with the Worker Protection Act. Some states have REIs, and they are not always listed on the label. It is your responsibility to determine if an REI has been established. It is illegal to ignore the REIs. If no REI appears on the label and none is set by your state, then you must wait at least until sprays are dried or dusts have settled before reentry or allowing others to enter a treated area without protective clothing.

Some labels for agricultural pesticides may list a *preharvest interval.* A preharvest interval is the time in days that must pass between the last application of a pesticide and harvesting for a food or feed crop. The interval varies with the pesticide and depends largely on its persistence on or in the crop, as well as the pesticide toxicity. If you do not obey the preharvest interval, your crop can be seized and destroyed and you can be fined. Most importantly, if crops are harvested and consumed before the preharvest interval has passed, people or animals may be poisoned. Table 7–1 is an example of preharvest intervals taken directly from a pesticide label.

Note: EPA PR Notice 92-4, effective October 9, 1992, states "A Material Safety Data Sheet may accompany a pesticide or device as long as it does not conflict with approved pesticide labeling. MSDSs may be attached to or shipped with pesticide containers so long as they do not obscure or conflict with the pesticide labeling."

Material safety data sheets may not all be organized the same, but they will basically cover the same information. The 10 sections contain the following:

I. **Identification of product.** This section includes the *manufacturer's name* and *address* and gives the *telephone number* of at least one office where additional information may be obtained. The 24-hour CHEMTREC telephone number is also included. The product name (*trade Name*) is the name used on the product label. The *chemical name and synonyms* entry generally includes readily recognized common names of active ingredients and at least one widely used chemical name for each. The *chemical family* entry indicates the chemical family to which the active ingredient(s) belongs, and in some cases, the type of use for which the product is intended (e.g., herbicide, bactericide, etc.).

II. **Hazardous ingredients of mixture.** This section usually includes a list of active ingredients contained in the product; whether or not each is considered hazardous; the CAS number (from Chemical Abstracts Service) which identifies the

TABLE 7-1 PREHARVEST AND LIVESTOCK GRAZING/FEEDING INTERVALS ON A PESTICIDE LABEL

Crops	Insects	Rates "Lannate" LV pints per acre	Last Application—Days		REI	Further use information
			To harvest	To livestock grazing/feeding		
Barley	Armyworms, cereal leaf beetle, aphids	$\frac{3}{4}$–$1\frac{1}{2}$	7	10	48 hours	Do not apply more than 1.8 lbs a.i./acre/crop Do not make more than 4 applications/crop
Beans (succulent) includes black-eyed peas and cowpeas)	Leafhopper, Mexican bean beetle	$\frac{3}{4}$–3	Succulent beans	3 vines	48 hours	Succulent beans— Do not apply more than 4.5 lbs a.i./acre/crop
	Fall armyworm, variegated cutworm	$1\frac{1}{2}$		7 hay		Do not make more than 10 applications/crop
	Beet armyworm loopers, corn earworm, saltmarsh caterpillar, yellowstriped armyworm western yellowstriped armyworm, lygus bugs, thrips, aphids, loopers.*	$1\frac{1}{2}$–3	$\frac{3}{4}$–$1\frac{1}{2}$ pt. —1, over $1\frac{1}{2}$ pt. —3			
	European corn borer (Ovicide & Larvicide) —Initiate when moth flights first appear and continue preventive treatments at 3–4 day intervals to control eggs and larvae					
	Spotted cucumber beetle	$\frac{3}{4}$–$1\frac{1}{2}$				

Crop	Pest	Rate	Days to harvest		REI	Restrictions
Beans (dry)	(same as succulent beans)	(same as succulent beans)	14†	14 vines† 14 hay†	48 hours	Do not apply more than 4.5 lbs a.i./acre/crop Do not make more than 10 applications/crop
Beets (table)	Imported cabbageworm, beet armyworm, cabbage looper, diamondback moth	$\frac{3}{4}$–3	0–roots		48 hours	Do not apply more than 3.6 lbs a.i./acre crop Do not make more than 8 applications/crop
	Cucumber beetle, variegated cutworm	$1\frac{1}{2}$	14–tops			
Bermudagrass pasture	Fall armyworm, armyworm	$\frac{3}{4}$–3		7 forage 3 dehydrated hay	48 hours	Do not apply more than 0.9 lbs a.i./acre/crop Do not make more than 4 applications/crop
Blueberries	Blueberry leafhopper, aphids, tussock moth, weevil, sharp-nosed leafhopper	$1\frac{1}{2}$				Do not apply during bloom Do not apply more than 3.6 lbs a.i./acre crop Do not make more than 4 applications/crop
	Cranberry fruitworm‡, cherry fruitworm‡	$1\frac{1}{2}$–3				
	Flea beetle (larvae), sawfly (larvae), blueberry leafroller	3				
	Blueberry maggot	$\frac{3}{4}$–$1\frac{1}{2}$				

* Not recommended in Al & GA.

† Do not apply within 14 days of cutting.

‡ For ground use only.

active chemical and the percent, approximate percent, or range of amount contained in the product; and the published Threshold Limit Values for each ingredient, where data are available.

III. **Physical data.** This section contains information on known physical constants. Where such constants are not known, an approximate or estimated value may be given.

IV. **Fire and explosion hazard data.** The *flash point* is given where applicable. *Flammable limits* are given where known, and refer to the limits of concentration in the atmosphere that would be likely to support combustion, or possibly explode if an ignition source were present. Products that flash at 100°F and below are DOT flammable; those that flash at 200°F and below are DOT combustible. *Fire extinguishing media* lists recommended fire extinguishing media to be used. Generally, water is not recommended unless used as a fog or spray to cool down containers, or to put out very small fires. The use of water may lead to environmental contamination, which may be extremely difficult and expensive to clean up. *Special fire fighting procedures* recommend safety equipment and protective equipment to be used, as well as any unusual procedures to be followed. *Unusual fire and explosion hazards* are pointed out for specific products.

V. **Reactivity data.** This section lists *stability and conditions to avoid* when handling or storing certain chemicals. It also lists some of the types of hazardous products that may be formed under a fire situation or, occasionally, under storage or use conditions.

VI. **Health hazard data.** Information is given on the *effects of overexposure,* including possible routes of entry. Recommendations are given for *emergency first aid procedures* and how to cope with medical emergencies caused by different routes of exposure. *Medical conditions aggravated by exposure* are pointed out, including certain preexisting illnesses that may be aggravated or made worse by exposure to the product. *Possible carcinogenicity* (cancer causing) information is included if such data is available.

VII. **Spill or leak information.** This section contains *steps to be taken in case a material is released or spilled.* Recommendations for containing the spill, use of absorbent material, picking up the absorbed material, and proper disposal are given. The *waste disposal method* recommended normally follows instructions given on the product label.

VIII. **Special protection information.** Recommendations are given for protecting users against excessive exposure. The entries are self-explanatory, and tend toward the philosophy of protecting against foreseeable exposure resulting from anticipated methods of use.

IX. **Special precautions.** Any special precautions to be followed in handling or storing the product are given, usually along with some "generic" statements about keeping the storage area clean, not contaminating water supplies, and reading the product label.

X. DOT information and date. This section gives appropriate DOT regulatory shipping information for highway shipments and the required DOT labeling. It also gives the date of preparation of the MSDS and indicates whether it is issued as a new MSDS or an update of a previous edition.

WHEN SHOULD YOU READ THE LABEL?

There are five times when you should read the pesticide label (Figure 7–3). Many people are guilty of only reading the label when they are going to apply the pesticide. Read the label as follows:

1. *First time.* Before you buy the pesticide, to determine:
   - If this is the best chemical for the job.
   - If the product can be used safely under your conditions.
   - If the concentration and amount of active ingredient are right for your job.
   - If you have the proper equipment to apply the pesticide.
2. *Second time.* Before you mix pesticides, to determine:
   - Protective equipment needed to handle the pesticide.
   - Specific warnings and first aid measures.
   - What the pesticide can be mixed with (compatibility).
   - How much pesticide to use.
   - Proper mixing procedure.
3. *Third time.* Before you apply the pesticide, to determine:
   - Safety measures for the applicator.
   - Where the pesticide can be applied (crops, structures, livestock, and so on).
   - When to apply the pesticide (remember to check waiting periods for crops and animals).
   - How to apply the pesticide.
   - Rate of application.

Figure 7–3   Read the label.

- Other restrictions on use.
- Any special instructions.

4. *Fourth time.* Before you store the pesticide, to determine:
   - Where and how to store the pesticide.
   - What incompatibility problems there may be.
   - Storage requirements (temperature, palleting, and the like).

5. *Fifth time.* Before you dispose of any excess pesticide or the container, to determine:
   - How to decontaminate and dispose of the pesticide container.
   - Where and how to dispose of surplus pesticides.
   - Any restrictions that may apply.

The importance of reading the label cannot be stressed too often. The information that appears on the label is put there for your information and your protection. If it is read and understood and all the directions are followed, the likelihood of misusing the material or of having an accident with the pesticide is remote. This is why we so often stress that *the most important few minutes in pest control are the time spent reading the label.*

WARNING CONCERNING OUT-OF-DATE SOURCES OF PESTICIDE INFORMATION

Accepting information about pesticides from unreliable sources is the surest way to create problems for yourself and others. One mistake can cost you money, injure someone, or cause you legal problems. Guard against the following:

1. Friends mean well but often cannot remember exactly the name of a product or, in fact, they may know very little about the problem or the product.
2. Don't be oversold by personnel in sales or others who recommend pesticides. The best source of information is from specialists who work directly with the class of chemical under consideration.
3. Interested bystanders may often offer suggestions, but if not from an authoritative source, the suggestions should be disregarded.
4. Old bulletins, circulars, and mimeographed information from state or federal sources should be disregarded. Only recommendations of a current year should be accepted as valid.
5. Sales catalogs may contain incomplete or out-of-date information. Doubtful recommendations should be checked against up-to-date information.
6. Memories should never be trusted. Too many pesticide chemical names sound nearly alike.
7. Trade names may not be consistent with the actual chemical ingredient.
8. Recommendations from other states may or may not apply in your state. Never accept recommendations on any pesticide without knowledge of the source and the validity of the source.

SPECIMEN LABELS

Pesticide manufacturers publish updated label books every year. The specimen labels on the following pages are from 1995 label books and are representative of the toxicity category signal words **CAUTION, WARNING, DANGER,** and **DANGER-POISON.** Take a few minutes to study these labels. Note that the specimen label bearing the signal word **CAUTION** is a biological insecticide. Note also that the specimen label bearing the signal word **WARNING** is also a restricted use pesticide. Remember too that the specimen label bearing the signal word **DANGER** has that designation because of the possibility of eye damage. See if you can find the various parts of the label as described on the previous pages. It is important to be aware that labels may change annually. Be alert in knowing that you are using the latest available labels. If you have doubts, contact the manufacturer.

BIOLOGICAL INSECTICIDE

FOR CONTROL OF INSECT PESTS OF VEGETABLES, FRUIT AND FIELD CROPS

SPECIMEN LABEL

ACTIVE INGREDIENT:
Bacillus thuringiensis, subspecies kurstaki,
  Potency of 52,863 Spodoptera Units
  (at least 60 million viable spores)
  per milligram* 6.4%
INERT INGREDIENTS:.................................. 93.6%
                          Total 100.0%

Potency units should not be used to adjust use rates beyond those specified in the Directions for Use section.

* Equivalent to 24.0 billion Spodoptera units per pound (17.0 billion International Units per pound).

EPA Reg No. 55947-136

KEEP OUT OF REACH OF CHILDREN
CAUTION
See additional precautionary statements

PRECAUTIONARY STATEMENTS
HAZARDS TO HUMANS
AND DOMESTIC ANIMALS
CAUTION

Harmful if inhaled or absorbed through the skin. Avoid breathing vapors or spray mist. Prolonged or frequently repeated skin contact may cause an allergic reaction in some individuals. Avoid contact with skin, eyes or clothing. Discontinue use if reaction occurs. In case of contact immediately flush eyes or skin with plenty of water. Get medical attention if irritation persists.

STATEMENT OF PRACTICAL TREATMENT
IF ON SKIN
Wash with plenty of soap and water. Get medical attention if irritation persists.

IF IN EYES
Immediately flush eyes with plenty of water. Get medical attention if irritation persists.

PERSONAL PROTECTIVE EQUIPMENT (PPE)
Applicators and other handlers must wear:

• Long-sleeved shirt and long pants

• Waterproof gloves

• Shoes plus socks

Follow manufacturer's instructions for cleaning/maintaining PPE. If no such instructions for washables, use detergent and hot water. Keep and wash PPE separately from other laundry.

ENGINEERING CONTROLS STATEMENTS
When handlers use closed systems, enclosed cabs, or aircraft in a manner that meets the requirements listed in the Worker Protection Standard (WPS) for agricultural pesticides [40 CFR 170.240 (d) (4-6)], the handler PPE requirements may be reduced or modified as specified in the WPS.

USER SAFETY RECOMMENDATIONS
Users should:

• Wash hands before eating, drinking, chewing gum, using tobacco or using the toilet.

• Remove clothing immediately if pesticide gets inside. Then wash thoroughly and put on clean clothing.

• Remove PPE immediately after handling this product. Wash the outside of gloves before removing. As soon as possible, wash thoroughly and change into clean clothing.

ENVIRONMENTAL HAZARDS

Do not contaminate water when disposing of equipment washwaters.

DIRECTIONS FOR USE

It is a violation of Federal Law to use this product in a manner inconsistent with its labeling.

For any requirements specific to your State or Tribe, consult the agency responsible for pesticide regulation.

AGRICULTURAL USE REQUIREMENTS

Use this product only in accordance with its labeling and with the Worker Protection Standard, 40 CFR part 170. This Standard contains requirements for the protection of agricultural workers on farms, forests, nurseries, and greenhouses, and handlers of agricultural pesticides. It contains requirements for training, decontamination, notification, and emergency assistance. It also contains specific instructions and exceptions pertaining to the statements on this label about personal protective equipment (PPE), and restricted-entry interval. The requirements in this box only apply to uses of this product that are covered by the Worker Protection Standard.

Do not apply this product in a way that will contact workers or other persons, either directly or through drift. Only protected handlers may be in the area during application.

Do not enter or allow worker entry into treated areas during the restricted entry interval (REI) of 12 hours.

PPE required for early entry to treated areas that is permitted under the Worker Protection Standard and that involves contact with anything that has been treated, such as plants, soil, or water is:

- Coveralls
- Waterproof gloves
- Shoes plus socks

Prosper with pesticides by using them properly! Read and follow label directions.

This labeling must be in the possession of the user at the time of pesticide application.

Notice: Read "Limitation of Warranty and Limitation of Liability" on the container or in this Directions for Use section before buying or using. If terms are not acceptable, return at once unopened.

Unless specifically stated, do not apply this product through any type of irrigation system.

JAVELIN® WG is a biological insecticide for the control of lepidopterous larvae (see RECOMMENDED APPLICATION RATES section).

JAVELIN WG attacks the larval gut and must be ingested by the insect to be effective.

JAVELIN WG may be applied up to and on the day of harvest.

GENERAL USE INSTRUCTIONS

For most consistent control, apply at first sign of newly hatched worms (1st and 2nd instar larvae). Instructions for specific crops are located in the ADDITIONAL INSTRUCTION sections under RECOMMENDED APPLICATION RATES.

Reapply as necessary under a pest management program that includes close scouting.

If rapid knockdown of heavy worm or non-lepidopterous populations is necessary, include an effective contact insecticide in combination with JAVELIN WG.

For heavy worm infestations, use the higher JAVELIN WG rate. During situations of dense foliage and/or rapid growth, increasing water carrier volumes will provide better crop coverage and improve JAVELIN WG performance.

Tank mix recommendations are for use only in states where the tank mix product and application site are registered.

Read and follow all label directions for use for other pesticides used as tank mix partners with JAVELIN WG for specific rate recommendations, application timing, and precautions.

MIXING

Fill spray or mixing tank $^3/_4$ full. Turn on agitation and pour JAVELIN WG into water while maintaining continuous agitation. Add other spray material (if any) and add balance of water. Agitate as necessary to maintain suspension. Do not allow diluted sprays to remain in the tank for more than 48 hours. JAVELIN WG is formulated to provide desirable coverage and adherence to leaf surfaces. Additional adjuvants, spreaders, or stickers may be added to improve product performance, especially under heavy dew or rainy conditions. Combinations with commonly used insecticides, fungicides, or other spray tank adjuvants are generally not deleterious to JAVELIN WG if the mix is used promptly. Before mixing in the spray tank, it is advisable to test physical compatibility by mixing all components in a small container in proportionate quantities.

GROUND APPLICATION

Unless otherwise stated, use recommended amount of JAVELIN WG in a minimum of 20 gallons of water per acre depending on type of crop and requirements of state regulations. If lower volumes are used, proper application equipment must be used to insure adequate coverage. Thorough and uniform crop coverage is required for adequate insect control.

AERIAL APPLICATION

Use recommended amount of JAVELIN WG in at least 3 gallons of water per acre. Applications at higher water volumes have demonstrated improved control of targeted pests. Apply early morning or evening when air is calm.

INSECTS CONTROLLED

When used as directed, JAVELIN WG will control the following insects.

COMMON NAME	SCIENTIFIC NAME
Alfalfa caterpillar	Colias eurytheme (Boisduval)
Almond Moth	Cadra cautella (Walker)
Armyworm	Pseudaletia unipuncta (Haworth)
Artichoke plume moth	Platyptilia carduidactyla (Riley)
Bagworm	Thyridopteryx ephemeraeformis (Haworth)
Banana skipper	Erionota thrax (Haworth)
Blueberry leafrollers	various
Blueberry spanworm	Itame argillacearia (Pack.)
Bollworm	Helicoverpa zea (Boddie)
California oak moth	Phrygnidia californica (Packard)
Cherry fruitworm	Grapholita packardi (Zeller)
Citrus cutworm	Xylomyges curialis
Codling moth	Cydia pomonella (Linnaeus)
Cotton leafperforator	Bucculatrix thurberiella (Busck)
Cotton leafworm	Alabama argillacea (Hubner)
Cutworm	various, family Noctuidae
Diamondback moth	Plutella xylostella (Linnaeus)
Douglas-fir tussock moth	Orgyia pseudotsugata (McDunnough)
Elm spanworm	Ennomos subsignaria (Hubner)
European corn borer	Ostrinia nubilalis (Hubner)
Fall cankerworm	Alsophila pometaria (Harris)
Fall webworm	Hyphantria cunea (Drury)
Filbert webworm	Melissopus latiferreanus (Walsingham)
Fruittree leafroller	Archips argyrospila (Walker)
Grape leaffolder	Desmia funeralis (Hubner)
Grapeleaf skeletonizer	Harrisina americana (Guerin)
Green cloverworm	Plathypena scabra (Fabricius)
Green fruitworm	Lithophane antennata (Walker)
Gypsy moth	Lymantria dispar (Linnaeus)
Helicoverpa spp.	Helicoverpa spp.
Heliothis spp.	Heliothis spp.
Hornworms	Manduca spp.
Imported cabbageworm	Pieris rapae (Linnaeus)
Jack pine budworm	Chloristoneura pinus (Freeman)
Loopers	various
Mimosa webworm	Homadaula anisocentra (Meyri)
Naval orangeworm	Amyelois transitella (Walker)
Obliquebanded leafroller	Choristoneura rosaceana (Harris)
Omnivorous leafroller	Platynota stultana
Omnivorous leaftier	Cnephasia longana (Haworth)
Orange tortrix	Argyrotaenia citrana (Fernald)
Orangedog	Papilio cresphontes (Cramer)
Oriental fruit moth	Grapholita molesta (Busck)
Peach twig borer	Anarsia lineatella (Zeller)
Pecan nut casebearer	Acrobasis nuxvorella (Neunzig)
Redbanded leafroller	Argyotaenia velutinana (Walker)
Redhumped caterpillar	Schizura concinna (J.E. Smith)
Rindworm complex	various
Roughskinned cutworm	Athetis mindara (Barnes & McDunnough)
Saltmarsh caterpillar	Estigmene acrea (Drury)
Sod webworm	Crambus mutabilis
Southwestern corn borer	Diatraea grandiosella (Dyar)
Spotted cutworm	Xestia spp.
Spring cankerworm	Paleacrita vernata (Peck)
Spruce budworm	Choristoneura fumiferana (Clemens)
Tent caterpillar	various, family Lasiocamidae
Tobacco budworm	Helicoverpa virescens (Fabricius)
Tobacco hornworm	Manduca sexta (Linnaeus)
Tomato pinworm	Keiferia lycopersicella (Walsingham)
Tropical sod webworm	Herpetogramma phaeopteralis (Guenee)
Tufted apple bud moth	Platynota idaeusalis (Walker)
Variegated leafroller	Platynota flavedana (Clemens)
Velvetbean caterpillar	Anticarsia gemmatalis (Hubner)
Western tussock moth	Orgyia vetusta (Boisduval)

RATE SELECTION CONSIDERATIONS

Rate recommendations are typically given as a range:

Lower rate ranges may be desired when tank mixing with contact insecticides labeled for worm control or under conditions of light worm infestations or when uniformly small worms are present.

Medium rate ranges may be desired when multiple worm life stages are present, continuous egg hatches are occurring or young or light armyworm infestations exist.

Upper rate ranges may be desired for heavy worm infestations, mature (larger) worms or for moderate to heavy infestations or armyworm, bollworm or other difficult to control worm species.

Unless otherwise stated, use recommended amount of JAVELIN WG in a minimum of 20 gallons of water per acre depending on type of crop and requirements of state regulations. Lower volumes may be used, but proper application equipment must be used to insure adequate coverage. Thorough and uniform crop coverage is required for adequate insect control.

RECOMMENDED APPLICATION RATES

Crops	Lbs./Acre
VEGETABLE CROPS	
Artichokes	0.50-1.25

ADDITIONAL INSTRUCTIONS:

Apply in a minimum of 100 gal. of water per acre with a spray interval of 10 days or less.

Apply 1 lb/A in combination with ASANA® XL, AMBUSH®, or SUPRACIDE® aerially to aid in resistance management of the artichoke plume moth. Use and follow all label directions of the tank mix partner regarding application, timing, gallonage, and schedules.

Crops	Lbs./Acre
Asparagus, Beans, (Green, Lima, Mung) Broccoli, Broccoli Raab (Rapini), Brussels Sprouts, Cabbage, Cardoon, Carrots, Cauliflower, Celeriac, Celery, Chick Peas, Chinese Broccoli, Chinese Cabbage, Collards, Cucumbers, Dry Bulb Onions, Eggplants, Garlic, Green Onions, Greens (Dandelion, Turnip, Mustard, Beet, China), Herbs (Basil, Cilantro, Dill, Oregano, Thyme, etc.), Horseradish, Kale, Kohl Rabi, Leeks, Lettuce (Endive, Romaine, Head Lettuce, Escarole, Butter Crunch, Leaf, etc.), Melons (Cantaloupe, Crenshaw, Honeydew, Muskmelon, Watermelon, etc.), Okra, Onions, Parsley, Parsnips, Peas, Peppers, Potatoes, Pumpkins, Radishes, Rutabaga, Salsify, Spinach, Squash (Summer and Winter), Sweet Corn, Sweet Potatoes, Swiss Chard, Table Beets, Tomatoes, Turnip Root, Watercress, Yams	0.12-1.50

ADDITIONAL INSTRUCTIONS:

Apply as necessary to maintain control.

Crops	Lbs./Acre
FIELD CROPS	
Alfalfa (Hay and Seed), Sudan Grass, Hay Crops & Other Forage Crops	0.25-1.50

ADDITIONAL INSTRUCTIONS:

Under conditions of rapid growth and rapidly increasing armyworm populations (10 worms or greater per 180° sweep) use the highest rate. Against heterogenous worm populations, where 4th and 5th instars are present, and continuous egg laying is occurring, applications may provide variable control. Under these conditions, the addition of a contact insecticide in combination with JAVELIN WG is recommended.

The addition of a spreader sticker to JAVELIN WG may provide improved performance.

Crops	Lbs./Acre
Canola and Evening Primrose	0.12-1.50

ADDITIONAL INSTRUCTIONS:

Apply as necessary to maintain control.

Crops	Lbs./Acre
Dry Beans and Peas, Lentils, Mint, Peanuts, Rice, Safflower, Soybeans, Sugar Beets, Sunflower, Sorghum	0.25-1.50

ADDITIONAL INSTRUCTIONS:

Apply as necessary to maintain control.

Crops	Lbs./Acre
Field Corn, Pop Corn, Seed Corn	0.50-1.50

ADDITIONAL INSTRUCTIONS:

Make initial application when economically damaging populations exist. Repeat as necessary to maintain control. Applications must be made to early instars prior to entering the ear or plant.

Crops	Lbs./Acre
Hops	0.25-1.00

ADDITIONAL INSTRUCTIONS:

Apply as necessary to maintain control.

Begin treatment as soon as possible after hatching and before larvae are protected by leaf folds.

Crops	Lbs./Acre
Jojoba	0.50-1.00

ADDITIONAL INSTRUCTIONS:

Apply in a minimum of 50 gallons of water per acre by ground equipment or a minimum of 10 gallons of water by aerial equipment. Thorough coverage of foliage is essential and dictates the minimum spray volumes necessary.

Crops	Lbs./Acre
Small Grains	1.00-1.50

ADDITIONAL INSTRUCTIONS:

Apply as necessary to maintain control.

Crops	Lbs./Acre
Tobacco	0.12-1.25

ADDITIONAL INSTRUCTIONS:

Apply as necessary to maintain control.

Crops	Lbs./Acre
Cotton	0.50-1.50

Including Arizona and California

Early and Mid-Season

ADDITIONAL INSTRUCTIONS:

Repeat as necessary throughout season to maintain control. If egg laying frequency indicates future moderate to heavy worm populations, time application spray to coincide with the 2nd instar larvae. During periods of high temperatures, worms will progress through 1st and 3rd instars very rapidly and early application timing is necessary for control.

To be effective, JAVELIN WG spray must be deposited at the larval feeding site. When plant cover is dense and worms are feeding in the lower ⅔ portion of the plant, aerial application of JAVELIN WG may not provide adequate control.

For the suppression of light to moderate infestations, apply at first sign of egg-laying or newly-hatched worms (1st instar larvae).

Crops	Lbs./Acre
Except Arizona and California	0.25-1.25

Early

ADDITIONAL INSTRUCTIONS:

For early season management of Helicoverpa and Heliothis species. Initiate applications when 50% of plants are at pinhead square cotton stage, independent of Helicoverpa and Heliothis egg and larval counts, or at 1st egg lay, whichever occurs earlier. Continue applications on 5-day spray interval up to synthetic pyrethroid spray window.

For added control of Helicoverpa and Heliothis, tank mixing of JAVELIN WG with a labeled ovicide, such as, methomyl (0.125 lb. a.i./acre), profenofos (0.25 lb. a.i./acre), or thiodicarb (0.125-0.25 lb.a.i./acre) is recommended.

Read and follow all directions for use, precautions and restrictions on tank mix product labels.

FRUIT, NUT & VINE CROPS

Crops	Lbs./Acre
Apples and Pears	0.50-4.00

ADDITIONAL INSTRUCTIONS:

Apply when newly hatched larvae appear and before leaves are rolled.

Continue applying as a part of the normal cover spray program until pest is adequately controlled. Apply when caterpillars are actively feeding (2nd-4th instars).

Crops	Lbs./Acre
Avocados	0.50-1.25

ADDITIONAL INSTRUCTIONS:

Apply as necessary to maintain control. Begin treatment as soon as possible after hatching and before larvae are protected by leaf folds.

(Amorbia [Mexican leafroller] is suppressed only).

Crops	Lbs./Acre
Bananas	0.50-1.00

ADDITIONAL INSTRUCTIONS:

Hawaii only. Use calibrated ground equipment with adequate water to apply to point of runoff.

Crops	Lbs./Acre
Citrus	0.25-1.50

ADDITIONAL INSTRUCTIONS:

Use 50-600 gallons of water per acre when using ground equipment and 10 gallons of water minimum per acre by air.

(Amorbia [Mexican leafroller] is suppressed only).

Blueberries, Caneberries, Currants, Kiwi	0.25-1.00

ADDITIONAL INSTRUCTIONS:

Apply by ground equipment only. Begin treatment as soon as possible after hatching. For leafrollers, apply before larvae are protected by leaf folds.

Grapes	0.50-1.25

ADDITIONAL INSTRUCTIONS:

Apply by ground equipment in up to 200 gallons total spray per acre to obtain thorough coverage of leaf surfaces. Start treating as soon as possible after hatching and before larvae are protected by leaf folds.

Almonds, Apricots, Cherries, Filberts, Nectarines, Peaches, Pecans, Persimmons, Plums, Pomegranate, Prunes, Walnuts	0.25-4.00

ADDITIONAL INSTRUCTIONS:

For leafrollers, start treating as soon as possible after hatching and before larvae are protected by leaf folds.

Apply when caterpillars are actively feeding (2nd to 4th instar).

Application timing is very important for good casebearer suppression. Consult your local university or extension agent for information concerning specific modeling that predicts egg lay, typical application dates, and scouting techniques for your area. JAVELIN WG must be present at egg hatch for best control. Make application when the majority of eggs are in the pink stage. For best control make two applications 7 days apart. If only one application is made, a minimum of 1 lb. should be applied.

Melons (Also see vegetables)	0.50-0.75

ADDITIONAL INSTRUCTIONS:

Apply at first sign of hatch before larvae enter fruit. Repeat as necessary to maintain control.

Strawberries	0.25-1.50

ADDITIONAL INSTRUCTIONS:

Apply as necessary to maintain control. Use 20 gallons water minimum per acre when using ground equipment and 5 gallons water minimum per acre by aircraft.

In a tank mix with contact insecticides, rates as low as 1/2 lb. of JAVELIN WG may be used for the control of armyworm.

Crops	Lbs./Acre
SHADE TREES AND ORNAMENTALS (Including Roses)	0.12-1.25

ADDITIONAL INSTRUCTIONS:

Apply when leaf expansion reaches 40% to 50% as infestation warrants. If eggs hatch over a long period of time, or if reinfestation occurs, spray about 14 days after first application.

Apply when most larvae are 3rd-4th instar. Also consider the opening of the bud cap to ensure foliage exposure. Apply after eggs have hatched and early instar larvae are feeding on exposed foliage.

TURF AND GRASS SEED PRODUCTION	1.00

ADDITIONAL INSTRUCTIONS:

Repeat as necessary throughout season to maintain control.

STORED SOYBEANS, GRAINS
(Indian Meal Moth, Almond Moth)
ADDITIONAL INSTRUCTIONS:

To control and prevent Indian Meal Moth and Almond Moth infestations of stored soybeans and grains, prepare a spray mixture which includes 1 gallon of water of every 1.5 oz. by weight of JAVELIN WG. The spray mixture may be applied either by treating the top 4 inches of grain as it is being augured into storage (applying 0.6 pint of mixture per bushel in the grain stream), or by treating the surface of grain after it is in the bin. The Table below can be used as a guide in determining the total amount of JAVELIN WG needed according to the bin diameter or the number of bushels to be treated.

Bin Diameter (Ft.)	Surface Area (Sq. Ft.)	Bushels (to 4 in. Depth)	JAVELIN WG Rate (by Weight)	
			Grams	Oz.
8	50	13	21	0.75
12	113	30	50	1.75
16	201	53	85	3.00
20	314	84	120	4.25
24	452	120	185	6.50
28	615	163	255	9.00
32	804	214	326	11.50

To insure thorough coverage when making applications to the grain surface after it is in the bin, apply spray mixture in three (3) applications. Mix the grain with a scoop or rake to a depth of four (4) inches after each application.

Stored grain may be treated anytime, but for best results, treat grain at the time it is placed into storage or shortly thereafter, or in the early spring prior to egg-laying. Full season control is normally experienced. Re-treat only if reinfestation occurs.

For the protection of bagged grain, apply spray mixture to entire grain mass, and mix thoroughly prior to bagging. JAVELIN WG at 6 oz. by weight per 10 gallons of water will treat approximately 100 bushels.

Treated grain may be used at any time after treatment.

Flowers and Ornamentals

JAVELIN WG may also be used on flowers and ornamentals outdoors and in the greenhouse at a rate of 0.25-1.50 lb. per 100 gallons of water for control of listed insects on this label.

Guide for Small Spray Volume Mixing

Rate Lbs./A	Conversion Rate* Teaspoons/Gallon
1/4	1/2
1/2	1
1	2

* Assumes Application to spray runoff.

STORAGE AND DISPOSAL

Do not contaminate water, food, or feed by storage and disposal.

STORAGE

Store in original container in a cool, dry place inaccessible to children and pets and away from heat and direct sunlight. Protect from freezing. Storage at temperatures above 90°F may impair effectiveness.

PESTICIDE DISPOSAL

Pesticide, spray mixture, or rinse water that cannot be used according to label instruction must be disposed of according to Federal, State, or Local procedures.

CONTAINER DISPOSAL

Completely empty container into application equipment. Triple rinse or equivalent. Then offer for recycling or reconditioning, or puncture and dispose of in a sanitary landfill or by incineration, or if allowed by state and local authorities, by burning. If burned, stay out of smoke.

LIMITATION OF WARRANTY AND LIMITATION OF LIABILITY

NOTICE: Read this Limitation of Warranty and Limitation of Liability before buying or using this product. If the terms are not acceptable, return the product at once, unopened, and the purchase price will be refunded.

It is impossible to eliminate all risks inherently associated with the use of this product. Crop injury, ineffectiveness, or other unintended consequences may result because of such factors as weather conditions, presence of other materials, or the manner of use or application, all of which are beyond the control of Sandoz or seller. All such risks shall be assumed by buyer or user.

Sandoz warrants that this product conforms to the chemical description on the label and is reasonably fit for the purposes stated in the Directions for Use, under normal use conditions, subject to the risks described above. **Sandoz makes no other express or implied warranty of fitness or of merchantability or any other express or implied warranty.**

In no event shall Sandoz or seller be liable for any incidental, consequential or special damages resulting from the use or handling of this product. **The exclusive remedy of the user or buyer, and the exclusive liability of Sandoz or seller for any and all claims, losses, injuries or damages (including claims based on breach of warranty, contract, negligence, tort, strict liability or otherwise) resulting from the use or handling of this product, shall be the return of the purchase price of the product or, at the election of Sandoz or seller, the replacement of the product.**

Sandoz and seller offer this product, and buyer and user accept it, subject to the foregoing limitations of warranty and limitation of liability, which may not be modified by any oral or written agreement.

REGISTERED TRADEMARKS

Ambush® is a registered trademark of ICI Group Companies.

Asana® XL is a registered trademark or E.I. duPont de Nemours and Co., Inc.

Supracide® is a registered trademark of Ciba-Geigy Corp.

JAVELIN® WG is a registered trademark of Sandoz Ltd.

⚠ SANDOZ
SANDOZ AGRO, INC.
1300 EAST TOUHY AVENUE, DES PLAINES, ILLINOIS 60018

SP-376

©1994 SANDOZ AGRO, INC.

July 1994
Des Plaines, IL

RESTRICTED USE PESTICIDE

This product is a restricted use herbicide due to reproductive and ground and surface water concerns. Users must read and follow all precautionary statements and instructions for use in order to minimize potential for Cyanazine to reach ground and surface water.

For retail sale to and use only by Certified Applicators or persons under their direct supervision and only for those uses covered by the Certified Applicator's certification.

Bladex® 4L

herbicide

FOR USE ON FIELD CORN, POPCORN, SWEET CORN, FIELD CORN GROWN FOR SEED, AND COTTON

LIQUID

Contains 4 lbs. active ingredient per gallon.

Active Ingredient	*By Weight*
Cyanazine	
2[[4-chloro-6-(ethylamino)-s-triazin-2-yl]	
amino]-2-methylpropionitrile	43%
Inert Ingredients	57%
TOTAL	100%

EPA Reg. No. 352-470

KEEP OUT OF REACH OF CHILDREN
WARNING   AVISO

Si usted no entiende la etiqueta, busque a alguien para que se la explique a usted en detalle. (If you do not understand this label, find someone to explain it to you in detail.)

STATEMENT OF
PRACTICAL TREATMENT

IF SWALLOWED, call a physician or poison control center. Drink 1 or 2 glasses of water and induce vomiting by touching back of throat with finger. Do not induce vomiting or give anything by mouth to an unconscious person.

IF IN EYES, flush with plenty of water. Get medical attention if irritation persists.

IF ON SKIN, wash immediately with plenty of soap and water.

IF INHALED, remove victim to fresh air. If not breathing, give artificial respiration, preferably mouth-to-mouth. Get medical attention.

For medical emergencies involving this product,
call 1-800-441-3637.

PRECAUTIONARY STATEMENTS
HAZARD TO HUMANS
AND DOMESTIC ANIMALS

WARNING! May be fatal if swallowed. Harmful if inhaled or absorbed through the skin. Causes temporary eye injury.

This product may be hazardous to your health. This product is classified "Restricted Use" because, at doses which caused serious maternal illness in laboratory animals, birth defects were present. Use of protective clothing and equipment and following the precautions below can reduce risk. Avoid breathing spray mist. Avoid contact with skin, eyes, or clothing. Do not get in eyes or on clothing.

Keep out of reach of domestic animals, particularly cattle. Consumption of this product, spray solutions, or water contaminated with product can result in serious illness or possible death of bovines.

PERSONAL PROTECTIVE EQUIPMENT
Applicators and other handlers must wear:
Long-sleeved shirt and long pants
Chemical-resistant gloves, such as barrier laminate or butyl rubber or nitrile rubber or polyvinyl chloride or viton or neoprene rubber.
Chemical-resistant footwear plus socks
Protective eyewear
Chemical-resistant apron when cleaning equipment, mixing or loading.

(continued on next page)

PRECAUTIONARY STATEMENTS
(continued)

PERSONAL PROTECTIVE EQUIPMENT (continued)

Discard clothing and other absorbent materials that have been drenched or heavily contaminated with this product's concentrate. Do not reuse them. Follow manufacturer's instructions for cleaning/maintaining PPE. If no such instructions for washables, use detergent and hot water. Keep and wash PPE separately from other laundry.

ENGINEERING CONTROL STATEMENTS

When handlers use closed systems, enclosed cabs, or aircraft in a manner that meets the requirements listed in the Worker Protection Standard (WPS) for agricultural pesticides [40 CFR part 170.240 (d)(4-6)], the handler PPE requirements may be reduced or modified as specified in the WPS.

USER SAFETY RECOMMENDATIONS

USERS SHOULD: Wash hands before eating, drinking, chewing gum, using tobacco or using the toilet. Remove personal protective equipment immediately after handling this product. Wash the outside of gloves before removing. As soon as possible, wash thoroughly and change into clean clothing .

ENVIRONMENTAL HAZARDS

Do not apply directly to water, or to areas where surface water is present, or to intertidal areas below the mean high water mark. Do not contaminate water by cleaning equipment or disposal of wastes. Cyanazine, the active ingredient in BLADEX has been detected in surface waters that receive run-off from treated areas. To minimize cyanazine run-off, follow the Best Management Practices outlined in the Directions For Use section of this label.

Cyanazine is a chemical which can move (seep or travel) through soil and can contaminate groundwater which may be used as drinking water. Cyanazine has been found in groundwater as a result of agricultural use. Groundwater contamination may be reduced by diking and flooring of permanent liquid bulk storage sites with an impermeable material. Users are advised not to apply BLADEX where the water table (groundwater) is close to the surface and where the soils are very permeable (i.e., well drained soils such as loamy sands). Your local agricultural agencies can provide further information on the type of soil in your area and the location of groundwater.

DIRECTIONS FOR USE

It is a violation of federal law to use this product in a manner inconsistent with its labeling. This labeling must be in possession of the user at the time of pesticide application.

Do not apply this product in a way that contacts workers or other persons, either directly or through drift. Only protected handlers may be in the area during application. For any requirements specific to your state or tribe, consult the agency responsible for pesticide regulation.

AGRICULTURAL USE REQUIREMENTS

Use this product only in accordance with its labeling and with the Worker Protection Standard, 40 CFR part 170. This Standard contains requirements for the protection of agricultural workers on farms, forests, nurseries, and greenhouses, and handlers of agricultural pesticides. It contains requirements for training, decontamination, notification, and emergency assistance. It also contains specific instructions and exceptions pertaining to the statements on this label about personal protective equipment(PPE) and restricted-entry interval. The requirements in this box only apply to uses of this product that are covered by the Worker Protection Standard.

Do not enter or allow worker entry into treated areas during the restricted entry interval (REI) of 12 hours.

PPE required for early entry to treated areas that is permitted under the Worker Protection Standard and that involves contact with anything that has been treated, such as plants, soil, or water, is:
Coveralls.
Chemical-resistant gloves, such as barrier laminate or butyl rubber or nitrile rubber or polyvinyl chloride or viton or neoprene rubber.
Shoes plus socks.
Protective eyewear

DuPont's BLADEX® 4L Herbicide is a liquid formulation that can be applied alone or in combination with other herbicides to control a wide variety of weeds on field corn, sweet corn, popcorn, field corn grown for seed, and cotton.

WEEDS CONTROLLED ON CORN

BLADEX effectively controls the following weeds on corn when used alone or in combination with other herbicides according to label directions:

Grasses

Annual bluegrass	Foxtail, giant
Annual fescues	Foxtail, green
Annual (Italian) ryegrass	Foxtail, yellow
Annual sedge	Goosegrass
Barnyardgrass*	Junglerice
Bullgrass	Stinkgrass (Indian lovegrass)
Crabgrass	Witchgrass
Fall panicum	

Broadleaves

Annual groundcherry	Mayweed
Annual morningglory	Nightshade (annual)
Black mustard	Pigweed*
Buffalobur	Pineappleweed
Buttercup (annual)	Plantain
Carpetweed	Poorjoe
Common chickweed	Prickly sida (teaweed)
Cocklebur†	Prostrate knotweed
Common groundsel	Prostrate spurge
Common mallow	Russian thistle
Common ragweed	Shepherdspurse
Common purslane	Smallflower galinsoga
Corn spurry	Smartweed (Pennsylvania)
Curly dock (seedling) Fiddleneck	Sunflower† (wild, annual, common)
Florida pusley (Florida purslane)	Tarweed cuphea (gumweed) Velvetleaf*
Hedge mustard	Wild buckwheat
Jimsonweed*	Wild mustard
Kochia	Wild radish
Ladysthumb	Wild turnip
Lambsquarters	

* Under conditions that delay germination of the seeds, such as low temperatures or lack of soil surface moisture, the effectiveness of BLADEX against these weeds may be impaired.

† The degree of weed control may be reduced if soil moisture and temperature conditions cause deep germination of these seeds.

CONVENTIONAL TILLAGE PREEMERGENCE OR PREPLANT INCORPORATED USES ON CORN

FIELD CORN, POPCORN, SWEET CORN, AND FIELD CORN GROWN FOR SEED

Apply BLADEX treatments just before, at, or after planting but before the crop has emerged. Do not remove treated soil from the seedrows prior to or during planting.

BLADEX may be applied early, prior to planting or in a split application, if pre-season weed control is desired. For split applications, do not exceed the total amount of BLADEX for the soil texture and organic matter shown in Table 1.

If BLADEX is applied early, more than 15 days before planting, a split application of BLADEX or some other herbicide treatment may be necessary at or after planting to provide additional length of weed control. Refer to the Conservation Tillage Preemergence Uses section of this label.

Rotary hoeing is recommended for preemergence applications that do not receive adequate rainfall or sprinkler irrigation to wet the top 2 in. of soil or depth of germinating weeds within about 10 days after application.

BLADEX alone or tank mix combinations should not be incorporated more than 3" deep to keep from burying the herbicide. Single or two-pass incorporation with a tool such as a field cultivator operated at 5 to 7 mph is acceptable. A spike-toothed harrow, deep tillage disk, or rolling basket device is not recommended for incorporating BLADEX.

Use Rates for BLADEX Applied Alone

- Use Table 1 for field corn, popcorn, and field corn grown for seed.

Use Rates for BLADEX in Combination with Other Herbicides

BLADEX can be tank mixed with Atrazine, Lasso 4EC[3], Dual 8E[4] Frontier[5], Surpass[6], Harness Plus[7], Sutan+[8], and Eradicane 6.7E[9] herbicides. Refer to the manufacturers' labels for proper use rates, rotational guidelines, and all other precautions. Follow the label with the most restrictive requirements.

- Use Table 3 for BLADEX tank mix rates with Atrazine on field corn, popcorn, and field corn grown for seed in all states except Kentucky, Missouri, Tennessee, and Kansas east of Highway 99.
- Use Table 4 for BLADEX tank mix rates with Atrazine on field corn, popcorn, and field corn grown for seed in all of Kentucky, Missouri, Tennessee, and Kansas east of Highway 99.
- Use Table 5 for BLADEX tank mix rates with "Lasso", "Dual", "Frontier", "Surpass", "Harness Plus", "Sutan+", or "Eradicane 6.7E" on field corn and popcorn.
- Use Table 6 for BLADEX tank mix rates with Atrazine and "Lasso", Dual, "Frontier", "Surpass", "Harness Plus", "Sutan+", or "Eradicane 6.7E" in field corn, popcorn, and field corn grown for seed.

BLADEX Plus "Eradicane 6.7E" or "Sutan+"

- Do not use BLADEX in tank mixes with "Eradicane 6.7E" or "Sutan+" on field corn grown for seed.
- Use 3.6 pints per acre of "Sutan+" or "Eradicane 6.7E". For loam soils with ≥5% organic matter or clay loams and clay soils with ≥4% organic matter, use 4.8 pints per acre.
- Apply tank mix combinations of BLADEX and "Sutan+" or "Eradicane 6.7E" before planting. Incorporate the mixture 2" to 3" deep immediately after application. Refer to the "Sutan+" and "Eradicane 6.7E" manufacturers' labels for appropriate incorporation methods. Do not incorporate the BLADEX deeper than 3" or weed control may be reduced.
- As an alternative, BLADEX may be applied preemergence, as an overlay over previously incorporated "Sutan+" and "Eradicane 6.7E".

- Existing stands of quackgrass and purple and yellow nutsedge must be turned under and thoroughly chopped up prior to chemical treatments.
- In addition to the weeds controlled by BLADEX, this tank mix will control or suppress the following weeds: shattercane, quackgrass, yellow and purple nutsedge, sandbur, Texas panicum, and wild proso millet. For fields with moderate to heavy infestations of these weeds, refer to the "Sutan+" or "Eradicane 6.7E" labels for appropriate rates.

Use Rates for BLADEX on Sweet Corn

BLADEX may be applied preemergence or preplant incorporated for the control of annual grasses and broadleaf weeds in sweet corn.

- BLADEX may cause injury or stand loss on new or "supersweet" varieties of sweet corn. Consult with Agricultural Extension Agencies and sweet corn seed suppliers about the sensitivity of new varieties to potential injury.
- Apply BLADEX treatments just before, at, or after planting, but before crop has emerged.
- Use Table 1 for use rates for BLADEX applied alone preemergence on sweet corn.
- Use Tables 7, 8, and 9 for BLADEX tank mix rates with other herbicides on sweet corn.

CONSERVATION TILLAGE PREEMERGENCE USES ON CORN

FIELD CORN, POPCORN, SWEET CORN, AND FIELD CORN GROWN FOR SEED

(30 days prior to planting until emergence)

BLADEX may be used for early preplant or preemergence weed control for land going into the production of corn under conservation tillage (including no-till) programs. Complete any planned early spring tillage prior to application. Tillage after application may reduce the effectiveness of the herbicide treatment. In corn planted in no-till stalk ground (corn, sorghum), stubble ground (soybean, small grains), and any minimum-till land, BLADEX, when used according to label directions, will

- kill most existing small weeds,
- suppress many emerged perennial weeds, and
- provide residual control of annual weeds.

A nitrogen solution or complete fertilizer solution may replace all or part of the water as a carrier. The spray gallonage and boom design must be adequate to give thorough, uniform coverage of the weed foliage. Follow the label requirements of all products used in tank mix combinations.

Use Rates

- Use Table 1 for field corn, popcorn, or field corn grown for seed with surface residue <30%.
- Use Table 2 for field corn, popcorn, or field corn grown for seed with residue >30%.
- Use Tables 7 - 9 for sweet corn.

Annual Grass and Broadleaf Weeds Up to 3"

- Use BLADEX alone and add 1-2 qt per acre of crop oil concentrate (COC) if weeds are emerged at the time of application.
- For best burndown results use a minimum of 20 gal per acre of liquid fertilizer as the carrier and replace COC with a nonionic surfactant.

Broadleaf Weeds Exceeding 3"

- If broadleaf weeds exceed 3 in. at application, add 2,4-D LV Ester and/or "Banvel" and non-ionic surfactant at recommended rates.
- Additional weeds controlled with 2,4-D are: wild buckwheat, dandelion, dock, giant ragweed, marestail, pennycress, prickly lettuce and tansy mustard.
- To control existing alfalfa, add 0.3 to 0.5 pint per acre of "Banvel" to the spray mixture of BLADEX plus 2,4-D. Apply before the alfalfa exceeds 6 in. in height.

Grass Weeds Exceeding 3"

- If grass weeds exceed 3 in. at application, add either Gramoxone Extra[10] or Roundup[11] to the tank at the recommended rates for these products.
- Add 1 to 2 pints of a non-ionic surfactant per 100 gal of spray.
- With "Gramoxone Extra," well established weeds over 6 in. tall will not be well controlled.
- Do not apply "Gramoxone Extra" in a suspension type liquid fertilizer containing clay.

Burn Down Under Dry Conditions for Control of Sod Grasses

For burndown of existing sod grasses such as orchardgrass, bromegrass, rye or timothy, or when conditions are very dry, add "Gramoxone Extra" to the tank mix at the recommended rate.

Perennial Grass Weeds

For improved control of perennial grasses such as johnsongrass or quackgrass, add "Roundup" at the recommended rate or follow with a postemergence application of DuPont's ACCENT® Herbicide.

Other Labeled Tank Mixes

BLADEX can be tank mixed with other labeled products according to the directions for the treatments explained in the Conventional Tillage section of this label.

Early preplant applications of BLADEX may be tank mixed with 2 pints per acre of Princep[12] 4L or 1.1 lb of Princep Caliber[13] 90. Apply 30 days or more prior to planting.

Sequential Treatments

If, due to weather conditions, corn is planted more than 30 days after application, a sequential herbicide treatment may be necessary to provide additional length of weed control. This may be a postemergence treatment with ACCENT, BLADEX 90DF, EXTRAZINE II, or some other herbicide treatment applied at or after planting.

USE RATE TABLES FOR CORN

BLADEX 4L ALONE, PREEMERGENCE

TABLE 1, FOR FIELD CORN, POPCORN, SWEET CORN, AND FIELD CORN GROWN FOR SEED

Early Preplant or Preemergence Broadcast Rates in Conventional Tillage with <30% Surface Residue

Soil Texture	%OM =	BLADEX 4L (quarts/acre)*					
		<1%	1%	2%	3%	4%	≥5%
Sand, Loamy sand		Do Not Use	1.3	1.5	2.3	2.8	3.3
Sandy loam		1.3	1.8	2.0	2.5	3.0	3.5
Loam, Silt loam, Silt		1.5	2.0	2.5	3.0	3.5	4.0
Sandy clay loam, Clay loam, Silty clay loam		2.0	2.5	3.0	3.5	4.0	4.5
Sandy clay, Silty clay, Clay		2.8	3.0	3.5	4.0	4.5	4.8

TABLE 2, FOR FIELD CORN, POPCORN, SWEET CORN, AND FIELD CORN GROWN FOR SEED

Early Preplant or Preemergence Broadcast Rates in Conservation or No-till Tillage with >30% Surface Residue

Soil Texture	%OM =	BLADEX 4L (quarts/acre)*					
		<1%	1%	2%	3%	4%	≥5%
Sand, Loamy sand		Do Not Use	1.6	1.9	2.9	3.5	4.1
Sandy loam		1.6	2.3	2.5	3.1	3.8	4.0
Loam, Silt loam, Silt		1.9	2.5	3.1	3.8	4.4	5.0
Sandy clay loam, Clay loam, Silty clay loam		2.7	3.1	3.8	4.4	5.0	5.6
Sandy clay, Silty clay, Clay		3.5	3.8	4.4	5.0	5.6	6.0

BLADEX 4L PLUS ATRAZINE 4L, PREEMERGENCE

TABLE 3, FOR FIELD CORN, POPCORN, AND FIELD CORN GROWN FOR SEED

Early Preplant or Preemergence Broadcast Rates: For Use in All States Except Kentucky, Missouri, Tennessee, and Kansas East of Highway 99

Soil Texture	%OM =	BLADEX 4L (quarts/acre)* plus ATRAZINE 4L (quarts/acre)†					
		<1%	1%	2%	3%	4%	≥5%
Sand, Loamy sand		Do Not Use	0.8 + 0.5	1.0 + 0.5	1.5 + 0.5	1.8 + 0.8	2.3 + 1.0
Sandy loam		0.8 + 0.5	1.0 + 0.5	1.5 + 0.5	1.8 + 0.8	2.3 + 1.0	2.8 + 1.3
Loam, Silt loam, Silt		1.0 + 0.5	1.5 + 0.5	2.0 + 0.8	2.3 + 1.0	2.8 + 1.3	3.3 + 1.3
Sandy clay loam, Clay loam, Silty clay loam 1.		1.5 + 0.5	2.0 + 0.8	2.3 + 1.0	2.8 + 1.2	3.3 + 1.3	3.5 + 1.3
Sandy clay, Silty clay, Clay		2.0 + 0.8	2.3 + 1.0	2.8 + 1.2	3.3 + 1.3	3.5 + 1.3	3.8 + 1.5

---

* Maximum rate limits per acre per year for all applications is 6.5 lb cyanazine (6.5 quarts BLADEX 4L) except on highly erodible land with less than 30% plant residue cover, where the rate limit is 3.0 lb cyanazine (3.0 quarts BLADEX 4L).

† If 90% Atrazine is used, multiply the Atrazine rates shown in this table by 1.11 to equal the appropriate poundage of an 90% Atrazine product. If Atrazine 80W is used, multiply the Atrazine rates shown in this table by 1.25 to equal the appropriate poundage of Atrazine 80W.

TABLE 4, FOR FIELD CORN, POPCORN, AND FIELD CORN GROWN FOR SEED

Early Preplant or Preemergence Broadcast Rates: For Use Only in All Kentucky, Missouri, Tennessee, and Kansas East of Highway 99

Soil Texture	%OM =	<1%	1%	2%	3%	4%	≥5%
		BLADEX 4L (quarts/acre)* plus ATRAZINE 4L (quarts/acre)†					
Sand, Loamy sand		Do Not Use	0.8 + 0.5	1.0 + 0.5	1.5 + 0.6	1.8 + 0.8	2.3 + 1.0
Sandy loam		0.8 + 0.5	1.4 + 0.7	1.6 + 0.8	1.8 + 0.9	2.2 + 1.1	2.8 + 1.3
Loam, Silt loam, Silt		1.4 + 0.7	2.0 + 1.0	2.2 + 1.1	2.4 + 1.2	2.8 + 1.3	3.3 + 1.3
Sandy clay loam, Clay loam, Silty clay loam		1.6 + 0.8	2.2 + 1.1	2.4 + 1.2	2.8 + 1.3	3.3 + 1.3	3.5 + 1.4
Sandy clay, Silty clay, Clay		2.0 + 1.0	2.4 + 1.2	2.8 + 1.3	3.3 + 1.3	3.5 + 1.4	3.8 + 1.5

BLADEX 4L PLUS TANK MIX PARTNERS, PREEMERGENCE

TABLE 5, FOR FIELD CORN, POPCORN, AND FIELD CORN GROWN FOR SEED

Early Preplant or Preemergence Broadcast Rates in Tankmix Combinations§ with "Lasso", "Sutan+", "Eradicane 6.7E", "Dual 8E", "Frontier", "Surpass", or "Harness Plus" in Conventional or Conservation Tillage

Soil Texture	%OM =	<1%	1%	2%	3%	4%	≥5%
		BLADEX 4L (quarts/acre)*					
Sand, Loamy sand		0.6‡	0.8	1.3	1.5	1.8	2.0
Sandy loam		0.8	1.3	1.5	1.8	2.0	2.3
Loam, Silt loam, Silt		1.3	1.5	1.8	2.0	2.3	2.5
Sandy clay loam, Clay loam, Silty clay loam		1.5	1.8	2.0	2.3	2.5	2.8
Sandy clay, Silty clay, Clay		1.8	2.0	2.3	2.5	2.8	3.0

TABLE 6, FOR FIELD CORN, POPCORN, AND FIELD CORN GROWN FOR SEED

Early Preplant or Preemergence Broadcast Rates in Tankmix Combinations§ with Atrazine 4L and "Lasso", "Sutan+", "Eradicane 6.7E", "Dual 8E", "Frontier", "Surpass", or "Harness Plus" in Conventional or Conservation Tillage

Soil Texture	%OM =	<1%	1%	2%	3%	4%	≥5%
		BLADEX 4L (quarts/acre)* plus ATRAZINE 4L (quarts/acre)†					
Sand, Loamy sand		0.4 + 0.2‡	0.5 + 0.3	0.8 + 0.5	1.0 + 0.5	1.3 + 0.5	1.3 + 0.8
Sandy loam		0.5 + 0.3	0.8 + 0.5	1.0 + 0.5	1.3 + 0.5	1.3 + 0.8	1.5 + 0.8
Loam, Silt loam, Silt		0.8 + 0.5	1.0 + 0.5	1.3 + 0.5	1.3 + 0.8	1.5 + 0.8	1.8 + 0.8
Sandy clay loam, Clay loam, Silty clay loam		1.0 + 0.5	1.3 + 0.5	1.3 + 0.8	1.5 + 0.8	1.8 + 0.8	2.0 + 0.8
Sandy clay, Silty clay, Clay		1.3 + 0.5	1.3 + 0.8	1.5 + 0.8	1.8 + 0.8	2.0 + 0.8	2.0 + 1.0

* Maximum rate limits per acre per year for all applications is 6.5 lb cyanazine (6.5 quarts BLADEX 4L) except on highly erodible land with less than 30% plant residue cover, where the rate limit is 3.0 lb cyanazine (3.0 quarts BLADEX 4L).

† If 90% Atrazine is used, multiply the Atrazine rates shown in this table by 1.11 to equal the appropriate poundage of an 90% Atrazine product. If Atrazine 80W is used, multiply the Atrazine rates shown in this table by 1.25 to equal the appropriate poundage of Atrazine 80W.

‡ Do not use in the light sandy soils of the Atlantic Coastal Plain.

§ Do not use BLADEX 90DF in tankmixes with "Eradicane 6.7E" and/or "Sutan+" on field corn grown for seed.

BLADEX 4L ON SWEET CORN, PREEMERGENCE

TABLE 7, FOR SWEET CORN

Early Preplant or Preemergence Broadcast Rates in Tankmix Combinations with Atrazine 4L

Soil Texture	%OM =	BLADEX 4L (quarts/acre)* plus ATRAZINE 4L (quarts/acre)†					
		<1%	1%	2%	3%	4%	≥5%
Sand, Loamy sand		Do Not Use	0.8 + 0.4	1.1 + 0.4	1.3 + 0.7	1.5 + 0.9	2.2 + 1.1
Sandy loam		Do Not Use	1.1 + 0.4	1.3 + 0.7	1.5 + 0.9	2.0 + 1.1	2.8 + 1.3
Loam, Silt loam, Silt		Do Not Use	1.3 + 0.7	1.5 + 0.9	2.0 + 1.1	2.5 + 1.3	3.2 + 1.3
Sandy clay loam, Clay loam, Silty clay loam		Do Not Use	1.5 + 0.9	1.8 + 1.1	2.5 + 1.3	3.2 + 1.3	3.4 + 1.6
Sandy clay, Silty clay, Clay		Do Not Use	1.8 + 1.1	2.8 + 1.3	3.2 + 1.3	3.4 + 1.6	3.6 + 1.8

TABLE 8, FOR SWEET CORN

Early Preplant or Preemergence Broadcast Rates in Tankmix Combinations with "Lasso", "Sutan+", "Eradicane 6.7E", or "Dual 8E"

Soil Texture	%OM =	BLADEX 4L (quarts/acre)*					
		<1%	1%	2%	3%	4%	≥5%
Sand, Loamy sand		Do Not Use	0.8‡	1.2	1.4	1.6	2.0
Sandy loam		Do Not Use	1.2	1.4	1.6	2.0	2.2
Loam, Silt loam, Silt		Do Not Use	1.4	1.6	2.0	2.2	2.6
Sandy clay loam, Clay loam, Silty clay loam		Do Not Use	1.8	2.0	2.2	2.6	2.8
Sandy clay, Silty clay, Clay		Do Not Use	2.0	2.2	2.6	2.8	3.0

TABLE 9, FOR SWEET CORN

Early Preplant or Preemergence Broadcast Rates in Tankmix Combinations with Atrazine 4L and "Lasso", "Sutan+", "Eradicane 6.7E", or "Dual 8E"

Soil Texture	%OM =	BLADEX 4L (quarts/acre)* plus ATRAZINE 4L (quarts/acre)†‡					
		<1%	1%	2%	3%	4%	≥5%
Sand, Loamy sand		Do Not Use	0.6 + 0.2	0.9 + 0.4	1.0 + 0.5	1.0 + 0.6	1.4 + 0.6
Sandy loam		Do Not Use	0.8 + 0.4	1.0 + 0.5	1.1 + 0.6	1.4 + 0.6	1.6 + 0.6
Loam, Silt loam, Silt		Do Not Use	1.0 + 0.5	1.2 + 0.6	1.4 + 0.6	1.6 + 0.6	1.8 + 0.9
Sandy clay loam, Clay loam, Silty clay loam		Do Not Use	1.2 + 0.6	1.4 + 0.6	1.6 + 0.6	1.8 + 0.9	2.0 + 0.9
Sandy clay, Silty clay, Clay		Do Not Use	1.4 + 0.6	1.6 + 0.9	1.8 + 0.9	2.0 + 0.9	2.0 + 1.1

* Maximum rate limits per acre per year for all applications is 6.5 lb cyanazine (6.5 quarts BLADEX 4L) except on highly erodible land with less than 30% plant residue cover, where the rate limit is 3.0 lb cyanazine (3.0 quarts BLADEX 4L).

† If 90% Atrazine is used, multiply the Atrazine rates shown in this table by 1.11 to equal the appropriate poundage of an 90% Atrazine product. If Atrazine 80W is used, multiply the Atrazine rates shown in this table by 1.25 to equal the appropriate poundage of Atrazine 80W.

‡ Do not use in the light sandy soils of the Atlantic Coastal Plain.

COTTON

IDLE SEASON OR EARLY PREPLANT USES IN COTTON

FOR USE ONLY IN CALIFORNIA

BLADEX may be used for burndown of small existing annual weeds and residual control of weeds during the winter and early spring season prior to planting cotton. Complete any planned tillage prior to application. Apply herbicide treatment before weeds germinate or before weed seedlings are more than 3" tall. Tillage after application may reduce the effectiveness of the herbicide treatment.

Apply BLADEX at least 30 days prior to planting. Apply the proper rate for the soil texture, organic matter, and time interval between application and planting, as shown in Table 10. Where existing weeds are present, add COC, surfactant, or emulsible vegetable oil at its recommended rate to aid in the burndown of small weeds.

Where existing weeds are greater than 3" in height, when very dry conditions exist or where volunteer grains are a major problem, tankmix BLADEX with 1 to 2 pints per acre of "Gramoxone Extra". Well-established weeds 6" or taller may not be well controlled.

Apply BLADEX plus "Gramoxone Extra" in at least 20 gal per acre of carrier by ground sprayer. (The volume of carrier and the application equipment must be adequate to give a uniform application.) Add nonionic surfactant at 1 to 2 qt per 100 gal of diluted spray (or other suitable surfactant at recommended rates) where "Gramoxone Extra" is used. COC or emulsible vegetable oil are not needed where "Gramoxone Extra" is used. Do not apply "Gramoxone Extra" combinations in suspension-type fertilizer.

BLADEX can also be tank mixed with Treflan[14] or Prowl[15] and incorporated for fall-listed cotton beds instead of surface-applied as described above.

Weeds Controlled

Grasses

Annual bluegrass	Rabbitsfoot grass
Annual ryegrass	Volunteer small grains
Barnyardgrass*	(suppression)
Bristly foxtail	Wild oat*
	Yellow foxtail

Broadleaves

Annual henbit	London rocket
Black nightshade	Marestail
Burclover	Miners lettuce
Cheeseweed*	Pineapple weed
Chickweed	Prickly lettuce
Fiddleneck	Shepherdspurse
Groundsel	Sowthistle
Knotweed	Wild mustard
Lambsquarters	Wild radish

* Under soil moisture conditions favoring deep germination, these species may not be completely controlled.

PREPLANT USES IN COTTON

FOR USE ONLY IN ARIZONA

BLADEX in tank mix combination with "Prowl" or a trifluralin herbicide product may be applied to land to be planted in cotton. Apply on the flat, incorporate and list. Irrigation may be applied preplant to beds or cotton may be planted in dry beds and irrigated up.

Carefully match the BLADEX rate with the soil texture as shown in Table 11. If in doubt about the soil texture, a composite sample of the soil should be tested for the average texture class. Do not use on fields where soil texture varies from coarse to fine.

While cotton exhibits tolerance to BLADEX, application rates in excess of label recommendations for a soil texture class can cause chlorosis and stunting in the crop and may result in stand reduction. Cool, wet weather conditions during seedling stage of cotton can also result in temporary yellowing, stunting, or stand reduction. These effects will be more pronounced in the presence of cotton seedling diseases. Avoid "irrigating back" until a stand of seedling cotton has been established. Do not use sprinkler irrigation during seedling stage of cotton on fields treated preplant with BLADEX.

Application Directions

Use Table 11 for use rates of BLADEX applied with "Prowl" or a trifluralin product. The correct amount of BLADEX must be uniformly suspended in the spray tank before adding "Prowl" or a trifluralin herbicide product. Sufficient jet or mechanical agitation to keep the spray mixture uniformly suspended must be provided while filling and applying.

Apply broadcast using equipment calibrated to give a uniform application at the correct dosage. Use a minimum of 20 gal of water per acre for ground application. Do not apply by air. Incorporate as soon as possible.

Refer to the "Prowl" or the trifluralin product label for the application-to-incorporation interval and description of incorporation equipment and methods. Refer to the "Application Information" section of this label for tank mix compatibility testing procedures.

Weeds Controlled

Barnyardgrass	Woolly morningglory*
Junglerice	Wright groundcherry
Palmer amaranth	

* Heavy infestations of this species may not be completely controlled.

CONSERVATION TILLAGE/FALLOW

BLADEX may be used for burndown of small existing annual weeds and residual control of weeds during the winter and early spring season prior to planting cotton. Complete any planned tillage prior to application. Apply herbicide treatment before weeds germinate or before weed seedlings are more than 3 inches tall. Tillage after application may reduce the effectiveness of the herbicide treatment.

For rates of 3 pts or over, apply BLADEX at least 30 days prior to planting. For the 1 pt rate, apply at least seven days prior to planting. Apply the proper rate for the soil texture, organic matter and time interval between application and planting as indicated in Table 12. Where existing weeds are present, add crop oil concentrate or surfactant at recommended rates.

For application with ground sprayers, use at least 20 gal/acre of carrier. For aerial application, refer to Supplemental Labeling BLADEX 4L Herbicide Aerial Application via Closed Loading Systems (H-13127). For aerial application, apply BLADEX 4L in a minimum of 3 gal. spray volume per acre. In all cases, application equipment and carrier volume must be adequate to give uniform application.

Weeds Controlled

Grasses

Annual bluegrass	Yellow foxtail
Annual ryegrass	

Broadleaves

Annual henbit (purpletop)	Lambsquarters
Black nightshade	Marestail
Burclover	Shepherd's-purse
Chickweed	Wild mustard
Cutleaf eveningprimrose	Wild radish

BLADEX 4L Herbicide plus "Gramoxone" Extra

BLADEX plus "Gramoxone" Extra is recommended for early preplant/postemergence application to control annual broadleaf and grass weeds.

BLADEX plus "Gramoxone" Extra tank mix may be used for burndown of existing annual weeds and residual control of weeds during early spring prior to planting conservation tillage cotton. Apply with crop oil concentrate at 1.0 % V/V or surfactant at 0.25-0.5 % V/V. See Table 12 for BLADEX rates and timings. Refer to "Gramoxone" Extra label for rates and timings.

Weeds Controlled

Grasses

Annual bluegrass	Volunteer wheat
Annual ryegrass	Yellow foxtail

Broadleaves

Annual henbit (purpletop)	Marestail
Black nightshade	Shepherd's-purse
Chickweed	Wild mustard
Cutleaf eveningprimrose	Wild radish
Lambsquarters	

PREEMERGENCE USES IN COTTON

BLADEX Alone

BLADEX is a selective preemergence herbicide for early season weed control in cotton. Supplemental practices (such as BLADEX applied directed postemergence) may be necessary to control late season weeds.

Carefully match the BLADEX rate (Table 13) with the soil texture. Do not use BLADEX on fields where the soil texture changes from coarse to fine. Avoid overlapping the spray pattern or overdosing the field with BLADEX. Application rates above those recommended for the soil texture can result in a yellowing or stunting of the cotton.

While cotton exhibits tolerance to BLADEX, adverse growing conditions such as excessive rains, standing water, or cold weather may result in stand reduction.

BLADEX Plus Zorial Rapid 80

BLADEX may be used in a tank mix combination with Zorial Rapid 80[16] on cotton. Apply BLADEX plus "Zorial Rapid 80" at the proper rate for the soil texture shown in Table 13. The soil must contain at least 1.0% organic matter. Seed placement should be 1/2" to 3/4" from the soil surface. Plant only cotton within six months after the last application of "Zorial Rapid 80"; other crops may be injured by residual herbicide in the soil.

Weeds Controlled

Annual morningglory	Prickly sida (Teaweed)
Cocklebur	Spurge

DIRECTED POSTEMERGENCE AND LAYBY USES IN COTTON

BLADEX may be applied alone and in tank mix combinations as a directed postemergence or layby treatment to cotton. These applications may be either a preemergence or postemergence treatment to weeds in all cotton growing states.

Apply BLADEX before weeds are more than 2" tall. For a directed postemergence treatment, apply BLADEX after the cotton is at least 6" tall. For layby treatment, apply BLADEX after the cotton is at least 12" tall.

Direct the spray mixture toward the soil around the base of the cotton plants. Direct contact of the spray mixture with the cotton leaves will injure the foliage. The use of leaf lifters or shields on application equipment is recommended to avoid spraying the cotton foliage.

BLADEX may be applied as a directed postemergence and/or layby treatment following a preemergence application of BLADEX. Apply no more than two directed postemergence applications plus one preemergence application (three applications total) to the same crop in any one year. If BLADEX is not used preemergence, apply no more than three directed postemergence applications, including layby treatment, to the same crop in any one year. (In California, apply no more than two directed postemergence applications, including layby treatment.)

When applied as a layby treatment before weeds emerge, the effectiveness of BLADEX depends on rainfall or irrigation to move the herbicide into the soil. The degree of preemergence control from a layby treatment will be reduced if soil moisture and temperature conditions cause deep germination of weed seed.When irrigation water is used to activate the herbicide, every row must be watered; for skip row cotton, all treated soil must be irrigated.

BLADEX Applied Alone

For a directed postemergence treatment, apply BLADEX at the rate shown in Table 14. For layby treatment, apply BLADEX at the rates for the soil texture indicated in Table 15. Add a nonionic agricultural surfactant suitable for use on growing cotton at the rate of 2 qt per 100 gal of spray mixture (or as directed by the manufacturer).

BLADEX Plus MSMA

Apply a tank mix combination of BLADEX plus MSMA and a surfactant after the cotton is 6" tall, but before it reaches the bloom stage. Apply no more than two applications of this mixture before the first bloom stage. Tank mix BLADEX plus MSMA at the rates indicated in Table 16. Add a nonionic surfactant at the rate of 2 qt per 100 gal of spray mixture (or as directed by the manufacturer).

Weeds Controlled

Annual morningglory	Palmer amaranth
Bristly starbur	Pigweed (Redroot and Spiny)
Cocklebur	Prickly sida (Teaweed)
Crotalaria	Sicklepod
Jimsonweed	Spurge
Lambsquarters	Tropic croton
Nightshade (annual	Wright groundcherry

PRECAUTIONS FOR COTTON

- Failure to wait the recommended time interval between application and planting may result in crop injury.
- At least 1" of rainfall or an equivalent irrigation that waters the surface of the soil after application must precede planting.
- The use of this treatment on calcareous or caliche soil outcroppings may result in crop injury.
- Do not graze or feed foliage from treated areas to livestock.
- Do not apply BLADEX to cotton land in irrigation water.
- Do not apply within 54 days of harvest.

USE RATE TABLES FOR COTTON

BLADEX 4L ALONE, IDLE SEASON OR EARLY PREPLANT - CALIFORNIA ONLY

TABLE 10, FOR COTTON

		BLADEX 4L (quarts/acre)*† Days Prior to Planting					
		30 Days		60 Days		90 Days	
Soil Texture	%OM =	Under 2%	Over 2%	Under 2%	Over 2%	Under 2%	Over 2%
Sand, Loamy sand		1.5	2	2.5	3	3	3.5
All Other Soils		2	2.5	3	3.5	3.5	4

* For the time intervals between those listed in this table, adjust the rates proportionately.

† The maximum rate limit per acre per year for all applications is 6.5 lb cyanazine (6.5 qts BLADEX 4L) except on highly erodible land with less than 30% plant residue cover, where the rate limit is 3.0 lb cyanazine (3 qts BLADEX 4L).

** The soil must contain at least 1% organic matter.

†† Do not use on coarse soils (sands and loamy sands).

USE RATE TABLES FOR COTTON

BLADEX 4L PLUS PROWL OR TRIFLURALIN, PREPLANT - ARIZONA ONLY

TABLE 11, FOR COTTON

| Soil Texture†† | BLADEX 4L† (quarts/acre) | + | "Prowl" | | or | Trifluralin | |
			4EC (qts/A)	3.3EC (qts/A)		5EC (qts/A)	4EC (qts/A)
Sand, Loamy sand (<0.5% OM)	Do Not Use		Do Not Use	Do Not Use		Do Not Use	Do Not Use
Sandy loam, Loam (>0.5% OM)	1-1.5		0.5	0.6		0.32	0.4
Silt loam, Silt, Sandy clay loam	1.5-2		0.75	0.9		0.4-0.5	0.5-0.6
Clay loam, Silty clay loam, Clay	2-2.5		1	1.2		0.5	0.6

"BLADEX" 4L FOR CONSERVATION TILLAGE/FALLOW - EARLY PREPLANT

TABLE 12, FOR COTTON

Soil Texture**	BLADEX 4L (quarts/acre)†				
	Days Prior to Planting*				
	7 Days	30 Days to 60 Days		More than 60 Days	
	Organic Matter	Organic Matter		Organic Matter	
	Over 1%	Under 2%	Over 2%	Under 2%	Over 2%
Loamy Sands	--	1.5	2	2.5	3
Med. to Heavy	0.5	2	2.5	3	3.5

BLADEX 4L ALONE AND PLUS "ZORIAL RAPID 80", PREEMERGENCE

TABLE 13, FOR COTTON

For Use in Alabama, Arkansas, Georgia, Louisiana, Mississippi, Missouri, and Tennessee

Soil Texture**	BLADEX 4L† Alone (quarts/acre)	BLADEX 4L† (qts/A)	+	"Zorial Rapid 80" (lb/A)
Sandy loam ††	0.5	0.5		0.8
Silt and Silt loam	0.6	0.6		1.3
Loam, Clay loam, Sandy clay loam, Sandy clay	0.9	0.9		1.3
Silty clay loam, Silty clay, Clay	1.2	1.2		1.6

---

*   For the time intervals between those listed in this table, adjust the rates proportionately.

†   The maximum rate limit per acre per year for all applications is 6.5 lb cyanazine (6.5 qts BLADEX 4L) except on highly erodible land with less than 30% plant residue cover, where the rate limit is 3.0 lb cyanazine (3 qts BLADEX 4L).

** The soil must contain at least 1% organic matter.

†† Do not use on coarse soils (sands and loamy sands).

USE RATE TABLES FOR COTTON

BLADEX 4L ALONE, DIRECTED POSTEMERGENCE AND LAYBY

TABLE 14, FOR COTTON

Directed Postemergence Rates		BLADEX 4L (quarts/acre)†	
			Banded 38" Row
Height of Cotton Band		Broadcast	12" Band 19"
6" or more	0.6-1 qts/A	0.2-0.3 qts/A	0.3-0.5 qts/A

Use the maximum rate when dry or arid conditions exist.

TABLE 15, FOR COTTON

Layby Rates Height of Cotton	Soil Texture	BLADEX 4L (quarts/acre)†
12" or more	Sandy loam, Silt, Silt loam	0.8 qts/A
	Loam, Clay loam, Sandy clay loam, Sandy clay	1.2 qts/A
	Silty clay loam, Silty clay, Clay	1.6 qts/A

BLADEX 4L PLUS MSMA, DIRECTED POSTEMERGENCE

TABLE 16, FOR COTTON

Product (quarts/acre)	Broadcast	Banded 38" Row	
		12" Band	19" Band
BLADEX 4L†	0.6-1 qts/A	0.2-0.33 qts/A	0.3-0.5 qts/A
+	+	+	+
MSMA (4 lb/gal)	2.0 qts/A	0.65 qts/A	1.0 qts/A
or	or	or	or
MSMA (6.6 lb/gal)	1.2 qts/A	0.4 qts/A	0.6 qts/A

* For the time intervals between those listed in this table, adjust the rates proportionately.

† The maximum rate limit per acre per year for all applications is 6.5 lb cyanazine (6.5 qts BLADEX 4L) except on highly erodible land with less than 30% plant residue cover, where the rate limit is 3.0 lb cyanazine (3 qts BLADEX 4L).

** The soil must contain at least 1% organic matter.

†† Do not use on coarse soils (sands and loamy sands).

WEATHER EFFECTS AND MODE OF ACTION

As a preemergence herbicide, BLADEX is active mainly through the roots. Its effect on weeds is dependent on adequate rainfall to move the herbicide into the root zone. The soil must be thoroughly wet throughout the zone where weed seeds germinate. (The soil should be too wet to cultivate.)

Under conditions that delay weed germination—such as low temperatures and lack of soil surface moisture—or when germination is extended over a long period, the effectiveness of the herbicide may be impaired. Rotary hoeing, shallow cultivation, or a postemergence herbicide treatment may be useful under these circumstances. Follow these guidelines:

- Rotary hoeing or shallow cultivation is recommended if there has not been adequate rainfall or sprinkler irrigation within 10 days after application of BLADEX, and if the herbicide was not incorporated at the time of treatment.

- If the crop is cultivated, tillage should be shallow to minimize diluting the herbicide in the soil.

- To enhance weed control in areas of less than 25" of rainfall or where long dry periods are common, these treatments may require shallow incorporation with a tool such as a field cultivator operated at 5 to 7 mph. Incorporation should not be more than 3" deep to avoid burying the herbicide. Do not use a spike-toothed harrow, deep tillage disk, or rolling basket device to incorporate BLADEX.

Heavy rainfall between planting and crop emergence may cause excessive concentrations of herbicide in the seed furrow, resulting in possible crop injury or stand loss. To prevent rainfall from pooling, level deep planter marks or seed furrows before application.

Rainfastness

BLADEX Herbicide is active throung both shoot and root uptake, therefore rainfastness is not as critical as with most postemergence herbicides. However, best results are obtained on emerged weeds when there is an interval of at least 4 hours between application and rainfall.

APPLICATION INFORMATION

This product may not be mixed/loaded or used within 50 ft of all wells including abandoned wells, drainage wells and sink holes.

Application Equipment

Nozzles

Use nozzles that provide accurate and uniform coverage. Ensure that the nozzles are the same size and are spaced uniformly. Calibrate the sprayer before use and check it frequently during use.

Pump

Use a pump with capacity to:

a. Maintain 35 to 40 psi at the nozzles.

b. Provide sufficient agitation in tank to keep mixture in suspension.

c. Provide a minimum of 20% bypass at all times.

In addition, use centrifugal pumps that provide sufficient shear action to disperse and mix this product. The pump should circulate at least 10 gal per min for every 100 gal in the tank through the jets of a correctly positioned sparger tube or jet agitator.

Nozzle Screens

To prevent the nozzles from clogging, place 10- to 16-mesh nozzle screens on the suction side of the pump. Do not place a screen in the recirculation line. Use a 40- to 50-mesh screen between the pump and boom. Check your equipment manufacturer's literature for specific recommendations.

General Mixing and Spraying Directions

The following general mixing instructions are recommended:

1. Unless otherwise specified, use at least 10 gal of water per acre for soil applications and at least 15 gal of water per acre for foliar applications.

   Note: Use sufficient carrier to ensure uniform application. Follow the label requirements of all products used in tank mix combinations.

2. A nitrogen solution or complete liquid fertilizer may replace all or part of the water as a carrier for preemergence or preplant application on corn. For best burndown, use a minimum of 20 gal per acre of liquid fertilizer as the carrier. Do not apply fertilizer mixtures after the crop emerges, since this may cause crop injury.

3. Always check the tank mix compatibility (TMC) of any formulation before mixing BLADEX with liquid fertilizer carriers or other formulations. A simple but generally reliable TMC evaluation procedure is explained in the Tank Mix Compatibility Evaluation Procedure section of these mixing instructions.

4. Start with thoroughly clean equipment. (See labels of the previous compounds used for cleaning instructions.)

5. Fill the tank at least 1/2 full with carrier. Start and maintain consistent agitation through all mixing and spraying procedures. Make sure the agitation system is working properly and creates a rippling or rolling action on the liquid surface.

6. Add the recommended amount of BLADEX 4L to the tank with agitation.

7. Fill the tank to 75% capacity with carrier. Filling bypass lines should be kept below the liquid surface. Increase tank agitation, as necessary, to maintain the rippling or rolling action on the liquid surface.

8. If desired, add the appropriate emulsible crop oil, crop oil concentrate, or other tank mix formulations. Slurry these additional ingredients before adding them to the tank, if the compatibility test shows it to be necessary.

9. Finish filling the tank, maintaining sufficient agitation at all times to ensure surface action. In both spray tanks and nurse tanks, ensure that the BLADEX is completely dispersed and in uniform suspension before applying it.

10. Tank mixtures should always be applied immediately after preparation. If, for any reason, this is not possible, agitate the mixture sufficiently to remix all products, and check it for complete resuspension before application.

11. When tank mixing with other formulations, empty the tank as completely as possible before refilling it to prevent buildup of oil or EC residue. Always maintain agitation so that the mixture does not separate. If an oil or EC film begins to build up, drain and clean the tank with a strong detergent solution or an appropriate solvent.

12.If any emulsible crop oil, crop oil concentrate, or other emulsible formulation has been used either alone or in tank mix combinations with other pesticide formulations, clean the sprayer thoroughly by flushing it with a detergent solution at the end of each work day. This ensures a clean sprayer and continued trouble-free operation.

Tank Mix Compatibility Evaluation Procedure

1. Add 1 pint of carrier liquid to each of two 1-qt jars. Mark the first jar "with" and the other "without."

2. Add 1/4 tsp of a suitable compatibility agent to the jar marked "with," cap the jar, and shake it gently for 5 to 10 seconds to mix (1/4 tsp per 1 pint of carrier = 2 pints per 100 gal of carrier).

3. Add the appropriate amount of herbicide to both jars, cap each jar, and shake them gently for 5 to 10 seconds.

Note: If problems are encountered in mixing wettable powder or dry flowable formulations into a liquid fertilizer, slurry these formulations in water before adding them to the liquid fertilizer.

The following chart shows the amount of BLADEX to use for the jar test, depending on the intended use rate (gal of liquid carrier per 1 qt of BLADEX).

Jar Test for BLADEX Compatibility

Gallons of liquid carrier per acre.	Add this many teaspoons of BLADEX 4L per pint of liquid carrier for the jar test.
4.0	6.0
7.5	3.2
15.0	1.6
20.0	1.2
25.0	1.0
30.0	0.8

When the intended use rate varies—that is, when the amount of BLADEX added to each gallon of liquid carrier is less than or greater than 1 qt—adjust the jar test proportionately. If the intended field use rate is 3 qt (rather than 1 qt) of BLADEX in 15 gal of carrier per acre, add 4.8 tsp (rather than 1.6 tsp) of BLADEX to the quart jars containing 1 pint of carrier (3 qt of BLADEX in 15 gal of carrier per acre = 4.8 tsp of BLADEX in 1 pint of carrier).

4. Let each jar stand one-half hour. If the mixture separates, agglomerates, or precipitates, shake the jar again for 10 to 15 seconds, and note whether any of the following occur:

   a. Separated phases do not remix uniformly.

   b. Lumps do not disperse.

   c. Precipitate does not resuspend readily.

   d. Precipitate sticks tenaciously to the glass.

5. If the mixture does not exhibit any of these problems in either jar, the herbicides can, in most cases, be safely used in that carrier without a compatibility agent.

6. If problem 4.a or 4.b occurs in the jar marked "without" but does not occur in the jar marked "with", the compatibility agent should be used.

7. If problem 4.a or 4.b is seen in both jars, then the herbicides and carrier are incompatible and should not be used in the same spray tank. Alternatively, a different tank mix compatibility agent can be evaluated.

8. If problem 4.c or 4.d occurs in the jar marked "without" but does not occur in the jar marked "with", the compatibility agent should be used unless constant, thorough agitation can be maintained and immediate clean-out of the spray system is performed.

9. If problem 4.c or 4.d is seen in the jar marked "with," the user proceeds with mixing and application at his own risk should the agitation in the system be insufficient or curtailed.

10. When the components of a mixture are determined compatible by this test, they should be mixed for application according to the General Mixing and Spraying section of this label.

   If a compatibility test indicates that components of a proposed mix are compatible, the applicator is still responsible for following all mixing directions prescribed on the labels of the herbicides or pesticides involved.

11. The following compatibility agents, noted by the various tank mix combinations, may improve compatibility in liquid.

Tank Mix Combination	Compatibility Agents
BLADEX /"Lasso" (liquid fertilizer grade)	Probably not needed in 28-0-0, 10-34-0. Compex may help in others.
BLADEX /"Sutan+" or "Eradicane 6.7E"	Probably not needed in 28-0-0. Incompatible in 10-34-0. Unite, Spray-Mate, Kem-Link may help in others.
BLADEX /"Dual 8E"	Probably not needed in 28-0-0. Unite, Spray-Mate, Ivory Liquid may help in others.

FERTILIZER IMPREGNATION, APPLICATION, AND CLEANOUT

BLADEX may be used to coat or impregnate dry granular fertilizer for early preplant, preemergence, or preplant incorporated weed control in field corn. All recommendations, cautions, and special precautions on this label must be followed, in addition to any state regulations for blending, impregnating, and labeling dry bulk fertilizer.

General Blending Directions

Dry bulk fertilizers may be coated or impregnated with BLADEX using tower blenders, rotary drum blenders, or blending augers or conveyors. Observe the following precautions when blending BLADEX with dry bulk fertilizers:

- Do not impregnate BLADEX—or tank mixes containing BLADEX—in or on fertilizers containing ammonium nitrate, potassium nitrate, or sodium nitrate.

- Do not use BLADEX on straight limestone, which cannot absorb the fertilizer; however, fertilizer blends containing limestone can be impregnated using BLADEX alone.

- Use 200 to 450 lb per acre of dry fertilizer.

- Use equipment that uniformly distributes the herbicide throughout each batch of impregnated fertilizer. Nonuniform impregnation can cause crop injury or unsatisfactory performance.

Impregnating Fertilizer with BLADEX Alone

1. Add BLADEX 4L when at least 1/2 the total fertilizer volume required is in the mixer. A minimum of 200 lbs per acre of an approved fertilizer should be used.
2. BLADEX 4L can be moved from the chemical holding tank to the mixer tank by using an air system or a liquid pump with the hose at least 1 inch in diameter.
3. Position the spray nozzles to achieve uniform coverage with the BLADEX on the dry fertilizer without spraying the walls of the mixer.
4. Flush spray lines with water and spray into the dry fertilizer.
5. Add remaining fertilizer, plus drying agents when necessary, and blend thoroughly for at least 3 minutes.
6. Add 2 to 5% of a suitable drying agent to ensure a herbicide/fertilizer mixture that will spread through air spreaders. The need for a drying agent is determined by the wetness of the fertilizer batch. Wetness can change with humidity, nitrogen content, fertilizer rates, and herbicide rates.

Impregnating Fertilizer with BLADEX in Tank Mixes with Other Dry Herbicides

1. While the fertilizer is blending, add the BLADEX first and the tank mix partner last.
2. Add any necessary drying agent to ensure a spreadable herbicide/fertilizer mixture and blend thoroughly for at least three minutes.
3. Follow the other appropriate use instructions found in Section A for use with BLADEX mixtures.

Application of Impregnated Fertilizer

Fertilizer that is impregnated or coated with BLADEX must be applied uniformly. Crop injury and/or poor weed control may result if impregnated fertilizer is not uniformly applied. To ensure uniform application:

- Calibrate the fertilizer applicator accurately.
- Do not apply the fertilizer mixture while turning at the ends of the fields; this may result in excessive application rates causing crop injury.
- Do not double-apply the fertilizer across the ends or sides of the field.
- Apply impregnated fertilizer in one pass using air-flow or augur-metered application equipment. If other equipment is used, apply 1/2 the recommended rate and overlap each pass by 50%, splitting the middle of each pass to obtain the best distribution pattern.

Apply the fertilizer immediately after impregnation. Impregnated fertilizer may become lumpy and difficult to spread if it is stored.

Equipment Cleanout

Equipment used to impregnate or apply fertilizer impregnated with BLADEX alone or combinations with other herbicides must be cleaned out if the next batch of material is to be applied to a crop for which BLADEX or a combination herbicide is not registered. To clean out impregnating equipment, run at least 1,000 lb of unimpregnated fertilizer through the equipment before using it to make another application.

ROTATIONAL CROPS

Use these guidelines to determine which rotational crops can be planted safely following use of BLADEX:

- If the crop stand is lost due to adverse weather, insects, etc., the field can be replanted to corn or sorghum.
- If the field is replanted to sorghum, allow at least a 30-day interval between treatment and planting. The sorghum plants may be injured if the full preemergence rate is used and adverse conditions exist for sorghum growth.

- Any rotational crop may be planted the fall or spring following treatment with BLADEX.
- When BLADEX is tankmixed with other herbicides, refer to the manufacturers' label and use the most restrictive crop rotation interval.

BEST MANAGEMENT PRACTICES FOR GROUND AND SURFACE WATER PROTECTION

Application Requirements

- Do not mix, load, or apply BLADEX within 50 ft of all wells, including abandoned wells, drainage wells, and sinkholes.
- Do not apply aerially or by ground within 66 ft of the points where field surface water runoff enters perennial or intermittent streams and rivers or within 200 ft of natural or impounded lakes and reservoirs.
- If this product is applied to highly erodible land, the 66 ft buffer or set-back from runoff points must be planted to crop or seeded with grass.

Mixing/Loading Requirements

Do not mix or load BLADEX within 50 ft of intermittent streams and rivers, natural or impounded lakes, and reservoirs.

Operations that involve mixing, loading, rinsing, or washing of this product into or from pesticide handling or application equipment or containers within 50 ft of any well are prohibited unless conducted on an impervious pad constructed to withstand the weight of the heaviest load that may be positioned on or moved across the pad. Such a pad shall be designed and maintained to contain any product spills or equipment leaks, container or equipment rinse or wash-water, and rainwater that may fall on the pad. Surface water shall not be allowed to either flow over or from the pad; the pad must be self-contained. The pad shall be sloped to facilitate material removal. An unroofed pad shall be of sufficient capacity to contain at least 110% of the capacity of the largest pesticide container or application equipment on the pad. A pad that is covered by a roof or sufficient size to completely exclude precipitation from contact with the pad shall have a minimum containment capacity of 100% of the capacity of the largest pesticide container or application equipment on the pad. Containment capacities as described above shall be maintained at all times. The above-specified minimum containment capacities do not apply to vehicles when delivering pesticide shipments to the mixing/loading site.

Some states may have in effect additional requirements regarding well-head setbacks and operational area containment.

Cyanazine Rate Limits

One quart of BLADEX 4L contains 1.0 lb cyanazine active ingredient (a.i.). Adhere to the use rate recommendations in this or other label. In addition, observe the following requirements:

- Do not apply more than 6.5 lb total cyanazine a.i. (all sources) per acre per year to any land.
- On highly erodible land, as defined by the Soil Conservation Service, if plant residue cover is less than 30%, do not apply more than 3.0 lb total cyanazine a.i. (all sources) per acre per year.

When state/local requirements regarding the use of cyanazine (including lower maximum rates and/or higher set-backs) differ from the label, the more restrictive/protective requirements apply.

SPRAY DRIFT MANAGEMENT

The interaction of many equipment and weather-related factors determines the potential for spray drift. The applicator is responsible for considering all these factors when making application decisions.
AVOIDING SPRAY DRIFT IS THE RESPONSIBILITY OF THE APPLICATOR.

Importance of Droplet Size

The most effective way to reduce drift potential is to apply large droplets (>150 - 200 microns). The best drift management strategy is to apply the largest droplets that provide sufficient coverage and control. The presence of sensitive species nearby, the environmental conditions, and pest pressure may affect how an applicator balances drift control and coverage. APPLYING LARGER DROPLETS REDUCES DRIFT POTENTIAL, BUT WILL NOT PREVENT DRIFT IF APPLICATIONS ARE MADE IMPROPERLY OR UNDER UNFAVORABLE ENVIRONMENTAL CONDITIONS! See **Wind**, **Temperature and Humidity**, and **Temperature Inversions** sections of this label.

Controlling Droplet Size - General Techniques

- **Volume** - Use high flow rate nozzles to apply the highest practical spray volume. Nozzles with higher rated flows produce larger droplets.
- **Pressure** - Use the lower spray pressures recommended for the nozzle. Higher pressure reduces droplet size and does not improve canopy penetration. WHEN HIGHER FLOW RATES ARE NEEDED, USE A HIGHER-CAPACITY NOZZLE INSTEAD OF INCREASING PRESSURE.
- **Nozzle Type** - Use a nozzle type that is designed for the intended application. With most nozzle types, narrower spray angles produce larger droplets. Consider using low-drift nozzles.

Boom Height

Setting the boom at the lowest labeled height (if specified) which provides uniform coverage reduces the exposure of droplets to evaporation and wind. For ground equipment, the boom should remain level with the crop and have minimal bounce.

Wind

Drift potential increases at wind speeds of less than 3 mph (due to inversion potential) or more than 10 mph. However, many factors, including droplet size and equipment type determine drift potential at any given wind speed. AVOID GUSTY OR WINDLESS CONDITIONS.
Note: Local terrain can influence wind patterns. Every applicator should be familiar with local wind patterns and how they affect spray drift.

Temperature and Humidity

When making applications in hot and dry conditions, set up equipment to produce larger droplets to reduce effects of evaporation.

Temperature Inversions

Drift potential is high during a temperature inversion. Temperature inversions restrict vertical air mixing, which causes small suspended droplets to remain close to the ground and move laterally in a concentrated cloud. Temperature inversions are characterized by increasing temperature with altitude and are common on nights with limited cloud cover and light to no wind. They begin to form as the sun sets and often continue into the morning. Their presence can be indicated by ground fog; however, if fog is not present, inversions can also be identified by the movement of smoke from a ground source or an aircraft smoke generator. Smoke that layers and moves laterally in a concentrated cloud (under low wind conditions) indicates an inversion, while smoke that moves upward and rapidly dissipates indicates good vertical air mixing.

Shielded Sprayers

Shielding the boom or individual nozzles can reduce the effects of wind. However, it is the responsibility of the applicator to verify that the shields are preventing drift and not interfering with uniform deposition of the product.

PRECAUTIONS

Use BLADEX only in field corn, popcorn, sweet corn, field corn grown for seed, and cotton.

BLADEX is not effective when used preemergence on peat or muck soils.

Do not apply this product through any type of irrigation system.

Do not apply this product with aerial application equipment.

In fields where triazine-resistant biotypes of weeds have been identified, BLADEX should be used in combination with or in sequence with other registered nontriazine herbicides. (Triazine-resistant biotypes of kochia and pigweed have been identified in some fields in the Western Great Plains, and triazine-resistant biotypes of pigweed and lambsquarters have been identified in some fields in various states). Consult with appropriate state agricultural extension service representatives for specific recommendations.

STORAGE AND DISPOSAL

STORAGE: Do not contaminate water, food or feed by storage or disposal. Do not use or store around the home environment. Avoid contact with water. In case of spill or leak, soak up with sand, earth or synthetic absorbant (do not use alkaline absorbants) and dispose of wastes in compliance with local, State and Federal regulations. Ground water contamination may be reduced by diking and flooring of permanent liquid bulk storage sites with an impermeable material.

PESTICIDE DISPOSAL: Pesticide, spray mixture or rinsate that cannot be used according to label instructions must be disposed of according to applicable Federal, State or local procedures.

CONTAINER DISPOSAL: Triple rinse (or equivalent). Then offer for recycling or reconditioning, or puncture and dispose of in a sanitary landfill or by incineration, or, if allowed by state and local authorities, by burning. If burned, stay out of smoke.

NOTICE OF WARRANTY

Du Pont warrants that this product conforms to the chemical description on the label thereof and is reasonably fit for the purposes stated on such label only when used in accordance with the directions under normal use conditions. It is impossible to eliminate all risks inherently associated with the use of this product. Crop injury, ineffectiveness or other unintended consequences may result because of such factors as weather conditions, presence of other materials, or the manner of use or application, all of which are beyond the control of Du Pont. In no case shall Du Pont be liable for consequential, special or indirect damages resulting from the use or handling of this product. All such risks shall be assumed by the buyer. DU PONT MAKES NO WARRANTIES OF MERCHANTABILITY OR FITNESS FOR A PARTICULAR PURPOSE NOR ANY OTHER EXPRESS OR IMPLIED WARRANTY EXCEPT AS STATED ABOVE.

1   "Banvel" is a registered trademark of Sandoz Crop Protection Corp.
2   "Marksman" is a registered trademark of Sandoz Crop Protection Corp.
3   "Lasso" is a registered trademark of Monsanto Agri. Co.
4   "Dual" is a registered trademark of CIBA-Geigy Corp.
5   "Frontier" is a registered trademark of Sandoz.
6   "Surpass" is a registered trademark of Zeneca, Inc.
7   "Harness Plus" is a registered trademark of Monsanto.
8   "Sutan+" is a registered trademark of ICI Americas, Inc.
9   "Eradicane" is a registered trademark of Zeneca, Inc.
10   "Gramoxone Extra" is a registered trademark of ICI Agrochemicals.
11   "Roundup" is a registered trademark of Monsanto Corp.
12   "Princep" is a registered trademark of CIBA-Geigy Corp.
13   "Caliber" is a registered trademark of CIBA-Geigy Corp.
14   "Treflan" is a registered trademark of DowElanco.
15   "Prowl" is a registered trademark of American Cyanamid Company.
16   "Zorial" is a registered trademark of Sandoz Crop Protection Corporation.

SL-161 9025 1/31/95

Funginex®

One Gallon
U. S. Standard Measure

FUNGICIDE

Active Ingredient: Triforine: (N,N′-[1,4-pipera-
zinediyl-bis (2,2,2-trichloroethylidene)]-bis-
[formamide]) 18.2%

Inert Ingredients:	81.8%
Total:	100.0%

EPA Reg. No. 100-721   EPA Est. 100-AL-1

KEEP OUT OF REACH OF CHILDREN.
DANGER/PELIGRO

Si usted no entiende la etiqueta, busque a
alguien para que se la explique a usted en
detalle. (If you do not understand the label, find
someone to explain it to you in detail.)

See additional precautionary statements and
directions for use inside booklet.
CGA 112L2C 074

ciba™

DIRECTIONS FOR USE AND CONDITIONS OF SALE AND WARRANTY

IMPORTANT: Read the entire **Directions for Use** and the **Conditions of Sale and Warranty** before using this product. If terms are not acceptable, return the unopened product container at once.

Conditions of Sale and Warranty

The **Directions for Use** of this product reflect the opinion of experts based on field use and tests. The directions are believed to be reliable and should be followed carefully. However, it is impossible to eliminate all risks inherently associated with use of this product. Crop injury, ineffectiveness, or other unintended consequences may result because of such factors as weather conditions, presence of other materials, or the manner of use or application all of which are beyond the control of Ciba-Geigy or the Seller. All such risks shall be assumed by the Buyer.

Ciba-Geigy warrants that this product conforms to the chemical description on the label and is reasonably fit for the purposes referred to in the **Directions for Use** subject to the inherent risks referred to above. **Ciba-Geigy makes no other express or implied warranty of Fitness or Merchantability or any other express or implied warranty. In no case shall Ciba-Geigy or the Seller be liable for consequential, special, or indirect damages resulting from the use or handling of this product.** Ciba-Geigy and the Seller offer this product, and the Buyer and user accept it, subject to the foregoing **Conditions of Sale and Warranty,** which may be varied only by agreement in writing signed by a duly authorized representative of Ciba-Geigy.

Funginex®

DIRECTIONS FOR USE

It is a violation of federal law to use this product in a manner inconsistent with its labeling.

Do not apply this product in a way that will contact workers or other persons, either directly or through drift. Only protected handlers may be in the area during application. For any requirements specific to your State or Tribe, consult the agency responsible for pesticide regulation.

AGRICULTURAL USE REQUIREMENTS

Use this product only in accordance with its labeling and with the Worker Protection Standard, 40 CFR part 170. This Standard contains requirements for the protection of agricultural workers on farms, forests, nurseries, and greenhouses, and handlers of agricultural pesticides. It contains requirements for training, decontamination, notification, and emergency assistance. It also contains specific instructions and exceptions pertaining to the statements on this label about personal protective equipment (PPE) and restricted-entry interval. The requirements in this box only apply to uses of this product that are covered by the Worker Protection Standard.

Do not enter or allow worker entry into treated areas during the restricted entry interval (REI) of 12 hours.

PPE required for early entry to treated areas that is permitted under the Worker Protection Standard and that involves contact with anything that has been treated, such as plants, soil, or water is:

- Coveralls
- Chemical-resistant gloves, such as barrier laminate or butyl rubber
- Shoes plus socks
- Protective eyewear

FAILURE TO FOLLOW ALL PRECAUTIONS ON THIS LABEL MAY RESULT IN POOR DISEASE CONTROL, CROP INJURY, OR ILLEGAL RESIDUES.

Chemigation

Apply this product only through the following type of system: sprinkler including center pivot, lateral move, end tow, side (wheel) roll, traveler, big gun, solid set, or hand move. Do not apply this product through any other type of irrigation system.

Crop injury, lack of effectiveness, or illegal pesticide residues in the crop can result from non-uniform distribution of treated water.

If you have questions about calibration, you should contact State Extension Service specialists, equipment manufacturers, or other experts.

Do not connect an irrigation system (including greenhouse systems) used for pesticide application to a public water system unless the pesticide label-prescribed safety devices for public water systems are in place.

A person knowledgeable of the chemigation system and responsible for its operation or under the supervision of the responsible person, shall shut the system down and make necessary adjustments should the need arise.

When mixing with other pesticides or fluid fertilizers, agitation is recommended for mixing.

Posting of areas to be chemigated is required when (1) any part of a treated area is within 300 feet of sensitive areas such as residential areas, labor camps, businesses, day care centers, hospitals, in-patient clinics, nursing homes, or any public areas such as schools, parks, playgrounds, or other public facilities not including public roads, or (2) when the chemigated areas are open to the public such as golf courses or retail greenhouses.

Posting must conform to the following requirements. Treated areas shall be posted with signs at all usual points of entry and along likely routes of approach from the listed sensitive areas. When there are no usual points of entry, signs must be posted in the corners of the treated areas and in any other location affording maximum visibility to sensitive areas. The printed side of the sign should face away from the treated area towards the sensitive area. The signs shall be printed in English. Signs must be posted prior to application and must remain posted until foliage has dried and soil surface water has disappeared. Signs may remain in place indefinitely as long as they are composed of materials to prevent deterioration and maintain legibility for the duration of the posting period.

All words shall consist of letters at least 2½ inches tall, and all letters and the symbol shall be a color which sharply contrasts with their immediate background. At the top of the sign shall be the words KEEP OUT, followed by an octagonal stop sign symbol at least 8 inches in diameter containing the word STOP. Below the symbol shall be the words PESTICIDES IN IRRIGATION WATER.

This sign is in addition to any sign posted to comply with the Worker Protection Standard.

Public water system means a system for the provision to the public of piped water for human consumption if such system has at least 15 service connections or regularly serves an average of at least 25 individuals daily at least 60 days out of the year.

Chemigation systems connected to public water systems must contain a functional, reduced-pressure zone, backflow preventer (RPZ), or the functional equivalent in the water supply line upstream from the point of pesticide introduction. As an option to the RPZ, the water from the public water system should be discharged into a reservoir tank prior to pesticide introduction. There shall be a complete physical break (air gap) between the outlet end of the fill pipe and the top or overflow rim of the reservoir tank of at least twice the inside diameter of the fill pipe.

The pesticide injection pipeline must contain a functional, automatic quick-closing check-valve to prevent the flow of fluid back toward the injection pump.

Funginex®

The pesticide injection pipeline must contain a functional, normally closed, solenoid-operated valve located on the intake side of the injection pump and connected to the system interlock to prevent fluid from being withdrawn from the supply tank when the irrigation system is either automatically or manually shut down.

The system must contain functional interlocking controls to automatically shut off the pesticide injection pump when the water pump motor stops, or in cases where there is no water pump, when the water pressure decreases to the point where pesticide distribution is adversely affected.

Systems must use a metering pump, such as a positive displacement injection pump (e.g., diaphragm pump) effectively designed and constructed of materials that are compatible with pesticides and capable of being fitted with a system interlock.

Do not apply when wind speed favors drift beyond the area intended for treatment.

Sprinkler Irrigation

The system must contain a functional check-valve, vacuum relief valve and low pressure drain appropriately located on the irrigation pipeline to prevent water source contamination from backflow.

The pesticide injection pipeline must contain a functional, automatic, quick-closing check-valve to prevent the flow of fluid back toward the injection pump.

The pesticide injection pipeline must also contain a functional, normally closed, solenoid-operated valve located on the intake side of the injection pump and connected to the system interlock to prevent fluid from being withdrawn from the supply tank when the irrigation system is either automatically or manually shut down.

The system must contain functional interlocking controls to automatically shut off the pesticide injection pump when the water pump motor stops.

The irrigation line or water pump must include a functional pressure switch which will stop the water pump motor when the water pressure decreases to the point where pesticide distribution is adversely affected. Systems must use a metering pump, such as a positive displacement injection pump (e.g., diaphragm pump) effectively designed and constructed of materials that are compatible with pesticides and capable of being fitted with a system interlock.

Do not apply when wind speed favors drift beyond the area intended for treatment.

Precautions: (1) Do not graze animals in treated orchards. (2) Because timing of fungicide applications for disease control varies due to climatic and other conditions, consult Agricultural Experiment Station or State Extension Service Specialist to determine appropriate treatment timing. (3) Do not use this material if it cannot be applied according to the use pattern on this label. (4) Do not mix Funginex with wetting agents, spreader-stickers, or other adjuvants. (5) Do not let spray mixture stand in tank overnight.

Almonds: Brown Rot Blossom Blight (*Monilinia* spp.) (CA Only)

Apply a mixed solution of 12 fl. oz. of Funginex per 100 gals. of water; spray to run-off. Or, for low volume application, apply a mixed solution of 36-48 fl. oz. of Funginex in 50-200 gals. of water per acre. Make the first application at pink bud and the second at 50-100% bloom. Do not exceed 2 applications. Do not apply after petal fall.

Apples: Scab (*Venturia inaequalis*), Powdery Mildew (*Podosphaera leucotricha*), and Rust (*Gymnosporangium* spp.)

For full coverage spray only, mix 10 fl. oz. of Funginex per 100 gals. of water and apply to run-off. For low volume sprayers, apply 36-40 fl. oz. of undiluted Funginex per acre per application in sufficient water (50-200 gals. of water per acre). For aerial application, apply 36-40 fl. oz. of Funginex in a minimum of 20 gals. of water per acre. Complete coverage is essential to insure adequate control. Make first application at ½ inch green tip and repeat every 7 days for a preventive control program. Do not apply after petal fall. Do not exceed a total of 5 applications per growing season. Consult Agricultural Experiment Station or State Extension Service Specialist for use of Funginex in an apple scab monitoring control program.

Apricots, Cherries, Nectarines, Peaches, Plums, and Prunes: Brown Rot Blossom Blight (*Monilinia* spp.)

For full coverage spray only, mix 12-16 fl. oz. of Funginex per 100 gals. of water and apply to run-off. For low volume sprayers, apply 36-48 fl. oz. of undiluted Funginex per acre per application in sufficient water (50-200 gals. of water per acre). For aerial application, apply 36-48 fl. oz. of Funginex in a minimum of 20 gals. of water per acre. Complete coverage is essential to insure adequate control. Make first application at early bloom (peaches, nectarines: pink bud; apricots: red bud; cherries, plums, prunes: white bud or popcorn). Repeat after 50% bloom. If necessary, depending upon the length of the bloom period and conditions favoring brown rot blossom blight development, make a third application at early petal fall. Alternately, if warm, wet weather prevails, apply the 2 or 3 applications at 2-4 day intervals beginning at early bloom, since blossom period will be shortened. Do not exceed 3 sprays of Funginex during the blossom period. The higher rate of Funginex is only necessary under conditions of severe disease pressure.

CA Only: For full coverage spray only, mix 12 fl. oz. of Funginex per 100 gals. of water and apply to run-off. For low volume sprayers, apply 36-48 fl. oz. of undiluted Funginex per acre per application in sufficient water (50-200 gals. of water per acre). For aerial application, apply only 36-48 fl. oz. of Funginex in a minimum of 20 gals. of water per acre. Complete coverage is essential to insure adequate control. Make first application on peaches and nectarines at pink bud to 5% bloom, on apricots at red bud, on cherries, plums, and prunes at popcorn or white bud, followed by a second application at 50-100% bloom. Do not exceed 2 sprays of Funginex during the blossom period.

Funginex®

Nectarines, Peaches: Brown Rot, Fruit Rot (*Monilinia* spp.)

For full coverage spray only, mix 12-16 fl. oz. of Funginex per 100 gals. of water and apply to run-off. For low volume sprayers, apply 36-48 fl. oz. of undiluted Funginex per acre per application in sufficient water (50-200 gals. of water per acre). For aerial application, apply 36-48 fl. oz. of Funginex in a minimum of 20 gals. of water per acre. Complete coverage is essential to insure adequate control. Make the first application 2-3 weeks before harvest and repeat in 5-10 days. Make a third application just prior to harvest. Do not exceed 3 sprays of Funginex during the pre-harvest period. The higher rate of Funginex is only necessary under conditions of severe disease pressure.

CA Only: For full coverage spray only, mix 12 fl. oz. of Funginex per 100 gals. of water and apply to run-off. For low volume sprayers, apply 36-48 fl. oz. of undiluted Funginex per acre per application in sufficient water (50-200 gals. of water per acre). For aerial application, apply 36-48 fl. oz. of Funginex in a minimum of 20 gals. of water per acre. Complete coverage is essential to insure adequate control. Make first application 2-3 weeks before harvest and repeat in 5-10 days. Do not exceed 2 sprays of Funginex during the pre-harvest period.

Asparagus: Asparagus Rust (*Puccinia asparagi*) (CA Only)

Apply 10-20 fl. oz. of Funginex per acre in 20-50 gals. of water for ground application or in 5 gals. of water for aerial application. Apply at 7-14 day intervals. Adjust application rate and intervals depending on the severity of rust infection and climatic conditions favorable for rust sporulation. For application through sprinkler irrigation systems, apply in 0.12 acre inch of water through sprinkler systems during the last few minutes of irrigation in 150-200 gals. of water per acre. Apply to asparagus ferns only. Do not make more than 7 applications per growing season. Do not harvest spears within 24 weeks of the last fern application.

Highbush Blueberries: Mummyberry Disease (*Monilinia vacciniicorymbosi*) (Pacific and Midwestern States)

Apply 24 fl. oz. of Funginex per acre in 20-50 gals. of water for ground application or in 5 gals. of water for aerial application. Make the first application at leaf bud break and repeat in 7-10 days. Make the third application at pink bud stage and repeat in 7-10 days at early bloom. For the last application, apply 16 fl. oz. of Funginex per acre in 20-50 gals. of water for aerial application. Make the last application between full bloom and early petal fall. Do not make more than 5 applications from leaf bud break to early petal fall.

Eastern Seaboard States (for primary infection only)

Apply 24 fl. oz. of Funginex per acre in 20-50 gals. of water for ground application or in 5 gals. of water for aerial application. Make the first application at leaf bud break and repeat in 7-10 days. Make the last application at pink bud stage. Do not make more than 3 applications from leaf bud break to pink bud stage. Application of Funginex during or beyond early bloom may result in fruit russetting.

Storage and Disposal

Pesticide Storage

Do not store below 32°F (0°C).

Keep out of reach of children and animals. Store in original container only. Store in a cool, dry place and avoid excess heat. Carefully open container. After partial use, replace lid and close tightly. Do not put concentrate or dilute material into food or drink containers. Do not contaminate other pesticides, fertilizers, water, food, or feed by storage or disposal.

Pesticide Disposal

Wastes resulting from the use of this product may be disposed of on site or at an approved waste disposal facility. Do not contaminate water, food, or feed by storage or disposal.

Container Disposal

Triple rinse (or equivalent). Then offer for recycling or reconditioning, or puncture and dispose of in a sanitary landfill, or incinerate, or burn, if allowed by state and local authorities. Stay out of smoke from burning container.

For minor spills, leaks, etc., follow all precautions indicated on this label and clean up immediately. Take special care to avoid contamination of equipment and facilities during cleanup and disposal of wastes. In the event of a major spill, fire, or other emergency, call 1-800-888-8372, day or night.

Funginex®

Precautionary Statements

Hazards to Humans and Domestic Animals

DANGER / PELIGRO

Corrosive, causes irreversible eye damage. Do not get in eyes, or on skin, or clothing. Harmful if swallowed, inhaled, or absorbed through skin.

Statement of Practical Treatment

If in eyes: Flush with plenty of water for 15 minutes and get medical attention.

If swallowed: Drink 1 or 2 glasses of water and induce vomiting by touching back of throat with finger or blunt object. Do not induce vomiting or give anything by mouth to an unconscious person. Get medical attention or call Poison Control Center.

If inhaled: Remove patient from contaminated area and get medical attention. If not breathing, give artificial respiration, preferably mouth-to-mouth.

If on skin: Wash skin with soap and water. Get medical attention.

Personal Protective Equipment

Some materials that are chemical-resistant to this product are listed below. If you want more options, follow the instructions for Category B on an EPA chemical resistance category selection chart.

Applicators and other handlers must wear:

- Long-sleeved shirt and long pants
- Chemical-resistant gloves, such as barrier laminate or butyl rubber
- Shoes plus socks
- Protective eyewear

Discard clothing and other absorbent materials that have been drenched or heavily contaminated with this product's concentrate. Do not reuse them. Follow manufacturer's instructions for cleaning/maintaining PPE. If no such instructions for washables, use detergent and hot water. Keep and wash PPE separately from other laundry.

Engineering Control Statements

When handlers use closed systems, enclosed cabs, or aircraft in a manner that meets the requirements listed in the Worker Protection Standard (WPS) for agricultural pesticides [40 CFR 170.240 (d) (4-6)], the handler PPE requirements may be reduced or modified as specified in the WPS.

User Safety Recommendations

Users should:

- Wash hands before eating, drinking, chewing gum, using tobacco, or using the toilet.
- Remove clothing immediately if pesticide gets inside. Then wash thoroughly and put on clean clothing.

Environmental Hazards

Do not apply directly to water, to areas where surface water is present, or to intertidal areas below the mean high water mark. Do not contaminate water when disposing of equipment wash waters.

Physical or Chemical Hazards

Do not use or store near heat or open flame.

Funginex® is a registered trademark

©1994 Ciba-Geigy Corporation

Ciba Crop Protection
Ciba-Geigy Corporation
Greensboro, North Carolina 27419
CGA 112L2C   074

Funginex®

One Gallon

U. S. Standard Measure

FUNGICIDE

Active Ingredient: Triforine: (N,N'-[1,4-piperazinediyl-bis (2,2,2-trichloro-ethylidene)]-bis-[formamide]) 18.2%

Inert Ingredients:	81.8%
Total:	100.0%

EPA Reg. No. 100-721   EPA Est. 100-AL-1

See directions for use in attached booklet.

AGRICULTURAL USE REQUIREMENTS

Use this product only in accordance with its labeling and with the Worker Protection Standard, 40 CFR part 170. Refer to supplemental labeling under "Agricultural Use Requirements" in the Directions for Use section for information about this standard.

KEEP OUT OF REACH OF CHILDREN.

DANGER / PELIGRO

Si usted no entiende la etiqueta, busque a alguien para que se la explique a usted en detalle. (If you do not understand the label, find someone to explain it to you in detail.)

Precautionary Statements

Hazards to Humans and Domestic Animals

Corrosive, causes irreversible eye damage. Do not get in eyes, or on skin, or clothing. Harmful if swallowed, inhaled, or absorbed through skin.

Statement of Practical Treatment

If in eyes: Flush with plenty of water for 15 minutes and get medical attention.

If swallowed: Drink 1 or 2 glasses of water and induce vomiting by touching back of throat with finger or blunt object. Do not induce vomiting or give anything by mouth to an unconscious person. Get medical attention or call Poison Control Center.

If inhaled: Remove patient from contaminated area and get medical attention. If not breathing, give artificial respiration, preferably mouth-to-mouth.

If on skin: Wash skin with soap and water. Get medical attention.

Environmental Hazards

Do not apply directly to water, to areas where surface water is present, or to intertidal areas below the mean high water mark. Do not contaminate water when disposing of equipment wash waters.

Physical or Chemical Hazards

Do not use or store near heat or open flame.

Chemigation

Refer to section entitled **Chemigation** in booklet for chemigation use directions. Do not apply this product through any irrigation system unless the directions for chemigation are followed.

Funginex® is a registered trademark

©1994 Ciba-Geigy Corporation

Ciba Crop Protection
Ciba-Geigy Corporation
Greensboro, North Carolina 27419
CGA 112L2C   074

Furadan® 4 F
INSECTICIDE-NEMATICIDE

RESTRICTED USE PESTICIDE
Due to acute oral and inhalation toxicity. For retail sale to and application only by certified applicators or personnel under their direct supervision.

ACTIVE INGREDIENT:
* Carbofuran 44.0%
INERT INGREDIENTS: 56.0%
TOTAL.. 100.0%
* 2–3-Dihydro-2, 2-dimethyl-7-benzofuranylmethylcarbamate.
This product contains 4 lbs. of carbofuran per gallon.
EPA Reg. No. 279-2876 ZC   EPA Est. 279-
Code 279

KEEP OUT OF REACH OF CHILDREN
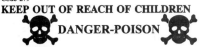
DANGER-POISON

PELIGRO
See below for additional precautionary information.
Si usted no entiende la etiqueta, busque a alguien para que se la explique a usted en detalle. (If you do not understand the label, find someone to explain it to you in detail.)

STATEMENT OF PRACTICAL TREATMENT
If swallowed: Drink 1 or 2 glasses of water and induce vomiting by touching back of throat with finger. Do not induce vomiting or give anything by mouth to an unconscious person. Get medical attention.
If inhaled: Remove to fresh air. Call a physician immediately.
If in eyes: Flush with plenty of water for at least 15 minutes. Get medical attention.
If on skin: Wash skin immediately with soap and water.

ANTIDOTE
NOTE TO PHYSICIAN: Carbofuran is an N-methyl carbamate and a reversible cholinesterase inhibitor. Do not use oximes such as 2-PAM. Start by giving 2 mg. atropine intramuscularly. According to clinical response, continue until signs of atropinization occur (dry mouth or dilated pupils). If in eye, instill one drop of homatropine.
For Emergency Assistance Call (800) 331-3148.

PRECAUTIONARY STATEMENTS

HAZARDS TO HUMANS (AND DOMESTIC ANIMALS)
DANGER
Poisonous if swallowed or inhaled. May be fatal or harmful as a result of skin or eye contact or by breathing spray mist. Causes cholinesterase inhibition. Warning symptoms of poisoning include weakness, headache, sweating, nausea, vomiting, diarrhea, tightness in chest, blurred vision, pinpoint eye pupils, abnormal flow of saliva, abdominal cramps, and unconsciousness. Atropine sulfate is antidotal.

Personal Protective Equipment
Some materials that are chemical-resistant to this product are listed below. If you want more options, follow the instructions for category C on an EPA chemical resistance category selection chart.

Applicators and other handlers must wear: Long-sleeved shirt and long pants; Chemical-resistant gloves, such as Barrier Laminate or Butyl Rubber, or Nitrile Rubber or Neoprene Rubber or Polyvinyl Chloride or Viton; Shoes plus socks; Protective eyewear when mixing or loading; For exposure in enclosed areas: A respirator with either an organic vapor-removing cartridge with a prefilter approved for pesticides (MSHA/NIOSH approval number prefix TC-23C), or a canister approved for pesticides (MSHA/NIOSH approval number prefix TC-14G); For exposure outdoors: Dust/mist filtering respirator (MSHA/NIOSH approval number prefix TC-21C).

Discard clothing and other absorbent materials that have been drenched or heavily contaminated with this product's concentrate. Do not reuse them. Follow manufacturer's instructions for cleaning/maintaining PPE. If no such instructions for washables, use detergent and hot water. Keep and wash PPE separately from other laundry.

When handlers use closed systems, enclosed cabs, or aircraft in a manner that meets the requirements listed in the Worker Protection Standard (WPS) for agricultural pesticides [40 CFR 170.240 (d) (4-6)], the handler PPE requirements may be reduced or modified as specified in the WPS.

The handler PPE requirements may be reduced or modified as specified in the WPS.

| User Safety Recommendations: |
Users should:
● Wash hands before eating, drinking, chewing gum, using tobacco or using the toilet.
● Remove clothing immediately if pesticide gets inside. Then wash thoroughly and put on clean clothing.
● Remove PPE immediately after handling this product. Wash outside of gloves before removing. As soon as possible, wash thoroughly and change into clean clothing.

ENVIRONMENTAL HAZARDS

This product is toxic to fish, birds and other wildlife. Birds feeding on treated areas may be killed. For waterfowl protection, do not apply immediately before or during irrigation, or on fields in proximity of waterfowl nesting areas, or on fields where waterfowl are known to repeatedly feed. Drift and runoff from treated areas may be hazardous to fish in neighboring areas. Do not apply directly to water.

Notice: It is a federal offense to use any pesticide in a manner that results in the death of a member of an endangered species.

The use of Furadan 4 F may pose a hazard to the following Federally designated endangered/threatened species known to be found in certain areas within the named locations.

> **Attwater's Greater Prairie Chicken**— Texas counties including: Aransas, Austin, Brazoria, Colorado, Galveston, Goliad, Harris, Refugio and Victoria
> **Aleutian Canada Goose**— California counties including: Colusa, Merced, Stanislaus and Sutter
> **Kern Primrose Sphinx Moth**— Walker Basin of Kern County, California

This product may not be used in areas where adverse impact on the Federally designated endangered/threatened species noted above is likely. Prior to making applications, the user of this product must determine that no such species are located in or immediately adjacent to the area to be treated. If the user is in doubt whether or not the above named endangered species may be affected, he should contact either the regional U.S. Fish and Wildlife Service office (Endangered Species Specialist) or personnel from the State Fish and Game office.

This product is highly toxic to bees exposed to direct treatment or residues on crops. Do not apply this product or allow it to drift to blooming crops or weeds if bees are visiting the treatment area. Protective information may be obtained from your Cooperative Agricultural Extension Service.

Carbofuran is a chemical which can travel (seep or leach) through soil and can contaminate groundwater which may be used as drinking water. Carbofuran has been found in groundwater as a result of agricultural use. Users are advised not to apply carbofuran where the water table (groundwater) is close to the surface and where the soils are very permeable, i.e., well-drained soils such as loamy sands. Your local agricultural agencies can provide further information on the type of soil in your area and the location of groundwater.

DIRECTIONS FOR USE

It is a violation of Federal law to use this product in a manner inconsistent with its labeling.

Do not apply this product in a way that will contact workers or other persons, either directly or through drift. Only protected handlers may be in the area during application. For any requirements specific to your State or Tribe, consult the agency responsible for pesticide regulation.

Do not apply this product through any type of irrigation system.

Do not use this product on Long Island, N.Y.

SHAKE WELL BEFORE USING

AGRICULTURAL USE REQUIREMENTS

Use this product only in accordance with its labeling and with the Worker Protection Standard, 40 CFR part 170. This Standard contains requirements for the protection of agricultural workers on farms, forests, nurseries, and greenhouses, and handlers of agricultural pesticides. It contains requirements for training, decontamination, notification, and emergency assistance. It also contains specific instructions and exceptions pertaining to the statements on this label about personal protective equipment (PPE), notification to workers, and restricted-entry interval. The requirements in this box only apply to uses of this product that are covered by the Worker Protection Standard.

Do not enter or allow worker entry into treated areas during the restricted entry interval (REI) of 48 hours (except for foliar applications to corn, sunflowers, or sorghum).

After foliar applications on corn, sunflowers, and sorghum, do not enter or allow worker entry into treated areas during the restricted entry interval (REI) of 14 days. Exception: for the last 12 days of the REI, workers may enter the treated area to perform hand labor or other tasks involving contact with anything that has been treated, such as plants, soil, or water, without time limit, if they wear the early entry personal protective equipment listed below.

PPE required for early entry to treated areas that is permitted under the Worker Protection Standard and that involves contact with anything that has been treated, such as plants, soil, or water, is: Coveralls, Chemical-resistant gloves, such as Barrier Laminate or Butyl Rubber, or Nitrile Rubber or Neoprene Rubber or Polyvinyl Chloride or Viton, and Shoes plus socks.

Notify workers of the application by warning them orally and by posting warning signs at entrances to treated areas.

STORAGE AND DISPOSAL

PESTICIDE STORAGE

Not for use or storage in or around the house.

Do not store below 35° F, (2° C).

Keep out of reach of children and animals. Store in original containers only. Store in cool, dry place and avoid excess heat. Carefully open containers. After partial use, replace lids and close tightly. Do not put concentrate or dilute material into food or drink containers. Do not contaminate other pesticides, fertilizers, water, food or feed by storage or disposal.

Bulk holding containers used for Furadan 4F, and associated equipment (e.g., pumps, meters, hoses, etc.), should be thoroughly rinsed to avoid cross contamination with materials subsequently introduced into the bulk system. Begin by draining any remaining product from tanks and equipment and storing in approved appropriately labeled containers. Thoroughly rinse down the inside of the tank and cycle the rinsate through the pump and metering system, repeating this rinsing and cycling procedure at least three times. Refer to the Pesticide Disposal statement for directions for disposal of rinsate.

In case of spill, avoid contact, isolate area and keep out animals and unprotected persons. Confine spills. Call FMC: (800) 331-3148.

To confine spill: If liquid, dike surrounding area or absorb with sand, cat litter or commercial clay. If dry material, cover to prevent dispersal. Place damaged package in a holding container. Identify contents.

PESTICIDE DISPOSAL

Pesticide wastes are acutely hazardous. Improper disposal of excess pesticide, spray mixture, or rinsate is a violation of Federal law. If these wastes cannot be disposed of by use according to label instructions, contact your State Pesticide or Environmental Control Agency, or the Hazardous Waste representative at the nearest EPA Regional Office for guidance.

CONTAINER DISPOSAL

Metal Containers: Triple rinse (or equivalent). Then offer for recycling or reconditioning, or puncture and dispose of in a sanitary landfill, or by other procedures approved by state and local authorities. Do not cut or weld metal containers.

Plastic Containers: Triple rinse (or equivalent). Then offer for recycling or reconditioning, or puncture and dispose of in a sanitary landfill, or incineration, or, if allowed by state and local authorities, by burning. If burned, stay out of smoke.

Returnable/Refillable Sealed Container: Do not rinse container. Do not empty remaining formulated product. Do not break seals. Return intact to point of purchase.

GENERAL INSTRUCTIONS

Do not plant with any crop other than alfalfa, artichokes, bananas, barley, coffee, corn (field, pop or sweet), cotton, cranberries, cucurbits (cucumbers, melons, pumpkins, squash), flax, grapes, non-bearing fruit, oats, ornamentals, peanuts, peppers, potatoes, rice, seed crops (Bermudagrass, spinach), sorghum, strawberries, soybeans, sugar beets, sugarcane, sunflowers, tobacco, or wheat for at least 10 months following use of this product.

Do not rotate with any crop on soil treated at greater than 3.0 pounds active ingredient per acre for at least 10 months.

ALFALFA: Alfalfa Weevil Larvae, Egyptian Alfalfa Weevil, Pea Aphid, and in N.Y. State for Snout Beetle control—Apply the amount of Furadan 4 F, indicated in the chart, when feeding is noticed or when insects appear. Alfalfa Weevil Adult—Apply 1 to 2 pints per acre when insects appear. Alfalfa Blotch Leafminer and Potato Leafhopper. Apply 1 to 2 pints per acre when insects appear. Lygus Bugs—Apply 2 pints per acre prior to bloom. Grasshoppers—Apply ¼ to ½ pint per acre when grasshopper feeding is noticed. For control of Blue Alfalfa Aphid (nymphs and wingless adults)—Apply Furadan 4 F at ½ to 1 pint per acre when insect feeding is noticed or when insects appear. Do not apply more than twice per season. Do not apply more than once per cutting. Do not use more than 1 pint per acre in the second application. Apply only to fields planted to pure stands of alfalfa. Do not move bees to alfalfa fields within 7 days of application. Observe the indicated number of days after application before cutting or grazing.

Pints of Furadan 4 F Per Acre	Do not cut or graze within
½	7 days
1	14 days
2	28 days

Minimum gallonage requirements. Ten gallons of finished spray per acre with ground equipment, two gallons per acre with aircraft, except Blue Alfalfa Aphid use 30 gallons of finished spray per acre with ground equipment, 5 gallons per acre with aircraft.

For waterfowl protection do not apply on fields in proximity of waterfowl nesting areas and/or fields where waterfowl are known to repeatedly feed.

COTTON: Thrips—Use Furadan 4 F at 2.5 fluid ounces per 1,000 linear feet of row (1 quart per acre with 40 inch row spacing). Apply in the seed furrow at planting. Furadan 4 F may be mixed with water or liquid fertilizer. When Furadan 4 F is used with liquid fertilizers, premix one part of Furadan 4 F with two parts of water. Check physical compatibility before mixing large quantities. Add this premix to the tank of fertilizer along with the rinsings from the premixing container. Maintain agitation in the tank after mixing and during application. Do not mix until ready to use. Do not feed cotton forage.

FIELD CORN, POPCORN, SWEET CORN—AT PLANTING: Corn Rootworms, Flea Beetles, and to aid in the control of first generation European Corn Borer—Use 2.5 fluid ounces of Furadan 4 F per 1,000 linear feet of row (1 quart per acre with 40 inch row spacing). Apply at planting, as a 7 inch band over the row or inject on each side of the row by mixing with water or liquid fertilizers.

Corn Rootworms, Flea Beetles, Seedcorn Maggot, Wireworms, and to aid in the control of first generation European Corn Borer and Armyworm for approximately 4 to 6 weeks after planting—Use 2.5 fluid ounces of Furadan® 4 F insecticide-nematicide per 1,000 linear feet of row (1 quart per acre with 40 inch row spacing). Apply at planting directly into the seed furrrow.

Furadan 4 F may be mixed with water or liquid fertilizers. If Furadan 4 F is used with liquid fertilizers, premix one part of Furadan 4 F with two parts of water. Check physical compatibility before mixing large quantities. Add this premix to the tank of fertilizer along with the rinsings from the premixing container. Maintain agitation in the tank after mixing and during application. Do not mix until ready to use. Do not feed forage within 30 days of last application.

FIELD CORN, POPCORN—POST PLANT: Corn rootworms (northern, southern and western)—use 2.5 fluid ounces of Furadan 4 F per 1,000 linear feet of row (1 quart per acre with 40 inch row spacing). Apply as a post emergent spray by banding over the row, or by side dressing or basal spraying both sides of the row after corn emerges. Control will generally be improved if the treatment is cultivated into the soil. Do not feed forage within 30 days of last application.

FIELD CORN—FOLIAR APPLICATION

European Corn Borer—Use 1½ to 2 pints of Furadan 4 F per acre as a foliar spray when corn borer eggs begin to hatch. Use the higher rate for heavier pest infestations. For treatment with aerial equipment, apply as a broadcast spray using a minimum of 1 gallon of finished spray per acre. For treatment with ground equipment, direct the spray into the corn whorl for first brood and into the ear zone for second brood, using a minimum of 10 gallons of finished spray per acre. Repeat if necessary. Observe all precautions listed below.

Southwestern Corn Borer—Use Furadan 4 F at 1 to 2 pints per acre as a broadcast foliar spray when eggs begin to hatch. Rate used will depend on desired residual activity. If infestation continues, retreat in 7 days after a 1 pint application and within 14 days after a 2 pint application. Apply in sufficient water for thorough coverage using a minimum of 1 gallon of finished spray per acre with air equipment or 10 gallons of finished spray per acre with ground equipment. Observe all precautions listed below.

Banks Grass Mites (suppression)—Furadan 4 F when applied at 2 pints per acre for the control of European or Southwestern Corn Borers will also suppress Banks Grass Mites.

Grasshoppers—Use Furadan 4 F as a foliar spray at a ¼ to ½ pint per acre when insects appear or feeding is noticed.

—Use the ¼ pint rate under light to moderate population levels (0 to 14 grasshoppers per sq. yd.).

—Use the ½ pint rate under more severe population levels (15 or more grasshoppers per sq. yd.).

Apply in sufficient water for thorough coverage using a minimum of 2 gallons of finished spray per acre with air equipment or 10 gallons of finished spray per acre with ground equipment. Observe all precautions listed below.

Do not make more than two applications per season at the 1½ to 2 pint use rate. Do not make more than four applications per season at the 1 pint use rate. Do not forage, cut or harvest within 30 days of last application. Do not apply on seed corn less than 14 days prior to detasseling or rogueing.

SWEET CORN—FOLIAR APPLICATION (Machine Harvested Only): European Corn Borer—For control of second generation borers apply 1 pint Furadan 4 F per acre. Make first application just prior to first silking and repeat at weekly intervals. Do not make more than four (4) applications per season. Do not apply within seven (7) days of harvest. If prolonged, intimate contact will result, do not reenter treated field within 14 days of application without wearing proper protective clothing. Do not graze or harvest stalks within 21 days of last application.

Do not make a foliar application if more than 12 ozs. of Furadan 10 G per 1,000 linear feet of row (10 lbs. per acre with 40" row spacing) or 8 ozs. of Furadan 15 G per 1000 linear feet of row (6.7 lbs. per acre with 40" row spacing) or 2.5 fluid ounces of Furadan 4 F per 1,000 linear feet of row (1 quart per acre with 40 inch row spacing) were used in an at-planting application.

Minimum gallonage requirements: 10 gallons of finished spray per acre with ground equipment, 2 gallons per acre with aircraft.

GRAPES (For use in California): Nematodes (Root Knot and Dagger) and Grape Phylloxera—Apply Furadan® 4 F insecticide-nematicide at 2½ gallons per acre as a broadcast treatment only, to the soil surface between the vine rows and immediately incorporate by mechanical means. Do not apply within 200 days of harvest. Remove dense weed growth prior to treatment. Do not use on soils of pH 8.0 or greater. Do not make more than one application per crop year.

ORNAMENTALS—CONTAINER GROWN

Root Weevil Larvae—Prepare a stock solution by thoroughly mixing 1 to 2 fluid ounces of Furadan 4 F per 100 gallons of water. Apply as a soil drench in sufficient volume to saturate the entire soil profile within each container. The following guidelines give approximate amounts to use on various sized containers. Make a single application when larvae are present (usually from July to mid-October in outdoor growing areas). Later application may be less effective due to lower temperatures and/or the presence of more mature larvae.

Container Size	Amount of Stock Solution Per Container
6 inches diameter	1 pint
8 inches diameter	1 to 2 pints
10 inches diameter	2 to 4 pints
12 inches diameter	3 to 6 pints
16 inches diameter	4 to 8 pints

Early marginal necrosis or leaf drop may occur on Hydrangea or Birch. Not all species or varieties of ornamentals have been tested. Before treating large numbers of plants of a particular variety, treat a few plants and observe prior to full scale application.

Application of Furadan 4 F through overhead sprinkler equipment is prohibited.

PINE SEEDLINGS: Pales Weevil and Pitch-Eating Weevil in pine plantations—Apply a 1% (W/W) active Furadan clay slurry (see following for preparation) to the roots of pine seedlings prior to transplanting. Treat seedlings by dipping roots or use any other suitable means which allows thorough coating. Keep roots moist until transplanted. Prepare the slurry as follows: Add 1.6 ounces (2½ tablespoons) of Furadan 4 F to ½ gallon of water. Mix thoroughly. Add 2 pounds of pulverized kaolin clay (pH 4.5) to this suspension. Mix thoroughly. This is sufficient to treat the roots of 150 to 200 seedlings. Adequate ventilation is required for indoor treatment.

POTATOES: Potato Tuberworm (Virginia only); Colorado Potato Beetle, European Corn Borer, Potato Flea Beetle, Potato Leafhopper—Use 1 to 2 pints of Furadan 4 F in sufficient water to treat one acre. Apply when insects first appear and repeat as necessary to maintain control. Do not make more than 8 foliar applications per season. Do not apply more than 6 pints to the foliage if either Furadan 10 G or Furadan 15 G was used at planting. Do not apply more than 2 pints per application. Do not apply within 14 days of harvest. Minimum gallonage requirements: 10 gallons of finished spray per acre with ground equipment and 3 gallons per acre with aircraft.

Do not use this product on Long Island, New York.

For waterfowl protection do not apply:

—immediately before or during furrow irrigation

—on fields in proximity of waterfowl nesting areas

—on fields where waterfowl are known to repeatedly feed

SMALL GRAINS (Wheat, Oats, Barley): Grasshoppers—Use Furadan 4F at ¼ to ½ pint per acre when insects appear. Cereal Leaf Beetle—Use ½ pint per acre when insects appear. Apply before heads emerge from boot. Do not make more than two applications per season. Do not feed treated forage to livestock.

Minimum gallonage requirements: Ten gallons of finished spray per acre with ground equipment, two gallons per acre with aircraft.

For waterfowl protection do not apply on fields in proximity of waterfowl nesting areas and/or on fields where waterfowl are known to repeatedly feed.

SOYBEANS—AT PLANTING APPLICATION: Nematodes (Root-Knot, Cyst, Stunt, Ring Sting Spiral Lesion, Lance Dagger, Stubby Root)—Use 3.75 to 5 fluid ounces of Furadan 4 F per 1,000 linear feet of row (3 to 4 pints per acre with 40-inch row spacing). Apply at planting as a 12-inch surface band. Soybeans may be grazed or cut for forage 30 days or later following an at planting application.

SOYBEANS—FOLIAR APPLICATION: Mexican Bean Beetle—Use Furadan 4 F at ½ to 1 pint per acre when insects appear. Grasshoppers—Use Furadan 4 F at ¼ to ½ pint per acre when insects appear. Do not use Furadan 4 F as a foliar application if Furadan 10 G, Furadan 15 G, Furadan 4 F was applied to soybeans at planting time. Do not make more than 2 foliar applications per season. Do not apply within 21 days of harvest. Do not graze or feed foliar-treated forage to livestock or cut for silage or hay. Minimum gallonage requirements: 20 gallons of finished spray per acre with ground equipment, 1½ gallons per acre with aircraft.

STRAWBERRY: (For use in Washington & Oregon) Root Weevils—Use Furadan 4 F at 2 to 4 pints per acre (2.6 to 5.1 fluid ounces per 1,000 linear feet of row for 42 inch row spacing). Apply as a 10 to 12 inch band over the row after last harvest but before October 1st. Do not make more than one application per season. Do not apply if berries are present.

SUGARCANE: Sugarcane Borer—Apply 1 to 1½ pints Furadan 4 F per acre using ground or aerial equipment. Check sugarcane fields weekly, beginning in early June and continuing through August. Make first application only after visible joints form and 5% or more of the plants are infested with young larvae feeding in or under the leaf sheath and which have not bored into the stalks. Repeat whenever field checks indicate the infestation exceeds 5 percent. Do not apply within 17 days of harvest. Do not use in Hawaii.

SUNFLOWERS (Confectionary and Oil)—At Planting

Sunflower Stem Weevil, Sunflower Beetle, Grasshoppers—Use 2.5 to 5.0 fluid ounces of Furadan 4 F per 1,000 linear feet of row (1.4 to 2.8 quarts per acre with 30 inch row spacing). Apply directly into the seed furrow—OR—Apply in a 7 inch band. Furadan 4 F may be mixed with water or liquid fertilizer. When Furadan 4 F is used with liquid fertilizers, premix one part of Furadan 4 F with two parts of water. Check physical compatibility before mixing large quantities. Add this premix to the tank of fertilizer along with the rinsings from the premixing container. Maintain agitation in the tank after mixing and during application. Do not mix until ready to use.

Use the higher rate when heavier insect infestations are anticipated.

SUNFLOWERS (Confectionary and Oil)—Foliar Application

Sunflower Moth, Banded Sunflower Moth, Stem Weevils, Seed Weevils—Use Furadan 4 F at 1 pint per acre. Grasshoppers—Use Furadan 4 F at ¼ to 1 pint per acre. Sunflower Beetle—Use Furadan 4 F at ¼ to ½ pint per acre. Apply as a foliar spray when insects appear. When a rate range is indicated, use higher rate for heavier insect infestations. Repeat applications as necessary but not more than four times per season. Apply a minimum of two gallons of finished spray per acre with aircraft and 10 gallons by ground equipment. Do not harvest crop within 28 days of last application.

TOBACCO—BURLEY: Flea Beetles—Use 1 gallon of Furadan 4 F per acre. Apply as a broadcast spray over the soil surface prior to transplanting and incorporate into the top 3 inches of soil with a suitable device.

TOBACCO—FLUE-CURED: Flea Beetles, Wireworms, and to aid in the control of Budworms and Root Knot Nematodes—Use 1½ gallons of Furadan 4 F per acre— OR—Use 1 gallon of Furadan 4 F per acre for control of Flea Beetles only. Apply as a broadcast spray over the soil surface prior to forming beds. Incorporate into the top 3 inches of soil. Form beds and plant as usual. This product may induce flecking of the bottom or lower leaves.

DEALERS SHOULD SELL IN ORIGINAL PACKAGES ONLY.

TERMS OF SALE OR USE: On purchase of this product buyer and user agree to the following conditions:

WARRANTY: FMC makes no warranty, expressed or implied, concerning the use of this product other than indicated on the label. Except as so warranted, the product is sold as is. Buyer and user assume all risk of use and/or handling and/or storage of this material when such use and/or handling and/or storage is contrary to label instructions.

DIRECTIONS AND RECOMMENDATIONS: Follow directions carefully. Timing and method of application, weather and crop conditions, mixture with other chemicals not specifically recommended, and other influencing factors in the use of this product are beyond the control of the seller and are assumed by buyer at his own risk.

USE OF PRODUCT: FMC's recommendations for the use of this product are based upon tests believed to be reliable. The use of this product being beyond the control of the manufacturer, no guarantee, expressed or implied, is made as to the effects of such or the results to be obtained if not used in accordance with directions or established safe practice.

DAMAGES: Buyer's or user's exclusive remedy for damages for breach of warranty or negligence shall be limited to direct damages not exceeding the purchase price paid and shall not include incidental or consequential damages.

Furadan and FMC—FMC trademarks   (279-9/29/92-A)
3/94

This is a specimen label. FMC is not responsible for the accuracy of the information contained herein. As labels are subject to revision, always carefully read and follow the label on the product container.

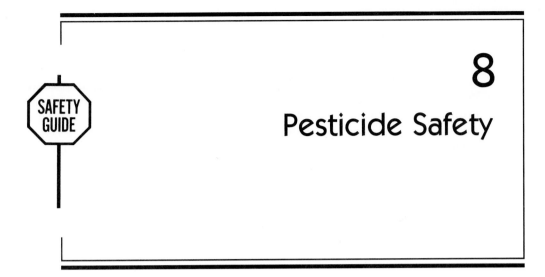

8

Pesticide Safety

GENERAL CONSIDERATIONS

Pesticide technology has come a long way since the days when diesel oil, lead arsenate, Bordeaux mixture, and nicotine sulfate were the mainstays of the industry. We are able to do much more with today's organic insecticides and weed killers to increase crop and livestock yields than we were 50 years ago when synthetic pesticides were just being discovered.

Many of our modern pest control chemicals are two-edged swords. We have not yet learned all the tricks of the pharmaceutical trade, for we have not yet devised many economic poisons that are selectively toxic only to their intended victims. Some of them are highly poisonous to people and animals. This situation is true today and is likely to remain so for a long time.

Pesticides, like drugs, are beneficial to people when properly used; misused they may be extremely dangerous. Most pesticides are designed to kill something—insects, mites, fungi, weeds, rodents, or other pests. Therefore, all pesticides should be handled as poisons.

With proper handling and application, pesticides rank among the safest production aids used. They undergo exhaustive manufacturer's tests and pass stringent label requirements before being approved and registered for market by the Environmental Protection Agency (EPA). Pesticide residues and their effects are carefully monitored or surveyed. Teams of scientists maintain surveillance on soils, crops, water, air, animals, and humans to ensure that dangerous levels of pesticides are not accumulating in the environment.

Pesticides have a wide safety margin when used properly. But, like automobiles, firearms, and medicines, they can be and sometimes are improperly used, causing accidents. National statistics show that approximately 20 children and adults die of pesticide poisoning annually. Though these accidental poisonings are fewer than deaths reported due to such items as aspirin, the number can and must be reduced.

It is entirely possible for a user to handle pesticides safely for many years with no obvious ill effects to himself or the environment. Illnesses resulting from overexposure to pesticides do occur among those who work with these substances, but these illnesses are preventable. They are caused by misuse of the chemicals, and misuse results from carelessness, which, in turn, results from ignorance. Well-informed pesticide users are more likely to take proper precautionary measures in handling toxic materials, and these persons play an important part in maintaining a good safety record for pesticide usage.

There is no reason for pesticide misuse because many sources of information are available as guidance for proper use. Manufacturers' labels and MSDSs, Cooperative Extension, EPA, USDA, Public Health Service, American Medical Association, and the various state agencies all provide reliable sources of information.

PESTICIDE SELECTION

Choosing the correct pesticide to use is one of the most important segments of carrying out an effective pest control program. The pesticide you choose will not only be instrumental in the effectiveness of your control program, but will also have a direct bearing on the hazards that you subject yourself to, as well as other persons and the environment.

Actually, a potential hazard may be present the moment a person purchases a pesticide. The selection of a type of pesticide, the formulation to be used, and even the container type may be factors contributing to a pesticide accident. Before making a selection, the pest problem should be identified by a competent person. Control measures should not be undertaken unless the pest problem is of economic importance, has a potential of developing into a problem, is of health importance, or is a nuisance.

After the pest has been properly identified, select a pesticide that will control the pest with a minimum of danger to other organisms. The pesticide must be one currently registered by the EPA and the State Department of Agriculture and recommended by the Cooperative Extension Service (Figure 8–1).

There are considerable differences in the safety of various pesticide formulations. Select the safest pesticide formulation whenever there is a choice. Granular-type formu-

Figure 8–1   Read the label.

Pesticide Safety   Chap. 8

lations are safer than sprays or dusts because they drift less. Formulations with greatest drift and dispersion potential are most likely to cause damage to desirable plants under unfavorable climatic conditions. These formulations can be of greater risk to the applicator if a highly toxic pesticide is being used (see Chapter 10). Emulsifiable concentrate pesticides with a petroleum-type carrier are generally more hazardous than water-soluble ones because they penetrate the skin more rapidly and are difficult to wash off.

Finally, estimate the amount of pesticide needed and purchase only enough for the particular job or for one season so that you do not have a storage or disposal problem. Smaller containers are easier to handle and present less chance of accidental spillage and contamination.

HANDLING AND MIXING PESTICIDES

Pesticides are toxic materials and extreme caution should be used in mixing and handling them. Of the various work activities associated with pesticide use, the mixing and loading operation is considered to be the most hazardous part of the spraying or dusting job. Obviously, many factors are involved that determine the degree of hazard of any operation. The mixing and loading of pesticides, however, will generally result in possibilities of exposure to spills, splashes, and the like when a mix is of higher concentration and in greater quantity. Although not minimizing the importance of protective measures during all work activities, the mixing and loading operation warrants special attention and care. The following general safety instructions should be observed during the mixing and loading of pesticides:

- Having selected the right pesticide for the job, read the label and carry out the necessary calculations for the required dilution of the pesticide. Obtain the proper equipment, including protective clothing and respirator if required, and have first aid equipment available.
- Never work alone when handling highly hazardous pesticides.
- Mix chemicals outside or in a well-ventilated area (Figure 8–2). Carefully open original concentrate containers. Never position any portion of the body directly over the seal or the pouring spout. The release of pressure may cause the liquid

Figure 8–2 Mix pesticides outdoors.

to be expelled from the container. Open sacks with a knife rather than tearing them because dry formulations such as dusts and powders can billow up in large concentrations. Always stand upwind when mixing or loading pesticides.

- When mixing chemicals, all quantities of the active ingredient should be measured with extreme accuracy. Have clean measuring and transfer containers available. Containers to accurately measure liquid in either or both avoirdupois (pints, quarts, gallons) or metric (milliliters, liters) measurements as well as scales for weighing dry materials should be kept in the area where pesticides are stored. The measuring containers should be thoroughly cleaned after each use.

- As concentrate containers are emptied and drained, water or other diluting material being used in the spray program should be used to rinse the container. Rinse three times, allowing 30 seconds for draining after each rinse. (Use 1 quart for each rinse of a 1-gallon can or jug; 1 gallon for each 5-gallon can; and 5 gallons for either 30- or 55-gallon drums.) Drain each rinse into the spray tank before filling it to the desired level (see also Chapter 14).

- Clean up spilled pesticides immediately. If the pesticide is accidentally spilled on skin, immediately wash it off with soap and water. Should the pesticide be spilled on clothing, change clothing as soon as possible and launder before wearing again. Do not store or wash contaminated clothing with other soiled clothing items.

- Protective gloves should be washed before removing them. Replace protective gloves frequently regardless of signs of wear or contamination.

Figure 8–3   Wash thoroughly.

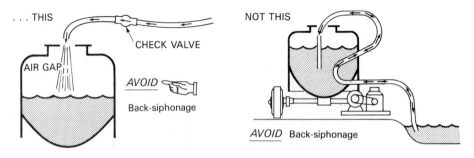

Figure 8–4   Avoid back-siphoning.

- Persons mixing, handling, or applying pesticides should never smoke, eat, or drink until they have washed thoroughly (Figure 8–3). It can be a means of swallowing pesticides that have accumulated on the mouth or hands.
- Do not use your mouth to siphon a pesticide from a container.
- When filling the spray tank, do not allow the delivery hose below the highest possible water surface in order to avoid back-siphoning (Figure 8–4).

APPLICATION OF PESTICIDES

Prior to pesticide application, the label should be referred to again even though the pesticide is familiar to you. Details are often forgotten and labels are frequently revised. Use protective clothing and equipment if prescribed no matter how uncomfortable it may be during hot weather. The discomfort must be endured to use the pesticide safely. New types of protective clothing have been developed that allow air to circulate without letting pesticides penetrate.

Do not apply higher dosages or rates than specified on the label or recommended by your state university Cooperative Extension personnel. To apply the correct pesticide amount, the application equipment must be calibrated properly. Knowledge of equipment output is essential to compute the amount of pesticide needed in the mixture to treat a given area (see Chapter 12).

Be sure the application equipment is clean, in good condition, and operating properly. Poorly operating equipment can cause unnecessary hazards to the user, as well as possible damage to the crop and the environment. Extra time spent during spraying to fix and adjust equipment that is in poor operating condition can cause excessive exposure to a pesticide. Do not blow out clogged hoses, nozzles, or lines with your mouth.

The application should be performed at the proper time using recommended dosages to prevent illegal residues from remaining in or on food, feed, or forage crops. Many pesticide labels state the number of days between the last pesticide application and the time that the crop can be harvested. These waiting intervals should be followed closely. For livestock, there is a waiting period between treatment and slaughter.

Precautions should be observed to guard against drift of pesticides onto nearby crops, pastures, or livestock. Extreme care should be observed to prevent applications from contaminating streams, ponds, lakes, or wells. If pesticides must be applied when there is a breeze, drive or spray into or at right angles to the breeze to prevent the pesticide from blowing into your face (Figure 8–5).

If you feel ill or notice any irregular body symptoms when applying pesticides, work should cease and efforts should be made to seek medical attention.

RESTRICTED ENTRY INTO TREATED FIELDS

Standards for restricted entry into treated fields (not to be confused with number of days from last spray until harvest) have been established for certain pesticides by the EPA. These standards apply particularly to agricultural uses and were established to protect field workers. Refer to Worker Protection Standards and restricted entry intervals in Chapter 6.

Restricted Entry Into Treated Fields

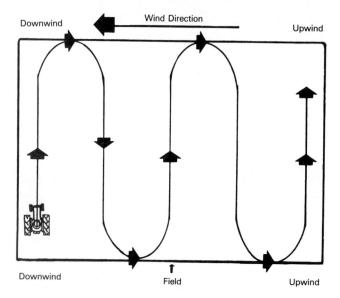

Downwind   Wind Direction   Upwind

Downwind   Field   Upwind

Figure 8–5 Spray at right angles to wind direction.

Although a commercial applicator may feel that he or she is not directly involved, there are three rules that must be followed:

1. No unprotected person may be in a field that is being treated with a pesticide.
2. No pesticide application is to be permitted that will expose any person to pesticides either directly or through drift, except those involved in the application, who should be wearing protective clothing as required.
3. The label restrictions and directions must be followed.

In addition to the federal restricted entry intervals that have been issued under the Worker Protection Standards, some states may impose certain requirements. It is your responsibility to keep abreast of label information and EPA requirements.

Anytime you apply a material with a restricted entry interval (REI), be sure to notify your client in order to avoid misunderstandings. Tell him or her the name of the material, time of application, and the required reentry period during which protective clothing must be worn so that the client may inform employees accordingly. If workers must enter fields prior to expiration of the required interval, growers are obligated to notify them of the date of treatment and furnish them with protective clothing. Protective clothing means at least a hat or other suitable head covering, a long-sleeved shirt and long-legged trousers or a coverall-type garment (all of closely woven fabric), shoes, and socks.

Warnings to workers may be given orally, or by posting signs at field entrances (Figure 8–6) and/or on bulletin boards where workers usually assemble. If a worker does not understand English, the warnings must be made in the appropriate language to ensure that the warning is clearly understood. Refer again to the Worker Protection Standards in Chapter 6. Fields treated with pesticides other than those with REIs may be reentered without protective clothing after the spray has dried or the dust has settled.

Figure 8–6 Pesticide warning sign.

Keeping everyone who is involved informed of your actions will be good public relations and could keep you from possible litigation. Some state agencies may impose and enforce standards for workers that are more restrictive than those specified in this chapter. Be sure you know the regulations for your state.

Don't confuse *days-to-harvest intervals* with restricted entry intervals. Days-to-harvest is the time that must pass between pesticide application and crop harvest to protect the consumer by allowing the pesticide residue to degrade to or below a specific tolerance level. The same pesticide will have different intervals for different crops, based on dosages necessary to control pests, different surfaces treated, whether edible parts are treated, stripping or washing outer leaves, and so on. Check the harvest interval on the label before an application is made.

TOXICITY OF PESTICIDES

Keep in mind when considering the toxicity of pesticides that any chemical substance is toxic or poisonous if absorbed in excessive amounts; therefore, the poisonous effect of a chemical depends on the amount consumed or absorbed. If enough common salt is consumed at one time, it is quite poisonous.

The assumption can be made, therefore, that all pesticides in sufficient amounts can be toxic or poisonous. They are designed to be poisons to kill the pest they are used against. There are, however, great ranges in the level of toxicity among different pesticides. It is important that persons working with pesticides have a broad general knowledge of the relative toxicity of at least the most common pesticides used.

Pesticide users should be concerned with the hazard associated with exposure as well as the toxicity of the chemical itself. *Toxicity* is the inherent capability of the substance to produce injury or death. *Hazard* is a function of *toxicity* and *exposure*. The hazard can be expressed as the probability that injury will result from the use of the pesticide in a given formulation, quantity, or manner. The hazard might be specific in nature, as posing a toxicity problem to humans or to animals, or in another instance causing injury to some plants.

Toxicity of Pesticides

A pesticide can be extremely toxic as a concentrate, yet possibly pose very little hazard to the user if (1) used in a very dilute formulation, (2) used in a formulation (granule) that is not readily absorbed through the skin or inhaled, (3) used only occasionally under conditions of no human exposure, or (4) used by experienced applicators who are properly equipped to handle the pesticide safely.

In another example, however, a pesticide may have a relatively low mammalian toxicity, but present a hazard because it is used in the concentrated form, which may be readily absorbed or inhaled. It could be hazardous to the nonprofessionals who are not aware of the possible hazards to which they are being exposed.

The toxicity value of a pesticide is at best just a relative measure to estimate its toxic effect on humans or other animals. Human toxicity ratings would be the best guide to a pesticide toxicity; however, no actual scientific tests can be conducted in which humans are subjected to lethal doses of pesticides. There are some fragmentary data in regard to human toxicity ratings based on accidental exposures and suicides caused by some pesticides, but essentially all pesticide toxicity ratings are based on animal tests.

Toxicities of pesticides are generally expressed as LD_{50} or LC_{50} values, which mean lethal dose or lethal concentration to 50% of a test population. To determine LD_{50} or LC_{50} values, the dosage of a particular pesticide necessary to kill 50% of a large population of test animals under certain conditions is computed. An example might be a pesticide that has an LD_{50} of 10, which would indicate that 10 milligrams of this pesticide given to animals that weigh 1 kilogram each would kill 50% of the population.

Graphs 8–1 through 8–4 illustrate the relative toxicity to humans of certain pesticides taken orally or by skin exposure. Note the longer bars indicating more relative safety to dermal exposure of the same insecticide than when taken orally. Although this is generally true, note also that some insecticides are equally as toxic on the skin as when swallowed.

The American Association of Pesticide Control Officials (AAPCO) has formulated and adopted certain regulatory principles relating to the determination of highly toxic pesticides. An economic poison that falls within any of the following categories when tested on laboratory animals (mice, rats, and rabbits) is **highly toxic** to humans within the meaning of these principles:

1. **Oral toxicity.** Those that produce death in half or more than half of the animals of any species at a dosage of 50 milligrams at a single dose, or less, per kilogram of body weight when administered orally to 10 or more such animals of each species.

2. **Toxicity on inhalation.** Those that produce death in half or more than half of the animals of any species at a dosage of 200 parts or less by volume of the gas or vapor per million parts by volume of air when administered by continuous inhalation for 1 hour or less to 10 or more animals of each species, provided such concentration is likely to be encountered by humans when the economic poison is used in any reasonably foreseeable manner.

3. **Toxicity by skin absorption.** Those that produce death in half or more than half of the animals (rabbits only) tested at a dosage of 200 milligrams or less per kilogram of body weight when administered by continuous contact with the bare

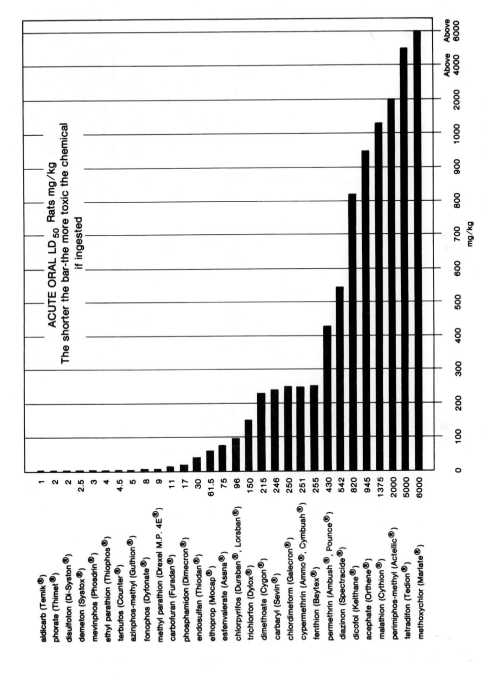

Graph 8–1 Relative toxicity to humans of certain common insecticides when ingested (swallowed).

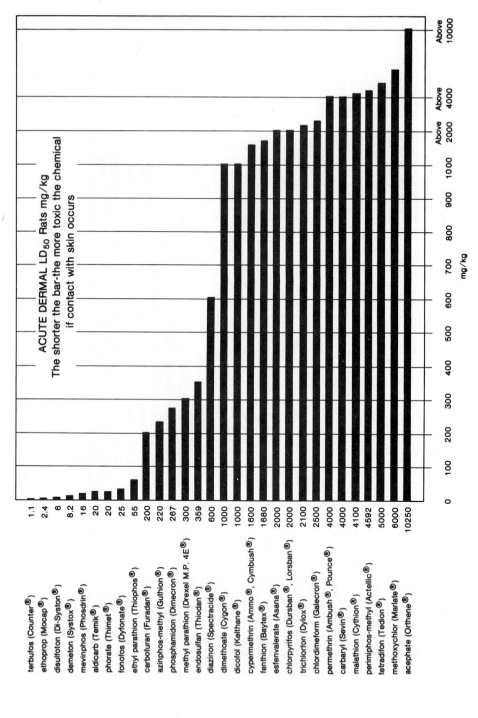

Graph 8–2 Relative toxicity to humans of certain common insecticides when exposure is dermal (skin contact).

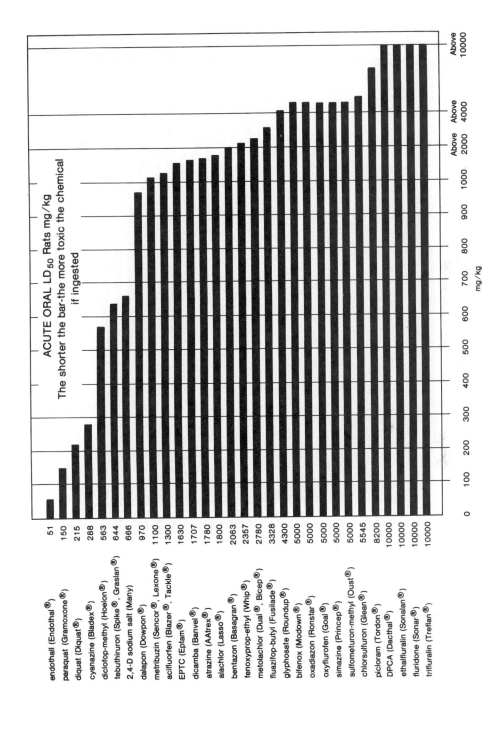

Graph 8–3 Relative toxicity to humans of certain common herbicides when ingested (swallowed).

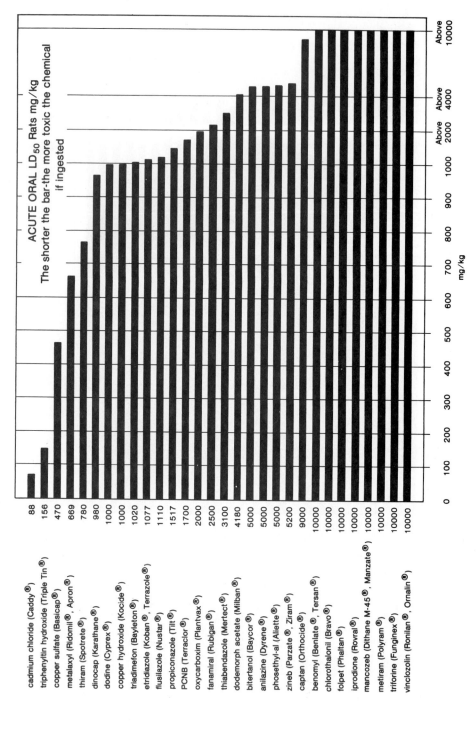

Graph 8–4 Relative toxicity to humans of certain common fungicides when ingested (swallowed).

196

skin for 24 hours or less to 10 or more animals. However, an enforcement official may exempt any economic poison that meets this standard but which is not in fact highly toxic to humans, and may after hearing designate as highly toxic to humans any economic poison which experience has shown to be so in fact.

Toxicity data based on LD_{50} values with animals involve several problems and should not be interpreted as exact values for human toxicity, but these values can be used as guides if one uses caution in considering the following four points.

1. The accidental hazard represented by any pesticide depends more on how it is used than how toxic it is.

2. Pesticide toxicity levels may vary according to the species of test animal, test method, sex of the species, whether the animals have been fasted or not, state of animal health, purity of the chemical tested, carrier in which the chemical is administered, route of administration, and length of time and frequency of exposure.

3. LD_{50} values are only a statistic. They provide no information on the dosage that will be fatal to a specific individual in a large population of animals. Statistically, however, the LD_{50} value is the most accurate means available to indicate the level of pesticide toxicity.

4. It must be remembered that LD_{50} values are usually expressed in terms of single dosages or exposures only, such as:
   • *Acute oral:* refers to a single dose taken or ingested by mouth.
   • *Acute dermal:* refers to a single dose applied directly to and absorbed by the skin.
   • *Inhalation:* refers to exposure through breathing or inhaling.

Thus LD_{50} values give little or no information about the possible cumulative effects of the pesticide.

TOXICITY CATEGORIES

Hazard indicators	I	II	III	IV
Oral LD_{50}	Up to and including 50 mg/kg	From 50 through 500 mg/kg	From 500 through 5000 mg/kg	Greater than 5000 mg/kg
Inhalation LD_{50}	Up to and including 0.2 mg/liter	From 0.2 through 2 mg/liter	From 2.0 through 20 mg/liter	Greater than 20 mg/liter
Dermal LD_{50}	Up to and including 200 mg/kg	From 200 through 2000	From 2000 through 20,000	Greater than 20,000
Eye effects	Corrosive; corneal opacity not reversible within 7 days	Corneal opacity reversible within 7 days; irritation persisting for 7 days	No corneal opacity; irritation reversible within 7 days	No irritation
Skin effects	Corrosive	Severe irritation at 72 hours	Moderate irritation at 72 hours	Mild or slight irritation at 72 hours

Signal Words Assigned by Levels of Toxicity

The EPA has published regulations for the use of human hazard signal words on pesticide labels:

Toxicity category I.   All pesticide products meeting the criteria of toxicity category I shall bear on the front panel the signal word DANGER. In addition, if the product was assigned to toxicity category I on the basis of its oral, inhalation, or dermal toxicity (as distinct from skin and eye local effects), the word POISON shall appear in red on a background of distinctly contrasting color and the skull and crossbones shall appear in immediate proximity to the word POISON.

Toxicity category II.   All pesticide products meeting the criteria of toxicity category II shall bear on the front panel the signal word WARNING.

Toxicity category III.   All pesticide products meeting the criteria of toxicity category III shall bear on the front panel the signal word CAUTION.

Toxicity category IV.   All pesticide products meeting the criteria of toxicity category IV shall bear on the front panel the signal word CAUTION.

Use of signal words.   Use of signal word(s) associated with a higher toxicity category is not permitted except when the agency determines that such labeling is necessary to prevent unreasonable adverse effects.

TOXICITY VALUES

As an aid to those unfamiliar with signal words and toxicity values, Table 8–1 groups and translates LD_{50} values into practical terms.

Toxicity classifications for insecticides, herbicides, fungicides, rodenticides, nematicides, fumigants, and bird repellents are presented in Appendix F, giving both the common and trade name as well as the producer of the particular chemical. LD_{50}

TABLE 8–1   ORAL, DERMAL, AND INHALATION RATINGS OF PESTICIDE GROUPS

Epa category-meaning	Pesticide label signal word[a]	Oral LD_{50} (mg/kg)	Dermal LD_{50} (mg/kg)	Inhalation LC_{50}[b] (μg/1)	Lethal oral dose for a 150-lb person
I. Highly hazardous	Danger-Poison	0–50	0–200	0–2000	Few drops to teaspoonful
II. Moderately hazardous	Warning	50–500	200–2000	2000–20,000	Teaspoonful to one ounce
III. Slightly hazardous	Caution	500–5000	2000–20,000	Over 20,000	One ounce to one pint or pound
IV. Relatively nonhazardous	Caution	Over 5000	Over 20,000	—	Over one pint or pound

[a]All pesticide labels carry a "Keep Out of Reach of Children" warning.

[b]Vapor or gas inhalation value may also be expressed in parts per million (ppm).

values are not shown in the appendix because chemical companies do not put them on their labels. As stated earlier, LD_{50} values are only a statistic that is used as an index to assign a signal word. Signal words are shown on every pesticide label and should be used as a guide to the toxicity of the pesticide and a signal to the possible hazards when mixing, loading, and applying the pesticide.

CHRONIC TOXICITY

When an organism is rapidly affected by a pesticide, this condition is referred to as *acute poisoning* or *acute toxicity*. Generally, this results from a single or a very few closely spaced exposures and is the kind of reaction most people equate with being poisoned.

Toxicity is the innate capacity of a chemical to be poisonous. It is important to remember that all poisoning is dose related. For any compound, the toxic response depends on the size of the dose in relation to the size of the victim. The more that is taken, the more serious the response.

Chronic toxicity is used to describe the effects of prolonged or repeated low-level exposures to a poisonous substance. It is dose related. Types of chronic effects include tumors and cancer, reproductive problems such as sterility and birth defects, damage to the nervous system, damage or degeneration of internal organs such as the liver, and allergic sensitization to particular chemicals. Any organ system in the body may show a significant injury after chronic exposure to a toxic material. In trying to determine the potential for chronic effects in people, toxicologists must use very high doses of a poison in animal tests compared to the probable human exposure. Such information is necessary to establish whether or not any risk level is possible, even as small as a one-in-a-million response. Remembering that chronic toxicity is dose related, does that mean that a real-world exposure will produce a lesser effect or just that it will take longer to show up? Decision makers must also consider whether there is such a thing as a lesser cancer or an acceptable birth defect.

Hazard is different than toxicity. Hazard is the degree of danger involved in using a pesticide. It is the chance or risk that poisoning could occur under certain circumstances of use. Hazard varies according to exposure and can be expressed as an equation of exposure times toxicity. The toxicity comes built into the pesticide. However, we can tilt the risk equation in our favor if we eliminate or minimize exposure. Even a compound with a high toxicity, acute or chronic, can be used with negligible risk if you virtually eliminate exposure to it. The more you expose yourself to it, the higher the risk. Preventing exposure is the key to the safe use of pesticides and is the reason why so much emphasis is placed on safety precautions during storing, handling, using, and disposing of them. With chronic toxicity, exposure is the cumulative effects from using pesticides, from residues in treated commodities that we eat and in the things our skin comes in contact with, and possibly from the water we drink. That is why it is especially important to pay attention to all label statements for groundwater protection, restricted entry intervals, preharvest intervals, disposal instructions, and requirements for protective safety equipment. The EPA uses these regulatory options to reduce the exposure and therefore the hazard of particular pesticides to a level acceptable under the law. If this option is lost because of general disregard by applicators, then the EPA will be required to use some other option, such as limiting or stopping further use.

Some pesticides bear warning statements on their labels due to possible carcinogenicity or reproductive toxicity. The imposition of a warning statement requirement for either cancer or birth defects is decided on a case-by-case basis. The following four factors are considered in this decision:

1. Weight of the evidence: How solid is the information that the pesticide is an oncogen or teratogen?
2. Is there a substantial magnitude of risk, taking into consideration potency and expected exposure levels?
3. Size of the exposed population: Is the number of people who (a) would read the label and (b) are affected by the risk sufficiently small that the statement would be misleading to all other users of the pesticide?
4. Nature of the exposure: Will the label statement be read by the exposed population? That is, if the exposure of concern is dietary, then a warning statement on the label would have no impact. If a statement is to be used, it must be because the population of concern would have access to the label (for example, the population of concern is that of mixers, loaders, and/or applicators).

The EPA's assumption is that imposition of a hazard warning statement will not reduce the number of people exposed to the pesticide; that is, the statement will not actually result in people using another product or no pesticide at all. Instead the statement is intended to provide informed consent and to minimize exposure. To further emphasize the care that needs to be taken with these pesticides, their use may need to be restricted to persons who are certified applicators.

A separate decision may also be made to classify these chemicals for restricted use. If the EPA believes that the extra care inherent in taking a certified applicator training course and reading a cancer or birth defects warning statement is necessary to ensure proper handling, then the product would also be classified as "restricted use."

Following is a list of pesticides for which warnings are required on the pesticide label due to possible carcinogenicity or reproductive toxicity. Those pesticides that have also been classified as "restricted use" are noted.

Cancer/Tumor Statement

This product contains [chemical name], which has been determined to cause tumors in laboratory animals. Risks can be reduced by closely following the use directions and precautions, and by wearing protective clothing specified elsewhere on the label.

Alachlor	Chlordane*
Aldrin*	Chlordimeform*
Amitraz	Diclofop-methyl*
Amitrole*	EDB*
Arsenic acid (nonwood uses)*	Formaldehyde
Captafol	

Heptachlor*   Pronamide*

Lindane*   Telone*

Linuron   TPTH*

Pentachlorophenol*

Birth Defects Statement

This product contains [chemical name], which has been determined to cause birth defects in laboratory animals. Risks can be reduced by closely following the use directions and precautions, and by wearing protective clothing specified elsewhere on this label.

Brodifacoum   Dinocap

Bromadiolone   Diphacinone

Chlorophacinone**   Pindone

Cyanazine*   Warfarin
   *Restricted use.
   **One restricted use.

PESTICIDE EXPOSURE: PROTECTIVE CLOTHING AND EQUIPMENT

Any time you or the people you supervise may be exposed to a pesticide, consider the need to use personal protective equipment (PPE)—clothing and devices that protect the body from contact with pesticides. The law requires pesticide users to follow all PPE instructions on the pesticide label. More protection may be a good idea in some situations.

Most pesticide handlers know to wear PPE during mixing, loading, and application, but many do not wear it at other times when they may be exposed. Wear PPE for any task that could cause pesticides to get on the skin or in the mouth, eyes, or lungs.

There are three ways that toxic chemicals may enter the human body to cause poisoning. The routes of entry include oral exposure (through the mouth by swallowing), dermal exposure (absorption through the skin and eyes), and inhalation or respiratory exposure (absorption by breathing). Figure 8–7 summarizes how the accidental exposures might occur.

Oral Exposure and Protective Measures

Although not considered a major source of occupational exposure to pesticides, oral exposure and subsequent absorption through the gastrointestinal tract is, with few exceptions, the most serious exposure because of the rapid internal absorption and possibility of quick death. Generally, when a pesticide is taken into the mouth in amounts sufficient to cause serious injury or death, it is consumed either by accident as a result of gross negligence or by intent to do self-inflicted injury. Accidental oral ingestion, in most cases, is a result of putting pesticides in unlabeled containers such as soft drink bottles or food containers for storage in an area where children or unknowing adults may consume it.

The most likely means of ingesting pesticides is through accidental splashing of liquid into the mouth or by wiping the face with a sleeve, cuff, or hand. Other means

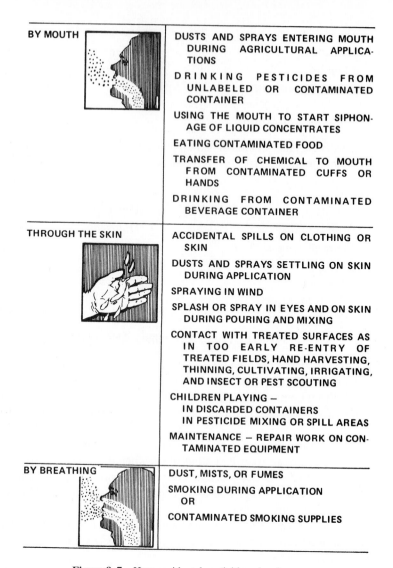

BY MOUTH	DUSTS AND SPRAYS ENTERING MOUTH DURING AGRICULTURAL APPLICATIONS
	DRINKING PESTICIDES FROM UNLABELED OR CONTAMINATED CONTAINER
	USING THE MOUTH TO START SIPHONAGE OF LIQUID CONCENTRATES
	EATING CONTAMINATED FOOD
	TRANSFER OF CHEMICAL TO MOUTH FROM CONTAMINATED CUFFS OR HANDS
	DRINKING FROM CONTAMINATED BEVERAGE CONTAINER
THROUGH THE SKIN	ACCIDENTAL SPILLS ON CLOTHING OR SKIN
	DUSTS AND SPRAYS SETTLING ON SKIN DURING APPLICATION
	SPRAYING IN WIND
	SPLASH OR SPRAY IN EYES AND ON SKIN DURING POURING AND MIXING
	CONTACT WITH TREATED SURFACES AS IN TOO EARLY RE-ENTRY OF TREATED FIELDS, HAND HARVESTING, THINNING, CULTIVATING, IRRIGATING, AND INSECT OR PEST SCOUTING
	CHILDREN PLAYING – IN DISCARDED CONTAINERS IN PESTICIDE MIXING OR SPILL AREAS
	MAINTENANCE – REPAIR WORK ON CONTAMINATED EQUIPMENT
BY BREATHING	DUST, MISTS, OR FUMES
	SMOKING DURING APPLICATION OR
	CONTAMINATED SMOKING SUPPLIES

Figure 8–7   How accidental pesticide poisonings occur.

of oral exposure to pesticides in occupational situations are food handled with contaminated hands, food exposed to pesticide sprays or dusts, contaminated drinking utensils, and attempting to clear nozzles by blowing through the nozzle openings or spray lines.

Oral exposure can be minimized by the following steps and protective measures:

1. Check the label for special instructions or warnings regarding oral exposure. Some pesticide labels will contain a precautionary illustration such as this one:
2. Never eat or drink while spraying or dusting.
3. Wash thoroughly with soap and water before eating or drinking.
4. Do not touch lips to contaminated objects or surfaces.

CAN KILL YOU IF SWALLOWED

This product can kill you if swallowed even in small amounts; spray mist or dust may be fatal if swallowed.

5. Do not wipe the mouth with hands, forearm, or clothing.

6. Do not expose lunch, lunch container, beverage, or drinking vessel to pesticides.

7. If you are involved in operations where highly toxic pesticides are used and the label has indicated a high danger of oral ingestion, it would be advisable to wear a full face plastic shield or mask when there is a possibility of the pesticide splashing.

Dermal Exposure and Protective Measures

Most sources of information indicate that pesticide absorption through the skin is the most common cause of poisoning in agricultural workers. The incidence of occupational poisoning by various pesticides is more closely related to their acute dermal toxicity in animals than to their acute oral toxicity, further substantiating the importance of dermal exposure as a significant route of entry into the body. Absorption through the skin and eyes may occur as a result of a splash, spillage, or drift when pesticides are being measured, mixed, loaded, or applied. Most pesticide handlers know to wear PPE during mixing, loading, and application, but many do not wear it at other times when they may be exposed. Wear PPE for any task that could cause pesticides to get on the skin or in the mouth, eyes, or lungs. For example:

- Disposing of pesticides or pesticide containers
- Transporting (or carrying) pesticide containers that are open or have pesticide spilled on the outside
- Helping with an application, such as scouting, monitoring, or checking pesticide coverage
- Flagging for aerial applications
- Cleaning, adjusting, or maintaining equipment that has pesticides on it
- Entering enclosed areas after fumigation to measure air levels or operate ventilation systems
- Entering treated areas after soil fumigation to adjust or remove coverings, such as tarpaulins
- Cleaning up spills
- Contact with a deposit or residue remaining on the treated crop or site

Several factors affect the dermal penetration of a pesticide:

- Physical and chemical properties of the pesticide
- Health and condition of the skin

Pesticide Exposure: Protective Clothing and Equipment

203

- Temperature
- Humidity
- Presence of other chemicals (solvents, surfactants, and others)
- Concentration of the pesticide
- Type of formulation
- Lack of eye protection

All these factors can affect penetration and absorption of a pesticide by the skin. The physical and chemical properties, concentration, formulation, and presence of other chemicals are established when a specific pesticide is selected for use. Temperature and humidity are the conditions existing at the time of application, which are, at the time, selected as weather conditions most favorable for effective pest control. The health and condition of skin may be variable at the time of pesticide application. Individuals with skin problems should avoid other than minimal exposure to pesticides unless extra precautions are observed. Skin cuts, abrasions, scratches, scuffs, or any other such damage or disruption can be sources of quick absorption of pesticides. Extra care should be taken to minimize pesticide exposure if these conditions exist.

Pesticides can be absorbed at different rates, and extra care must be taken to protect certain areas of the body. An index number of 100 indicates the most rapid absorption (Figure 8–8).

Wind, type of activity, application method, rate of application, and duration of exposure are other factors that may affect skin and eye exposure. All these factors must be taken into consideration when selecting protective clothing and equipment to protect against exposure to pesticides.

The type and amount of protective clothing needed depends on the job being done and the type of pesticide being used. Several factors are important for the operator to consider when determining protective needs:

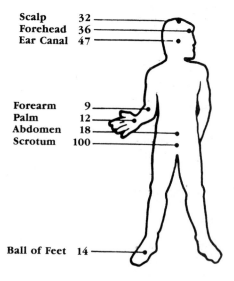

Scalp	32
Forehead	36
Ear Canal	47
Forearm	9
Palm	12
Abdomen	18
Scrotum	100
Ball of Feet	14

Figure 8–8   Pesticide absorption rates.

Pesticide Safety   Chap. 8

- Toxicity, concentration, and vapor action of the pesticide being used
- Degree of expected exposure during application
- Length of expected exposure during application
- Extent to which the pesticide can be absorbed through the skin

These requirements may vary in a number of situations. The following seven suggested procedures should be observed, depending on an evaluation of all the factors.

1. Observe all recommended protective measures specifically mentioned on the label of the pesticide to be used. Some pesticide labels may contain an illustration as follows:

CAN KILL YOU BY SKIN CONTACT

This product can kill you if touched by hands or spilled or splashed on skin, in eyes or on clothing (liquid goes through clothes).

2. Cover up before exposure, not after (Figure 8–9). Putting on protective clothing after you have been exposed to pesticides on the skin will only hasten the absorption rate.

3. For any pesticide handling task, wear at least a long-sleeved shirt and long-legged pants. The pants and shirt should be made of sturdy material. Fasten the shirt collar to protect the lower part of the neck. Skin contamination can be reduced by wearing coveralls that afford the entire body protection. Several different types of disposable coveralls are available. Coveralls made of Tyvek® are quite good and can be worn several times before disposal. If the situation is such that coveralls would be wet through mist spray or spillage, a chemical-resistant apron or waterproof rainsuit should be used. Coveralls, aprons, and rainsuits should be thoroughly washed with soap and water after each use. Contaminated PPE should be handled as carefully as the pesticide itself. Protective rubberwear should be periodically checked for cracks and holes.

4. Efforts should be made to protect the hair, skin about the head, eyes, and neck from pesticides. Headgear should include a waterproof rain hat or a safety hard hat. Waterproof or water-repellent parkas offer good protection for the head and neck area at the same time. Goggles or a face shield should be worn to protect the eyes. Old felt hats or other types of absorbent headgear should not be worn because they absorb pesticides, especially in the area of the sweat band. Once contaminated, they can provide a source of continuous and very dangerous skin contact.

5. Natural rubber gloves, at least 14 ml thick, should generally be used when handling organophosphates, carbamates, and other pesticides that are readily absorbed through the skin. Some fumigant-type pesticides cause rubber gloves to disintegrate, and polyethylene gloves must be used in this case. The type of

PROTECTIVE CLOTHING FOR HAZARDOUS PESTICIDES

(Pesticides with labels marked: DANGER POISON, or WARNING)

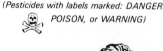

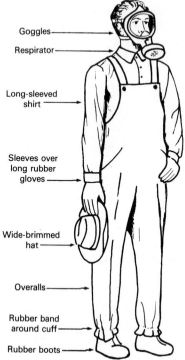

Goggles

Respirator

Long-sleeved shirt

Sleeves over long rubber gloves

Wide-brimmed hat

Overalls

Rubber band around cuff

Rubber boots

Drawn by Dr. James R. Baker
North Carolina State University

Figure 8–9 Protective clothing necessary for safe pesticide application.

glove to use is usually suggested by the pesticide manufacturer. Select a lightweight, unlined, natural rubber glove with long enough gauntlets to protect the wrist area, but flexible enough to allow freedom of finger movement. Several types of disposable gloves are presently manufactured that can be used when applying certain types of pesticides. Gloves should be checked very carefully at regular intervals for holes or other signs of wear. It is probably a good idea to replace the gloves periodically rather than chance an accidental exposure. Leather and cotton gloves should never be used when handling or applying pesticides. They can absorb the material and become a constant source of pesticide exposure. They could then become more of a potential hazard to poisoning than if no gloves were worn.

6. Waterproof boots or footgear should be worn during most types of pesticide application. Foot protection is most important when hand applications of pesticides are made. Only natural rubber boots should be worn when handling or spraying pesticides. Leather and canvas shoes can absorb the pesticide and hold

it in contact with the wearer, thus making this type of footwear undesirable when applying pesticides.

7. Eye protection is most often needed when measuring or mixing pesticide concentrates, as well as when spray or dust drifts might be a problem. Protective shields or goggles should be used whenever there is a chance of a pesticide coming into contact with the eyes. These pieces of equipment should be kept clean at all times and should be handled the same as protective clothing when being washed or repaired, keeping in mind that they should be handled as carefully as the pesticide itself.

Table 8–2 ties together the label signal word with various pesticide formulations and suggests the proper clothing and equipment that should be used, depending on the toxicity of the formulation.

Respiratory Exposure and Protective Measures

Pesticides are sometimes inhaled in sufficient amounts to cause serious damage to nose, throat, and lung tissues. The potential hazard of respiratory exposure is great, due to what is considered to be near complete absorption of pesticides through this route. Vapors and extremely fine particles represent the most serious potential for respiratory exposure.

Some means of protection for the respiratory route is needed when toxic dust and vapors or small spray droplets are prevalent. Pesticide dusts, aerosol, fog, fume, smoke, and certain mists represent a high potential for respiratory exposure.

Pesticide exposure is usually relatively low when dilute sprays are being applied with conventional application equipment. This is due primarily to the larger droplet sizes produced. When low-volume equipment is being used, respiratory exposure is increased by the smaller droplets or particles being produced. Application in confined spaces also contributes to an increased potential respiratory exposure.

A respiratory device is one of the most important pieces of PPE both for commercial and private applicators when they are applying toxic pesticides. Respirators are sometimes uncomfortable to wear, particularly in hot and dusty conditions, but it is important that the applicator fully realize the need for protection or serious injury may occur.

Several kinds of respiratory devices are available to protect applicators from breathing dust and chemical vapors. Pesticide applicators need to know what types are available and the hazards they will protect against. A number of manufacturers make various respiratory safety devices. These are available from many supply houses and farm equipment and pesticide dealers. Look in the yellow pages of your telephone book for dealers in your area.

Some pesticide labels may contain an illustration as follows:

CAN KILL YOU IF BREATHED

This product can kill you if vapors, spray mist, or dust are breathed.

TABLE 8–2 PROTECTIVE CLOTHING AND EQUIPMENT GUIDE

	Label signal word		
Formulation	Caution: Slightly toxic	Warning: Moderately toxic	Danger-Poison: Highly toxic
Dry	Long-legged trousers, no cuffs, long-sleeved shirt, shoes, socks, hat or head covering.	Long-legged trousers, long-sleeved shirt, shoes, socks, wide-brimmed hat, and gloves.	Long-legged trousers, long-sleeved shirt, shoes, socks, hat, gloves, and cartridge or canister respirator if dust in air or label precautionary statement says: "Poisonous or fatal if inhaled."
Liquid	Long-legged trousers, long-sleeved shirt, shoes, socks, and hat.	Long-legged trousers, long-sleeved shirt, shoes, socks, wide-brimmed hat, and water-impermeable gloves. Respirator if required by label precautionary statement: "Do not breathe vapors or spray mists" or "Poisonous if inhaled."	Long-legged trousers, long-sleeved shirt or disposable coveralls, water-impermeable boots, wide-brimmed hat, water-impermeable gloves, and goggles or face shield. Canister respirator if label precautionary statement says: "Do not breathe vapors or spray mists" or "Poisonous if inhaled."
Liquid (when mixing)	Long-legged trousers, long-sleeved shirt, shoes, socks, hat, gloves, and water-impermeable apron.	Long-legged trousers, long-sleeved shirt, shoes, socks, wide-brimmed hat, water-impermeable gloves, goggles or face shield, and water-impermeable apron. Respirator if label precautionary statement says: "Do not breathe vapors or spray mist" or "Poisonous (or fatal or harmful) if inhaled."	Long-legged trousers, long-sleeved shirt, water-impermeable gloves, goggles, water-impermeable apron, and canister respirator or disposable coveralls.
Liquid (prolonged exposure to spray or application in enclosed area)	Long-legged trousers, long-sleeved shirt, water-impermeable boots, water-impermeable gloves, and water-impermeable hat.	Water-repellent, long-legged trousers, long-sleeved shirt, water-impermeable gloves, water-impermeable apron, waterproof wide-brimmed hat, face shield, and cartridge or canister respirator.	Waterproof suit, water-impermeable boots, water-impermeable gloves, waterproof hood or wide-brimmed hat, face shield, and canister respirator.

Adapted from U.S. Environmental Protection Agency and Department of Agriculture, *Applying Pesticides Correctly.*

The pesticide label will contain information concerning the proper respirator to use for the pesticide you will be applying. Since April 1994, the Worker Protection Standards have required that pesticide container labels specify the type of respirator needed for mixing, handling, or application of that particular pesticide. The label contains a

NIOSH/MSHA "TC" approval number. The prefix of this approval number indicates the minimum level of respiratory protection that can be used. Under the WPS, certain low-toxicity pesticides will allow a TC-21C type dust/mist respirator. Certain high-toxicity pesticides may require a TC-14G gas mask. The Worker Protection Standards require that employers (1) provide employees with the respirator type stated on the pesticide label, (2) see that cartridge/filters are replaced according to certain guidelines, (3) make sure the respirator fits the wearer correctly, (4) maintain respirators in a clean and operating condition, with daily inspection for worn parts, (5) provide a storage area for respirators, and (6) take steps to prevent heat illness while respirators are being worn. Be sure to use the respirator designed for the pesticide and the job. Never try to use one type of respirator for all kinds of hazards.

There are primarily two types of respiratory devices for use by pesticide applicators when handling toxic pesticides. These are the chemical cartridge respirator and the canister-type respirator or gas mask. Supplied-air respirators and self-contained breathing equipment are also available when oxygen is deficient or highly toxic gases are so concentrated that the applicator must have his or her own air supply. Types of protective devices are illustrated and discussed next.

Cartridge respirator (Figure 8–10).   Protects against certain pesticide dusts, mists, and fumes. Check the label to see if this type of respirator is sufficient for protection against the pesticide you are using. Some cartridge respirators can be purchased with a full face piece that would eliminate the need for protective eyewear.

Canister-type (gas mask) (Figure 8–11).   Protects the lungs and the eyes against certain chemical dusts, mists, and fumes. Be sure to check the label to see if this type of protective device is required for the pesticide you are using.

This kind usually has a half-face mask that covers the nose and mouth, but does not protect the eyes. It has one or two cartridges attached to the face-piece. An absorbent material such as activated charcoal, plus dust filters, purify the air you breathe. **DOES NOT PROTECT from lack of oxygen.**

Figure 8–10   Cartridge respirator.

This covers the entire face. The face-piece may hold a canister directly (chin-type), or connect by a flexible hose to a canister carried on your chest or hip. The "gas mask" canister contains more absorbing material and longer-life filter than the cartridge-type respirator. **DOES NOT PROTECT from lack of oxygen.**

Figure 8–11   Canister-type respirator.

Pesticide Exposure: Protective Clothing and Equipment

Brings in air through a hose, from a safe distant supply. The operator's hood is connected by hose to a source of outside fresh air, usually pumped by a blower. Types are:

- Air-line respirators with constant flow, demand flow, or pressure demand flow.
- Hose masks, with or without blower.

Figure 8–12   Supplied-air respirator.

Lets you take your air supply with you for short-term jobs. The mask is connected to an air or oxygen cylinder, usually carried on operator's back.

(Certain gases, like ammonia, can cause harm through the skin. You may need protective clothing as well as oxygen.)

Figure 8–13   Self-contained breathing apparatus.

Supplied-air respirator (Figure 8–12).   Used in toxic atmospheres where oxygen is at least 19.5%. Analyze your own spray situation to determine if you need this type of equipment.

Self-contained breathing apparatus (Figure 8–13).   Can serve the same functions as a supplied-air respirator, but is approved for atmospheres that are immediately dangerous to life or health (IDLH). Mobility is provided with this type of equipment.

CARE AND MAINTENANCE OF RESPIRATORS AND GAS MASKS

No matter how well the respiratory device is designed and made, unless it is properly cared for and maintained, it may fail to provide protection. The two most common failings are said to be (1) failure to occasionally wash the facepiece with soap and water and (2) neglecting to change the filter cartridges or canisters regularly.

To minimize respiratory exposure and to be certain that protective devices are working properly, the instructions on the following page should be observed.

PESTICIDE POISONING EFFECTS AND SYMPTOMS

The ways in which pesticides affect humans and other mammals are commonly referred to as modes of action. The modes of action of a number of pesticides in use today are either not known or in some instances are only partially understood. Even

INSTRUCTIONS FOR USE

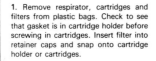

1. Remove respirator, cartridges and filters from plastic bags. Check to see that gasket is in cartridge holder before screwing in cartridges. Insert filter into retainer caps and snap onto cartridge holder or cartridges.

2. FIT RESPIRATOR ON FACE with narrow end over nose. Adjust headband around neck and crown of head, snug enough to insure a tight but comfortable seal.

3. TEST FOR TIGHTNESS: Place the palm of the hand or thumb over the valve guard and press lightly. Exhale to cause a slight pressure inside facepiece. If no air escapes, respirator is properly fitted. If air escapes, readjust respirator and test again.

4. FILTERS (a) REPLACE when breathing becomes difficult. Generally the filter discs should be changed after eight hours of dusty exposure. (b) REPLACE CARTRIDGES when any leakage is detected by taste or smell.

5. CLEAN AND SANITIZE YOUR RESPIRATOR after each day's use. First remove filters and cartridges, then wash other parts thoroughly with warm soapy water and/or sanitize with Willson's Germisol.®

6. The cartridge holders are keyed to assure their correct positioning and maintain the proper balance of the device. Make sure they are properly positioned and seated.

7. KEEP RESPIRATOR CLEAN when not in use. Store in container provided. Replace worn or faulty parts immediately, and order by part number.

8. FOR YOUR PROTECTION the DUST FILTERS and CHEMICAL CARTRIDGES must be assembled tightly, and changed frequently, according to exposure.

9. Many chemicals can be absorbed through the skin. Wear protective clothing when necessary.

TAKE CARE OF YOUR RESPIRATOR AND YOUR HEALTH

CAUTION
Respirators and canister gas masks are not to be worn in atmospheres immediately dangerous to life or health or in atmospheres containing less than 19.5% oxygen.

---

though it may not be known exactly how a pesticide poisons the body in all cases, some of the signs and symptoms resulting from such poisoning are quite well known. These warnings (signs and symptoms) of the body in response to poisoning should be recognized by those who use pesticides. It must be remembered, however, that everyone is subject to various sicknesses or diseases. Therefore, it cannot be assumed that because an individual is in the vicinity where pesticides are or have been used that certain signs or symptoms are the result of pesticide poisoning. But if any signs appear after contact with pesticides, call your physician and advise him or her of the nature of the pesticide involved.

Pesticide applicators should be alert to the early stages of signs and symptoms of poisoning and immediately and completely remove the source of exposure, such as contaminated clothing. Quick action may prevent additional exposure and help to minimize injury. In some instances, early recognition of the signs and symptoms of poisoning and immediate complete removal of the source of exposure may save the applicator's life.

Pesticide Poisoning Effects and Symptoms

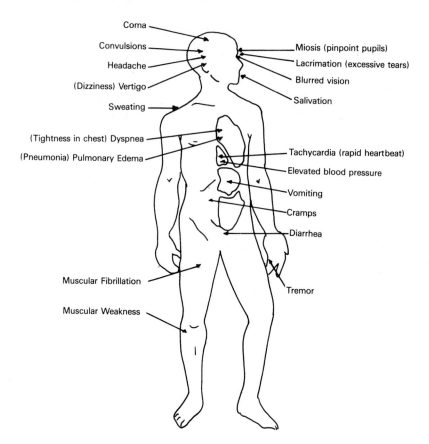

Coma

Convulsions

Headache

(Dizziness) Vertigo

Sweating

(Tightness in chest) Dyspnea

(Pneumonia) Pulmonary Edema

Muscular Fibrillation

Muscular Weakness

Miosis (pinpoint pupils)

Lacrimation (excessive tears)

Blurred vision

Salivation

Tachycardia (rapid heartbeat)

Elevated blood pressure

Vomiting

Cramps

Diarrhea

Tremor

Figure 8–14 Symptoms of pesticide poisoning.

Learn how to recognize poisoning symptoms (Figure 8–14). The following are some of the more important hazardous pesticide groups and some of the ways they affect humans and other animals.

1. *Organophosphorus pesticides.* The organophosphorus chemicals are one of the largest groups of pesticides presently being used. The group includes insecticides such as parathion, malathion, phorate, mevinphos, and diazinon. Toxicity values of these pesticides range from high toxicity for parathion to low toxicity in the case of malathion. The organophosphorus pesticides can be absorbed dermally, orally, or through inhalation of vapors.

The pesticides in this group attack a chemical in the blood, cholinesterase, that is necessary for proper nerve functioning. Because this group of pesticides affects the enzyme cholinesterase, they are sometimes referred to as cholinesterase inhibitors or anticholinesterase compounds. Organophosphates can inhibit, or poison, the cholinesterases by forming chemical combinations with them that prevent them from

PHYSIOLOGICAL ACTION OF ACETYLCHOLINE

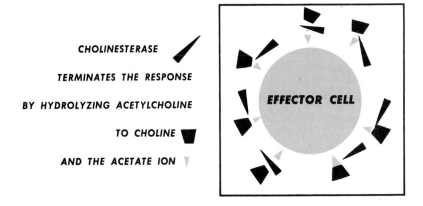

ACETYLCHOLINE,

LIBERATED BY NERVE

IMPULSE, ACTS DIRECTLY UPON

EFFECTOR CELLS TO PRODUCE THEIR

CHARACTERISTIC RESPONSES

EFFECTOR CELL
Examples are:

GLAND   MUSCLE   NERVE

PHYSIOLOGICAL ROLE OF CHOLINESTERASE

CHOLINESTERASE

TERMINATES THE RESPONSE

BY HYDROLYZING ACETYLCHOLINE

TO CHOLINE

AND THE ACETATE ION

EFFECTOR CELL

Figure 8–15 Role of acetylcholine and cholinesterase.

doing their work in the nervous system. (Refer to Figures 8–15 and 8–16 for a graphic explanation of the role of cholinesterase and how it is affected by organophosphates.) When the enzymes are poisoned, nerve impulse transmission races out of control because of a buildup of acetylcholine at the ends of nerve fibers. Muscle twitchings referred to as tremors or fibulations then become noticeable. Convulsions or violent muscle actions result if the tremors become intense.

Regular testing of cholinesterase levels in the blood of applicators is a good way to monitor exposure to organophosphate and carbamate pesticides. Pesticides that can

MECHANISM OF TOXIC ACTION OF PHOSPHATE ESTER INSECTICIDES

PHOSPHATE ESTER

COMPOUNDS ATTACH

A PHOSPHORYL GROUP

TO CHOLINESTERASE AND THEREBY

RENDER THE ENZYME UNABLE

TO PERFORM ITS FUNCTION

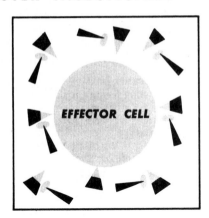

MECHANISM OF PROTECTIVE ACTION OF ATROPINE

ATROPINE

BLOCKS THE ACTION OF

ACETYLCHOLINE BY INTER-

FERING WITH THE ABILITY

OF THE CELL TO RESPOND

TO THIS CONTINUING STIMULUS

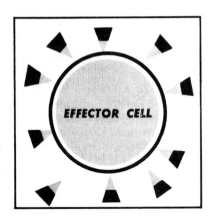

Figure 8–16 Mechanisms of action for organophosphate pesticides and atropine.

be tracked this way include malathion, parathion, and acephate (Orethene®) among the organophosphates, and aldicarb (Temik®), carbaryl (Sevin®), and benomyl (Benlate®) among the carbamates.

By periodically checking cholinesterase levels in the blood of pesticide applicators, chronic or low-level exposure and accumulation of pesticides in the body can be detected. Baseline levels should be established 60 days before pesticide exposure. Blood sampling should be repeated weekly during the exposure period.

Additional symptoms and signs of organophosphorus poisoning include headache, giddiness, nervousness, blurred vision, dizziness, weakness, nausea, cramps, diarrhea, and chest discomfort. Other symptoms and signs might be sweating, pinpoint eye pupils, watering eyes, excess salivation, rapid heartbeat, excessive respiratory secretions, and vomiting. Advanced stages of poisoning usually result in convulsions, loss of bowel control, loss of reflexes, and unconsciousness. Quick action and proper medical treatment can still save persons in the advanced stages of poisoning, even though they may be near death.

2. *Carbamate pesticides.* This group of pesticides includes such insecticides as aldicarb (Temik®), carbaryl (Sevin®), and carbofuran (Furadan®); herbicides such as cycloate (Ro-Neet®) and diallate (Avadex®); and such fungicides as benomyl (Benlate®) and ferbam (Fermate®).

The mode of action of these compounds is very similar to that of the organophosphorus compounds in that they inhibit the enzyme cholinesterase. However, they differ in action from the organophosphorus compounds in that the effect on cholinesterase is brief, because the carbamates are broken down in the body rather rapidly. Carbamates are therefore referred to as "rapidly reversing inhibitors" of cholinesterase. The reversal is so rapid that, unless special precautions are taken, measurements of blood cholinesterase of human beings or other animals exposed to carbamates are likely to be inaccurate and always in the direction of appearing to be normal. Carbamate pesticides can be absorbed through the skin and by breathing or swallowing. Symptoms and signs of carbamate poisoning are essentially the same as those caused by organophosphorus pesticides.

3. *Bipyridyliums.* This group of herbicides, which includes diquat and paraquat, may be fatal if swallowed and harmful if inhaled or absorbed through the skin. Lung fibrosis may develop if these materials are taken by mouth or inhaled. Prolonged skin contact will cause severe skin irritation. Signs and symptoms of injury may be delayed, but there are no adequate treatments and effects are generally irreversible.

4. *Anticoagulants.* This group of pesticides is commonly used as rodenticides. Warfarin (Prolin®), coumafuryl (Fumarin®), brodifacoum (Talon®), bromadiolone (Maki®), and diphacinone (Diphacin®) are representative examples of this group. These materials are of danger only when taken orally, and large doses would be required to cause human death.

Anticoagulants reduce the body's ability to produce blood clots and sometimes damage capillary blood movement in the body. Bleeding can occur with the slightest body damage, resulting in a severe nosebleed or massive bruises. Severe back and abdominal pains have been noted in persons attempting suicide.

5. *Botanical pesticides.* Botanical pesticides are manufactured from plant derivatives. They vary greatly in chemical structure and toxicity to humans, ranging from pyrethrum, one of the least toxic to mammals of all the insecticides in use currently, to strychnine, which is extremely toxic.

- *Pyrethrins (pyrethrum):* These botanical materials are used against a wide variety of insect pests. Most symptoms in humans as a result of exposure to pyrethrum have been reported to be skin allergies, sneezing, runny nose, and in some cases stuffiness of the nose. Most have been reported to be minor; however, some individuals are known to be more sensitive than others.
- *Strychnine:* The mode of action of this pesticide is not completely understood. However, it is known to act on the nervous system. Symptoms and signs are nervousness, stiff muscles in the face and legs, cold sweats, and fits. From one to ten violent attacks may occur before the patient either recovers or dies as a result of the inability to breathe. Without medical attention, death commonly occurs within 3 to 5 hours after exposure.
- *Nicotine (Black Leaf 40®):* This is one of the most toxic of all poisons and its action is very rapid. Poisoning can be caused by oral, dermal, and respiratory exposure. Signs and symptoms of nicotine poisoning include local skin burns and irritations. If nicotine is absorbed through the skin or taken in through the mouth, the patient becomes highly stimulated and excitable. This is generally followed by extreme depression. In fatal cases of nicotine poisoning, death is usually rapid, nearly always within 1 hour and occasionally within 5 minutes due to paralysis of respiratory muscles.
- *Rotenone:* Relatively few cases of poisoning of people from rotenone have been reported. Direct contact with rotenone will result in mild irritation of the skin and eyes in some individuals. Most of the symptoms reported have come from animal studies. In these studies, the animals exhibited numbness of the mouth, vomiting, and pains in the stomach and intestines. The inability of a test animal to coordinate its activities may result in muscle tremors, followed by chronic convulsions. Initially breathing is often very rapid changing to extremely slow. In cases of animal death, it has generally been due to the inability to breathe. If rotenone dust is inhaled, severe irritation of the inside of the nose, throat, and lungs may occur.

6. *Fumigation materials.* Most fumigation materials are highly toxic and are extremely dangerous when inhaled. Because of the nature of their use, they are extremely hazardous in enclosed areas.
- *Methyl bromide (Bromo Gas®):* The mode of action of this compound is to affect the protein molecules in certain cells of the body. The signs and symptoms produced by this compound include severe chemical burns of the skin and other exposed tissue, chemical pneumonia (which produces water in the lungs), severe kidney damage, and extreme nervousness. Any of these effects can be fatal. A person who inhales smaller amounts of methyl bromide may exhibit mental confusion, double vision, tremors, lack of coordination, and slurred speech. This sometimes produces effects that give the appearance of alcoholic intoxication, and victims have been jailed or sent to mental hospitals by mistake.
- *Chloropicrin (Picfume®):* This is also a highly hazardous chemical, but unlike methyl bromide, it gives off a gas with an odor and it is very irritating to the eyes. It is sometimes mixed with methyl bromide as a warning agent.

- *Carbon tetrachloride:* This chemical affects the nerves and also severely damages the cells in the kidneys and liver.

7. *Synthetic pyrethrins (pyrethroids):* Pyrethroid insecticides are fairly new. They are similar to the naturally occurring pyrethrins. Examples of early pyrethroids are fenvalerate (Pydrin®, Ectrin®) and permethrin (Ambush®, Pounce®, Ectiban®, Atroban®). These compounds are active at very low rates of application and, like natural pyrethrins, they are also low in toxicity to humans. The latest pyrethroids include cypermethrin (Ammo®, Cymbush®, Demon®, and Ripcord®), flucythrinate (Pay-off®), and fluvalinate (Mavrik®). These are stable compounds that are effective against a wide range of insect pests.

FIRST AID FOR PESTICIDE POISONING

IMPORTANT! *The information in this section may help to save a life—yours or your friend's. Even if this book is close by when an accident occurs, there may not be time to find and read this section before you are obliged to take action. This information should be in your head—not just in the book.*

It is essential that pesticide poisoning incidents be recognized immediately because prompt treatment may mean the difference between life and death. Do not substitute first aid for professional treatment. First aid is only to relieve the patient before medical help can be reached.

If you are alone with the victim:

1. See that the victim is breathing; if not, give artificial respiration.
2. Decontaminate immediately; that is, wash the victim off thoroughly.
3. Call a physician.

If another person is with you and the victim:

1. Speed is essential; one person should begin first aid treatment while the other calls a physician.
2. The physician will give you instructions. She or he will very likely tell you to get the victim to the emergency room of a hospital. The equipment needed for proper treatment is there. Only if this is impossible should the physician be called to the site of the accident.

General

1. Give mouth-to-mouth artificial respiration if breathing has stopped or is labored.
2. Stop exposure to the poison. If poison is on the skin, cleanse the person, including hair and fingernails. Do not induce vomiting unless instructed to do so by a physician.
3. Save the pesticide container and any remaining material; get readable label or name of chemical(s) for the physician. If the poison is not known, save a sample of the vomitus.

Specific

Poison on skin

1. Drench skin and clothing with water (shower, hose, faucet).
2. Remove clothing.
3. Cleanse skin and hair thoroughly with soap and water; speed in washing is most important in reducing the extent of injury.
4. Dry the victim and wrap him or her in a blanket.

Poison in eye

1. Hold eyelids open; wash eyes with a gentle stream of clean running water immediately. Use large amounts. Delay of a few seconds greatly increases the extent of the injury.
2. Continue washing for 15 minutes or more.
3. Do *not* use chemicals or drugs in washwater. They may increase the extent of injury.

Inhaled poisons (dusts, vapors, gases)

1. If the victim is in an enclosed space, do *not* go in without an air-supplied respirator.
2. Carry the patient (do not allow to walk) to fresh air immediately.
3. Open all doors and windows, if any.
4. Loosen all tight clothing.
5. Apply artificial respiration if breathing has stopped or is irregular.
6. Call a physician.
7. Prevent chilling (wrap the patient in blankets but don't overheat).
8. Keep the patient as quiet as possible.
9. If the patient is convulsing, watch breathing and protect him or her from falling and striking the head on the floor or wall. Keep the patient's chin up so the air passage will remain free for breathing.
10. Do *not* give alcohol in any form.

Swallowed poisons

1. Call a physician immediately.
2. Do not induce vomiting if:
   - Patient is in a coma or unconscious.
   - Patient is in convulsions.
   - Patient has swallowed petroleum products (that is, kerosene, gasoline, lighter fluid).
   - Patient has swallowed a corrosive poison (strong acid or alkaline products); symptoms are severe pain and a burning sensation in the mouth and throat. A corrosive substance is any material that in contact with living tissue will cause destruction of tissue by chemical action, such as lye, acids, or Lysol®.

3. If the patient can swallow after ingesting a corrosive poison, give the following substances by mouth.
   - *For acids:* milk, water, or milk of magnesia (1 tablespoon to 1 cup of water).
   - *For alkali:* milk or water; for patients 1 to 5 years old, 1 to 2 cups; for patients 5 years and older, up to 1 quart.

Chemical burns of skin

1. Wash with large quantities of running water.
2. Remove contaminated clothing.
3. Immediately cover with a loosely applied clean cloth; any kind will do, depending on the size of the area burned.
4. Avoid use of ointments, greases, powders, and other drugs in first aid treatment of burns.
5. Treat shock by keeping the patient flat, warm, and reassured until the physician arrives.

HOW TO INDUCE VOMITING WHEN A NONCORROSIVE SUBSTANCE HAS BEEN SWALLOWED

- Induce vomiting by placing the blunt end of a spoon (not the handle) or your finger at the back of the patient's throat.
- When retching and vomiting begin, place the patient facedown with the head lowered, thus preventing vomitus from entering the lungs and causing further damage. Do not let the victim lie on his or her back.
- Do not waste excessive time in inducing vomiting if the hospital is a long distance away. It is better to spend the time getting the patient to the hospital where drugs can be administered to induce vomiting and/or stomach pumps are available.
- Clean vomitus from the person. Collect some in case the physician needs it for chemical tests.

ARTIFICIAL RESPIRATION

In many conditions where breathing has ceased or apparently ceased, the heart action continues for a limited time. If fresh air is brought into the lungs so that the blood can obtain the needed oxygen from the air, life can be sustained. This can be accomplished in many instances by artificial respiration.

Certain general principles must always be kept in mind in applying any method of artificial respiration. Time is of prime importance; seconds count. Do not take time to remove the victim to a more satisfactory place unless the place is unsafe for the victim or rescuer; begin at once. Do not delay resuscitation to loosen the victim's clothes. These are secondary to the main purpose of getting air into the victim's lungs.

If possible, place the victim so the head is slightly lower than the feet to permit better drainage of fluid from respiratory passages. Remove from the victim's mouth all foreign bodies, such as false teeth, tobacco, or gum, and see that the head is tipped back to maintain an open airway; loosen any tight clothing about the victim's neck, chest, or waist.

Keep the victim warm by covering with blankets, clothing, or other material; if possible, the victim's underside should also be covered. Continue artificial respiration rhythmically and uninterrupted until spontaneous breathing starts or a doctor pronounces the patient dead.

If the victim begins to breathe of his or her own accord, adjust your timing. Do not fight the attempt to breathe. A brief return of natural respiration is not a signal for stopping the resuscitation treatment. Not infrequently, a patient, after a temporary recovery of respiration, stops breathing again. The patient must be watched; if natural breathing stops, resume artificial respiration at once.

Always treat the victim for shock during resuscitation, and continue such treatment after breathing has started. Do not give any liquids whatever by mouth until a patient is fully conscious. If it is necessary (due to extreme weather or other conditions) to move a patient before he or she is breathing normally, continue artificial respiration while the patient is being moved.

Mouth-to-Mouth or Mouth-to-Nose Method of Resuscitation (Adult)

If foreign matter is visible in the mouth or if the victim vomits, roll him on his side and prop him up with your knee behind his shoulders with his head resting on an extended arm. Turn the head slightly downward and wipe the mouth out quickly with your fingers or a cloth wrapped around them. Refer to Figure 8–17:

1. Tilt the head back so the chin is pointing upward (A). If this does not open the airway, pull the jaw into a jutting-out position (B and C). These maneuvers

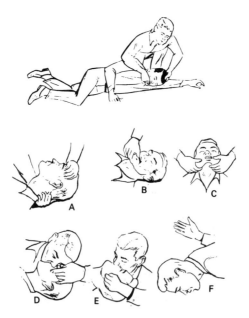

Figure 8–17 Mouth-to-mouth air exchange.

should relieve obstruction of the airway by moving the base of the tongue away from the throat.

2. Open your mouth wide and place it tightly over the victim's mouth. At the same time pinch the victim's nostrils shut (D) or close the nostrils with your cheek (E). Or close the victim's mouth and place your mouth over the nose. Blow into the victim's mouth or nose. (Air may be blown through the victim's teeth, even though they may be clenched.) The first blowing efforts should determine whether or not obstruction exists.

3. Remove your mouth, turn your head to the side, and listen for the return of air. Watch rise and fall of chest. Repeat the blowing effort. For an adult, blow vigorously at the rate of about 12 breaths per minute (for infants, 20 breaths per minute).

4. If you are not getting air exchange, recheck the head and jaw position (A or B and C). If you still do not get air exchange, quickly turn the victim on his side and administer one or two sharp blows between the shoulder blades in the hope of dislodging foreign matter (F).

Again sweep your fingers through the victim's mouth to remove foreign matter. Those who do not wish to come in contact with a person may place a cloth over the victim's mouth or nose and breathe through it. The cloth does not greatly affect the exchange of air. If the aider has false teeth, they should be removed for obvious reasons.

Manual Means of Artificial Respiration

If the nature of the injury is such that mouth-to-mouth resuscitation cannot be used, the rescuer should apply the manual method. It has already been pointed out that the base of the tongue tends to press against and block the air passage when a person is unconscious and not breathing. This action of the tongue can occur whether the victim is in a facedown or faceup position.

The following is an accepted manual method of artificial respiration (Figure 8–18). Kneel on one or both knees at the victim's head. Place one hand on each side

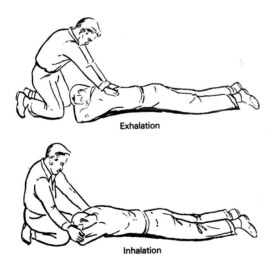

Exhalation

Inhalation

Figure 8–18 Manual artificial respiration procedure.

Artificial Respiration

of the spine (midback) approximately 2 inches from the spine (according to size of the victim), just below the shoulder blades. Do not have thumbs on the spine. Rock forward and exert slow, steady, moderate pressure until firm resistance is met. Gradually release pressure, rock backward; place your hands beneath the victim's arms close to elbows, slowly drawing the elbows toward you and upward as you rock backward until firm resistance is met. Gently lower arms to ground. Repeat this process at the rate of approximately 12 complete cycles per minute. Back pressure decreases the chest cavity, producing active exhalation. The arm lift and stretch increase the chest cavity and induce active inhalation.

DO'S AND DON'TS! ARTIFICIAL RESPIRATION

DO:

1. Start artificial respiration as quickly as possible.
2. Use the method best suited in each particular case.
3. Maintain the airway in all cases (clean out air passages and tip head back).
4. Give respiration: apply 12 to 15 times per minute for adults, 16 to 20 times for children, and 20 to 24 times for infants under 1 year of age.
5. Treat for shock.
6. Loosen tight clothing.
7. Administer oxygen if available.
8. After the victim starts breathing, stand by in case you are needed.

DO NOT:

1. Move the victim unless necessary to remove from danger.
2. Wait or look for help.
3. *Give up.*

Administration of Oxygen

It is advisable to supplement artificial respiration by administering oxygen, especially when the quantity of fresh air supplied to the patient by artificial respiration is inadequate, as is likely if the patient has been breathing poisonous gases. However, no time should be lost waiting for oxygen before artificial respiration is begun. When oxygen is administered, the manual treatment should not be stopped but should continue as long as there is hope of reviving the patient or until she or he begins to breathe normally. Oxygen is frequently available in compressed form in an oxygen bottle or cylinder. The valve should be opened slightly, with the flow directed away from the patient. After a moderate flow has been established, the oxygen should be directed into a cap, hat, or piece of cloth used as an improvised mask to confine the oxygen to the face of the victim; an oxygen inhalator aids in its administration. To eliminate explosive hazards, regulator control valves should be retained on oxygen bottles.

First Aid for Gas Poisoning

The steps in first aid treatment of poisoning by toxic or noxious gases are:

1. Always remove the person to fresh air as quickly as possible.
2. Obtain medical aid.
3. If breathing has stopped, start artificial respiration at once and continue until normal breathing is resumed or a doctor pronounces the victim dead. When giving artificial respiration, always administer oxygen if available.
4. Keep the patient at rest, lying down, to avoid any strain on the heart. Later give the patient plenty of time to rest and recuperate.

Inhalations of oxygen, when administered immediately, will greatly reduce the severity of carbon monoxide poisoning, as well as decrease the possibility of serious aftereffects. Give oxygen whether the patient is conscious or unconscious.

All industries in which gas poisoning may be a danger should provide apparatus for efficient administration of oxygen. Such apparatus should be placed at convenient points and employees should be trained in its use.

ANTIDOTES: THEIR USE BY NONMEDICAL PEOPLE

An antidote is a remedy used to counteract the effects of a poison or prevent or relieve poisoning. Therefore, good judgment and safety practices, including the use of protective clothing, safety devices, and knowledge of first aid, are in a sense antidotes since they can and frequently do prevent or relieve poisoning. However, the general public is inclined to think of antidotes almost exclusively in terms of special chemicals that must be purchased from a druggist or prescribed by a physician. The following brief discussion of some of the more common antidotes will group these materials as to internal or external use and general mode of action.

Antidotes for External Use

Clean water dilutes and washes away poisons. It is recommended for poison in the eyes or on skin and other tissues. Always have at least several gallons of clean water readily available for emergency use when you are handling pesticides.

Soaps, detergents, or commercial cleansers and water dilute and wash away poisons. They are recommended for poison on skin, hair, under fingernails, and on other external tissues not irritated by soap. Soap and water should always be readily available for emergency use when handling pesticides.

Note: In an emergency, use any source of reasonably clean water, such as irrigation canals, lakes, ponds, or water troughs. Don't let the victim die while you worry how dirty the water is.

Antidotes for Internal Use

Check first aid instructions before giving anything to a person by mouth!

- Clean water dilutes poison.
- Milk dilutes poison and helps counteract acid or alkali poisons.
- Syrup of ipecac is used to promote vomiting. Use only as directed by a physician or poison control center. It is poisonous if improperly used.

NOTIFY YOUR PHYSICIAN

Most medical doctors are not well informed as to the symptoms and treatment of pesticide poisoning. This is due to the few cases that they treat. Pesticide poisoning symptoms are similar to those of other illnesses and poisonings. You, the pesticide applicator, should tell your doctor which chemicals you use. Then she or he will know the symptoms and treatment and have the antidotes on hand (Figure 8–19).

Warning: Neither atropine nor 2-PAM should be used to prevent poisoning. Workers should not carry either antidote for first aid purposes. They should be given only under a doctor's directions.

POISON CONTROL CENTER

All states have some form of consultant service for diagnosis and treatment of human illness resulting from toxic substances. Some states have regional offices or at least several locations that can be contacted for information. Make sure you and your physician know the telephone number of the Poison Control Center nearest you.

MEDICAL SUPERVISION FOR PESTICIDE APPLICATORS

Persons handling or applying cholinesterase-inhibiting pesticides such as organophosphates or carbamates throughout the application season may wish to obtain regular cholinesterase blood tests under medical supervision. It is valuable to have preseason cholinesterase tests performed at a time when there has been no contact with these insecticides for an extended period. This way a baseline value may be obtained for each person at a time when he or she should be free from evidence of exposure. Cholinesterase tests should be obtained at weekly intervals during spray season to see

Figure 8–19 Keep your doctor informed.

if the cholinesterase levels are normal. If cholinesterase values for any person drop below 50% of the baseline value, that individual should be removed from contact with organophosphate and carbamate insecticides until his or her work habits can be reviewed carefully and cholinesterase levels have returned to normal. Many commercial companies follow these precautions with employees who are frequently exposed to pesticides.

AERIAL APPLICATION SAFETY

Aerial application of pesticides has developed into an extensive skilled profession. Various hazards are associated with this type of work, but these can be minimized by observing safety precautions. Although the previous sections provide general pesticide safety information common to all types of pesticide use, the trend to more pesticides being aerially applied each year creates a need for additional safety information for pesticide applicators engaged in agricultural aviation.

Equipment

The aircraft is the most important item to consider. The type of aircraft selected should depend on the work it will be used for. Large areas with long runs are generally more suited to fixed-wing aircraft application. Helicopters are generally safer and more maneuverable for treating small areas, particularly those that have obstacles.

Aircraft should be designed and built specifically for aerial application to protect the pilot and increase operating efficiency. Aerial application aircraft should be equipped with numerous safety features not generally found in conventional aircraft. These features should include (1) ease of control during slow flight and partial aileron control during stalls, (2) special ventilation that reduces pilot exposure to pesticides, (3) provisions for pilot comfort to keep fatigue at a minimum, and (4) good visibility in all directions.

Fixed-wing aircraft that have boom and nozzle equipment should not have booms that are longer than three-quarters of the wing span. Longer booms that extend near the wing tips result in spray being picked up by the wing tip vortices, causing uneven spray patterns and in some cases increasing drift potential. Aircraft should be equipped with a positive cutoff valve to prevent leaking when the spray is shut off at the end of swath runs.

Ground Crew

Members of ground crews, including mixers, loaders, and flag persons, must have access to protective clothing and equipment. In addition, they must have immediate access to an adequate supply of soap and water. (Refer to Chapter 6 for further information.) Ground crew personnel should bathe and change clothes each day or immediately if a concentrate or toxic pesticide is accidentally spilled on their clothing or skin.

It is important that ground crews be able to recognize symptoms of pesticide poisoning. Many times poison symptoms are more obvious to others than to the victim. Any abnormal reaction may indicate the onset of pesticide poisoning.

When water is drawn from streams or ponds for pesticide application purposes, ground crews should be sure that no pesticide flows back into the water source. Accidentally spilled pesticides should be immediately cleaned up by ground crews to protect persons and animals from coming into contact with them (see Chapter 14).

Pilot Safety

The conduct of a safe aerial application program is primarily the responsibility of the pilot. The pilot must be aware of any particular condition that may affect her or him personally, as well as drift and other hazards that could be detrimental to humans, livestock, crops, wildlife, and any other environmental situation.

With proper precautions, the pilot should experience less actual exposure to pesticides than the ground crew. The following 10 points should be considered for pilot safety:

1. The pilot should not assist the ground crew when toxic pesticides are being mixed or loaded. The pilot should remain on the windward side of the loading operation to avoid inhalation or exposure to toxic materials. The pilot can remain in the aircraft if closed system loading is being used.

2. The pilot should anticipate greater than normal hazards when applying a very toxic or highly concentrated pesticide.

3. The pilot should anticipate increased potential for hazards when fatigue results from flying for long periods.

4. The pilot should never turn or fly through the path of the previous swath.

5. The pilot should be constantly alert to detect pesticide leaks or accidental spillage inside the aircraft. The airplane cockpit should be tight to prevent pesticide spray or dust particles from entering during application. Cockpits should be checked and cleaned frequently.

6. A safety helmet and shoulder harness should always be worn during flight.

7. When applying toxic pesticides, the proper type of protective equipment should be used.

8. It is of utmost importance that the pilot be able to recognize symptoms of pesticide poisoning such as dizziness, blurred vision, watering of the eyes, and nausea. Flying should cease as soon as possible if any of these symptoms are evident.

9. Clothing worn during application should be laundered each day.

10. Two-way radio equipment to enable the pilot to keep in touch with the ground crew is a valuable asset.

Drift Control

Complete elimination of drift during aerial application is nearly impossible, but there are several factors that help to keep it at a minimum.

- Wind velocity is of utmost importance. When there are excessive winds, pesticides cannot be applied without drift. When wind velocity creates a drift hazard, the application operation should cease immediately.

- Granular materials are less likely to drift than sprays or dusts.

- Fine spray droplets will drift easier and farther than coarse spray droplets. Pesticides applied in oil will tend to drift farther than those applied in water.
- The extent of drift can be reduced by keeping the altitude of flight as low as possible. Precautions should be observed to maintain effective pesticide dispersal and to consider the safety of the operation.
- Dispersal equipment should be calibrated precisely to produce the desired rate of application. Considerations should be given to height of flight and swath coverage. Sometimes, under some conditions, temperature variations between air and ground will cause inversions that will prevent the pesticide application from settling to the ground. If this occurs, pesticide applications should be postponed until conditions are more favorable.
- When aerial pesticide applications are made over rough terrain, the downhill movement of surface air can carry a spray a considerable distance outside of the target area. The pilot should make allowances for this air movement. Aerial applications should not be attempted on very small fields where the hazard of drift to adjoining fields is unavoidable.
- Treatment of fields near canals, streams, lakes, or ponds should be made with extreme caution. The swaths should be parallel to the water so that turns will not be made across it.
- Pilots should not prime the application equipment or test the flow rate while ferrying between the airstrip and the area to be treated.
- It is the pilot's responsibility to see that the aircraft and its equipment are cleaned daily or after each use to prevent pesticide spray from building up.

Safety Management and Supervision

The overall responsibility of a safe pesticide application operation belongs to the managers or owners of the business. They should have all safety devices needed in the aircraft to protect the pilot. Aircraft dispersal equipment should be in good working order, delivering the desired rate and swath pattern. Loading equipment should be up-to-date to provide efficiency and safety for the ground crew operation. Pilots and workers exposed to organophosphate pesticides throughout the spray season should be placed under a medical program and required to take blood cholinesterase tests frequently.

Immediate supervisors should be sure all workers understand the nature of the pesticide being used and know what to do if poisoning occurs. The names of physicians and the hospital to contact should be immediately obtainable if an emergency arises. All workers, including the pilot, must be required to wear protective equipment when needed. At least two persons should work together when toxic pesticides are being mixed and loaded. It is the supervisor's responsibility to clear the area being treated and, if the material is toxic to bees, to notify beekeepers.

CRITICAL TIME PERIODS

There are three periods during a chemical application season that present the greatest likelihood for overexposure of chemical handlers and users:

1. *Early spring when new and inexperienced crews begin handling chemicals for the first time.* Overexposures are more likely to occur among new crew members until they gain some training and experience.

2. *The first prolonged hot spell of the summer.* Crew members may become tired of using uncomfortable protective clothing and respirators during hot weather and thus may run greater risk of overexposure. In addition, workers are more likely to become dehydrated during hot weather and then may be more susceptible to the effects of some chemicals.

3. *Late in the season, just before or during harvest time, when the workload has been prolonged and heavy for both workers and equipment.* At this time, individuals who have experienced repeated small exposures to pesticides sufficient to cause cholinesterase depression, but not severe enough to cause symptoms, may acquire small additional exposures whose effects are cumulative with the previous ones and result in overt symptoms. Equipment and protective devices that were functioning well earlier in the season may become faulty by the season's end and further contribute to this danger period.

GREENHOUSE OPERATIONS

Applying pesticides in greenhouses presents special problems. In normal greenhouse operations, employees must work inside. Space is generally limited, and personal contact with plants and other treated surfaces is almost a certainty. In addition, unauthorized persons may attempt to enter the premises. Ventilation in greenhouses is frequently kept to a minimum to maintain desired temperatures and, as a result, fumes, vapors, mists, and dusts may remain in the air for considerable periods of time, creating hazards.

Pesticides applied to plants and other surfaces in greenhouses do not generally break down as rapidly as they do outside. This is due to a reduced amount of moving air and lack of rain to wash the chemicals off, dilute them, or combine with them chemically to break them down further. In addition, the glass in the greenhouse filters out the ultraviolet light that would normally contribute to the degradation of certain chemicals.

Checklist for Greenhouse Applications

- Select chemicals that are most effective for pest control and present the least hazard to humans and animals.
- When using toxic pesticides, especially fumigants, applicators should use gas masks and waterproof protective clothing.
- Post warning signs on the outside of all entrances to the house when fumigants or other highly toxic pesticides are being applied (Figure 8–20). Follow label instructions. There are specific requirements under the Worker Protection Standards for greenhouse workers. Be sure you are aware of these.
- Do not enter the building without a gas mask or permit others to do so until it has been aired for the length of time recommended on the pesticide label.
- All possible skin contact with treated plants and other surfaces should be avoided by workers and others to minimize skin irritation, sensitization, and absorp-

Figure 8–20 Post warning signs in conspicuous places when using pesticides inside greenhouses or other enclosed areas.

tion of chemicals through the skin. Where this is impossible, workers should wear clean, dry protective clothing and wash frequently.

- It is suggested that handles and special wrenches for steam lines be kept in custody of authorized persons so that unauthorized persons will not inadvertently use the wrong line. These lines may be labeled by name with water-insoluble ink or color-coded blue for water, green for fertilizer, red for pesticide, and black for steam.

SAFETY CHECKLIST

Think and stay alert when working around pesticides. Here are 20 pesticide safety tips to think about.

1. Read the label often and follow the precautions shown.
2. Never work alone; always wear the specified protective clothing.
3. Don't let children or unauthorized people remain where pesticides are being mixed, loaded, or applied.
4. Don't drink intoxicating beverages before working with pesticides.
5. Make sure your equipment is clean, calibrated, and working properly.
6. Mix pesticides outdoors or, if you must work indoors, make sure that the area you use is well lit and ventilated.
7. Try not to destroy the label when opening bags or cans containing pesticides. Replace caps and close containers tightly after you finish.
8. Avoid eating, drinking, smoking, and touching your face while mixing or handling materials.
9. Measure materials correctly for the recommended rate. Don't mix or pour chemicals at eye level or on the edge of a table or pickup bed.
10. Pour liquid, powder, or dust slowly to avoid any splash, spill, or drift. When transferring concentrates from drums, use drum pumps.
11. If pesticides are splashed or spilled, immediately remove contaminated clothing; thoroughly wash skin with soap and water; put on fresh, clean protective clothing; clean up the spilled material.

12. Always apply pesticides under appropriate weather conditions.

13. Carry at least 5 gallons of clean water on tractors and other application equipment for use in washing eyes and skin in case of emergencies.

14. After handling pesticides, always wash thoroughly *before,* you eat, drink, smoke, or use the restroom.

15. Never leave pesticides unattended in a truck, field, or operation site.

16. Store only in the original container and keep tightly closed.

17. Avoid contamination of fish ponds and water supplies. Cover feed and water containers when treating around livestock or pet areas.

18. Keep separate equipment for use with hormone-type herbicides to avoid accidental injury to susceptible plants. Also avoid applications under wind conditions that could create drift to nontarget areas.

19. Rinse empty containers three times before disposing of them. Add the rinse to the spray tank and dispose of containers according to local regulations to avoid hazard to humans, animals, and the environment.

20. Plan ahead. Discuss with your physician the materials you will be using during the season so that she or he can be prepared to provide the appropriate treatment in case of accidental exposure. If symptoms of illness occur, call the physician or get the patient to a hospital immediately. Always provide medical personnel with as much information as possible.

9

Pesticides and Environmental Considerations

GENERAL CONSIDERATIONS

The quality of the environment has become a major issue, and with it an urgency for protecting the nation's air, water, land, and wildlife. Many who use pesticides regard them as a means of preserving the environment, but those who are most often quoted by the popular press place such materials near the top of the list of pollutants. The point at which a pesticide is considered a beneficial tool or a pollutant is at times hard to distinguish. Perhaps, as a weed is a "plant out of place," a pesticide pollutant is simply a "resource out of place."

Not only will problems of pollution grow with our expanding population, but sensitivity to and awareness of these problems will also increase. It appears inevitable that as our agriculture becomes more intensive, pollution will become a greater hazard to water and air and thus affect the sensitivity of more people. It is equally apparent that agriculture will be bound by rules and regulations to safeguard the environment, regulations based not so much on the need for a pesticide as on that for preventing pollution.

Pesticides become problems when they are dispersed beyond the target area. The characteristics of nonbiodegradability and accumulation of some pesticides have emphasized the problem. Conditions under which pesticides can become pollutants are leaching and contamination of groundwater from treated soil; when they are carried in surface runoff from agricultural land and urban areas; misuse or accident; pesticide drift; and careless disposal of surplus materials and empty containers. Concerns with pesticides are worldwide and we must do everything possible to keep pollution to a minimum and protect the environment (Figure 9–1).

Figure 9–1 Protect the environment!

Pesticides are a help to the environment when they are used carefully and wisely. For years they have been used to control pests that are harmful to humans. Rats carrying plague or mosquitos carrying malaria are good examples. These control programs are still necessary today, especially in crowded cities and countries with large numbers of people. With the help of pesticides, more food per acre can be produced. Diseases, insects, and other plant pests can be greatly reduced. There can be higher yields and better crop quality. Less land is needed to be tilled to feed people, thus allowing more land for wildlife and recreational purposes.

Pesticides can be a big help in our environment when used to protect outdoor activities in our parks and camping areas. Fly and mosquito control programs give relief from these annoying insects. Pesticides also protect livestock from harmful and annoying pests. The quantity and quality of livestock products (milk, meat, and the like) are improved when pests are controlled. Pesticides also aid in controlling insects or diseases that get into an area for the first time. Often these pests have few natural enemies in a new environment. Without a good pesticide program, they can rapidly overrun an area. Gypsy moths and Japanese beetles are good examples of this type of pest.

There are six major areas in our environment, besides humans, that require protection. These areas are soil, water, air, beneficial insects, plants, and wildlife. Each of these areas will be explored in more detail in the following paragraphs.

Soil

Soil has become more important as the need for food increases. To feed increasing numbers of people, fertile and healthy soil is an absolute requirement. Poor soil practices and misuse can cause poor yield and second-class crops, especially if root vegetables or forage crops are being planted.

Pesticides often become attached to fine soil particles, and when the surface soil erodes, the chemical is carried along (Figure 9–2). With some pesticides this attachment is so strong that the pesticide does not move down through the soil layers for any important distance. Overdoses of pesticides that remain for long periods in the soil may limit planting to only a few crops that will not be harmed by the chemicals. Most pesticides that become attached to soil probably stay near the point of application of the material. Here bacteria and other microscopic organisms in the soil are important

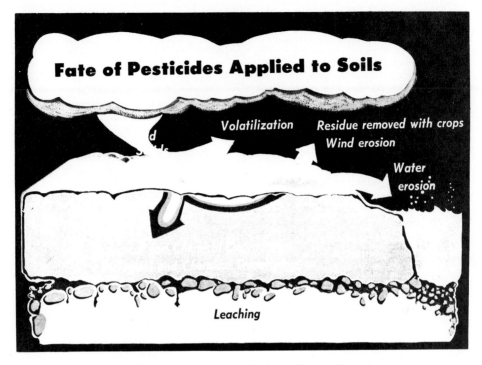

Figure 9–2 Fate of pesticides applied to soils.

in the chemical process of breaking down the pesticide into other chemicals and eventually into the simplest chemical building blocks.

Movement of pesticides in soil. Table 9–1 lists the relative mobility (movement) of certain pesticides in soils. These pesticides are listed by common names with some commonly used trade names in parentheses.

Pesticides with very low or low movement in soil tend to remain close to where they were originally placed. They may be a threat to sensitive crops planted in the area, but are not generally a threat to groundwater. Pesticides with moderate to high movement in soil are of greatest concern and must be kept out of groundwater. The EPA monitors groundwater sources for contamination by pesticides and industrial chemicals. Strict regulatory action is called for in cases of surface and groundwater pollution.

Water

Water is one of our greatest resources. Its unusual properties and abundance make it necessary for all life. Polluted water can fill many of our needs but not the most basic ones. Humans and wildlife need *clean* water for drinking and eating. Most fish and other marine life can survive only slight changes in their water environment. Farmers must use uncontaminated water for their livestock and irrigation processes to prevent plant or animal poisoning or illegal residues. Clean water is essential.

TABLE 9–1 MOVEMENT OF PESTICIDES IN SOILS

Very low or no movement	Low movement
Ametryn (Evik®, others)	Atrazine (AAtrex®, others)
Benefin (Balan®, others)	Azinphosmethyl (Guthion®, others)
DCPA (Dacthal®)	Butylate (Sutan®)
Diquat (several trade names)	Chlorpropham (Furloe®, Bud Nip®, others)
DSMA (Ansar®, others)	Dichlobenil (Casoron®, others)
Endrin (several)	Disulfoton (Di-Syston®, others)
Glyphosate (Roundup®)	Diuron (several)
Isopropalin (Paarlan®)	Linuron (several)
Lindane (several)	Prometon (Pramitol®)
MSMA (many)	Siduron (Tupersan®)
Oryzalin (Surflan®)	Vernolate (Vernam®, Surpass®)
Paraquat (Gramoxone®, others)	
Permethrin (Ambush®, Pounce®, others)	
Prometryn (Caparol®, others)	
Propazine (Milogard®, others)	
Simazine (many)	
Trifluralin (Treflan®, others)	

Moderate movement	High movement
Alachlor (Lassor®)	Aldicarb (Temik®)
Bromacil (Hyvar®, others)	Asulam (Asulox®)
Carbofuran (Furadan®)	Dalapon (Dowpon®, others)
Chloramben (Vegiben®, others)	Dicamba (Banvel®)
Chlorfenvinphos (Supona®, others)	Hexazinone (Velpar®)
Chlorsulfuron (Glean®, others)	Methomyl (Lannate®, Nudrin®)
EPTC (Eptam®)	Metribuzin (Lexone®, Sencor®)
Ethoprop (Mocap®)	Picloram (Tordon®)
Fenamiphos (Nemacur®)	TCA (several)
MCPA (Weedone®, others)	Tebuthiuron (Spike®, Graslan®)
MCPB (Can-Trol®, others)	
Metolachlor (Bicep®, Dual®, others)	
Monocrotophos (Azodrin®, others)	
Terbacil (Sinbar®)	
2, 4-D (many names)	

Pesticides get into water in several ways. They may be applied directly for control of aquatic plants or animals, they may reach the water by accident when the nearby land has been sprayed, or they may enter as spray drift from nearby applications. Pesticides attached to soil may wash into streams, and a certain amount of some chemicals may enter the water by being washed out of the air by rain. Pesticides that are dissolved in the water can be picked up directly by fish and other aquatic animals as the water passes over their gills. Therefore, it is essential that we be very careful in the application of pesticides where they may eventually find their way into water sources or systems.

Groundwater makes up 96% of the world's total water resources. Ninety percent of rural residents and 50% of the people in the United States depend on groundwater as their source of drinking water. By 1995 more than 30 pesticides were found in the groundwater in 34 states. Although only very small amounts of pesticides were detected in most situations, we must reduce the likelihood of pesticides entering groundwater and surface water so future water quality problems are not encountered.

Groundwater is primarily stored in aquifers, geological formations of permeable saturated zones of rock, sand, or gravel that contain enough water to yield usable amounts to wells and springs. In general, groundwater moves very slowly, sometimes only inches per year. Aquifers are recharged (replenished) by rainfall seeping through the ground or by surface waters with which they are interconnected. In many parts of the country, groundwater recharge areas are close to the surface and may be significantly affected by agricultural, residential, or industrial activities. The depth of groundwater below the earth's surface, the depth and type of soils above the aquifer, and many other factors affect the potential for contamination.

Once contaminated, groundwater is difficult or impossible to clean. Because groundwater moves slowly, contaminants do not spread or mix quickly. Contaminants remain concentrated in slow-moving masses of water (blooms) and are typically present for many years. If groundwater becomes contaminated, the contamination may eventually appear in surface water.

Once thought to be safe from contamination, groundwater is now a threatened resource. As more incidents of groundwater pollution are discovered across the country, the public is growing increasingly aware of the potential problem of groundwater contamination. As public concern has increased, so have demands for expanded protection of this vital resource. In October 1991, the Environmental Protection Agency published its final version of *Agricultural Chemicals and Groundwater: Pesticide Strategy.* The EPA's intent is to develop regulatory guidelines that will help it regulate groundwater and control the use of pesticides in certain areas. Its deliberations are not complete at the time of this writing, but it is expected that all pesticides will be examined for their *leachability,* and those that pose a problem to groundwater will be regulated through pesticide label requirements. It is possible that labels or accompanying literature will list counties or areas along with maps showing areas where certain pesticides cannot be used because of groundwater contamination dangers.

Many labels now state whether the pesticide can be dispensed through irrigation systems. Chemigation, by which pesticides are primarily dispensed through overhead sprinklers, is a common practice in various areas of the United States. The injection of pesticides into irrigation water can be a dangerous practice, especially when there is any possibility of back-siphoning of water-pesticide mixtures back down the well to the water source. Many states have passed legislation requiring safety equipment on all wells being used for chemigation.

Following are eight ways in which pesticide users can help prevent groundwater and surface water contamination:

1. *Follow label directions exactly.* The pesticide label will give valuable information concerning what that pesticide's potential dangers are for contaminating water and the environment. When applying a pesticide, the timing and placement instructions on the label must be followed correctly to ensure that the pesticide is applied properly. Following such steps is critical for the pesticide to work properly, as well as to protect the water, soil, air, wildlife, and other

environmental life. For example, applying a pesticide when heavy rains are predicted could lead to water contamination. Likewise, placing a pesticide on top of the soil when it should be incorporated not only cuts down on pest control but could lead to unnecessary runoff into surface water and perhaps groundwater.

2. *Prevent spills and back-siphoning.* When pesticides spill near wells or sinkholes, they can move into surface water and groundwater. Avoid mixing and loading pesticides near wells and other water sources. Use a long hose from the well to the sprayer so that if any spills occur they will be farther away from the clean water supply. If a spill takes place, be sure to clean it up and move the contaminated soil to a place where it will not seep into the water or otherwise harm the environment. When mixing pesticides, keep the end of the fill hose above the water level in the spray tank. Failure to do this can lead to a backflow situation where pesticides may be siphoned back into the water supply. Backflow siphoning can also occur if the pump quits. Use anti-back-siphoning valves on all chemigation equipment. Many states require this by law.

3. *Dispose of waste properly.* It is not only illegal to dispose of pesticides improperly, it is also very dangerous. If pesticides, their containers, or other hazardous materials are discarded where they can contaminate the water supply or environment, you and your family could drink pesticide-contaminated water. This contamination could move into your neighbor's or livestock's water supply, as well as affect wildlife and conditions of soil and air. Your responsibility for disposing of hazardous materials includes using proper disposal methods. The guidelines for disposing of such materials can be found on the pesticide label.

4. *Be aware of surface water control.* If there is more water in the soil than the soil can absorb, water (with pesticides in it) may flow into the groundwater or run off into streams, rivers, and lakes. Prolonged heavy rains and too much irrigation will also produce excess surface water. The use of weather forecasts, proper irrigation scheduling, and strip crops will cut down on potential surface water problems.

5. *Know the land characteristics.* The geology of the land plays a key role in controlling the groundwater and surface water. If the groundwater is within a few feet of the soil surface, pesticides are much more likely to reach groundwater. Irrigation or moderate rainfall may carry some pesticides directly into the groundwater. You must select pesticides carefully when either groundwater is close to the surface or soil permeability is great.

6. *Know the soil characteristics.* Soil texture (sand, silt, clay), soil permeability, and soil organic matter all play a major part in pesticide movement. Soils containing large amounts of organic matter and clay, for example, will hold (adsorb) some pesticides before they reach groundwater. Pesticides are more likely to move into groundwater through sandy soils and loose, porous soils.

7. *Use IPM practices.* Integrated pest management (IPM) practices, such as crop rotation and cover crops, not only reduce pest populations, but they also maintain and improve good soil and water conditions. Pest-resistant crop varieties and careful pest monitoring will also ensure that pesticides are used only when needed. Good IPM practices will reduce the expense of pest control from the use

of pesticides and will also help to preserve the use of pesticides for future situations when they may be vitally needed.

8. *Be familiar with pesticide characteristics.* Some pesticides move into the soil easier than others. Those with high water solubility are more likely to seep into the soil than pesticides with extremely low solubility. Table 9–1 gives the relative mobility (movement) of certain pesticides in soils.

Air

Air must be available for any plant or animal to live. It is a source of oxygen for breathing and receives the carbon dioxide waste. Air has the ability to move particles a long way before letting them go. Most of the time this ability aids pesticide application. Unfortunately, however, this same ability is the cause of drift. Pesticides carried by the air may be harmful to humans' and wildlife's health and safety. Pesticides in the air are not controllable and may settle into waterways, wooded areas, and heavily populated areas. Pesticide drift cannot be tolerated and must be avoided.

A major source of air pollution by pesticides is direct ground and aerial application. The amount of pesticides that enter the atmosphere is determined by wind, temperature inversions, pesticide formulations, particle size, drift of the compound, and type of compound applied.

The following factors affect drift:

Type of pesticide and formulation. Pesticides that are highly volatile drift great distances because of the release of vapors. Always attempt to use low- or nonvolatile pesticides. The formulation of the pesticide will determine its potential for drift. In general, dusts are most susceptible to drift, followed by sprays and granules, which are least likely to drift.

Spray additives. Carriers such as water or oil are different in potential for drift. Water-type sprays will drift more easily than oil. Spray additives may either increase or decrease drift.

Particle size. The smaller the particle size is, the greater the drift hazard. Particle size is related to the pesticide formulation and the type of application equipment.

Method of application. Aerial applications cause a greater drift problem than ground operations. This is caused by differences in equipment design and in volume of spray dispersed. Drift problems become intensified with low-volume and ultralow-volume spray techniques.

Rate of application. Larger spray operations have a greater potential for drift than small operations. In addition, higher application rates increase the amount of pesticide residue at a given distance downwind.

Climatic conditions. High or shifting winds can cause uneven distribution of the pesticide, and allow it to be carried away from the target area. The applicator should also consider both wind gusts and direction. Warming of the treatment area by sunlight causes convection currents to rise from the ground. This results in increased turbulence, causing drift. To minimize drift, spray during early morning hours, before sunrise, and after sunset.

High temperatures.   Excessive heat can cause increased vaporization of a pesticide. High temperatures can also affect the toxic activity of the pesticide.

Beneficial Insects*

Honey bees (Figure 9–3) and other pollinators are necessary for good farming and food production. In many cases when there is no pollination there is no crop. Honey bees produce honey and beeswax valued between $130 million and $140 million annually. Much of this honey comes from cultivated crops. The annual value of crops benefited by insect pollination, most of which is performed by honey bees, exceeds $10 billion. The farmer and the beekeeper therefore depend on each other.

Honey bees may be killed when crops are treated with pesticides. When this occurs, both the farmer and the beekeeper suffer a loss. For this reason, they need to cooperate fully in protecting the bees from pesticide damage.

Research to resolve the problem of bee losses due to pesticides has been under way since 1881 when damage to bees by lead arsenate was first reported. More than a century later, there still is no solution to this problem, although intensive study is continuing. Modern agriculture depends on bees for crop pollination and on chemicals for pest control. Unfortunately, in the United States pesticides annually destroy or damage about one million bee colonies or 20% to 25% of our honey bee colonies each year. This destruction has a significant economic impact on beekeepers and farmers alike. Yields from crops that require bees for pollination are frequently reduced due to the loss of bees. The absence of bees or inadequate use of pesticides reduces agricultural productivity, which results in higher food costs for consumers.

Whenever it is determined that a pesticide application is necessary, the following precautions should be taken:

- Use the proper dosage of the safest material (on bees) that will give good pest control.
- Tell the beekeeper what will be used and when it will be applied.
- Read the label and follow approved local, state, and federal recommendations.

Figure 9–3   Honey bee gathering nectar.

*The information presented in this section concerning honey bees has been adapted from USDA leaflet No. 563 *Pesticide and Honey Bees* and from University of Wisconsin leaflet No. A3086 *Protecting Honey Bees from Pesticides in Wisconsin.*

- Remember that the time the pesticide is applied, depending on the blooming period and attractiveness of the crop, makes a big difference in the damage to the bees; so treat the fields when the plants are least attractive to bees.
- Do not spray or dust chemicals over colonies, especially in hot weather when the bees cluster outside the hive.
- Apply chemicals at night or during early morning hours before bees forage.
- Do not spray or dust bee-visited plants in bloom, and do not let insecticides drift to plants in bloom.
- Remember that treating a nonblooming crop when weeds and wildflowers are in bloom in the field or close by can cause bee losses.
- Make as few treatments as possible, because repeated applications greatly increase the damage to colonies.
- Do not treat an entire field or area if local spot treatments will control harmful pests.
- Sprays do not drift as far as dust and, consequently, are less likely to harm bees.
- Granules are usually the safest and least likely to harm bees.

We must strive for a balance between the use of pesticides and the preservation of bees. Bee losses are frequently due to the inappropriate or careless application of pesticides, improper timing of application, and dumping of unused materials. Such losses can be minimized by keeping pesticide applicators and crop producers informed of the need to protect bees and the methods of doing so. Although it is unlikely that bee losses can be totally avoided when pesticides are used (many of the most effective insecticides are also very toxic to bees), they can be reduced. Remember, cooperation between pesticide applicators and beekeepers is essential. Cooperation will help to prevent many unnecessary bee losses as well as lawsuits and hard feelings.

In general, dusts are more toxic to bees than liquid sprays, and wettable powders are more toxic than emulsifiable concentrate sprays. Microencapsulated pesticides pose the greatest hazard to bees. The tiny capsules, adhering to foraging bees, are packed along with pollen into the pollen baskets and carried back to the hive where they are eaten by bees rearing the brood. Granular formulations and soil applications of pesticides are usually not harmful to bees. The hazard of some systemic pesticides is reduced because they are rapidly absorbed by the plant. At least one formulation, systox, has tentatively been reported to be a bee repellent.

Table 9–2 lists pesticides according to their relative hazard to bees. Table 9–3 gives common symptoms of pesticide poisoning in bees. Usually not all are evident at any one time.

Plants

Pesticides are used on plants to protect them from pests. Insecticides and fungicides are often used on ornamental and forest trees to control serious insect pests and diseases. These chemicals aid in keeping forests, parks, and lawns green and enjoyable. However, pesticides can injure plants. Injury from volatility and drift of herbicides to sensitive crops has led to specific legislation regulating the use of those materials in some states.

TABLE 9–2 PESTICIDES GROUPED ACCORDING TO THEIR RELATIVE HAZARDS TO HONEY BEES

Group 1: Highly hazardous

Severe losses may be expected if these pesticides are used when bees are present at treatment time or within a day thereafter.

Acephate (Orthene®)
Aldicarb (Temik®)
Arsenicals
Azinphosmethyl (Guthion®)
Carbaryl (Sevin®)
Carbofuran (Furadan®)
Chlorpyrifos (Lorsban®, Dursban®)
Diazinon (Spectracide®)
Dichlorvos (DDVP)
Dicrotophos (Bidrin®)
Dimethoate (Cygon®, DE-FEND®)
EPN
Famphur (Warbex®)
Fenitrothion (Cytel®)
Fensulfothion (Dasanit®)
Fenthion (Baytex®)
Lindane
Malathion (Cythion®)
Malathion ULV
Methyl parathion
Methamidophos (Monitor®)
Methidathion (Supracide®)
Methiocarb (Mesurol®)
Methomyl (Lannate®, Nudrin®)
Mevinphos (Phosdrin®)
Monocrotophos (Azodrin®)
Naled (Dibrom®)
Parathion
Phosmet (Imidan®)
Phosphamidon (Dimecron®)
Propoxur (Baygon®)
Resmethrin (Pyrethroid)
Tetrachlorvinphos (Appex®)

Group 2: Moderately hazardous

These can be used around bees if dosage, timing, and method of application are correct, but should not be applied directly on bees in the field or at the colonies.

Demeton (Systox®)
Disulfoton (Di-Syston®)
Endosulfan (Thiodan®)
Ethoprop (Mocap®)
Formetanate (Carzol®)
Leptophos (Phosvel®)
Oxamyl (Vydate®)
Oxydemeton methyl (Metasystox-R®)

TABLE 9–2 *(Continued)*

Group 2: Moderately hazardous

Phorate (Thimet®)
Phosalone (Zolone®)
Pyrazophos (Afugan®)
Ronnel
Temephos (Abate®)
Terbufos (Counter®)
Trichloronate (Agritox®)

Group 3: Relatively nonhazardous

These can be used around bees with a minimum of injury.

Insecticides and Acaricides

Allethrin
Bacillus thuringiensis
Diflubenzuron (Dimilin®)
Dinocap (Karathane®)
Dioxathion (Delnav®)
Ethion (ETHION-EC®)
Heliothis polyhedrosis virus
Methoxychlor (Marlate®)
Oxythioquinox (Morestan®)
Dienochlor (Pentac®)
Chlordimeform (Fundal®, Galecron®)
Chlorobenzilate (Acaraben®)
Cryolite (Kryocide®)
Dicofol (Kelthane®)
Propargite (Omite®, Comite®)
Pyrethrins (natural)
Rotenone
Tetradifon (Tedion®)
Trichlorfon (Dylox®, Dipterex®)

Herbicides, Defoliants, and Desiccants

Alachlor (Lasso®)
Amitrole (Weedazole®)
AMS (Ammate®)
Atrazine (AAtrex®)
Bifenox (Modown®)
Bromacil (Hyvar®)
Cacodylic acid (Phytar 560®)
Chloramben (Amiben®)
Copper sulfate (monohydrated)
Cyanazine (Bladex®)
2,4-D
2,4-DB
Dalapon
Dazomet (Mylone®)
Dicamba (Banvel®)
Dichlobenil (Casoron®)   *(continued)*

TABLE 9–2 *(Continued)*

Group 3: Relatively nonhazardous

Herbicides, Defoliants, and Desiccants

Diquat
Propazine (Milogard®)
Diuron (Karmex®)
Dazomet (Mylone®)
DSMA (Methar®)
Endothall (Endothal®)
EPTC (Eptam®)
Fluometuron (Cotoran®)
Fluorodifen (Preforan®)
Linuron (Lorox®)
MCPA (Weedar®)
Methazole (Probe®)
Metribuzin (Sencor®)
MSMA (Daconate®)
Naptalam (Alanap®)
Paraquat
Phenmedipham (Betanal®)
Picloram (Tordon®)
Prometon (Pramitol®)
Prometryn (Caparol®)
Pronamide (Kerb®)
Propachlor (Ramrod®)
Propanil (Stam®)
Propham (Chem-Hoe®)
Simazine (Princep®)
Terbacil (Sinbar®)
2,4,5-T
Terbutryn (Igran®)

Fungicides

Anilazine (Dyrene®)
Benomyl (Benlate®)
Captan
Carboxin (Vitavax®)
Copper oxychloride sulfate
Mancozeb (Dithane M-45®, Fore®)
Maneb (MANEB 80®)
Metiram (Polyram®)
Dodine (Cyprex®)
Folpet (Phaltan®)
Oxycarboxin (Plantvax®)
Sulfur
Thiram (Spotrete®)
Zineb

TABLE 9-3 COMMON INSECTICIDE POISONING SYMPTOMS

- Excessive numbers of dead bees in front of hive.
- Dead bees on top bars in hive.
- Lack of housecleaning by hive bees.
- Dying larvae crawling out of cells.
- Dead and dying bees, including many young (fuzzy) bees, from feeding on contaminated pollen.
- Break in brood cycle.
- Honey and pollen stored in brood area.
- Queen supercedure.
- Bees dying after being installed in hives stored over winter (caused by pesticides Sevin® and Penncap-M® retained in stored pollen for several months).
- Queenlessness within 30 days.

Organophosphate Insecticide Poisoning (for example, parathion dimethoate CYGON®):

- Bees regurgitate; dead bees stick together.
- Tongues of dead bees extended.
- Bee abdomen distended.
- Erratic attempts at self-grooming.
- Gentle bees become aggressive.
- Dying bees appear disoriented or paralyzed; bees are often spinning or crawling awkwardly.
- Wings held away from the body but remain hooked together.
- Dead bees are primarily field bees except when microencapsulated materials are responsible. Then the dead bees are primarily young bees, and a break in the brood-rearing cycle results.
- Loss of ability to maintain hive temperature.

Carbamate Insecticide Poisoning (for example, Furadan®, Lannate®, Sevin®):

- Affected bees become aggressive and erratic and often are unable to fly.
- Bees often die slowly; they may live up to 3 days.
- Bees appear stupefied as if chilled; paralysis is usually evident.
- Dead brood in front of hive.

Injury can range from slight burning or browning of leaves to death of the whole plant. This injury is called *phytotoxicity.* Phytotoxicity accidents can result from carelessness from the use of a pesticide that is highly hazardous to trees and smaller plants.

All kinds of pesticides (especially insecticides, fungicides, and herbicides) may injure or kill plants. Herbicides are especially hazardous because they are designed to kill or control plants. Nontarget plants can be severely damaged from drift or misuse of herbicides. Always be careful when herbicides are applied near desirable plants.

In addition, some pesticides and some formulations tend to move off target readily. These chemicals can be a great threat to desirable crops, other plants, and trees. Some are carried off target by rain and runoff water, and may injure plants in the water's path. Other pesticides may move through the soil to surrounding areas and cause phytotoxicity there. If plant injury could be a possible problem in your spray operation, try to use a pesticide and formulation that tends to remain on the target area. Be especially careful not to overdose when plant injury could be a problem. Remember that persistent pesticides can be very useful for long-term insect, disease, or weed control programs. But be sure to follow label directions carefully if future crops or other plants or trees will be planted in the area.

General Considerations

Wildlife

Fish, birds, and mammals are assets to humans (Figure 9–4). Land that is used only as farmland does not have to be a wildlife refuge. However, care should be taken to protect surrounding wooded areas and waterways when applying any pesticide. Fishing and hunting are very popular sports. If pesticides are carelessly used, these sports can disappear. Pesticide kills of mammals, birds, and fish in large numbers have been reported on occasion and have hampered fishing and hunting activities in some areas. Pesticides can be helpful to wildlife by controlling dangerous or annoying pests that could harm the animals. The toxicity of every chemical to every animal is not known. A pesticide that is only slightly toxic to one living thing may be very toxic to another. Check the label for specific instructions concerning the toxicity of a pesticide to wildlife.

Birds and mammals can be killed outright by insecticide applications. Although they may absorb chemicals through their skin or inhale them in sufficient quantities to be affected, the usual means of poisoning is by eating contaminated food. Plants and seeds, or areas treated with insecticides can be hazardous to wildlife through direct exposure to the chemicals. Animals that eat other animals for food may absorb pesticides from their prey and thus develop large loads of pesticides in their own systems. Wildlife may be killed or affected in other ways. Changes in behavior in birds and mammals, hatching failure in birds, and reduced reproduction in mammals may occur. The growth and survival rates of the young produced may be reduced. For mammals, these changes in reproduction may affect both males and females. For birds in the field, reproductive effects usually occur in females.

Most animals store certain kinds of pesticides in their body fat. Some animals such as ducks and bats use their fat quickly when they go without food for any period of time. They may then be poisoned by pesticides that were stored in their fat systems several months earlier. Over a long period of time, the most affected bird populations have been the bird eaters, such as falcons and some hawks; and the fish eaters, such as eagles, ospreys, and pelicans. Local and regional populations of these birds have on occasion declined because of the effect of certain insecticides. Similar widespread and

Figure 9–4 Wildlife is an asset to people.

long-term effects are not known to have occurred in mammals, although they can be killed by heavy applications of the more toxic insecticides.

Because the importance of wildlife losses cannot be expressed in economic terms, personal values must be used in evaluating losses. The loss of even one animal may be considered great by some. No estimate has been made of the annual loss of wildlife due to pesticide use. However, biologists agree that the loss of habitat is the most serious threat to wildlife in the United States today. Changes in land use, such as housing developments and highways, have greatly altered local wildlife populations. In the last 200 years, the activities of people have caused the extinction of about 60 species of birds, mammals, fishes, reptiles, and amphibians in the United States.

Accidental losses from pesticides have occurred in the past, although probably less than 100 such cases have been reported in the United States. Of greater concern are the possible long-term effects to wildlife populations, since some chemicals are known to reduce reproduction or affect defensive behavior.

Individual citizens making everyday choices may affect wildlife in unintended ways. When deciding about pesticide use, we are also deciding about adding to the pesticide burden of the environment. Each person's decisions about pesticides are, therefore, important.

On March 14, 1991, the EPA published a notice in the *Federal Register* regarding its "may affect" determinations for 32 pesticides. This notice was forwarded to the U.S. Fish and Wildlife Service. The pesticides for which determinations were made are as follows:

EPA "MAY AFFECT" PESTICIDE LIST FOR ENDANGERED SPECIES

Acephate	Magnesium phosphide
Aldicarb	Methyl bromide
Aluminum phosphide	Naled
Azinphos methyl	Parathion
Bendiocarb	Permethrin
Brodifacoum	Phorate
Bromadiolone	Pindone
Bromethalin	Potassium nitrate
Carbofuran	Sodium cyanide
Chlorophacinone	Sodium fluoroacetate
Chlorpyrifos	Sodium nitrate
Diphacinone	Terbufos
Endosulfan	Trifluralin
Fenthion	Vitamin D-3
Fenvalerate	Warfarin
s-fenvalerate	Zinc phosphide

Hazard is related to conditions existing at the time and place of application and to the fate of the chemical in the environment. The degree of exposure of wild animals is usually difficult to control, yet may be as important a factor in wildlife safety as using one of the insecticides having a low toxicity rating. Some chemicals may move in the environment and be accumulated by animals in amounts sufficient to destroy their reproductive capacity, make them more vulnerable to predators, kill them, or kill their predators.

There are reasons for optimism about the future concerning wildlife and pesticides. Some chemicals harmful to wildlife are beginning to lessen in the environment. Some scientists believe that environmental contamination by pesticide users is decreasing. New pesticides are being tested more carefully for effects on other animals.

Where wildlife values take priority over other values, avoid using pesticides. Determine whether the treatment is really necessary. If you feel you must use a pesticide, choose it carefully to pose the minimum hazard under your conditions of use, and follow label directions carefully. Select a pesticide that will last only a short time in the environment. Avoid using insecticides during bird migration and nesting periods. Whenever possible, insecticide treatment should be avoided during the nesting season of pheasants, quail, prairie chickens, grouse, and doves that nest on the ground at field edges or in fields (fallow, stubble, alfalfa, etc.). Always avoid treating over or near streams, lakes, and ponds; fish are often more sensitive to pesticides than are warm-blooded animals. Do not puddle sprays during application or when cleaning spray equipment because birds may drink or bathe there within minutes. Be very careful about safe storage and safe disposal of unused chemicals and their containers. Never throw "empty" containers into water or discard them where animals have access to them.

The encouragement and maintenance of wildlife on farms where various cropping practices are conducted depends on the pesticide applicator's attitude, appreciation of various wildlife, and willingness to provide a habitat for wildlife coexistence with farming practices. It is possible to maintain a fairly large variety of wildlife within a farm setting, provided the pesticide applicator is aware of various options that will help to protect wildlife, and is willing to time some operations so that they will be the least disruptive to certain species, especially pheasants.

Some pesticides are beneficial to wildlife in that they are used to eliminate undesirable species or to enhance a setting, making it more conducive to the propagation and growth of desirable species. A farmer who is conscientious about cropping practices is usually careful to ensure that these practices are complementary to and not competitive with the wildlife species that coexist on the farm.

ENDANGERED SPECIES

Certain plants and animals (including insects) have been identified as endangered or threatened species. Every effort must be made to avoid causing harm to these populations. Since all living things are part of a complex, delicately balanced network, the removal of a single species can set off a chain reaction affecting many others. It has been estimated, for example, that a disappearing plant can take with it up to 30 other species that depend ultimately on that plant, including insects, higher animals, and even other plants. The full significance of an extinction is not always readily apparent; much remains to be learned, and the full long-term impact is difficult to predict.

By definition, an *endangered species* is one on the brink of extinction throughout all or a significant portion of its range. A *threatened species* is one likely to become endangered in the foreseeable future. The reasons for a species' decline are usually complex, and thus recovery is difficult. A major problem for most wildlife is the destruction of habitat, usually the result of industrial, agricultural, residential, or recreational development.

The EPA in conjunction with the Federal Fish and Wildlife Service is charged with the responsibility of protecting endangered species of plants and animals from pesticides. At the time of this writing, the final rules and regulations for implementation of the Endangered Species Act were not completed, and indications are that they may not be fully implemented until 1996 or later. The essence of the requirement is that no pesticides will be permitted to be used where they will endanger any of a number of listed species. It is important to be aware of the Endangered Species Act and its requirements. Contact your local county cooperative extension office or the regional EPA office for more information.

PESTICIDE PERSISTENCE

All pesticides degrade or break down chemically into other related chemicals and, eventually, into the simple building blocks of which the whole world is made. This process occurs at very different rates for different pesticides. In some, the changes occur rapidly (in hours or in a few days); these materials are referred to as short-lived. In others, the changes are slower and the pesticides may be present for relatively long periods of time; these are known as persistent pesticides. The rates of breakdown or degradation of any chemical may change with differing conditions of temperature, sunlight, air, and location.

The environmental difficulty with persistent pesticides is that once released into the ecosystem they remain in the original chemical form long enough so that, if they have the other properties of moving readily and of being stored in animal tissues, they can spread to a distance and be concentrated at some other place than where they were applied. While it is true that the persistence of a pesticide may give it practical advantages for the control of the pest against which it is used, that same persistence means that, once such a material is released, its environmental damage cannot be controlled.

PESTICIDE ACCUMULATION

Some pesticides can build up in the bodies of animals (including humans). These are called *accumulative* pesticides. The chemicals can build up in an animal's body until they are harmful to it. These pesticides also accumulate in the food chain (Figure 9–5). Meat-eaters feeding on other animals with built-up pesticides may receive high doses of pesticides. If they feed on too many of these animals, the meat-eater can be poisoned without ever directly contacting the pesticide. The buildup through the food

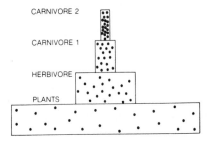

Figure 9–5 Concentration in the food chain.

chain can injure animals that aid humans. In fact humans, as one of the meat-eaters at the top of this food chain, could get very high doses of pesticides in this way.

Some pesticides do not build up in the body of animals or in the food chain. These are called *nonaccumulative* pesticides. These chemicals usually break down rapidly into other, relatively harmless materials. Organophosphate pesticides, for example, have high toxicity at first and could be potentially dangerous to wildlife, but they do not accumulate so they are not as dangerous to the environment in general. Usually, pesticides that break down quickly in the environment are the least harmful to it.

PEST RESISTANCE TO PESTICIDES

Pest resistance to pesticides has been of concern to pesticide applicators for a number of years. In recent years there have been more new reports of insects, mites, plant disease pathogens, nematodes, weeds, and rodents developing resistance to various pesticides. Because researchers started recording insect resistance to insecticides over the last 50 years or so, about 450 different insects have been found to be resistant to one or more insecticides. Most of these have been found in the last 20 years, and the overall number has doubled in that time period. Almost 60% of the 450 different insects that have developed resistance are of importance in agriculture.

During the last 20 years, more than 150 herbicide-resistant weeds have been reported as resistant to one or more herbicides, while at least 100 species of plant disease pathogens and three species of nematodes have also been identified as resistant. The Colorado potato beetle, for example, is probably the most resistant pest in the United States. Farmers in the northeast states have tried just about every pesticide against the Colorado potato beetle with very poor results.

The number of resistant weed species has grown alarmingly since 1980. Many of the weeds with known herbicide resistance are tolerant to triazines; however, resistance to 2,4-D, trifluralin, paraquat, and a few urea products has also been reported. One of the most recent resistance problems has arisen with chlorsulfuron (Glean®), to which several weed species have been reported as being resistant. Plant disease organisms show a similar pattern. Almost 84% of the species of fungi reportedly carrying resistance resist benomyl fungicides.

Scientists suspect that overuse or overexposure of a pesticide probably accelerates resistance problems. More judicious use of still-effective compounds may extend their useful life.

Six suggestions for reducing chemical resistance are as follows:

1. Use pesticides only on an as-needed basis. Fields should be scouted to determine pest counts and economic thresholds. Cost-return potentials should be evaluated before treating, and "insurance" applications should be avoided.
2. Rotate chemicals. If alternatives are available, products should be switched to guard against selective control. Whenever chemicals are used over and over, resistant, hard-to-kill strains survive, ultimately leading to highly resistant populations. By alternation of products, the applicator may extend the useful life of each product and help to prevent selective survival.

3. Use cultural practices that reduce pest numbers and minimize the need for chemicals. Keeping pests from building up to economic infestation levels through the use of crop rotations will help break the life cycles of many insects, weeds, and disease pathogens.

4. Monitor fields, golf courses, and other green areas closely so that new problems can be detected as they are developing. If control is necessary and other methods have not worked, a pesticide may be required to solve the problem. On the other hand, if a pesticide has not controlled the pest the way it should, determine if another method or product would be warranted.

5. Good IPM programs have been developed for a number of crops and situations. Use resistant varieties whenever possible to take advantage of genetic resistance. Use beneficial insects and biological controls whenever they are available.

6. Learn as much as you can about the pests you are managing and the methods that can best be used to control them. Keep up on the latest research and be aware of what is happening at your state university.

WHAT PESTICIDE APPLICATORS CAN DO TO PROTECT THE ENVIRONMENT

Commercial pesticide applicators are usually hired to apply a specific chemical to a particular area at a given rate within a given time period. Therefore, it may not be practical for you to assure yourself that there is a real need for pesticide use or to select a chemical that is the least dangerous to reduce hazards to a minimum. However, there are procedures that applicators can follow to help prevent damage to the total environment:

- Calibrate equipment carefully. A small increase in dosage rates may mean the difference between severe effects and no effects on fish and wildlife.
- Mix pesticides at the correct rate. Too much pesticide in the spray tank can be more harmful than a poorly calibrated spray machine.
- Be sure that you hit the designated target. Use care in developing ground application and flight patterns. Avoid any overlap in spray swaths, especially near water areas. Avoid applying spray materials directly to water areas.
- Spray after irrigating if possible. Spraying while irrigating or just before may result in the spray materials entering waters containing fish. Chemigate only with approved pesticides.
- Spray under favorable weather conditions to prevent drift into water courses or wildlife habitats.
- Learn what you can about the possible effects on fish and wildlife of the chemicals that you are using.
- Use alternate pest controls whenever possible.
- Do your part to aid the environment; your surroundings are worth protecting.

10

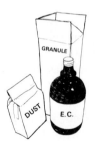

Pesticide Formulations and Adjuvants

The active ingredient in a pesticide is the chemical that controls the target pest. A pesticide chemical can seldom be used as originally manufactured, and the product you purchase is rarely made up only of active ingredients. Usually the basic chemical cannot be added directly to water or mixed in the field with solids, so the manufacturer must modify the product by combining it with other materials such as solvents, wetting agents, stickers, powders, or granulars. This mixture of active and inert (inactive) ingredients is called a pesticide formulation. Some formulations are ready to use while others must be further diluted with a solvent, water, or air before they are applied.

A single active ingredient often is sold in several formulations. For example, some pesticides can be purchased as emulsifiable concentrates, wettable powders, dusts, and granules. Most pesticides, however, are not available in such a wide range of formulations. When applicators have the opportunity to select from several formulations, they should choose the formulation that will best meet the requirements for a particular job. Considerations in making a choice include:

- Effectiveness against the pest
- Habits of the pest
- Plant, animal, or surface to be protected
- Application equipment available and best suited for the job
- Danger of drift or runoff
- Possible injury to the protected surface (discoloration or pitting of the surface)
- Safety to the applicator, helpers, and other organisms likely to be exposed

TYPES OF FORMULATIONS

Liquid Formulations

Aerosols (A). These formulations contain one or more active ingredients and a solvent. Most aerosols contain a low percentage of active ingredient. There are two types of aerosol formulations: the ready-to-use type, and those made for use in smoke or fog generators.

Ready-to-use aerosols are usually small, self-contained units that release the pesticide when the nozzle valve is triggered. The pesticide is driven through a small opening by an inert gas under pressure, creating fine droplets. These products are used in greenhouses, in small areas inside buildings, or in localized outdoor areas. Commercial models hold 5 to 10 pounds of pesticide, and these are usually refillable.

Advantages

- Ready to use
- Easily stored
- Convenient way of buying small amount of a pesticide
- Retain their potency over fairly long time

Disadvantages

- Expensive
- Practical for very limited uses
- Risk of inhalation injury
- Hazardous if punctured, overheated, or used near an open flame
- Difficult to confine to target site or pest

Formulations for smoke or fog generators are not under pressure. They are used in machines that break the liquid formulation into a fine mist or fog (aerosol) using a rapidly whirling disk or heated surface. These formulations are used mainly for insect control in structures such as greenhouses and warehouses, and for mosquito and biting fly control outdoors.

Advantages

- Easy method of filling entire space with pesticide

Disadvantages

- Highly specialized use
- Fairly expensive for pounds of active ingredient per gallon
- Difficult to confine to target site or pest
- Risk of inhalation injury

Emulsifiable concentrates (EC or E). An emulsifiable concentrate formulation usually contains the active ingredient, one or more petroleum solvents, and an emulsifier, which allows the formulation to be mixed with water. Each gallon of EC usually contains 2 to 8 pounds of active ingredient. ECs are among the most versatile formulations. They are used against agricultural, ornamental and turf, forestry, structural, food processing, livestock, and public health pests. They are adaptable to many types of application equipment, including small, portable sprayers, hydraulic sprayers, low-volume ground sprayers, mist blowers, and low-volume aircraft sprayers.

Advantages

- High concentration means price per pound of active ingredient is relatively low and product is easy to handle, transport, and store
- Little agitation required; not abrasive; will not settle out or separate when equipment is running
- Little visible residue on fresh fruits and vegetables and on finished surfaces

Disadvantages

- High concentration makes it easy to overdose or underdose through mixing or calibration errors
- Phytotoxicity hazard usually greater
- Easily absorbed through skin of humans or animals
- Solvents may cause rubber or plastic hoses, gaskets, and pump parts and surfaces to deteriorate
- May cause pitting or discoloration of painted finishes
- May be corrosive

Flowables (F or FL).   Some active ingredients are insoluble solids. These may be formulated as flowables in which the finely ground active ingredients are mixed with a liquid, along with inert ingredients, to form a suspension. Flowables are mixed with water for application and are similar to EC formulations in ease of handling and use. They are used in the same types of pest control operations for which ECs are used.

Advantages

- Seldom clog nozzles
- Easy to handle and apply

Disadvantages

- Require moderate agitation
- May leave a visible residue

Fumigants.   Fumigants are pesticides that form poisonous gases when applied. Sometimes the active ingredients are gases that become liquids when packaged under high pressure. These formulations become gases when released during application. Other active ingredients are volatile liquids when enclosed in an ordinary container and so are not formulated under pressure. They become gases during application. Others are solids that release gases when applied under conditions of high humidity or in the presence of water vapor.

Fumigants are used for structural pest control, in food and grain storage facilities, and in regulatory pest control at ports of entry and at state and national borders. In agricultural pest control, fumigants are used in soil, greenhouses, granaries, and grain bins.

Advantages

- Toxic to a wide range of pests
- Can penetrate cracks, crevices, wood, and tightly packed areas such as soil or grains
- Single treatment will usually kill most pests in treated area

Disadvantages

- Target area must be enclosed or covered to prevent the gas from escaping
- Highly toxic to humans; specialized protective equipment, including respirators, must be used with fumigants

Invert emulsions. This unusual mixture contains a water-soluble pesticide dispersed in an oil carrier. Invert emulsions require a special kind of emulsifier that allows the pesticide to be mixed with a large volume of petroleum carrier, usually fuel oil. When applied, invert emulsions form large droplets that do not drift easily. Invert emulsions are most commonly used in vegetation control along rights-of-way where drift to susceptible nontarget plants is a problem.

Solutions (S). A few pesticide active ingredients dissolve readily in water. Formulations of these pesticides contain the active ingredient and one or more additives. When mixed with water, they form a solution that will not settle out or separate. Solutions may be used in any type of sprayer indoors or outdoors.

Advantages

- No agitation necessary

Disadvantages

- Very few formulations of this type available

Low concentrate solutions (S). These formulations, usually solutions in petroleum solvents, contain small amounts (usually 1% or less) of active ingredients per gallon. They are designed to be used without further dilution. Low concentrate solutions are used for the following:

- Structural and institutional pests
- Clothes moths
- Livestock and poultry pests
- Space sprays in barns and warehouses
- Mosquito control

Types of Formulations

Advantages

- No mixing necessary
- Household formulations have no unpleasant odor; do not stain fabric

Disadvantages

- Expensive
- Limited number of uses

Ultralow-volume concentrate solutions (ULV). ULV concentrate solutions contain eight or more pounds of active ingredient per gallon. They may approach 100% active ingredient. ULV concentrates are designed to be used as is or to be diluted with only small quantities of specified solvents. These special-purpose formulations must be applied with highly specialized spray equipment. They are mostly used in outdoor applications such as agricultural, forestry, and ornamental, and in mosquito control programs. The advantages and disadvantages are similar to those for emulsifiable concentrates.

Dry Formulations

Baits (B). A bait formulation is an active ingredient mixed with food or another attractive substance. The bait attracts the pests, which are then killed by eating the pesticide it contains. The amount of active ingredient in most bait formulations is quite low, usually less than 5%. Baits are used inside buildings to control ants, roaches, flies, and other insects, and for rodent control. Outdoors they are sometimes used to control slugs and some insects, but their main use is for control of vertebrate pests such as birds, rodents, and other mammals.

Advantages

- Ready to use
- Entire area need not be covered, since pest goes to bait
- Control pests that move in and out of an area

Disadvantages

- Often attractive to children and pets
- May kill domestic animals and nontarget wildlife outdoors
- Pest may prefer the crop or other food to the bait
- Dead pests may cause odor problem
- Other animals feeding on the poisoned pests may also be poisoned
- Application costs are high

Dusts (D). Most dust formulations are ready to use and contain a low percentage of active ingredient (usually 1% to 10%), plus a very fine, dry inert carrier made from talc, chalk, clay, nut hulls, or volcanic ash. The size of individual dust particles is variable.

Dust concentrates contain a greater percentage of active ingredient. These must be mixed with dry inert carriers before they can be applied.

Dusts are always used dry and easily drift into nontarget areas. They sometimes are used for agricultural applications. In structures, dust formulations are used in cracks and crevices, and for spot treatments. They are widely used in seed treatment. Dusts are also used to control lice, fleas, and other parasites on pests, domestic animals, and poultry.

Advantages

- Usually ready to use, with no mixing
- Effective where moisture from a spray might cause damage
- Require simple equipment
- Effective in hard-to-reach indoor areas

Disadvantages

- Drift hazard high
- Expensive because of low percentage of active ingredient

Granules (G). Granular formulations are similar to dust formulations except that granular particles are larger and heavier. The coarse particles are made from an absorptive material such as clay, corn cobs, or walnut shells. The active ingredient either coats the outside of the granules or is absorbed into them. The amount of active ingredient is relatively low, usually ranging from 1% to 200%.

Granular pesticides are most often used to apply chemicals to the soil to control weeds, nematodes, and insects living in the soil. They also may be used as systemics, formulations that are applied to the soil and then absorbed into the plant through the roots and carried throughout the plant. They are sometimes used in airplane or helicopter applications because drift is minimal. Granular formulations are also used to control larval mosquitoes and other aquatic pests. Granules are used in agricultural, ornamental, turf, aquatic, right-of-way, and public health (biting insect) pest control operations.

Advantages

- Ready to use; no mixing
- Drift hazard is low; particles settle quickly
- Low hazard to applicator; no spray, little dust
- Weight carries the formulation through foliage to soil target
- Simple application equipment; often seeders or fertilizer spreaders
- May be more persistent than wettable powders (WPs) or ECs

Disadvantages

- Does not stick to foliage
- More expensive than WPs or ECs
- May need to be incorporated into soil
- May need moisture to activate pesticidal action

Microencapsulation.   Microencapsulated formulations are particles of pesticides (either liquid or dry) surrounded by a plastic coating. The formulated product is mixed with water and applied as a spray. Once applied, the capsule slowly releases the pesticide. The encapsulation process can prolong the active life of the pesticide by providing a timed release of the active ingredient.

Advantages

- Increased safety to applicator
- Easy to mix, handle, and apply

Disadvantages

- Constant agitation necessary in tank
- Some bees may pick up the capsules and carry them back to the hives where the released pesticide may poison entire hives

Pellets (P or PS).   Pellet formulations are very similar to granular formulations; the terms are often used interchangeably. A pellet, however, is a formulation manufactured to create a pellet of specific weight and shape. The uniformity of the particles allows them to be applied by precision applicators, such as those being used for precision planting of pelleted seed.

Soluble powders (SP).   Soluble powder formulations look like wettable powders. However, when mixed with water, soluble powders dissolve readily and form a true solution. After they are thoroughly mixed, no additional agitation is necessary. The active ingredient in soluble powders ranges from 15% to 95% (usually over 50%).

Soluble powders have all the advantages of the wettable powders and none of the disadvantages except the inhalation hazard during mixing. Few pesticides are available in this formulation, because few active ingredients are soluble in water.

Water-dispersible granules (WDG), also known as dry flowables (DF).
Water-dispersible granular formulations are like wettable powder formulations, except the active ingredient is prepared as granule-sized particles. Water-dispersible granules must be mixed with water to be applied. The formulation requires constant agitation to keep it suspended in water. Water-dispersible granules share the advantages and disadvantages of wettable powders except:

- They are more easily measured and mixed.
- They cause less inhalation hazard to the applicator during pouring and mixing.

Wettable powders (WP or W).   Wettable powders are dry, finely ground formulations that look like dusts. They usually must be mixed with water for application as a spray. A few products, however, may be applied either as a dust or as a wettable powder; the choice is left to the applicator. Wettable powders contain 5% to 95% active ingredient (usually 50% or more). Wettable powder particles do not dissolve in water. They settle out quickly unless constant agitation is used to keep them suspended.

Wettable powders are one of the most widely used pesticide formulations. They can be used for most pest problems and in most types of spray machinery where agitation is possible.

Advantages

- Low cost
- Easy to store, transport, and handle
- Lower phytotoxicity hazard than ECs and other liquid formulations
- Easily measured and mixed
- Less skin and eye absorption than ECs and other liquid formulations

Disadvantages

- Inhalation hazard to applicator while pouring and mixing the concentrated powder
- Require good and constant agitation (usually mechanical) in the spray tank
- Abrasive to many pumps and nozzles, causing them to wear out quickly
- Residues may be visible

PACKAGING

Most pesticides are packaged in either metal containers (from pint size to 55-gallon drums), glass containers (usually pint size up to gallon size), plastic containers (from pint size up to 5-gallon containers), and paper bags or cardboard boxes that contain granules or wettable powders. Some pesticides may be designated in metric measurements.

Premeasured doses of pesticides, usually wettable powders, are packaged in dissolvable bags. This technique involves the use of a dissolvable polymer bag material in which the correct dose for a specified unit is encased. The applicator merely drops the bag into a specified amount of water and the bag dissolves, releasing the material into the water. This procedure is designed to protect the pesticide handler as much as possible.

Safer handling and fewer container disposal problems are just two reasons why water-soluble packages (WSP) have become more prevalent in the retail industry in recent years. Another reason is the Gel-Tec® technology, developed by Rhone-Poulenc, which allows emulsifiable concentrates (ECs) to be sold in WSPs instead of only water-dispersible granules from larger containers. Figure 10–1 illustrates Rhone-Poulenc's mixing instructions for Buctril Gel®, but should apply to other products in water-soluble packaging as well. Following the instructions should eliminate problems with WSPs mixing thoroughly and completely.

Mini bulk containers are also popular. These are owned either by the applicator or by the chemical dealer. The containers have transfer systems that meter the pesticide into the spray tank. This type of container is reusable (for the same material) and eliminates the disposal problem of a number of smaller containers. These are usually rented from the dealer, who keeps them in proper operating condition and refills them for future rental.

MIXING INSTRUCTIONS
BUCTRIL GEL-PAKS

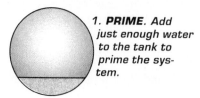

*1. **PRIME**. Add just enough water to the tank to prime the system.*

*2. **SOAK**. Add Gel-Paks and other products in WSP. Allow to soak for 2 minutes.*

*3. **AGITATE**. Turn on the agitation for 2 minutes.*

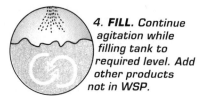

*4. **FILL**. Continue agitation while filling tank to required level. Add other products not in WSP.*

Figure 10–1 Mixing instructions for water-soluble packets.

SPRAY ADJUVANTS

Adjuvants are chemicals added to pesticide spray mixtures to enhance and modify the physical properties of the pesticide for the spray mix. Adjuvants are specialized chemicals that must be matched to a particular pesticide, the problem, and environmental conditions to ensure that they enhance and do not detract from the effectiveness of the pesticide. Adjuvants are added to the spray solution in two ways: (1) they are included as part of the pesticide formulation as sold to the applicator, and (2) they are added by the applicator to the spray solution as a tank mix. Adjuvants can be described either by their physical characteristics or by what they are intended to do in the pesticide mixture. Based on intended use or type of action, adjuvants used in a mixture can be divided into three broad categories of compounds: activators, spray modifiers, or utility modifiers.

Activator Adjuvants

An activator adjuvant increases the biological activity of the pesticide beyond that usually obtained without the material added. An activator adjuvant is particularly useful when applying herbicides. Activator adjuvants are usually classified by their physical characteristics. There are four major classes of activator adjuvants: surfactants, vegetable oils, crop oils, and crop oil concentrates.

Surfactants. A surfactant is a compound that improves the emulsifying, dispersing, spreading, wetting, or other surface-modifying properties of liquids. The term *surfactant* is derived from the words *surface-active agent.* It is important to note that, while surfactants make up a very large class of adjuvants, not all adjuvants are surfactants. A surfactant may be the sole ingredient in an adjuvant, or surfactants may be used in conjunction with other ingredients.

Surfactant molecules are comprised of two segments: the strong polar group that is attracted to water and a nonpolar group that is attracted to nonaqueous materials such as oil. Surfactants are classified into four types based on the polar portion of the molecule:

1. Anionic
2. Cationic
3. Ampholytic
4. Nonionic

Of these four types, the nonionic surfactants are most often used in agricultural sprays. They are usually chemically inactive when mixed with pesticides.

Surfactants have several different functions in spray solutions, depending on the type and concentration at which they are added. At low concentrations, surfactants are used primarily as *wetting agents.* Wetting agents reduce the surface tension of the spray droplet, reducing the tendency of the spray droplet to bead up on the leaf and increasing the leaf area covered by each droplet. At higher concentrations, surfactants can also act as *emulsifiers.* An emulsion is one liquid dispersed in another liquid, each retaining its original identity (oil in water for example). If oil is added to water and shaken vigorously, the oil is momentarily dispersed as small droplets in the water. However, if allowed to stand, the oil and water will separate unless an emulsifier is added. An emulsifiable concentrate is a pesticide formulated in an oil-based solvent that contains the necessary emulsifiers to allow this oil solution to disperse as finely divided oil droplets in water.

Vegetable oils as a carrier or adjuvant. The use of vegetable oils as an additive or complete carrier has gained considerable interest. This interest is primarily due to the high cost in time and money of applying conventional volumes of water, the development of equipment for low-volume application (such as rotary nozzles and electrostatic units), and the development of new pesticides (such as synthetic pyrethroid insecticides and postemergence herbicides).

Available research results indicate that vegetable oil can be substituted for petroleum-based oil concentrate additives with equal weed control in most cases. Studies by pesticide manufacturers indicate that weed control increases with the addition of a surfactant to the vegetable oil. Vegetable oils, as additives, are generally considered most important in enhancing weed control in low-humidity and high-temperature environments. Thus, oil additives generally provide protection against a loss of weed control with an herbicide under adverse environmental conditions.

In addition to using vegetable oils as an additive, vegetable oil alone can be used as the complete carrier. The development of equipment for low-volume application has made it feasible to use vegetable oil as a carrier at up to 2 gallons per acre. The use of vegetable oil in place of water has potential for improving the efficiency of low-volume applications because the drops are relatively nonevaporative, and may allow more uniform coverage and canopy penetration. In addition, pesticide-oil mixtures may be less subject to rain wash-off from plant surfaces, and an increased spread factor may increase the rate of absorption into the plant. The use of vegetable oils will increase particle drift, however, due to (1) smaller droplets and (2) lighter weight droplets.

Most of the research to verify the claimed advantage of vegetable oils has been with aerial application of synthetic pyrethroid insecticides. Work with low-volume ground applications of herbicides has not produced conclusive results. Initial studies with preemergence and postemergence herbicides indicate that soybean oil carriers can be very effective. Changing physical properties of herbicide–vegetable oil mixtures make it difficult to develop techniques for applying herbicides in vegetable oil carriers. In addition to rotary nozzles, experimental low-volume air-assist nozzles show promise of maintaining good atomization characteristics over a range of oil properties.

Crop oils. Crop oils are nonphytotoxic mineral or vegetable oils. They are commonly sold as combinations with 1% to 2% surfactant. Their function is similar to that of surfactants.

Crop oil concentrates. These are mineral or vegetable oils that contain up to 20% surfactant. Their principal function is to aid in cuticle penetration and reducing the surface tension of spray droplets. Crop oil concentrates also appear to be effective in increasing spray retention on certain weed species and prolonging drying time, thus allowing more time for water-soluble herbicides to penetrate the leaf.

Methylated seed oils. In recent years the use of the methyl esters of seed oil fatty acids has been investigated as additives or adjuvants with herbicides. Methylation of the fatty acids alters the properties of the starting materials. The main changes will most likely alter the water-oil affinity relationship. This in turn apparently can cause an increase in the herbicide absorption or uptake. The methylated seed oils are combined with additional surfactant-emulsifiers to further enhance their ability to mix with water and to promote penetration. Research will have to continue in order to more precisely define (a) what kind of seed oils function best after methylation, (b) what levels of methylation provide the best functionality, (c) what kinds and how much additional surfactant-emulsifier needs to be added to the methylated seed oil, and (d) with what herbicide chemistry and on what target weed species will the maximum effect of methylated seed oils be found.

Spray Modifiers

Spray modifiers are adjuvants that modify the spray solution in the tank on the way to the target or on the target itself. This group of adjuvants includes spreaders, stickers, spreader-stickers, foaming agents, and thickening agents.

Spreaders. Spreaders are added to an aqueous spray solution to increase the surface area of the spray droplet on the target. Spreaders are usually nonionic surfactants.

Stickers. Stickers are added to an aqueous spray solution to cause the spray droplets to adhere to the target surface. The primary function is to decrease wash-off during rainfall.

Foaming agents. Foaming agents are used to enhance herbicide action, reduce drift of sprays, and mark spray swath widths.

Thickening agents. Thickeners are materials added to spray solutions to increase the solution viscosity. These adjuvants are used to reduce spray drift from a target area.

Utility Modifiers

Utility modifiers are adjuvants that widen the range of conditions under which a given pesticide formulation is useful. These include antifoam, compatibility, and buffering agents.

Antifoam agents. Antifoam agents are added to a spray solution to decrease or prevent foaming when the solution is agitated or sprayed.

Compatibility agents. Compatibility agents are added to a spray solution to permit or maintain the emulsification of two or more ingredients that would otherwise separate when mixed. The most common use of these agents in agriculture is in applying pesticides in combination with liquid nitrogen and fertilizer.

Buffering agents. Buffering agents are added to an aqueous spray solution to increase the dispersion and/or solubility of a pesticide. Buffering agents are used in areas of extremely acid or alkaline water.

WHAT DOES THE LABEL SAY?

No single adjuvant can perform all adjuvant functions. However, different compatible adjuvants can be combined to perform multiple functions simultaneously. The need for adjuvants is, for the most part, determined by the requirements, recommendations, or suggestions on pesticide labels. The pesticide label should be consulted before any adjuvant use has been determined. Many pesticides already have an adjuvant added, and it is not necessary to add additional materials. There are, however, a fairly large number of EPA-registered pesticides that have very specific recommendations on their labels for the use of one or more adjuvants. Adjuvant information will usually fall into the following categories.

- Pesticide labels that specifically require the use of adjuvants
- Pesticide labels that suggest the use of adjuvants
- Pesticide labels that specifically prohibit the use of adjuvants
- Pesticide labels that neither require nor suggest adjuvant usage (These labels are without adjuvant information but do not prohibit adjuvant usage.)
- Pesticide labels that contain a combination of some or all of the previously listed information
- Pesticide labels that when used alone are void of adjuvant recommendations or prohibitions, but when tank mixed with other pesticides have suggestions or requirements for adjuvants. (Where two or more pesticides containing adjuvant recommendations are tank mixed, the one with the most restrictive labeling takes precedence.)

Although the information on pesticide labels concerning adjuvants can be confusing, it is important to take the time to understand this information because most EPA-registered pesticide labels have a statement on them that reads, "It is a violation of federal law to use this product in any manner inconsistent with its labeling." Adjuvant usage in conflict with the label can constitute such a violation. Other sources of adjuvant information include technical literature, technical data sheets, material safety data sheets (MSDSs), supplemental labels, and promotional literature.

FUNCTIONS OF ADJUVANTS

The spray solution may have one or more of the following nine functions to perform in order to provide a safe and effective application:

1. *Wetting of foliage and/or pest.* Adequate wetting is required to provide good retention and coverage of spray solutions. A suitable surfactant, at the proper concentration, will normally suffice, although certain plants and pests may have special requirements.

2. *Modifying rate of evaporation of spray.* The need for reducing the rate of evaporation of a spray solution applied at 2 to 3 gallons per acre in a hot, dry area is obvious. However, need may be equally great in the application of a concentrate spray in an orchard. Once the spray has been applied, it may be desirable to have the spray dry as rapidly as possible. Both functions can be performed by the proper adjuvant.

3. *Improving weatherability of spray deposits.* Resistance to heavy dews, rainfall, and sprinkler irrigation can mean the difference between successful control and failure of a fungicide application, for example.

4. *Enhancing penetration and translocation.* Many chemicals perform most effectively when they have been absorbed by the plant and transported to areas other than the point of entry. Systemic pesticides have this ability. Their absorption can be enhanced and certain nonsystemic chemicals can be made to penetrate plant cuticles through the use of a suitable adjuvant.

5. *Adjusting pH of spray solutions and deposits.* Many currently used pesticides (primarily organic phosphates and some carbamates) degrade rapidly under even mildly alkaline conditions found in some natural waters and on certain leaf surfaces. Buffering adjuvants can prolong the effective life of alkaline-sensitive chemicals under these conditions.

6. *Improving uniformity of deposit.* It is almost axiomatic that with nonsystemic pesticides the quality of performance of the pesticide can be no better than the quality of the spray deposit. This is particularly true of most fungicides, which require complete and uniform coverage.

7. *Compatibility of mixtures.* With the savings in labor costs to be obtained from doing more than one job with a single application, the effort is made frequently to mix various combinations of pesticides or pesticides with liquid fertilizers in the spray tank for simultaneous application. The attendant compatibility problems frequently can be corrected with the proper adjuvant.

8. *Safety to the crop.* Certainly, we do not wish to harm the crop we are trying to protect. This often happens, however, with chemicals that are potentially phytotoxic. The hazard can be increased through the use of the wrong adjuvant or substantially reduced through the choice of a proper one.

9. *Drift reduction.* No method currently in use for reducing drift of pesticide sprays is entirely satisfactory. The most promising of the newer approaches to drift reduction is the use of special adjuvants applied through conventional aerial or ground equipment. These products are primarily polymer spray thickeners or viscosity enhancers.

EPA Approval

The labeling of spray adjuvants is basically unregulated, and there is little consistency from one label format to another. The one regulation that applies uniformly across the country is that all chemicals to be sprayed into the environment must have a tolerance established by the EPA or be classified "tolerance exempt." Otherwise, they are illegal in pesticide sprays.

PRACTICAL CONSIDERATIONS

Whenever possible, use the exact adjuvant specified on a pesticide label. When a specified brand of adjuvant is not available, make sure that the substituted product has the same general characteristics. For a surfactant, check the following:

1. Does it match the ionization characteristics of the recommended brand (nonionic, anionic, and so on)?

2. What percent of surfactant is in the product? Ranges of 50% to 100% surfactant are typical in commercial products, the remainder being water or alcohol. If the concentration is different from the recommended brand, you will have to adjust the rate you use accordingly. One quart of 50% surfactant is equal to 1 pint of a 100% product, and so on.

If the product is a crop oil or crop oil concentrate, check the following:

1. Is the source oil the same as that specified on the label (vegetable, mineral, and so on)?
2. Is the amount of emulsifier the same as in the specified product? Many pesticide labels classify only the type of oil source and not the brand of adjuvant or the required percentage of emulsifier. When this is the case:
   - Buy a reputable brand from a reputable dealer.
   - Use a product that has 15% to 20% emulsifier by weight. Products with less emulsifier will cost less but may not work as well. For many vegetable oil–based concentrates, the difference between 10% and 15% emulsifier is the difference between no added performance and enhanced performance.
   - When only a type of adjuvant, and not a brand, is specified, it may pay you to make comparisons between several reputable brands. Small differences in an adjuvant can make large differences in adjuvant-pesticide performance. It is possible that there would be enough difference between brands to make comparison testing worthwhile.
   - Be careful when mixing adjuvants. When mixing more than one adjuvant with a pesticide for the first time (such as a surfactant as an activator and another product as a thickener to reduce the risk of drift), it is desirable to test the mixture before using it on large areas. First, do a standard compatibility test with the pesticide and both adjuvants. Then apply the mixture and check spray characteristics and performance on the intended crop.
   - If the label does not call for an adjuvant, think twice about using one. If the label says don't use an adjuvant, don't. You will just be spending money for something you do not need.
   - Be careful about adjuvant rates. Proper adjuvant rates are just as critical to performance as proper pesticide rates. We know that the characteristics of some adjuvants change markedly as the concentration and solution change. Adding more is not always better, and a little less may not do anything at all.
   - Remember that different environmental conditions and pests will most likely change the way the adjuvant affects the performance of the pesticide. What helps during cool weather may not be necessary during hot, dry conditions, or vice versa. Make sure the label calls for an adjuvant for the particular pest you are trying to control. If the label does not differentiate between weeds, for example, observe the results carefully and make adjustments as required to control the most difficult or serious weeds.

PROBLEM AREAS

Table 10–1 is a summarization of problem areas encountered with agricultural chemicals, and suggestions for adjuvant use in solving the problems. It may be of benefit in the identification of what product to use under what circumstance. This chart is only a guide. The first source for information is the pesticide label. The label recommendations will supersede any general recommendations for use on adjuvant labels.

TABLE 10–1 PROBLEM AREAS ENCOUNTERED WITH USE OF AGRICULTURAL CHEMICALS

Phase/Stage	Problem	Cause	Impact/Ramification	Solution
I. Mixing and storage: Normal sequence of events as the product is mixed and applied. Prolonged storage of mixed chemical is not part of the normal routine. Occasionally conditions will prevent timely application (weather, equipment breakdown, etc.).	Foaming	Excess or vigorous agitation. Emulsifying, wetting, or dispersant system of the pesticide. Soft water.	Foam (trapped air bubbles): can affect mixing, calibration, and distribution. Additionally, can cause overflow and contamination of the outside of equipment, exposing people to unnecessary hazard and requiring cleanup.	Utilize an antifoamer/defoamer material. Antifoam will break foam after it forms. Defoamer will prevent its formation. If materials are known to have foaming problems, product can be used as a preventative.
	Settling/Suspension	Primarily a problem with wettable powders, flowables, and water-dispersible granules. Particles are insoluble and will settle if they are large or if agitation is poor or is stopped for any length of time.	Nonhomogenous mixtures can produce erratic control, plugged screens and/or nozzles.	Utilize a thickening or suspension agent. These are usually clays, chlorides, or gums. They will extend the period of suspension. If settling does occur when these products are used, resuspension will be much easier.
	Incompatibility	Undesirable chemical reactions in combinations of pesticides and/or liquid fertilizer materials; these reactions produce gelling, curdling, or separation of products.	Nonhomogenous mixtures can produce erratic results, plugged screens and nozzles. Severe curdling or gelling may even prevent application entirely.	Know if mixtures to be used are compatible. Check the label. Perform a "jar test." If incompatibility occurs, determine if the use of a compatibility agent will make the mixture usable.
	Degradation	The breakdown of chemicals due to high pH of the water carrier is called alkaline hydrolysis.	Breakdown of the pesticide may happen to the extent that effectiveness may be reduced. The extent of the problem depends on susceptibility of pesticide, alkalinity of water, and time in contact with the water. Most water in the United States is alkaline.	Know the pH of water used as a carrier. Know the susceptibility of the chemical. If there is a potential problem, lower the pH of the water prior to mixing. Use an approved acidifier material.

(continued)

265

TABLE 10-1 *(Continued)*

Phase/Stage	Problem	Cause	Impact/Ramification	Solution
II. Application period: The time that elapses from discharge from spray nozzle until it contacts the target.	Drift	Wind, speed of application vehicle, spray pressure, nozzle design; small particles move from target area.	Spray droplets missing target are wasted; loss of control may pose hazard to people, livestock, or to nontarget crops and vegetation.	New nozzle design and orientation, rotary nozzles. Chemical thickening agents to reduce the number of very fine droplets. Usually effective reduction of drift will employ several countermeasures, not just a chemical additive.
III. Spray droplets in contact with target.	Coverage	Coverage of the spray droplets can be reduced by "beading" of the individual drops on the leaf surface. The main causes are surface tension, waxy-oily leaf surfaces, and hairy or pubescent leaves.	Lack of contact area translates to lower control levels for insecticides, fungicides, and herbicides.	Lower the surface tension of the spray so that the droplets will lie down or collapse to cover a wider area. Also, this will allow droplets to penetrate through hair structures to the leaf surface. Use a surfactant or wetting agent.
	Adherence	Surface tension of droplets cause "bounce" or leaf angle allows runoff. Rainfall or sprinkler irrigation may wash off applied pesticide. Wind and plant growth can also remove chemical deposits.	Lack of adherence translates to physical loss and reduces performance of the pesticide.	Utilize a "true sticker" or adhesion agent to firmly stick the pesticide droplet to the leaf.
	Penetration	Waxy-oily surfaces along with thick leaf cuticles can slow or resist penetration of chemicals that must be taken into the plant to be effective. Most herbicides, along with systemic insecticides and fungicides, fall into this category. Foliar nutrients.	Lack of penetration most likely results in reduced effectiveness of the pesticide.	Utilize a "true penetrant" additive. These are usually special surfactant combinations and not just general-purpose surfactant or wetting agents. They must be specifically designed to disrupt the covering on leaves to allow penetration to take place.

266

	Degradation	Natural phenomena such as temperature, moisture, sunlight, ultraviolet radiation, and bacteria, all work to break down or degrade the pesticide.	Under extremes of the factors listed under causes, breakdown rate can be fast enough to reduce effectiveness.	Utilize some form of protectant or extender. Film formers and encapsulants are currently the most popular. These products are usually combined with "sticker" properties.
	Attractiveness to insect pest	Odor or taste or lack thereof may make acceptance by insect feeders marginal.	Lower levels of insect control. Some insecticides (biologicals) require that they be consumed.	Add flavor or odor attractants to stimulate feeding by insects.
IV. Equipment after use.	Residues, rust, scale	Buildup of chemical residues over time. Lack of regular sanitation. Rust and scale can form also over time.	Residues that persist from one spray application to the next may cause damage to crops. New herbicides are now exceptionally concentrated, with rates as low as 0.1 oz being used. With these concentrated materials it is extremely important to thoroughly clean application equipment between uses. Scale and rust can plug screens and nozzles.	Clean equipment with a specifically designed cleaner neutralizer; follow pesticide label for guidelines on cleaning equipment.
	Stains	Certain chemicals, usually herbicides, can stain equipment.	Sight tubes and gauge faces can become obscured. Other strains probably have impact for esthetic reasons.	Utilize cleaner with stain-removal properties.
	Dirt, grease, oil	Buildup over time.	Removal more important for esthetics, but can be important for proper operation and ease of maintenance.	Cleaners used with pressure washing equipment.

WATER pH: HOW IT AFFECTS PESTICIDE STABILITY AND EFFICACY

The measure of the acidity or alkalinity (the pH) of the water you are using can greatly influence how pesticides and other products perform in the spray tank. The pH scale extends from 0 (extremely acidic) to 14 (extremely alkaline), with the middle value of 7 being exactly neutral at 25°C.

The pH of most well, lake, or stream waters falls within the range of 4 to 9. Most waters are slightly basic because of the presence of dissolved carbonate and bicarbonate salts. A deviation from the norm is caused when acidic or basic industrial, municipal, or natural waste discharges enter the water supply.

Physical or chemical incompatibility can arise from the following:

- Salts and other inorganic or organic compounds dissolved in the water
- Chemicals discharged from an industrial or municipal plant
- Products applied to the target pest

When the pH of the water and/or spray mix is strongly acidic or basic, you may have problems receiving satisfactory pest control.

The breakdown or hydrolysis of pesticides is measured in terms of half-life. For example, if a product is 100% effective when first added to water and it has a half-life of 4 hours, the effectiveness is cut in half (to 50%) in this time period. During the next 4 hours it is halved again. The shorter the half-life, the greater the effect of alkaline water. Many organophosphate insecticides are rapidly broken down into two or more smaller inactive chemicals when mixed with water that is alkaline. For example, azinphosmethyl (Guthion®), has a half-life of 17.3 days at a pH of 5, a half-life of 10 days at a pH of 7, and a half-life of only 12 hours at a pH of 9. Acephate (Orthene®) has a half-life of 65 days at a pH of 3 and only 16 days at a pH of 9. Carbamate insecticides, which include the widely used carbaryl (Sevin®), are also affected by the pH of water. The half-life of Sevin® is 100 to 150 days at a pH of 6.0, but only 24 hours at a pH of 9.0

Certain fungicides and herbicides are also affected by alkaline conditions and undergo hydrolysis in higher pH levels. Refer to Appendix G for further information listed under "Stability of Agri-Chemicals with Respect to pH Carriers/Diluents."

PESTICIDE COMPATIBILITY

Two or more chemicals frequently are combined:

- *To increase the effectiveness of one of the chemicals.* The manufacturer usually wants to increase the activity of its pesticide against specific pests. The manufacturer does this by adding a chemical referred to as a *synergist,* which is not an active pesticide but increases the effectiveness of the pesticide. Synergists usually increase the toxicity of the pesticide so that a smaller amount is needed to bring about the desired effect. This may reduce the cost of the application and also reduce the hazard, as less of the active material is used.
- *To provide better control than that obtained from one pesticide.* Applicators sometimes combine two or more active pesticides to kill a pest that has not been effectively controlled by either chemical alone. Many combinations are quite

effective, but in most cases it is not known if the improved control is a result of a synergistic action or an additive effect of the several chemicals on different segments of the pest population.

- *To control different types of pests with a single application.* Frequently, several types of pests need to be controlled at the same time, for example, insects, diseases, and mites. Generally, it is more economical to combine the pesticides needed and make a single application, in this case an insecticide, a fungicide, and an acaricide. However, the compatibility of the various chemicals must be known before the materials are combined.

Compatibility Problems in Certain Combinations of Chemicals

When two or more pesticides can be safely mixed together or used at the same time, they are said to be compatible. When they cannot be mixed, used together, or used at the same time, they are said to be incompatible. Some pesticides are incompatible because chemically they will not mix. Some pesticides will mix together well but the results are not the same as when they are used alone. Some combinations of chemicals result in mixtures that produce the opposite effect of synergism, known as *antagonism*. Antagonism, or incompatibility, may result in chemical reactions that cause the formation of new compounds or a separation of the pesticide from the water or other carrying agent. If one of these reactions occurs, one of the following may result:

- Reduced effectiveness of one or both compounds
- Precipitation, clogging screens and nozzles of application equipment
- Various types of plant injury (phytotoxicity), for example, russeting of fruits or vegetables, stunting of plants, and reduction of seed germination and production
- Excessive residues
- Excessive runoff

Some pesticides will not mix (are not compatible) with other pesticides or with liquid fertilizers in spray tank mixtures. For example, wettable sulfur cannot be mixed with Lorsban® or Morestan®. Some herbicides are not compatible with liquid fertilizers and herbicide oils. Anytime you plan to mix two or more pesticides, first make sure they are compatible. Follow the specific directions on the label to test for compatibility if you still have questions. Also remember that a pesticide may mix physically with another pesticide, but the activity of one or both may be altered (based on their chemical or biological incompatibility).

Formulation Sequence

If you use more than one pesticide formulation (WP, WDG, DF, L, EC, and so on) in a spray tank, there is a proper order for adding them.

1. Dry materials go into the spray before liquids. If a wettable powder (WP) is used, put it in first as follows: Make a slurry with the WP by adding a small amount of water to it until it forms a gravylike consistency. Slowly add this slurry to the tank with the spray tank agitator (mixer) running.

2. Dry flowables (DF) or water-dispersible granules (WDG) go in second. Flowables should be premixed (1 part flowable to 1 part water) and poured slowly into the tank.

3. Liquid flowables (F or FL) should be added third. (*Exception: When using Furadan 4F*®, this material should be put in last.) Liquids should also be premixed (1 part liquid chemical to 2 parts water or liquid fertilizer) before blending in the tank. Many labels will give you the proper mixing sequence.

4. Liquids, like emulsifiable concentrates (EC), should be combined last.

Potentiation

Another less familiar but extremely important undesirable effect of combining certain pesticides is potentiation. Some of the organophosphorus pesticides potentiate or activate each other as far as animal toxicity is concerned. In some cases, the combination increases the toxicity of a compound that is normally of very low toxicity to one that is highly toxic to people and other animals.

Phytotoxicity, excessive residue, or poisoning of livestock can also occur when one chemical is applied several days after the application of a different chemical. Check the pesticide label carefully for such warnings.

Before you mix pesticides, or a pesticide and another material such as a fertilizer, you should be certain that once they are mixed you will get the best benefit from both. One reason for combining pesticides is to save time. However, if the two materials are incompatible, you may lose valuable time. Besides time, you may also lose the chemicals (which are costly), or you may get poor results or injure the crop.

The pesticide label will sometimes indicate incompatibility problems. Some pesticide formulations are prepared for mixing with other materials and are registered for premixes or for tank mixes. If this is true it will be so indicated on the label. Make sure materials are compatible and cleared for the combination before mixing or using them together.

Because of the risks involved, combinations of pesticides and/or other agricultural chemicals should not be used unless a specific combination has been proved to be effective, tested for side effects, and accepted for registration by the EPA and the state Department of Agriculture.

11

Pesticide Application Equipment

GENERAL INFORMATION

Proper application is the key to the success of any pesticide treatment. Simply stated, the application process is getting the pesticide to the target. This process usually involves a carrier, which may be liquid, dry, or air, to transport the pesticide to the intended surface or target. Application may range from the simple act of spraying a repellent on our skin or painting a surface with a brush to the use of very elaborate and expensive application equipment.

Many factors affect our ability to place a pesticide on the target in the manner and amount for most effective results, with the least undesirable side effects, and at the lowest possible cost. Certainly the selection and use of equipment are of utmost importance and deserve major emphasis when considering pesticide application. However, without proper consideration to calibration, formulation, adjuvants, compatibility, and use records, successful application is not very probable. Successful application must also involve principles of proper timing and drift control.

Drift

The control of drift has become an important item of concern with the custom applicator as well as the private applicator. To be effective, the pesticide must be applied precisely on target at the correct rate, volume, and pressure.

Why be concerned with drift? There are many reasons why the pesticide applicator should be concerned with drift and ways to minimize damage that it may cause.

- Pesticides are expensive. Do not waste them by letting them get off target.
- Pesticides are undesirable in the general environment for many reasons: smell, appearance, and danger to wildlife and nontarget plants. Groundwater must be guarded against contamination.
- Pesticides can damage sensitive crops in surrounding areas. Recent federal laws place rigid penalties on both private and custom applicators for misapplication. Thus, an applicator who permits chemicals to drift off the target area to injure another crop may be faced with a lawsuit that requires many hours of litigation, or may even put him or her out of business.

What is drift? Particle drift is the movement of spray particles or droplets away from the spray site before they reach the target plant or ground surface. This is usually physical movement. Vapor drift can also occur, and is the movement of chemical vapor out of the target area.

Some chemicals vaporize at ordinary temperatures or on exposure to air. Physical drift is normally due to wind movement. However, a temperature inversion or convection currents, which usually occur in temperatures above 85°F, can prevent the settling of very small droplets and allow their movement away from the spray site during relatively calm conditions.

In agriculture, drift is only of concern when pesticides are deposited other than where we intended them to be. Drift is influenced by many factors. Spraying pressure, particle size, specific gravity, nozzle design, evaporation rate, height of release, horizontal and vertical air movements, temperature, and humidity are among the important considerations. Since so many variables play an active role in drift, some of them are discussed in detail to provide insight into drift control.

- *Pressure.* Some equipment operators attempt to correct drift by varying pressure. Droplet size is influenced by pressure, but rate and pattern configuration are affected even more. The higher the pressure, the smaller the droplet size. With smaller droplets, better coverage is gained, resulting in higher chemical performance at the expense of drift control. Reducing pressure will enhance drift control because of larger droplets, but it will also reduce coverage. Evaluations have demonstrated that a constant pressure setting produces the best results.
- *Nozzles.* In terms of drift control, nozzle selection is probably more important than pressure. Nozzle tips that give the largest droplet size and application rate acceptable with the proper pressure range are generally recommended. The larger droplet size will help reduce drift and deliver chemicals on target. Position and orientation of nozzles on the spray boom are also important. Droplet size distribution can be greatly affected by nozzle orientation.
- *Specific gravity.* Small, lightweight particles fall much slower than large droplets. For example, a 5-micron ($1 \mu = 0.00004$ inch) drop of water would require $5 \frac{1}{2}$ hours to fall 50 feet in still air as compared to only 55 seconds for a 100μ drop under the same conditions. It is the small droplets of lower specific gravity that will drift from the target area. Furthermore, water droplets and oil droplets react differently in the presence of wind movement. Because they are much lighter, oil spray droplets remain airborne longer.

Although a coarse spray will drift less than a fine spray, the coarse spray results in narrower swaths with fewer drops per unit of area. In applying certain herbicides, this may be beneficial, as thorough coverage may not be critical as long as the material reaches the target area. However, for more complete coverage, a finer spray may be necessary. When the average drop size is doubled, the number of drops produced from a gallon of spray will be reduced to one-eighth the original number. Generally, spray droplet size should be no finer than is necessary to do an effective job.

- *Height of nozzles.* Nozzles positioned too high will disperse spray over a wider area, but increase the likelihood of drift. Spray particles must fall a greater distance. The applicator must determine the desired swath width by striking a balance among nozzle size, pressure, and height above the target. For example, if an operator increases the application pressure and thereby increases the rate but maintains the swath width, then he or she should lower the nozzle while staying within manufacturer specifications to compensate for the increased pressure. If the pressure is increased correspondingly, there will be an increase in the application rate, and also an increased probability of drift.

- *Horizontal and vertical air movement.* Unless it is calm, most fields are subjected to variable air movement. Unpredictable changes in air movement may occur at any time to cause drift of spray being applied by ground or aerial sprayers. Thus, weather conditions directly affect the direction, amount, and distance of drift. Furthermore, as the ground warms up during the day, ground-air temperature increases rapidly. This warmer air rises and sets up convection and thermal air currents to lift small spray particles suspended above the target area. Horizontal air movement then carries these droplets to some distant point before they can settle out. The air temperature differential between the ground and 10 to 20 feet up is considerably less during early morning and late evening hours than during the middle of the day. On the other hand, during a temperature inversion the air near the ground is cooler than the air above, and minute spray droplets remain suspended in the layer of cold, undisturbed air. Eventually these mists move out of the area before coming to rest. This kind of drift is difficult to control.

- *Temperature and humidity.* The rate of droplet evaporation is determined to a great degree by temperature and humidity. As the diameter of the droplet decreases, the ratio of surface area to volume increases and evaporation occurs at a faster rate. The time of exposure to evaporation conditions also increases with decreasing droplet size because of slower fall.

Vapor drift, unlike spray or dust drift, is related directly to the chemical properties of the pesticide. Most problems are the result of accidental or unintentional misuse. If the applicator understands the chemical properties, she or he can control and avoid damage from vapor drift.

Vapor drift can be due to vapor leakage. Vapor leakage can be stopped by properly sealing fumigants or other volatile materials after they are applied and by applying these materials with vaportight equipment. Many volatile pesticides cause safety problems from vapor drift. For example, ester formulations of phenoxy ester herbicides may volatilize and drift under high temperature conditions. Amine or acid formulations should be used where vapor drift might cause problems.

The careful use of pesticides is of prime concern to everyone today. Many factors interact to influence the distance material will drift from the target area. Even when common sense and good application technology are adhered to, drift continues to present a problem for the applicator. Label instructions must be followed and strict attention must be given to the control of pesticide drift. To minimize spray drift, the following precautions should be taken:

- Use lower spraying pressures to reduce the number of fine spray droplets produced.
- Keep boom height as low as possible while maintaining uniform coverage.
- Use nozzles that are more resistant to drift, such as extended-range flat sprays, low-pressure flat sprays, or wide-angle, full-cone spray tips.
- Replace and clean faulty or plugged nozzles that may cause fogging.
- Use spray additives to increase the liquid viscosity at the nozzle orifice so that fewer fine droplets are produced.
- Spray when wind speeds are 5 mph or less; early morning or early evening are often good choices.

TYPES OF APPLICATION EQUIPMENT

Spray application equipment of the future will be complex, interconnected systems of mechanical, electronic, and computerized components. Integrated computer-controlled spray systems and electronically synchronized controls will soon replace the simple on/off switches of the present. Computer-aided spray equipment will allow applicators to apply products in a site-specific manner and enable them to meet future environmental recordkeeping requirements.

Most application equipment can be used for several different kinds of pest problems. By choosing the type of equipment best suited for the type of operation, the applicator can be assured of doing a good job. The agricultural applicator's equipment differs greatly from that of a structural pest control operator. Even when specializing in a specific type of pest control, the pesticide applicator will need to make a choice of equipment. The choice will depend on working conditions, pesticide formulations, type of area treated, and possible problems that will be encountered. While large power equipment may be desirable for some jobs, other problems may be best handled by using small portable or hand equipment.

Aerial Equipment*

Fixed-wing aircraft (Figure 11–1). Approximately 7500 aircraft are used each year in the United States for aerial application. Most of them are single-engine aircraft and are either high-wing monoplanes, low-wing monoplanes, or biplanes.

---

*See additional information at the end of this chapter.

Figure 11–1   Fixed-wing aircraft.   **Figure 11–2**   Helicopter.

These type of aircraft are used on smaller jobs. Multiengine aircraft are being used on large areas such as forests and rangelands and also for forest fire control.

Advantages.   Aerial application offers a fast, convenient method for pest control, especially when quick action must be taken. Aerial application also allows for treatment of pests when a field is too wet or muddy to allow ground application equipment to operate.

Disadvantages.   Fixed-wing aircraft generally cannot treat small fields and are difficult to operate in areas where many hazards are present, such as power lines and tall trees. Application costs are generally higher with aircraft than with ground equipment, but the speed of application and timeliness may offset this cost differential.

Helicopters (Figure 11–2).   The helicopter has shown substantial increase in use in recent years, but the total flight time in comparison to other aircraft is still quite small.

Advantages.   Helicopters in agriculture offer certain advantages over fixed-wing aircraft, such as slower speeds, safety, accuracy of swath, coverage and placement of chemical, and ability to operate without an airport.

Disadvantages.   The complicated construction means higher initial cost and maintenance and therefore a higher per acre application charge. Here again, however, this may not be a disadvantage if quick control of the pest problem is essential.

Ground Equipment

Low-pressure boom sprayers (Figure 11–3).   These sprayers are usually mounted on tractors, trucks, or trailers. They are designed to be driven over fields or large areas of turf, applying the pesticides in swaths to the crop. Low-pressure sprayers generally use a relatively low volume of dilute spray ranging from 5 to 40 gallons per acre applied at 30 to 60 pounds of pressure. These sprayers are designed to handle most of the spraying needs on general farms, and there are perhaps more low-pressure boom sprayers in use than any other type of equipment. They usually have centrifugal-type pumps that limit their pressure to about 100 pounds per square inch. Handguns can be attached for remote spraying for spot treatment and patches of weed infestations.

Figure 11-3 Boom sprayer.

Figure 11-4 High-pressure sprayer.

Advantages. Low-pressure sprayers are relatively inexpensive and light-weight, are adapted to many uses, and can cover large areas rapidly. They are usually low volume so that one tankful will cover a large area.

Disadvantage. Low-pressure sprayers cannot adequately penetrate and cover dense foliage because of their low pressure and gallonage rate.

High-pressure sprayers (Figure 11–4). High-pressure sprayers are often called hydraulic sprayers. They operate with dilute sprays, and pressures can be regulated up to several hundred pounds. They are used for spraying shade trees and ornamentals, livestock, orchards, farm buildings, and unwanted vegetation where dense foliage requires good penetration.

Advantages. High-pressure sprayers are useful for many different pest control jobs. They have enough pressure to drive a spray through heavy brush, thick cow hair, or to the tops of tall shade trees. Because they are strongly built, they are long lasting and dependable. Diaphragm pumps are generally used and resist wear by gritty or abrasive materials. Mechanical agitators are standard and keep wettable powders well mixed in the tank. With a long hose and handgun, trees, shrubs, or other targets in hard-to-reach places can be treated. This kind of attachment is commonly used by commercial sprayers for applications to ornamentals.

Disadvantages. High-pressure hydraulic sprayers have to be strongly built and so are heavy and costly. They generally use large amounts of water and thus require frequent filling.

The National Arborist Association, Inc., has formulated suggestions for the proper and most efficient use of high-pressure sprayers. Their suggestions are offered here:

- High-pressure, high-volume hydraulic sprayers are used to apply pesticides to tall shade trees as well as landscape plants. When properly adjusted, and when used with the appropriate hose and nozzle, such equipment is an effective and efficient delivery system.

- The tops of tall trees are reached by maximizing the volume of material that comes from the nozzle, *not* by increasing the pressure. If the pressure is increased beyond the capacity of the hose and/or nozzle, the spray will come out as a fine mist which will not travel very high and will be subject to any breeze.

- If the inside diameter of the hose, its fittings, or the orifice of the nozzle are too small for the volume of material that the pump is delivering, the same things will occur. If a swivel is used between the nozzle and the hose, its inside diameter must be the same as the spray hose.
- Use of the proper combination of pump, pressure, hose, fittings, and nozzle will provide maximum volume of delivery at the nozzle, permitting a higher, straighter trajectory of the material. This will result in a large droplet size pattern, avoiding premature atomization (small droplet size) of the material, thereby minimizing drift.

The following recommendations should be considered:

1. Hydraulic sprayers
   - Hydraulic sprayers used for spraying trees more than 60 feet in height should have a pump capacity of not less than 35 gallons per minute.
   - Hydraulic sprayers used for spraying trees less than 60 feet in height must have a pump capacity sufficient to reach the tops of the trees to be sprayed.
   - Tank capacities should not be less than 400 gallons to provide sufficient material at the job site.
   - Spray tanks must have either mechanical or jet agitation.
   - To provide maximum height and a minimum of spray drift, the pump should operate at such pressure (may be up to 800 psi) to produce a pressure of about 400 psi when the gun is open and equipped with the largest size spray tip (disc) that the pump will support at this pressure.
2. Spray hose
   - Spray hose should have sufficient inside diameter to deliver a minimum of two-thirds of the gallons per minute capacity of the pump.
   - Spray hose should have a minimum burst pressure of not less than two times the maximum operating pressure of the pump.
3. Spray nozzles
   - Spray nozzles (guns) (Figure 11–5) should have a sufficient capacity to deliver up to the gallons per minute rating of the pump.
   - Spray nozzles (guns) should be adjustable from a straight stream, to a fan, to shutoff.
   - Spray nozzles should have interchangeable tips to allow for increasing or decreasing volume of material.
   - Tips should be checked for wear and replaced as required.

The following are pesticide application techniques for hydraulic sprayers:

1. The applicator should position himself to take advantage of air movement with each tree or shrub; treat only designated trees or shrubs; treat so that the material carries into the tree being treated; when spraying a shade tree, start at the top working downward from side to side and gradually reduce the volume of spray as the lower extremities are reached; apply a sufficient amount of material to provide thorough coverage; avoid drenching; be sure that the pesticide reaches all of the leaves or bark, top or bottom, as the case requires; when treating plants near buildings, position himself so as to spray away from the building.

Types of Application Equipment

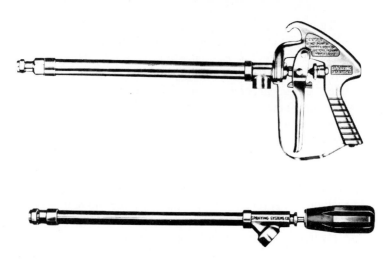

Figure 11–5 Hand spray guns.

2. When spraying a tall shade tree there are two very important techniques that must be considered.
- In order to compensate for the force of gravity, the proper distance can be determined by first aiming at the top of the tree and gradually raising the nozzle until the moving leaves indicate the top has been reached.
- Air movement can be used to drive the column of material up into the top of a tree rather than through it. By pointing the nozzle back over your shoulder and slightly away from the target, the air will push the material up and into the crown of the tree. (This principle is similar to that which applies when you are trying to water a garden that is 20 feet away and you only have 15 feet of hose. Raising the nozzle a few degrees enables you to reach the target.)

Air-blast sprayers. Practically all spraying in commercial orchards and much of the spraying on shade trees is with air-blast sprayers (Figure 11–6). Air-blast sprayers are primarily designed to carry pesticide-water mixtures under pressure from a pump through a series of nozzles into a blast of air that blows into the tree by means of a fan. High-volume fans supply the air, which is directed to one or both sides of the sprayer as it moves between rows of trees. Nozzles operating at low, moderate, or high pressure deliver the spray droplets into the high-velocity airstream. The high-speed air aids in breaking up larger droplets and transporting these smaller droplets for thorough coverage. Agitation of the spray material in the tank is usually accomplished with a mechanical agitator.

Advantages. A small amount of water covers a large area and very little operating time is lost in refilling. They are usually less tiring to operate than hydraulic sprayers and are particularly adapted to applying sprays over a large area.

Disadvantages. Since the pesticide is carried by an air blast, these types of equipment must operate under calm conditions. Windy conditions interfere with the

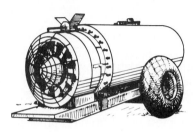

Figure 11–6 Typical air-blast orchard sprayer.

normal pattern of application of the blower. Larger models may not be able to treat hard-to-reach areas.

Low-volume air sprayers (mist blowers). Mist blowers are characterized by high air velocities and somewhat lower water volumes than conventional air-blast sprayers. This type of sprayer depends on a metering device, which may or may not be a conventional nozzle, that operates at low pressures and depends on the high-speed air for liquid breakup.

Advantages. A considerable savings in time and labor is possible with low-volume sprayers because less water is handled than with conventional air-blast sprayers. Small units that can be carried on a person's back are available for spraying small and difficult-to-reach areas.

Disadvantages. Calibration is more critical, and favorable weather for spraying is more essential. Coverage with very low volumes on some crops may be less satisfactory than with normal volumes delivered by conventional air-blast sprayers.

Ultralow-volume sprayers (ULV). Ultralow-volume spraying is accomplished by applying the chemical concentrate directly without the use of water or any other liquid carrier. Many ultralow-volume ground sprayers use a fan that delivers high-speed air to help break up and transport the spray droplets.

Advantage. The main advantage of ULV sprayers is the labor and time saved due to the elimination of water.

Disadvantages. There is an increased risk to the applicator from handling and spraying the concentrated pesticide. In addition, there are only a limited number of pesticides cleared for ULV application at the present time.

Aerosol generators (foggers). Aerosol generators and foggers break certain pesticide formulations into very small, fine droplets (aerosols). One droplet cannot be seen. But when large numbers of droplets are formed, they can be seen as fog or smoke. In some foggers, heat is used to break up the pesticide. These are called thermal aerosol generators. Other foggers break the pesticide into very fine particles by such means as rapidly whirling discs, air-blast breakup, or extremely fine nozzles. Foggers are usually used to completely fill an area with a pesticidal fog, whether it be a greenhouse, warehouse, or open recreational grounds. Insects and other pests in the treated area will be controlled when they come in contact with the aerosol fog.

Advantages. The droplets produced by foggers are so fine that they do not stick to surfaces within the area. Therefore, foggers using fairly safe formulations can be used in populated areas for mosquito or other insect control without leaving unsightly residues. The droplets float in the area and penetrate tiny cracks and crevices, or go through heavy vegetation to reach pests in hard-to-reach places. Because they blanket an area, it is difficult for pests to escape exposure.

Disadvantages. Since most of the droplets produced by foggers do not stick, little if any residual control is possible. As soon as the aerosol moves out of an area, other pests can move back in. Also, the droplets produced are so fine that they drift for long distances and may cause unwanted contamination or injury. Most aerosol generators require special pesticide formulations. When foggers are used, the weather conditions must be just right. For example, if an area is being treated for mosquitoes, rising air currents will carry the fog harmlessly over the pests and out of the area.

Dusters. Dusters blow fine particles of pesticide dusts onto the target surface. They may be very simply constructed. Dusters are used mostly by home gardeners, pest control operators, and truck gardeners for individual spot treatment of plants over small areas.

Advantages. Dusters are lightweight, relatively cheap, and fast acting. They do not require water.

Disadvantages. Dusts are highly visible, drift easily, and are difficult to control; therefore, dusters are less desirable for most crops or larger outdoor jobs.

Granular applicators. Granular equipment is designed to apply coarse, dry particles that are fairly uniform in size to soil, water, and in some cases foliage. Spreaders may work in several different ways including air blast, whirling discs, multiple gravity-feed outlets, and soil injectors. Broadcast or band spreaders can be used depending upon the needs. Many growers have turned to the Lock'n Load® closed handling system developed by American Cyanamid (Figure 11–7). Since its introduction in 1990, Lock'n Load® has set the standard as the simplest, most effective way to virtually eliminate exposure to insecticide dust. This has been a savior to users of insecticides such as Dyfonate®, Force®, Lorsban®, Counter®, and Thimet®.

DuPont, Zeneca, and Ingersoll-Dresser Pump Company are cooperatively developing the Smart Box®. It is a closed system, returnable-refillable container-applicator with electronic metering for granular soil insecticides. Testing showed that the Smart Box® accurately delivered 3 to 4 ounces of granules per 1000 row feet and could go lower with modifications. Most conventional applicators deliver 6 to 10 ounces of granules per 1000 row feet. With each Smart Box® containing enough granules to treat 15 acres, a set of boxes supplies a 12-row planter for a day's planting of 150 to 200 acres.

Advantages. Granular equipment, like dusting equipment, is light and relatively simple, and no water is needed. Because granules are uniform in size, flow easily, and are relatively heavy, seeders and fertilizer spreaders can be used to apply granules, often without any modification. Specially designed granular applicators such as the Lock'n Load® system and the Smart Box® are used to apply granules for soil insects.

Container as purchased

Container locked on
granule spreader box

Figure 11–7 Lock'n Load® granule spreader system.

Disadvantages. Because granular materials do not generally stick well to foliage, granule spreaders are not usually used on aboveground parts of plants. Therefore, the applicator will need other machinery for controlling most leaf-feeding insects and most plant diseases.

Soil injectors. Soil-injection equipment is frequently used to apply fumigation materials to the soil to control nematodes and other soilborne pathogens or insects. The most common method of soil application involves the use of chisel cultivators or shanks that have a liquid or granular tube down the back of the shank permitting materials to be placed in the soil to a depth of a foot or more. With volatile materials, the shanks may be spaced at 12 inches or more, and a continuous effective band or coverage will be obtained. Sweep-type elevator shovels, with a series of nozzles on the trailing edge, may be used to apply a single band or continuous cover beneath the soil surface.

Advantages. Since soil-applied materials are not as likely to cause phytotoxic damage as foliar applications, and since applications can be much more precise when injected into the soil, it has long been common practice to use undiluted technical or minimum dilution materials at very low rates of a gallon or less per acre.

Disadvantages. Because the pressure orifice or low application rates must be quite small, these are difficult to keep from plugging. This can be overcome, however, through the use of gravity-feed systems or the use of positive-displacement-type pumps, which are usually driven from a ground wheel of the applicator machine.

Hand Equipment

Hand-operated sprayers and dusters are most commonly used by individuals for their own relatively small pest problems. The custom applicator, however, will often find it convenient and efficient to have hand sprayers for small jobs that do not require larger powered equipment or that require only a small amount of spray. They are especially helpful on small jobs in hard-to-reach areas where spray equipment must be carried in.

Types of Application Equipment

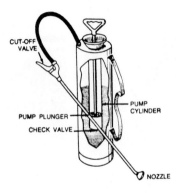

Figure 11–8 Compressed-air hand sprayer.

There are several types of hand application equipment:

- **Intermittent discharge sprayers,** which spray material with each stroke of the pump.
- **Continuous pressure sprayers,** which discharge spray as long as the pump is being operated.
- **Aerosol bombs,** which have pressurized cans or tanks with a discharge valve and nozzle and are essentially self-contained.
- **Knapsack granular applicators,** which are carried on the front of the applicator and are crank operated, spinning disc types.
- **Knapsack hand sprayers,** which are carried on the back with a capacity up to 5 gallons. A hand-operated piston provides the pressure.
- **Compressed-air hand sprayers** (Figure 11–8), in which air is pumped into a pressurized tank. These are usually designed to hold 1 to 5 gallons of spray mixture. Pressure can also be supplied with a carbon dioxide cylinder.
- **Hand dusters** range from small, self-contained units to larger wheelbarrow size units. Air velocity for dispensing the dust is created by a plunger, hand crank, or belt attached to a fan or blower.
  - *Advantages.* Hand sprayers and dusters are economical, uncomplicated, and lightweight, and yet will do a surprising amount of work and adapt to many different problems. The spray and dust are easily controlled as to direction and drift because relatively little material is used at low pressure.
  - *Disadvantages.* Hand sprayers and dusters are efficient and practical for small jobs only. Wettable powders tend to clog regular sprayer nozzles and agitation is frequently poor.

SPRAY EQUIPMENT PARTS

To do a proper job of applying pesticides, an operator must have the correct equipment and operate it correctly. A good spraying system consists of a number of parts assembled together in a manner that permits all parts to work together and function as a unit. Figure 11–9 illustrates the typical parts of a sprayer. Selection of a good quality sprayer will depend on careful attention to each of the following parts.

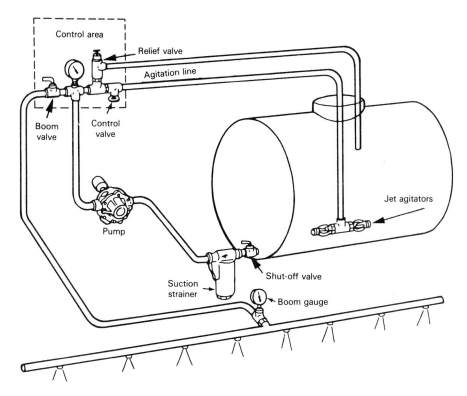

Figure 11–9 Schematic outline of a sprayer system.

Tanks and Agitators

Tanks. Tanks should be made of stainless steel or rust-resistant galvanized steel, fiberglass, or polyethylene to avoid rust, sediment, plugging, and restriction problems. Tanks from 55 to 150 gallons are available on mounted sprayers and 200 gallons or larger on pull-type sprayers. Saddle tanks are popular and give good agitation. Tanks need to have a large covered opening in the top with a removable strainer to make filling, inspection, and cleaning easy. Tanks should have a built-in sump with a drain plug in the bottom for complete drainage when cleaning.

Tank agitators. The return flow from the regulator valve usually gives sufficient agitation for solutions or emulsions. Wettable powder materials usually need more agitation to keep them in suspension. Agitation can be accomplished by mechanical or hydraulic means. Mechanical agitation (Figure 11–10) is more desirable for suspension sprays, but this type of agitation is difficult to secure in tractor- or trailer-mounted sprayers that use a tractor power takeoff to operate the pumps.

Hydraulic or jet agitation (Figure 11–10) is most commonly used on low-pressure sprayers. The fluid is circulated by returning a portion of the pump output to the tank and discharging it under pressure through holes drilled in a pipe running the entire length of the tank or through special agitator nozzles. Jet agitation

Spray Equipment Parts

283

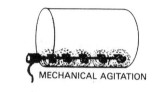

MECHANICAL AGITATION

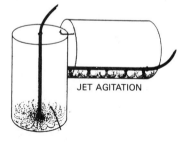

JET AGITATION

Figure 11–10 Mechanical and jet (air) agitation.

is simple and effective provided that the device is installed correctly and there is sufficient flow. The agitator orifices should receive fluid from a separate line on the discharge side of the pump and not merely from the bypass line.

The amount of flow needed for agitation depends on the chemical used, as well as on the size and shape of the tank. For a simple orifice jet agitator, a flow of 6 gallons per minute (gpm) for every 100 gallons of tank capacity is usually adequate. Several types of siphon attachments are available that will help stir the tank with less flow. If these are used, the agitator flow from the pump can be reduced to 2 to 3 gpm for every 100 gallons of tank capacity. Foaming can occur if the agitation flow rate remains constant as the tank empties. This condition can be prevented by using a control valve to gradually reduce the amount of agitator flow.

The jet stream coming from a jet agitator should pass through at least 1 foot of liquid before striking the side or bottom of the tank. Abrasive materials under high velocity will eventually wear a hole in the tank unless it passes through sufficient solution. If the sprayer is stopped, even briefly, it may be necessary to mechanically work the powders back into suspension so that the jet agitator can keep them there.

Pumps

The pump is the heart of the sprayer. There are many types of pumps on the market, with advantages and disadvantages specific to each pump. The five most important factors to consider in selecting a pump are as follows:

1. *Capacity.* The pump should have sufficient capacity to supply the boom output and provide for bypass agitation. The boom output is calculated from the number and size of nozzles used. Hydraulic agitation, if used, should be added to the boom output. Select the pump that will provide the boom output plus about 50% more for agitation through the bypass return. Pump capacities are given in either gallons per hour (gph) or gallons per minute (gpm).

2. *Pressure.* The pump must be able to produce a desired operating pressure at the capacity required for the spraying job to be done. The amount of pressure is

indicated in pounds per square inch (psi). Some pumps designated for low pressures can produce high pressures but will wear out rapidly if operated under high-pressure conditions.

3. *Resistance to corrosion and wear.* The pump must be able to handle the chemical spray materials without excessive corrosion or wear. Some pumps will handle abrasive materials like wettable powders with much less wear than others. Chemical reaction and corrosion affect certain materials more than others.

4. *Repairs.* Pumps should be designed so that repairs can be made economically and quickly.

5. *Type of drive.* The pump should be readily adaptable to the available power source. Most farm tractors are designed to operate at a power takeoff speed of 500 to 1000 revolutions per minute (rpm). High-speed pumps will require an auxiliary power source or speed step-up mechanisms. Roller pumps can be attached directly to the tractor power takeoff (PTO) but should be secured with a chain to prevent the case from rotating.

Figure 11–11 illustrates the five most popular pumps and compares their features and advantages and disadvantages.

Centrifugal and turbine pumps. Centrifugal pumps are the most popular type for new low-pressure sprayers. They are durable and simply constructed and can readily handle wettable powders and abrasive materials. Because of the high capacity of centrifugal pumps (to 120 gpm or more), hydraulic agitators can be used to agitate spray solutions even in large tanks.

Pressures up to 100 psi are developed by centrifugal pumps, but discharged volume drops off rapidly above 30 to 40 psi. This *steep performance curve* is an advantage, because it permits controlling pump output without a relief valve. Centrifugal pump performance is very sensitive to speed, and inlet pressure variations may produce uneven pump output under some operating conditions.

Centrifugal pumps should operate at speeds of about 3000 to 4500 rpm. When driven with the tractor PTO, a speedup mechanism is necessary. A simple inexpensive speed increaser is a belt and pulley assembly. Some pumps use a planetary gear system. The gears are completely enclosed and mounted directly on the PTO shaft. Centrifugal pumps can be driven by a direct-connected hydraulic motor and flow control operating off the tractor hydraulic system. This allows the PTO to be used for other purposes, and a hydraulic motor may maintain a more uniform pump speed and output with small variations in engine speed. Pumps may also be driven by a direct-coupled gasoline engine, which will maintain a constant pressure and pump output independent of tractor engine speed.

Centrifugal pumps should be located below the supply tank to aid in priming and maintaining a prime. Also, a pressure-relief valve is not needed with centrifugal pumps. The proper way to connect components on a sprayer using a centrifugal pump is shown in Figure 11–12. A strainer located in the discharge line protects nozzles from plugging and avoids restricting the pump input. Two control valves are used in the pump discharge line, one in the agitation and the other to the spray boom. This permits controlling agitation flow independent of nozzle flow. The flow from

TYPES OF PUMPS

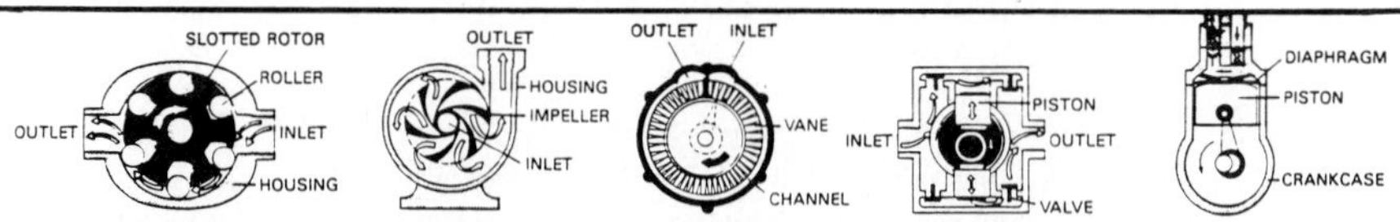

	ROLLER	CENTRIFUGAL	TURBINE	PISTON	DIAPHRAGM
MATERIALS HANDLED	Emulsions and nonabrasive materials	Any liquid	Most; some turbines may be damaged by abrasives	Any liquid	Most; some chemicals may damage diaphragm
RELATIVE PURCHASE PRICE	Low	Medium	Medium	High	Medium
DURABILITY	Pressure decreases with wear	Long life	Long life	Long life	Long life
PRESSURE RANGES — (psi)	0-300	0-75	0-60	0-1000	0-850
OPERATING SPEEDS — (rpm)	300-1000	2000-4500	600-1200	600-1800	200-1200
FLOW RATES — (gpm)	1-50	0-120	10-80	5-60	1-60
ADVANTAGES	• Low cost • Easy to service • Operates at PTO speeds • Medium volume • Easy to prime	• Handles all materials • High volume • Long life	• Can run directly from 1000 rpm PTO • High volume	• High pressures • Wear resistant • Handles all materials • Self-priming	• Wear resistant • Medium pressures
DISADVANTAGES	• Short life if material is abrasive	• Low pressure • Not self-priming • Requires speed-up drive or high-speed hydraulic motor	• Low pressures • Not self-priming • Requires speed-up drive for 540 rpm PTO shafts	• High cost • Needs surge tank	• Low volume • Needs surge tank

Figure 11–11 Comparison of features and advantages and disadvantages of the five most commonly used pumps.

centrifugal pumps can be completely shut off without damage to the pump. Spray pressure can be controlled by a simple gate valve, eliminating the pressure-relief valve with a separate bypass line. A special throttling valve to control the spray pressure and agitation flow is normally used. Electrically controlled throttling valves are popular for remote pressure control.

A boom shutoff valve allows the sprayer boom to be shut off while the pump and agitation system continue to operate. Electric solenoid valves eliminate the need for chemical-carrying hoses to be run through the cab of the vehicle. A switch box that controls the electric valve is mounted in the vehicle cab. This provides a safe operator area if a hose should break.

To adjust for spraying with a centrifugal pump, open the boom shutoff valve, start the sprayer, and open the throttling valve to the desired spraying pressure. Then adjust the control valve for proper agitation. If spraying pressure drops, readjust the control valve to restore desired pressure. Check to be sure flow is uniform from all nozzles.

Pesticide Application Equipment   Chap. 11

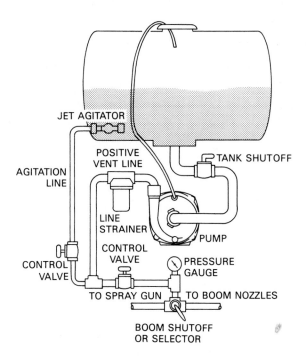

JET AGITATOR

POSITIVE
VENT LINE

TANK SHUTOFF

AGITATION
LINE

LINE
STRAINER

PUMP

CONTROL
VALVE

CONTROL
VALVE

PRESSURE
GAUGE

TO SPRAY GUN   TO BOOM NOZZLES

BOOM SHUTOFF
OR SELECTOR

Figure 11–12 Centrifugal pump and typical system hookup.

Turbine pumps are also available for low-pressure sprayers. These pumps are similar to centrifugal pumps except that they can provide the normal capacity and pressures up to 60 psi when mounted directly on a 1000-rpm PTO shaft, eliminating the need for step-up mechanisms.

Roller pumps. Roller pumps are popular due to low initial cost, compact size, and efficient operation at tractor PTO speed. They are positive-displacement pumps and self-priming.

Roller pumps have a slotted rotor that revolves in an eccentric case. Rollers in the slots seal the spaces between the rotor and the wall of the case. As the rollers pass the pump inlet, the spaces around the rollers enlarge and are filled with liquid drawn through the suction hose. When the rollers approach the pump outlet, the fluid is forced through the outlet. Pump output is determined by the length and diameter of the rollers and case, eccentricity, and speed of rotation. Roller pumps can develop pressures up to 300 psi and capacities to 50 gpm.

Material options for roller pumps include cast-iron or corrosion-resistant housings; and nylon, Teflon®, or rubber rollers. Viton®, rubber, or leather seals are also used. Nylon or Teflon® rollers are resistant to most farm chemicals and are recommended for multipurpose sprayers. Abrasive material will cause extreme wear and failure of roller pumps. Roller pumps should have factory-lubricated sealed ball bearings, stainless steel shafts, and replaceable shaft seals.

The recommended hookup for roller pumps is shown in Figure 11–13. A control valve is placed in the agitation line so that the bypass flow is controlled to regulate spraying pressure. Systems using roller pumps contain a pressure-relief valve. These

Spray Equipment Parts

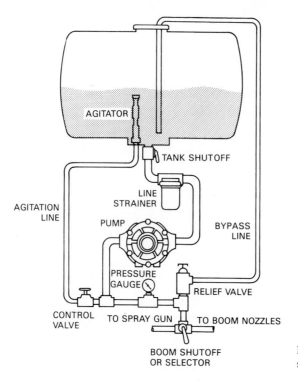

AGITATOR

TANK SHUTOFF

LINE
STRAINER

AGITATION
LINE

PUMP

BYPASS
LINE

PRESSURE
GAUGE

RELIEF VALVE

CONTROL
VALVE

TO SPRAY GUN

TO BOOM NOZZLES

BOOM SHUTOFF
OR SELECTOR

Figure 11–13   Roller pump and typical system hookup.

valves have a spring-loaded ball, disc, or diaphragm that opens with increasing pressure so that excess flow is bypassed back to the tank, preventing damage to sprayer components when the boom is shut off.

The agitation control valve must be closed and the boom shutoff valve must be opened to adjust the system. Start the sprayer, making sure flow is uniform from all spray nozzles, and adjust the pressure-relief valve until the pressure gauge reads about 10 to 15 psi above the desired spraying pressure. Slowly open the control valve until the spraying pressure is reduced to the desired point. Replace the agitator nozzle with one having a larger orifice if the pressure will not come down to the desired point. Use a smaller agitation nozzle if insufficient agitation results when spraying pressure is correct and the pressure-relief valve is closed. This will increase agitation and permit a wider open control valve for the same pressure.

Piston pumps.   Piston pumps are positive-displacement pumps, where output is proportional to speed and independent of pressure. Piston pumps work well for wettable powders and other abrasive liquids. They are available with either rubber or leather piston cups, which permit the pump to be used for water- or petroleum-based liquids and a wide range of chemicals. Lubrication of the pump is usually not a problem due to the use of sealed bearings.

The use of piston pumps for farm crop spraying is limited partly by relatively high cost, but a piston pump may be the best buy for the commercial applicator. The piston pump has a long life, which makes it economical for continuous use. Larger pis-

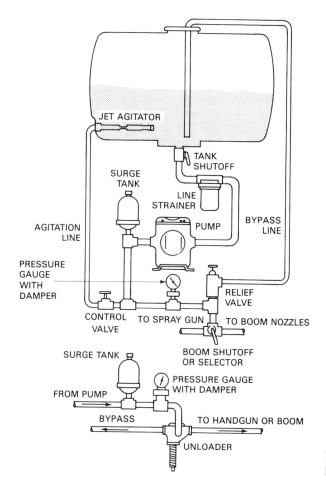

Figure 11–14 Piston or diaphragm pump and typical system hookup.

ton pumps have a capacity up to 60 gpm and are used at pressures up to 1000 psi. This high pressure is useful for higher-pressure cleaning, livestock spraying, or crop insect and fungicide spraying. A piston pump requires a surge tank at the pump outlet to reduce the characteristic line pulsation.

The connection diagram for a piston pump is shown in Figure 11–14. It is similar to a roller pump except that a surge tank has been installed at the pump outlet. A damper is used in the pressure gauge system to reduce the effect of pulsation. The pressure-relief valve should be replaced by an unloader valve (also shown in Figure 11–14) when pressures above 200 psi are used. This reduces the pressure from the pump when the boom is shut off, so less power is required. If an agitator is used in the system, agitation flow may be insufficient when the valve is unloading.

Open the control valve and close the boom valve to adjust for spraying. Then adjust the relief valve to open at a pressure 10 to 15 psi above spraying pressure. Open the boom control valve and make sure flow is uniform from all nozzles. Then adjust the control valve until the gauge indicates the desired spraying pressure.

Spray Equipment Parts

Diaphragm pumps.   Diaphragm pumps are positive-displacement and excellent general-purpose sprayer pumps. They are capable of producing high pressure (to 850 psi) as well as high volume (60 gpm). The price of diaphragm pumps is relatively high. High pressures and volume are needed when applying some pesticides such as fungicides. Diaphragm pumps are excellent for this job. The spray system hookup for diaphragm pumps is the same as for piston pumps (Figure 11–14). Be sure the controls and all hoses are large enough to handle the high flow, and all hoses, nozzles, and fittings must be capable of handling high pressures.

Strainers

Filter screens.   Filter screens are used in sprayers to prevent foreign materials from entering and wearing precision parts of the sprayer. These strainers, which give uniform flow, also prevent nozzle tips from clogging. Screens are normally placed at the entrance to the pump intake line (Figure 11–15), in the line from the pressure regulator to the boom, and in each nozzle. The mesh or size of the screens on the filters should be large enough to allow passage of wettable powders (50 mesh) and emulsifiable liquid concentrates (100 mesh). *Mesh* refers to the number of openings per square inch of the screen material. Usually 12- to 50-mesh screens are used in the tank strainers, 25- to 50-mesh screens in the in-line hose strainer, and 50- to 100-mesh screens in the nozzle tips. The screens in the nozzle tips (Figure 11–16) should be sized according to the opening of the nozzle. Screens should be checked and cleaned often to prevent poor coverage and loss of pressure. Self-cleaning strainers are now being offered by some manufacturers.

Flow-control Equipment

Pressure regulator.   The pressure regulator or relief valve maintains the required pressure in the system. It is a spring-loaded valve that opens to prevent excessive pressure in the line and to return some of the solution to the tank. Most pressure regulators are adjustable (Figure 11–17) to permit changes in the working pressure if desired.

Because the output of a centrifugal pump can be completely closed without damage to the pump, a pressure-relief valve and separate bypass line are not needed. The spray pressure can be controlled with simple gate or globe valves. It is preferable, however, to use special throttling valves that are designed to control the spraying pressure accurately. Electric-controlled throttling valves are becoming popular for remote pressure control. These valves are especially useful for enclosed cabs.

Figure 11–15   Intake filter screen.

Figure 11–16   Nozzle tip screen.

Pesticide Application Equipment   Chap. 11

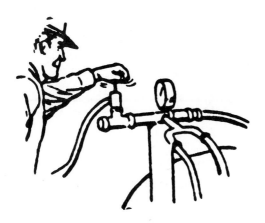

Figure 11–17 Adjusting pressure.   **Figure 11–18** Standard pressure gauge.

Pressure gauge.   Accurate pressure measurement is important since pressure is one factor that can be controlled and determines the amount of liquid being sprayed. Spray nozzles are designed to operate within certain pressure limits, and you should not exceed them. High pressures can cause dangerous fogging and drift, while low pressures may increase droplet size to the point that improper coverage is obtained.

Check pressure gauges periodically to determine their accuracy. The pressure gauge (Figure 11–18) is a delicate instrument and should be handled with care. It will indicate malfunctions by showing fluctuations in pressures. Pressure gauges should be selected to give accurate readings within the range of pressures normally used in the spraying system. For example, a gauge reading up to 500 psi would not be satisfactory for operating at pressures of 30 to 40 pounds because it would not be sensitive enough to register significant pressure changes.

Control valves.   Quick-acting control valves should be installed between the pressure regulator and the boom to control the flow of spray materials. One valve may be used to cut off the flow to the entire system, or a combination of two or three valves may be used to control flow to one or more sections of the boom. Special selector valves are available that control the flow of spray materials to any section of the boom. These controls should be located in a place where you can handle them without changing position or taking your eyes off your driving.

Hoses and fittings.   Hoses should be of neoprene or other oil-resistant materials and strong enough to withstand peak pressures. Hose test pressure should be twice the operating pressure. The suction hose should be made of two-ply fabric, wire-reinforced to prevent collapsing. It should be larger than the pressure hoses because it must provide the total pump flow through the suction hose, suction strainer, and valves.

Pressures are sometimes encountered that are much higher than the average operating pressures. Hoses on the pressure side of the system should be reinforced to prevent accidental breakage, which could subject the operator to dangerous spray materials. Fittings, including clamps, should be designed for quick, easy attachment and removal.

Spray Equipment Parts

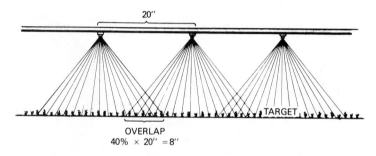

Figure 11–19 Typical boom setup with proper pattern overlap.

Nozzles

The nozzle and its adaptation to the spray application requirement are important in the effective use of agricultural chemicals. The nozzle helps control the rate, uniformity, thoroughness, and safety of pesticide application. All your efforts in identifying a pest, selecting a material to control that pest, and obtaining a proper piece of spray equipment for applying the pesticide can be a total waste if nozzles are not properly selected, installed, and maintained. Nozzles are the prime elements in the system.

Nozzle performance is the key to total system performance in most sprayer systems. Factors to consider are nozzle type, size, condition, height, orientation (position on boom), and spacing on the boom. Figure 11–19 illustrates the proper setup for 20-inch nozzle spacing.

The most frequent errors encountered when checking the accuracy of sprayers include wrong boom height, one end of the boom being too high or too low, misaligned nozzles on the boom, and the use of several different-sized nozzles together on the same boom. These errors are illustrated in Figure 11–20. The effects of worn nozzles are shown in Figure 11–21.

With several manufacturers producing a variety of nozzles specifically designed to do any of a wide range of jobs, precision application is within the grasp of any applicator who sets out to do a good job. The problem lies in the applicator's ability to select the proper nozzle, use it correctly, and maintain the system at peak efficiency.

Significant changes have been made in new nozzles. These changes can be found in the nozzle body. Traditional nozzles had a straight bore that carried liquids out of the boom down to the tip. In Spraying Systems' new design, called a "pre-orifice," fluid coming from the boom enters through a small hole and flows into a pocket machined into the nozzle body. The fluid then flows from this pocket or chamber into a short passage and out to the discharge orifice.

Another internal design change is the incorporation of protuberances in the liquid stream to increase turbulence. This mixing action helps decrease the percentage of undersized particles that emerge from the nozzle.

These internal changes are said to let the nozzle produce a uniform droplet size distribution over a wider range of line pressures than older designs. As a side benefit, the nozzle's internal passages are said to stay cleaner longer.

Boom too low or too high will give uneven patterns

Boom should be level to sprayed surface for uniform coverage

Nozzles should be aligned parallel with the boom

Do not use 80° and 65° nozzles together

Figure 11–20   Examples of problems with boom or nozzle alignment.

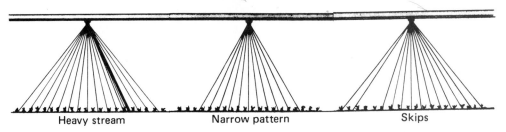

Heavy stream   Narrow pattern   Skips

Figure 11–21   Symptoms of worn nozzles.

A recent design from Spraying Systems, the Turbo Teejet® (Figure 11–22), uses these internal changes and has a new external look; the discharge orifice is in a barrel that looks like a small flooding nozzle instead of the narrow slit traditionally seen on a flat-fan spray tip. The company says the design maintains a consistently tight range of drift-resistant droplet sizes at fluid pressures of 15 to 90 psi. By contrast, the company's standard flat-fan Teejet® operates in a 30 to 60 psi range. Although nozzles exist for virtually every purpose, a few standard ones are basic to agricultural spray application. They are discussed next.

Regular flat-fan nozzles (Figure 11–22) are used for most broadcast spraying of herbicides and for certain insecticides when foliar penetration and coverage are not required. These nozzles produce a tapered-edge, flat-fan spray pattern. They are available in several selected spray-fan angles. Eighty-degree spray angle tips are most commonly used although 110-degree nozzles are fast gaining in popularity. The nozzles

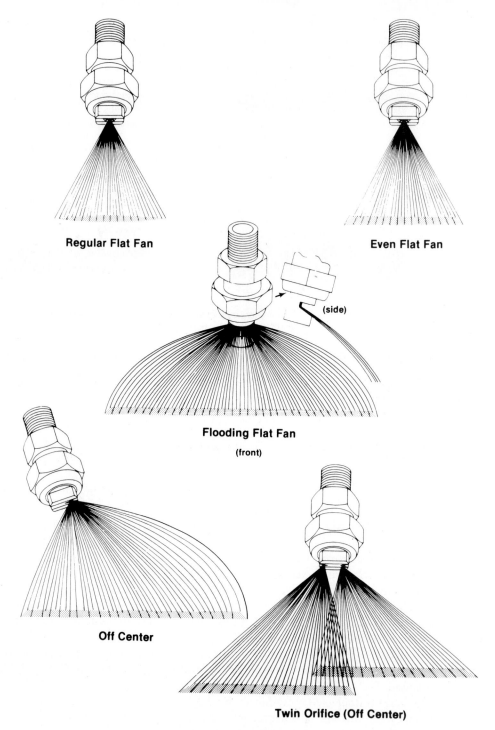

Regular Flat Fan

Even Flat Fan

(side)

Flooding Flat Fan

(front)

Off Center

Twin Orifice (Off Center)

Figure 11–22 Nozzle spray patterns.

are usually on 20-inch centers at a boom height of 10 to 23 inches. The boom heights for various spray angles are:

Spray angle (degrees)	Boom height, 20-inch spacing (inches)
65	21–23
73	20–22
80	17–19
110	10–12

When applying herbicides with standard flat-fan nozzles, keep the operating pressure above 25 psi to minimize spray angle collapse. If lower pressures are desired, Spraying Systems XR tips should be used. At these pressures, flat-fan nozzles produce medium-to-coarse drops that are not as susceptible to drift as the finer drops produced at pressures of 40 psi and higher. Regular flat-fan nozzles are recommended for some foliar-applied herbicides at pressures from 40 to 60 psi. These high pressures will generate fine drops for maximum coverage on the plant surface.

Because the outer edges of the spray pattern have tapered or reduced volumes, adjacent patterns along a boom must overlap in order to obtain uniform coverage. For maximum uniformity, this overlap should be about 40% to 50% of the nozzle spacing.

The low-pressure (LP) flat-fan nozzle is available from Spraying Systems. This nozzle develops a normal fan angle and distribution pattern at a spray pressure 15 psi. Operating at a lower pressure results in larger drops and less drift than the regular flat-fan nozzle designed to operate at pressures of 30 to 60 psi.

Even flat-fan nozzles (Figure 11–22) apply uniform coverage across the entire width of the spray pattern. They should be used only for banding pesticides over the row and should be operated at 30 psi. Pressures lower than 30 psi will cause an angle that is below rating and the coverage (band) will be narrower. Band width is determined by adjusting nozzle height. The band widths for various nozzle heights are shown as follows.

Band width (inches)	Nozzle height (inches)		
	40-degree series	80-degree series	95-degree series
8	10	5	4
10	12	6	5
12	14	7	6
15	19	9	8

Off-center, flat-fan nozzles (Figure 11–22) are used when extended coverage is needed at the end of a boom. These nozzles produce an off-center spray pattern extending in one direction from the nozzle tip. The coverage is relatively uniform when the nozzles are mounted at the proper height and operated within a pressure range of 15 to 40 psi. When mounted on a double swivel, the nozzles may be used as small boomless sprayers.

Spray Equipment Parts

Twin orifice (off-center) nozzles (Figure 11–22) produce a 150-degree, fan-shaped pattern with tapered edges that overlap to provide uniform coverage. They are used with "drops" between rows for herbicides and other applications requiring a wide coverage, fan-shaped spray. The spray pattern is comparable to that produced by two regular flat-fan nozzles.

Flooding flat-fan nozzles (Figure 11–22) produce a wide-angle, flat-fan pattern and are used for applying herbicides and mixtures of herbicides and liquid fertilizers. The nozzle spacing for applying herbicides should be 60 inches or less. The nozzles are most effective in reducing drift when they are operated within a pressure range of 8 to 25 psi. Pressure changes affect the width of spray pattern more with the flooding flat-fan nozzle than with the regular flat-fan nozzle. In addition, the distribution pattern is usually not as uniform as that of the regular flat-fan tip. The best distribution is achieved when the nozzle is mounted at a suitable height and angle to obtain at least double coverage or 100% overlap.

Flooding nozzles can be mounted so that they spray straight down, straight back, or at any angle in between. Position is not critical as long as double coverage is obtained. You can determine nozzle position by rotating the nozzle to the angle required to obtain double coverage at a practicable nozzle height.

Hollow-cone (disc and core type) nozzles (Figure 11–23) are used primarily when plant foliage penetration is essential for effective insect and disease control and when drift is not a major concern. At pressures of 40 to 80 psi, hollow-cone nozzles produce small drops that penetrate plant canopies and cover the underside of leaves more effectively than other nozzles. If penetration is not required, the pressure should be limited to 40 psi or less. The most commonly used hollow-cone nozzle is the two-piece, disc-core, hollow-cone spray tip. The core gives the fluid a swirling motion before it is metered through the orifice disc, resulting in a circular hollow-cone spray pattern.

Whirl chamber hollow-cone nozzles (Figure 11–23) have a whirl chamber above a conical outlet. These nozzles produce a hollow-cone pattern with fan angles up to 130 degrees and are used primarily on herbicide incorporation kits. The recommended pressure range is 5 to 20 psi.

RD and RA Raindrop® hollow-cone nozzles (Figure 11–23) have been designed by the Delavan Corporation to produce large drops in a hollow-cone pattern at pressures of 20 to 60 psi. The RD Raindrop® nozzle consists of a conventional disc-core, hollow-cone nozzle to which a special Raindrop can be added. The RA Raindrop® nozzles are used for herbicide incorporation, and the RD Raindrop® nozzles are used for foliar spraying. When used for broadcast application, these nozzles should be rotated 30 to 45 degrees to obtain uniform distribution.

Broadcast (boomless) nozzles (Figure 11–23) are used to spray swaths up to 30 feet or more. The spray is more susceptible to drift than that from nozzles mounted on a spray boom, and the distribution across the swath is not as uniform as with the other nozzle types. Broadcast nozzles produce small droplets immediately under the nozzle, but extremely large droplets are deposited at the outer edge of the swath.

Broadcast nozzles are used for boomless sprayers, to extend the effective width on the end of the boom, or when thorough coverage of the surface is not necessary. They are not recommended for field crops, but are suggested for spraying pastures, fencerows, and roadsides where obstructions make a boom undesirable.

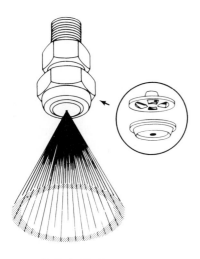

Hollow Cone

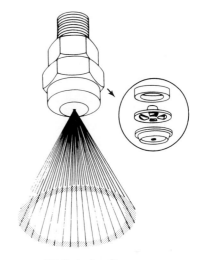

RD Raindrop®

Whirl Chamber

RA Raindrop®

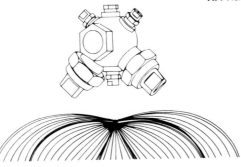

Broadcast (Boomless)

Figure 11–23 Nozzle spray patterns.

Spray Equipment Parts

Spraying Systems Company has some specialized spray tips that are worth mentioning. One is the XR Teejet® (Figure 11–24), which has an extended range that permits varying the size of the droplet by varying the spray pressure. The DG Teejet® (Figure 11–24) is a drift guard tip that produces large droplets to reduce drift. The Turbo Floodjet® (Figure 11–24) produces an even larger droplet to reduce drift, and operates at spray pressures from 10 to 40 psi.

Some new innovative spray tips were recently marketed by Spraying Systems Company. The Turbo Teejet® flat spray tip (Figure 11–24), with a considerably different look, is designed to provide drift control, excellent spray distribution, and long life in a flat spray tip. A wide pressure range makes it excellent for use with automatic rate controllers. The Drift Guard® Even Flat Spray Tip (Figure 11–24) is said to give operators the ability to work closer to the target with larger droplets. The pre-orifice design provides large droplets and the 95-degree spray angle allows a low spray height. The spray distribution is even across the entire band. The Turbo Floodjet® flat spray tip (Figure 11–24) is said to be ideal for lawn and garden sprayers, backpack units, and more. This tip is available in both stainless steel and polymer. It is great for drift control, plus gives excellent spray distribution, long wear, and a nonclogging design. This tip also maintains its spray angle over a wide range of pressures making it excellent for use with automatic rate controllers.

Delavan and Hardi are companies that also manufacture excellent spray tip nozzles as well as Spraying Systems Company. Catalogs are available from all three companies, and the author makes no endorsements.

Nozzle Tip Materials

Nozzle tips are available in a wide variety of materials, including ceramic, hardened stainless steel, stainless steel, polymer, and brass. Ceramic is the most wear-resistant material, but it is also the most expensive. Stainless steel tips have excellent wear resistance with either corrosive or abrasive materials. Although polymer and other synthetic plastics are resistant to corrosion or abrasion, they are subject to swelling when exposed to some solvents. Ceramic tips are extremely durable and will handle abrasive, corrosive materials. They are slightly more expensive than hardened stainless steel. Brass tips are the most common, but they wear rapidly when used to apply abrasive materials such as wettable powders and are corroded by some liquid fertilizers. Polymer tips are the most economical and wear longer than brass. Stainless steel is still a bargain.

Quick Attach Nozzles

Several companies offer quick attach tip and cap assemblies, which allow for quick changing of nozzle tips, screen cleaning, etc., while keeping self-alignment of flat spray patterns. The caps can be color coded so that one nozzle tip size can be assigned a certain color to prevent intermingling of several sizes.

ITEMS TO REMEMBER

- Select a nozzle that provides the desired droplet size, volume of flow, and pattern.
- Provide the nozzle with a liquid spray free of foreign particles, and provide it at a properly regulated pressure.

| AT LOW PRESSURE | AT HIGH PRESSURE | | |

- Uniform coverage at lower pressures
- Smaller droplets at high pressures for thorough coverage
- Nozzle spacing–20 inches
- Spraying pressure– 15-60 psi

- Large droplets to reduce drift
- Removable pre-orifice
- Nozzle spacing–20 inches
- Spraying pressure– 30-60 psi

- Uniform coverage along boom
- Pre-orifice design produces large droplets to reduce drift
- Nozzle spacing–30-40 inches
- Spraying pressure–10-40 psi

> Relative droplet size for each spray tip is shown in each pattern.

Turbo TeeJet® Flat Spray Tip - Drift control, excellent spray distribution and long life in a flat spray tip. This patented nozzle with built-in pre-orifice produces a wide spray pattern. Droplet size is larger than the XR TeeJet® or DG TeeJet® tips but smaller than the Turbo FloodJet® to ensure coverage in foliar applications. Wide pressure range makes it excellent for use with automatic rate controllers.

Drift Guard Even Flat Spray Tip - The DG TeeJet® Drift Guard Even Spray tip gives operators the ability to work closer to the target with larger droplets. The pre-orifice design provides large droplets and the 95° spray angle allows a low spray height. The spray distribution is even across the entire band. The pre-orifice can be removed to clean the tip if necessary.

Turbo FloodJet® Polymer Flat Spray Tip - Ideal for lawn and garden sprayers, backpack units and more. Like the stainless steel version, this tip is great for drift control, plus you get excellent spray distribution, long wear, and a non-clogging design. This patented tip maintains its spray angle over a wide range of pressures making it excellent for use with automatic rate controllers.

Figure 11–24 Nozzle spray patterns and innovative tips.

Spray Equipment Parts

- Mount the nozzles so that their location, relative to the target, remains constant and proper.
- Move the nozzles through the field at constant speed to ensure uniform application rates. Ground-driven systems and equipment with monitors do not require constant speed.
- After installation, properly calibrate the system to ensure that, in all respects, it is applying uniform coverage at the desired rates.
- Maintain the system in peak condition by periodic inspections and calibrations with particularly detailed attention to the performance of each nozzle.

A word about nozzle tip numbers. Nozzle tip numbers on tips manufactured by Spraying Systems Company have a meaning that will help you understand their use and output. The first two numbers refer to the angle of spray discharge. The last four numbers refer to the gallons per minute of the nozzle tip at 40 psi. For example, nozzle tip number 650067 means a nozzle tip with a 65-degree angle of fan discharge delivering 0.067 gpm at 40 psi. The decimal point is placed by counting three figures from the left side; for example, a Teejet® 8004 nozzle tip delivers 0.4 gpm. For spraying heights where the boom is 15 to 18 inches from the surface to be sprayed, 110-degree nozzle tips are recommended; for 17 to 19 inches, 80-degree nozzle tips; for 19 to 21 inches, 73-degree nozzle tips; and for 21 to 23 inches, 65-degree nozzle tips. Risk of drift is generally greater with the wider nozzle tip angle.

Spray discs are used in handguns. The number on the cap represents the diameter of the orifice (opening) in the disc in increments of 1/64 inch. A number 3 disc is 3/64 inch and a number 10 is 10/64. The larger the disc number, the coarser the droplet size is and the more gallons per minute that are delivered at a fixed pressure. Refer to manufacturers' charts to purchase the proper disc for your operation and equipment.

Nozzle configuration for band and directed spraying. *Band application* is applying a chemical in parallel bands, leaving the area between bands free of chemical. *Directed spraying* is application of a chemical to a specific area such as a plant canopy, a row, or at the base of the plants.

Several nozzle configurations are often used when foliage penetration or row crop height presents a problem. Figure 11–25 shows several commonly used nozzle configurations. The two- and three-nozzle configurations give better bottom leaf coverage than a single nozzle. This can be important with many pesticides. Drop nozzles are useful for herbicide application in taller row crops to reduce the risk of crop injury. In smaller row crops, a single-nozzle band configuration using a nozzle with a uniform pattern, for example, even flow flat-fan, should be adequate.

Spray Markers

A helpful accessory to aid uniform spray application is a foam or dye marker system to mark the edge of the spray swath (Figure 11–26). This mark shows the operator where to drive on the next pass to reduce skips and overlaps and is a tremendous aid

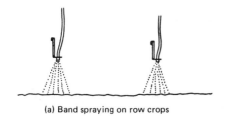
(a) Band spraying on row crops

(b) Directed spraying with one nozzle over the row

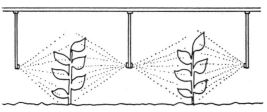
(c) Directed spraying with two nozzles

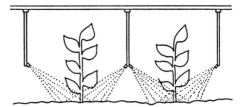
(d) Directed spraying at base of plants

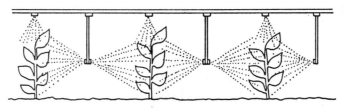
(e) Directed spraying with three nozzles often used for insecticides and fungicides

Figure 11–25   Nozzle placement for band and directed spraying.

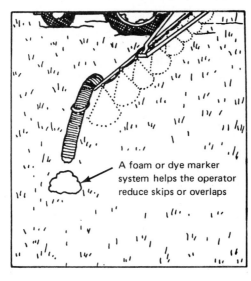
A foam or dye marker system helps the operator reduce skips or overlaps

Figure 11–26   Foam marker.

Spray Equipment Parts

in nonrow crops such as small grains. The mark may be continuous or intermittent. Typically, 1 to 2 cups of foam are dropped every 25 feet. The foam or dye requires a separate tank and mix, a pump or compressor, a delivery tube to each end of the boom, and a control to select the proper boom end. Another type of marker is the paper type. This unit drops a piece of paper intermittently the length of the field.

Three new marker dye products developed by Loveland Industries, Inc., can help applicators improve the visibility and accuracy of their applications. Marker Dye WSP® is a water-soluble packet of blue dye that is added to the mixing tank water for agricultural and other applications of pesticides and liquid fertilizers. The new dye enables the applicator to see exactly what areas have been treated, avoiding costly overlaps as well as skips and off-target applications. The dye dissolves quickly and completely and leaves no residue to clog nozzles. Marker Dye® is a liquid formulation available in one-quart bottles ideal for use in applications with tank or knapsack sprayers. The third marker dye is known as Foam Dye® and is a high-visibility foam colorant that leaves a noticeable color on crop residue upon dissipation of foam marker applications. It is claimed not to harm soil or crops at labeled application rates and will not affect the expansion capability of foam solutions.

OTHER APPLICATION TECHNIQUES

Several types of equipment have been developed in recent years that differ from the conventional equipment.

Rope wick applicators (Figure 11–27).   Sometimes known as carpet applicators, these applicators are popular for use against weeds that have grown higher than the crop. The wick or wipe applicators are tractor mounted and driven through the field as the herbicide is brushed off on the unwanted weeds at a height above the crop. Wick applicators are made with a reservoir boom made of various-sized poly vinyl chloride (PVC) with upright openings on one end for easy fill-up, and a drain for easy cleaning. Most of the PVC booms are of 3-inch diameter.

The height of the front-mounted booms can be hydraulically adjusted to best fit the situation with each crop. There is no spray drift problem because the only chemical that is used is that which is wiped on the weeds by the wick. There are a number of applicators on the market. Although several herbicides show promise with this type of application, Roundup® has been the most successful and is still the most popular.

Electrostatic sprayers.   This old concept is making a comeback with new technology. The electrostatic sprayer produces negatively charged spray droplets that are attracted to the crop. The electrical force is 40 times greater than the pull of gravity. The most significant part of this idea is that droplet size and movement are both controlled so that drift is virtually eliminated. The technique achieves up to 70 times greater coverage on undersides of leaves than conventional sprayers. When compared to conventional sprayers and noncharged air-assist units, electrostatic air-assist sprayers produce dramatically improved results with insecticides and fungicides, even at reduced rates and lower volumes per acre.

Figure 11–27 The rope wick applicator shown here is only one of many different designs and makes. Some wick applicators use ropes interlaced between two horizontal bars, while others use tufts of material other than rope. The ropes themselves are made of different materials, but work much the same way as a wick in a kerosene lamp.

The heart of this technique is a charged nozzle through which the liquid flows by gravity, picking up its electric charge on the way. The uniform, charged droplets move along specific lines of the electric field existing between the positive nozzle and the negatively charged target plant. This electric field envelopes the entire plant (Figure 11–28). Since all droplets carry the same charge and are mutually repellent, each follows a specific trajectory. The result is an even distribution of chemical over the upper and lower surfaces of all the leaves and stems.

Because droplets are physically attracted to the target at quite high velocity, drift is reduced and spraying is sometimes possible under conditions that would rule out the use of conventional equipment. Manufacturers include Electrostatic Spraying Systems of Watkinsville, Georgia, and GerVan Company of Modesto, California.

Air-assisted sprayers. These sprayers achieve improved coverage and better penetration of dense crops. The transfer vehicle is air moving at speeds of 80 to 150 miles an hour. Two types of units are available: (1) air curtain sprayers, which use normal hydraulic spray tips with a "curtain" of air blasting droplets to the target, and (2) air atomizers, which meter spray solution into an orifice, where an air blast produces droplets and drives them into the crop. Droplets bounce when they hit foliage or the

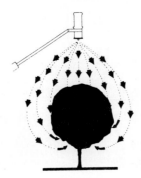

Figure 11–28 Electrostatic coverage.

ground. This provides thorough coverage of lower leaf surfaces. Effectiveness of some fungicides and insecticides has been said to be improved remarkably with this technique.

At least eight different companies offer such units, while Willmar Manufacturing offers a unit incorporating the first air-shear system to be used commercially in the United States. The liquid is atomized, or "sheared" into tiny droplets, about 100 to 125 microns in size, that are carried into the crop by the airstream. The combination of the forceful airstream and the angled nozzles creates turbulence that opens the crop canopy, improves penetration of the spray mixture, and reduces carrier volume, drift, and runoff.

Covered boom sprayers. Various forms of shrouding have been used on agricultural sprayers for decades. Most are basically inverted funnels or simple shields that keep spray aimed below the top of the crop canopy. The upper levels of vegetation block the wind and help trap small particles that might drift.

The patented Windfoil® sprayers from Rogers Innovative have been marketed for about 10 years, both as larger units for agricultural, and as smaller models for turf maintenance. Besides reducing the need for protective gear for the operator, Rogers says the sealed spray system permits use of a finer spray that may give more complete coverage of individual turf leaves. Rogers also suggests that sealed spray might achieve effective pest control while using lower chemical rates than open-air spraying systems.

Competing covered boom designs are being marketed by others. One such company, Environmental Technologies Equipment Corporation (ETEC), like Rogers is a Canadian company. ETEC recently introduced a 72-inch spray unit that replaces the cutting deck of front-mounted mowers. Besides helping the operator see exactly where the spray is being applied, the company says the arrangement may reduce public concerns because most people perceive the machine as a mower rather than a sprayer.

Subsurface injection of pesticides. Research on subsurface injection of pesticides has been under way for several years. Most of this work has been in relation to turfgrass pest control. Several new kinds of technology have been developed to enable turf managers to apply pesticides below the surface of the thatch, so that materials are applied directly at the soil-thatch interface, or at least farther into the thatch than with a conventional surface application. Some of the earliest field efforts of subsurface injection were conducted on warm-season grasses with equipment delivering up to 1000 psi through nozzles placed just above the turf. These units were developed in the early 1980s and were tested on white grubs and mole crickets in a variety of turf and pasture settings.

Further refinements at the Cross Equipment Company in Albany, Georgia, in the mid-1980s led to equipment that delivers liquid formulations at pressures up to 2000 psi. These units have nozzles at 2- or 3-inch spacing, placed just above the turf and aligned over drag bars. The orifices of the nozzles are extremely small, enabling the system to deliver extremely high pressures. The units do not slice the turf in any way, but simply drive the material into the thatch at high pressures. The Cross Equipment unit delivers a steady stream of material at a constant pressure.

The use of high pressure to apply liquid formulations has been expanded recently. The Toro Hydroject® unit, which was initially designed to fracture the soil and serve as a liquid aerator, provides penetration of the soil to depths of several inches. This unit delivers pulses of liquid at high pressure. Other companies are looking at developing units that can generate similar levels of pressure (up to 5000 psi) to penetrate the thatch and soil. For example, the Rogers Innovative Company of Saskatchewan, Canada, has developed a unit that can handle pesticides of various liquid formulations and deliver them through nozzles into the thatch and soil. Some of these units, like the Toro unit, which operate by delivering pulses of high pressure, are capable of providing penetration 2 inches into the soil.

Most turf managers can see the value of placing materials directly at the soil-thatch interface, at least when they are trying to control soil insects, such as white grubs or mole crickets. There is also evidence that the technology might be useful for improving the control of some root diseases, such as summer patch. Studies in Massachusetts showed that the use of high-pressure liquid injection reduces the surface residue of some insecticides 30% to 80% compared to conventional applications. Although subsurface technology appears to reduce surface exposure, and may reduce surface runoff following heavy rains, there is concern regarding the possibility of increased leaching. Studies at Penn State seem to suggest that leaching potential is not increased significantly under normal rainfall or irrigation conditions. There needs to be more research in this regard because people in areas that are particularly sensitive want to see more data or testimonials before they commit to the technology.

PRECISION APPLICATION DEVICES

Electronic monitors, sensors, and automatic control systems. Advances in electronics are having an impact on pesticide application techniques. Agriculture is one of the leaders in adopting electronics. Farmers and custom applicators have readily accepted the use of onboard sensors, monitors, and control systems to ensure a more precise and less costly application.

Several companies make spray rate monitors. The systems have a meter to sense total flow to the spray boom, a transducer to sense travel speed, a control to dial in the effective width sprayed, and a panel that continuously displays the spray rate during operation. Some units have such additional features as displays of nozzle flow, travel speed, area covered, total volume sprayed, and amount of solution remaining in the tank. Early problems with flow meters have been solved by the manufacturers, so most monitors are maintaining good reliability in the field. Spray rate monitors have the potential of eliminating many of the current errors in application accuracy.

Some manufacturers add a control system to their spray rate monitor. A microprocessor-controlled servo valve assembly automatically regulates the flow in proportion to travel speed to maintain a constant spray rate. Several control systems have such features as alarms, manual override, and individual boom controls.

Studies are also being conducted to develop sensors that continuously monitor soil organic matter content during field application of pesticides. In the research studies, reflected light passes through optical interference filters into phototransistors, which transform the reflected light into an electrical signal proportional to soil organic matter content. In a field unit an onboard microprocessor would deliver the properly amplified signal to a stepping motor for automatic control of application rate. The concept needs a great deal more development before becoming commercially available.

Weed sensing systems control sprayers, applying postemergence herbicides on a spot basis rather than continuously. Research is being conducted to develop spot-directed applications. It might be possible to ensure that the sensor sees right up to the crop, without sensing and spraying the crop. Bowman Manufacturing, Inc., of Newport, Arkansas, markets the Scan Ray II®, which senses and spot sprays weeds growing above the crop.

Marking swaths is a major problem when applying pesticides, especially with high-speed sprayers that have wide booms. Overlaps and skips could be reduced with accurate swath-marking systems or some way to determine the sprayer's exact position. A major breakthrough is global positioning systems (GPS), a military system consisting of 24 satellites orbiting the earth that can determine an individual's or object's location by sending out encoded radio signals. According to Dan Glickman, present Secretary of Agriculture, "technology that allows a military rescue team to find captain Captain Scott O'Grady in the middle of a Bosnian forest, land safely and unannounced within yards of his position, and pluck him out of harm's way, is the same technology which can guide a tractor through a 500-acre field to the precise location of insect infestations or to a spot where plant nutrient levels are below average." Linked with geographic information systems (GIS), a computer software program, the computer can decipher the remote-sensed information. Using this technology, pesticide applicators can pinpoint precisely where they are in a field. Variable rate technology (VRT) enables the spray operator to vary the application rate per acre, instead of varying rates per field.

Direct injection systems. Direct injection is an onboard, "on-the-go" mixing (injecting) and application of one or more formulated chemicals in the field. Tank mixing at the dealership or farm is eliminated. Full-strength pesticides and a carrier (water or fertilizer) are transported in separate tanks on the spray rig. Computerized controls programmed by the operator tell the pump how much chemical to mix with the carrier to achieve a particular application rate.

Because direct injection eliminates the need to tank-mix chemicals, pesticide compatibility problems are virtually eliminated. Chemicals and carrier are mixed only seconds before they are pumped into the boom and onto the field. More important are the environmental advantages. Having to dispose or reuse leftover tank solution or rinsate, especially in large volumes, has always been an application headache. Use of separate chemical and carrier tanks in direct injection systems solves this dilemma. There is no need to rinse the large spray tank and dispose of the rinsate since the spray tank usually only carries water or fertilizer. After application, unused product in the chemical tank can be transferred back into the original container. The tank needs to be rinsed only when a different chemical is used. Rinsate is minimal because the chemical tanks are small, usually 30 gallons or less.

Accuracy is inherent to direct injection systems. Application rates are electronically set and monitored through console computers connected to vehicle speed sensing devices and the pump. Operators can monitor rates and area covered continually on the console screen. Precise amounts of chemicals are delivered to the carrier and can be changed on-the-go via an override switch to spot treat areas that need more or less chemical. Chemical injection also maintains the spray pattern configuration during spraying because water spray pressure and volume are kept constant through changes in field conditions.

Currently, direct injection systems are commercially available from Raven Industries of Sioux Falls, South Dakota, and Midwest Technologies (Mid Tech) of Springfield, Illinois. Other companies are starting to offer direct injection as an option on their field applicators.

Recirculating ground sprayers. These sprayers have been developed to catch excess spray as the spray rig moves through the field. This technique is not perfected yet, but does have the advantages of preventing excess spray from reaching the ground, and the recirculation of that spray is a savings that allows for more acreage covered with less spray and less environmental contamination.

Closed systems. These systems minimize, if not eliminate, human contact with concentrated pesticides during mixing and loading activities. Closed systems (Figure 11–29) meter and transfer pesticide products from the shipping container to mixing or application tanks, and often rinse the emptied containers with a minimum of human exposure to the concentrate pesticide. In addition to reducing the risk of human exposure in metering and transferring pesticides, closed systems have the following advantages:

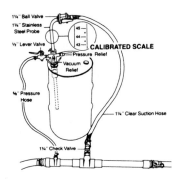

Figure 11–29 A simple closed transfer system.

- Provide greater accuracy in measuring dosage
- Reduce or eliminate fill site contamination
- Reduce the possibility of water source contamination
- Reduce the need for wearing hot, uncomfortable, protective clothing (wearing protective gloves and clean clothing remains advisable)

If provision for rinsing emptied containers is made, closed systems also:

- Decontaminate the container by rinsing to make the container easier to handle and dispose of
- Avoid wasting pesticide remaining in unrinsed containers
- Can reduce the time and effort required for rinsing and mixing

Dripless chemical transfer is just one of the many improved techniques available for handling crop chemicals in bulk. The technological development that makes this possible is dry break coupling, which is the heart of a true closed system. The dry break makes it possible to disconnect transfer hoses without dripping and spillage. The dry break was first used in other industries, but was introduced to agriculture in 1992 by Ingersoll-Dresser Pump/Scienco Products of Memphis, Tennessee. The company's Clean-Lock® dry break coupling, designed specifically for agricultural chemicals, is used in transfer systems offered by several major ag chem manufacturers, including FMC, Monsanto, Zeneca, and others. Two other suppliers of dry break systems to the ag market are Parker Hannifin Corporation of Minneapolis and Micro Matic of Northridge, California.

Returnable, refillable minibulks are popular, and while numerous systems are available, the Starr System®, introduced by Sandoz in 1994, has some unique features. The sumped bottom discharge eliminates waste that occurs with flat or grooved containers. And the stackable, palletized Starr tanks have a low center of gravity, which reduces the chance of tipping over. The biggest advantage of this system, however, is the Tru-Force® measuring chamber. Chemical concentrate is measured in the chamber by volume, eliminating the meter. Meters can vary in accuracy with temperature fluctuations, and they also require extensive maintenance. In the days when product rates were measured in terms of pints or quarts per acre, minor errors made little or no difference. With many of the newer products, rates are stated in ounces or grams per acre, and slight variations may result in significant error.

Pressure-free, volumetric chemical transfer is a way to move chemical concentrate from standard jugs, reusable containers, or minibulks to a calibrated measuring chamber by vacuum, instead of pressure. The obvious advantage is that if a hose or fitting should fail, the concentrate is not under pressure. This minimizes the chance of handler exposure. One such system is Vac-U-Meter® manufactured by S & N Sprayer Company of Greenwood, Mississippi. Another system is Empty-Clean®, manufactured by Empty-Clean Corporation of Cordele, Georgia. Empty-Clean is a handheld device with a suction probe that removes chemical concentrate from containers by vacuum, eliminating pouring. No additional power source is required since the unit operates from the discharge hose attached to the nurse tank pump. The chemical is transferred to the sprayer tank, and clean water from the nurse tank is used to rinse containers. Then the rinsate is automatically transferred into the sprayer tank.

Closed systems are required by law in California for handling highly toxic liquid pesticides. Other states may take similar action. At least one pesticide is labeled and marketed for use only in closed systems to reduce possible health hazards.

Chemigation

Many growers are using their irrigation systems to apply not only water but also a variety of agricultural chemicals. Chemigation, as this practice is commonly called, can be defined as the application of a chemical via an irrigation system by introducing or injecting the chemical into the water flowing through the system. Chemicals applied by this technique include fertilizers, herbicides, fungicides, insecticides, nematicides, growth regulators, and a few so-called biorational pesticides. Total chemigation technology has been developed, refined, and applied with a great degree of success over the past dozen years, especially by interdisciplinary teams of scientists at the University of Georgia and the University of Nebraska.

It has been estimated that approximately 15 million acres are now under sprinkler irrigation. The most commonly used system is the center pivot, of which there are approximately 75,000 systems in operation in the United States. Other systems include hand-move laterals, towline laterals, side-roll laterals, hose-drag traveling systems, and continuously moving linear/lateral systems.

The advantages of chemigation are as follows:

- Excellent uniformity of chemical application
- Prescription application of chemicals at the time they are needed
- Easy chemical incorporation and activation
- Reduction of soil compaction
- Reduction of mechanical damage to the crop
- Reduction of operator hazards
- Potential reduction of chemical requirements
- Potential reduction of environmental contamination
- Economical application
- Proven effectiveness

The disadvantages of chemigation are as follows:

- Additional capital outlay for injection equipment and a backflow prevention system
- Safety considerations to groundwater, humans, and the environment
- Management requirements to ensure that the equipment is working properly at all times

Chemigation is an effective crop production management tool that is particularly well adapted to center pivots and other continuously moving lateral sprinkler irrigation systems.

There has been a great deal of concern about the use of pesticides in irrigation water because of the possibility for miscalculation of pesticide amounts; for drift problems with center pivot sprinklers, especially when end guns do not shut off when the sprinkler system is near roads, farmhouses, or other areas that should not be sprayed; and the potential for groundwater contamination. The possibility for back-siphoning of pesticides into wells is a distinct possibility unless anti-back-siphoning equipment is installed. Without proper safety and antipollution devices installed and functioning, there is a potential for surface and groundwater contamination. One possibility is chemical injection system failure. This could allow water to backflow through the injection system, overflowing the chemical tank. The result would be contamination of the injection site. Mechanical or electrical failure that shuts down the irrigation pump could allow water and chemicals to flow back directly into the water source. Finally, the chemical injection system could continue to function after irrigation pump shutdown. Chemicals would then be pumped into the irrigation pipeline and flow into the water source. Serious contamination could result.

A few years ago, various individuals and groups began calling for regulation of the application of pesticides through irrigation systems. Some states have passed laws governing the application of pesticides through irrigation systems. Additional states may be considering legislation, although it may not be as necessary now that the Environmental Protection Agency has issued requirements for label statements on pesticides intended for sprinkler chemigation. These statements include requirements for safety devices on the sprinkler equipment. Figure 11–30 illustrates the minimum requirements for antipollution devices and the arrangement of the equipment for applying pesticides through irrigation systems.

Federal chemigation requirements

Generic label statements for chemigated products

- "Crop injury, lack of effectiveness, or illegal pesticide residues in the crop can result from nonuniform distribution of treated water."
- "If you have questions about calibration, you should contact State Extension Service specialists, equipment manufacturers, or other experts."
- "Do not connect an irrigation system (including greenhouse systems) used for pesticide application to a public water system unless the pesticide label-prescribed safety devices for public water systems are in place."
- "A person knowledgeable of the chemigation system and responsible for its operation, or under the supervision of the responsible person shall shut the system down and make necessary adjustments should the need arise."

Label statements for chemigated toxicity category I products. In addition to generic label statements, the labels of toxicity category I products (those with the label signal word DANGER) that allow chemigation must include the statements:

- "Posting of areas to be chemigated is required when (1) any part of a treated area is within 300 feet of sensitive areas such as residential areas, labor camps, businesses, day care centers, hospitals, inpatient clinics, nursing homes, or any public areas such as schools, parks, playgrounds, or other public facilities not including public roads, or (2) when the chemigated area is open to the public such as golf courses or retail greenhouses."

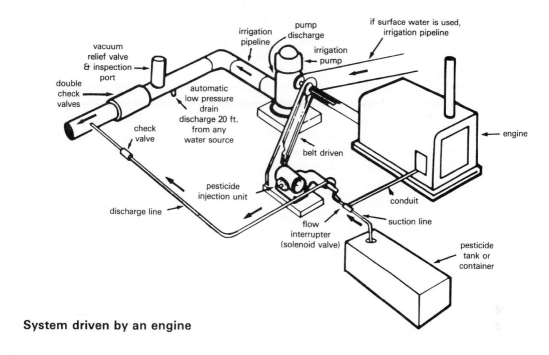

System driven by an engine

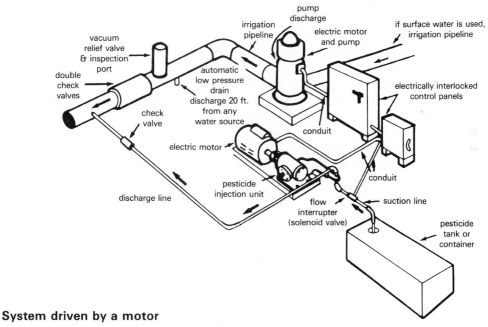

System driven by a motor

Figure 11–30   Minimum requirements for antipollution devices and arrangement of the equipment for applying pesticides through irrigation systems (chemigation).

Precision Application Devices   **311**

- "Posting must conform to the following requirements. Treated areas shall be posted with signs at all usual points of entry and along likely routes of approach from the listed sensitive areas. When there are no usual points of entry, signs must be posted in the corners of the treated areas and in any other location affording maximum visibility to sensitive areas. The printed side of the sign should face away from the treated area toward the sensitive area. The signs shall be printed in English. Signs must be posted prior to application and must remain posted until foliage has dried and soil surface water has disappeared. Signs may remain in place indefinitely as long as they are composed of materials to prevent deterioration and maintain legibility for the duration of the posting period."
- "All words shall consist of letters at least $2\frac{1}{2}$ inches tall, and all letters and the symbol shall be a color that sharply contrasts with their immediate background. At the top of the sign shall be the words KEEP OUT, followed by an octagonal stop sign symbol at least 8 inches in diameter containing the word STOP. Below the symbol shall be the words PESTICIDES IN IRRIGATION WATER."

Posting required for chemigation does not replace other posting and reentry interval requirements for farmworker safety.

Label statements for chemigation systems connected to public water supply systems. The labels of pesticide products that allow chemigation through systems connected to public water systems must include the following statements:

- "Public water system means a system for the provision to the public of piped water for human consumption if such system has at least 15 service connections or regularly serves an average of at least 25 individuals daily at least 60 days out of the year."
- "Chemigation systems connected to public water systems must contain a functional, reduced-pressure zone, backflow preventer (RPZ), or the functional equivalent in the water supply line upstream from the point of pesticide introduction. As an option to the RPZ, the water from the public water system should be discharged into a reservoir tank prior to pesticide introduction. There shall be a complete physical break (air gap) between the outlet end of the fill pipe and the top or overflow rim of the reservoir tank of at least twice the inside diameter of the fill pipe."
- "The pesticide injection pipeline must contain a functional, automatic, quick-closing check valve to prevent the flow of fluid back toward the injection pump."
- "The pesticide injection pipeline must contain a functional, normally closed, solenoid-operated valve located on the intake side of the injection pump and connected to the system interlock to prevent fluid from being withdrawn from the supply tank when the irrigation system is either automatically or manually shut down."
- "The system must contain functional interlocking controls to automatically shut off the pesticide injection pump when the water pump motor stops, or, in cases where there is no water pump, when the water pressure decreases to the point where pesticide distribution is adversely affected."

- "Systems must use a metering pump, such as a positive-displacement injection pump (e.g., diaphragm pump), effectively designed and constructed of materials that are compatible with pesticides and capable of being fitted with a system interlock."
- "Do not apply when wind speed favors drift beyond the area intended for treatment."

Specific label statements for sprinkler chemigation

- "The system must contain a functional check valve, vacuum relief valve, and low-pressure drain appropriately located on the irrigation pipeline to prevent water source contamination from backflow."
- "The pesticide injection pipeline must contain a functional, automatic, quick-closing check valve to prevent the flow of fluid back toward the injection pump."
- "The pesticide injection pipeline must also contain a functional, normally closed, solenoid-operated valve located on the intake side of the injection pump and connected to the system interlock to prevent fluid from being withdrawn from the supply tank when the irrigation system is either automatically or manually shut down."
- "The system must contain functional interlocking controls to automatically shut off the pesticide injection pump when the water pump motor stops."
- "The irrigation line or water pump must include a functional pressure switch that will stop the water pump motor when the water pressure decreases to the point where pesticide distribution is adversely affected."
- "Systems must use a metering pump, such as a positive-displacement injection pump (e.g., diaphragm pump), effectively designed and constructed of materials that are compatible with pesticides and capable of being fitted with a system interlock."
- "Do not apply when wind speed favors drift beyond the area intended for treatment."

SPRAYER MAINTENANCE AND CLEANING

Most trouble with sprayers can be traced to foreign matter that clogs screens and nozzles and sometimes wears out pumps and nozzles. Pump wear and deterioration are brought about by ordinary use, but they are also accelerated by misuse. The following suggestions will help minimize labor problems and prolong the useful life of the pump and sprayer.

1. *Use clean water.* Use water that looks clean enough to drink. A small amount of silt or sand particles can rapidly wear pumps and other parts of the sprayer system. Water pumped directly from a well is best. Water pumped directly from ponds or stock tanks should be filtered before filling the tank.
2. *Keep screens in place.* A sprayer system usually has screens in three places: a coarse screen on the suction hose, a medium screen between the pump and the

CLEAN NOZZLE
TIPS WITH A
SOFT BRUSH

Figure 11–31 Nozzle cleaning procedure.

boom, and a fine screen in the nozzle. The nozzle screen should be fine enough to filter particles that will plug the tip orifice.

3. *Use chemicals that the sprayer and pump were designed to use.* For example, liquid fertilizers are corrosive to copper, bronze, ordinary steel, and galvanized surfaces. If the pump is made from one of these materials, it may be completely ruined by just one application of the liquid fertilizer. Stainless steel is not adversely affected by liquid fertilizers, and pumps made from this substance should be used for applying these types of fertilizers.

4. *Never use a metal object to clean nozzles.* To clean, remove the tips and screens and clean them in water or a detergent solution using a soft brush (Figure 11–31). The orifice in a nozzle tip is a precision machine opening. Cleaning with a pin, knife, or other metallic object can completely change the spray pattern and capacity of the tip.

5. *Flush sprayers before using them.* New sprayers may contain large amounts of metallic chips and dirt from the manufacturing process. Sprayers that have been idle for awhile may contain bits of rust and dirt. Remove the nozzles and flush the sprayer with clean water. Clean all screens and nozzles thoroughly before trying to use the sprayer.

6. *Clean sprayer thoroughly after use.* After each day's use, thoroughly flush the sprayer with recommended sprayer decontamination/cleaning products, inside and out, to prevent corrosion and accumulation of chemicals. (Refer to Chapter 14 for further information.) Be sure to collect rinsate and dispose of or reuse it according to approved practices.

SPRAYER STORAGE

Be sure to prepare the spray system for off-season storage. When changing chemicals or when finished spraying for the season, clean the sprayer thoroughly both inside and out. Some chemicals, such as 2,4-D, are particularly persistent in the sprayer and must be removed completely to prevent possible crop damage from other spraying operations. For thorough cleaning between chemicals or at the end of the season, the following procedure is recommended:

1. Remove and clean all screens and tips in kerosene or a detergent solution using a soft brush.

2. Mix one box (about ½ pound) of detergent or the appropriate amount of commercially available sprayer decontamination/cleaning product with 30 gallons of water in the tank. Circulate the mixture through the bypass for 30 minutes, then flush it out through the boom.

3. Replace the screens and nozzle tips.

4. For sprayers having held 2,4-D or other phenoxy-type herbicides, fill the tank about one-third to one-half full of water; then add 1 quart of household ammonia or the appropriate amount of commercially available sprayer decontamination/cleaning product to each 25 gallons of water. Circulate the mixture for 5 minutes, allowing some to go out through the nozzles. Keep the remainder of the solution in the system overnight, and then run it out through the nozzles the following morning.

5. Flush the system with a tank full of clean water by spraying through the boom with nozzles removed.

6. When the pump is not in use, fill it with a light oil and store it in a dry place. If the pump has grease fittings, lubricate them moderately from time to time. Overlubrication can break seals and cause the pump to leak.

7. Remove nozzles and screens and place them in a light oil for storage.

8. Drain all parts to prevent freeze damage.

9. Cover openings so that insects, dirt, and other foreign material cannot get into the system.

10. Store the sprayer, hoses, and boom in a dry storage area.

Warning: Never use a pocket knife or other metal object to clean a nozzle. It will damage the precisely finished nozzle edges and ruin the nozzle performance. A round wooden toothpick is much better. Better still, remove the nozzle tip and back-flush it with air or water first before trying anything else. Don't blow through it by mouth. Chemicals are poisons!

AERIAL APPLICATION EQUIPMENT

Aerial dispersal equipment must be lightweight to provide for maximum payloads, strongly built, and mounted to withstand the wear and the loads that they encounter. Operating controls must be simple. All components should be designed and installed to permit thorough cleaning between jobs in order to avoid residual contamination from previous uses.

A clean aircraft is not only good advertising, it makes daily inspections quicker and easier. Well cared for equipment will inspire the customer's confidence in the operator.

Spray Dispersal Equipment

Spray dispersal equipment consists of the following items: tank(s), pump, pressure regulator, line filter, flow-control valve, boom, and nozzles. On fixed-wing aircraft the tank is built into the fuselage. The pump, pressure regulator, line filter, and control valve are generally mounted under and outside the fuselage. The boom and nozzles are usually attached under the trailing edge of the wing.

On rotary-wing aircraft, the tanks are mounted on the sides of the frame to keep the load in line with the main rotor shaft. The tanks are coupled at their bases with a crossover pipe that feeds the pump. The pump is driven by the engine. The filter, regulator, and valve are attached to the lower frame of the fuselage. The spray boom and nozzles are mounted to the toe of the landing skids or to the frame under the main rotor. The toe mounting puts the boom and nozzles where the pilot can see them. Mounting the boom under the main rotor gives a slightly wider swath.

Spray equipment provides effective swath widths of 40 to 60 feet in the range of 1 to 10 gallons per acre when flown at heights of 5 to 8 feet above the ground or crop. Special applications such as ULV can give much wider swaths, but they are flown at 15- to 25-foot altitude.

Tanks. Top-loading tank openings should be large enough to permit pouring materials into them. All tanks should be fitted with emergency dump valves, and the valve action should be checked each time the equipment is flushed. They should be fitted with vents to permit free flow of the materials at emergency dump rates. Bottom-loading systems should have capacity enough to match the loading pump. Spray tanks need agitation that reaches the bottom to maintain suspensions and mixture of materials. If hydraulic agitation (pumping the material back into the tank) is used, a rule-of-thumb flow rate is 10 gpm of agitation flow for every 100-gallon capacity of the tank. The pump outlet and the emergency dump should connect to the bottom of the tank to permit complete emptying.

Pumps. All water-based sprays can be handled with centrifugal pumps. Centrifugal pumps are mechanically simple, lightweight, flexible in their output, and have a maximum pressure determined by design. Normal power sources for these pumps are as follows:

- Wind-driven fan, using the forward motion of the aircraft
- Direct coupling to the aircraft engine, with or without a clutch
- Hydraulic or electric drive from the aircraft engine to a hydraulic or electric motor coupled to the pump

The size of centrifugal pumps is selected to provide adequate flow (gpm) to handle the maximum nozzle capacity (and agitation, if included) and the maximum pressure (psi) to give the needed boom pressure and to take care of the pressure loss (about 5 psi) in the pipes, valve, and filter. Specialized applications (ULV, bifluid, microbial sprays) call for special pumps.

Controls. Controls should include a shutoff valve and a pressure regulator between the pump and the boom. The valve should be a quick-acting gate or ball valve. The pressure regulator should be a bypass type that returns the relief liquid to the tank(s). If the pump is powered by a wind-driven fan, a brake should be installed to stop the spray pump when not in use.

Filters. Filter screens at the nozzles prevent the spray tips from plugging with the sediment that normally collects in a spray system. It is wise to have a line filter installed behind the pump so that sediment is not forced into the parts of the valve and the regulator, creating unnecessary wear. Screen size should be 50 or 100 mesh, depending on the size of the nozzle tips. A pressure gauge coupled to the line beyond the filter will indicate (by low pressure) when the line filter is starting to get clogged. Clean the line filter daily during spray operations.

Plumbing.   To prevent excessive pressure losses:

- For application rates over 2 gallons per acre, all main piping and fittings should be at least $1\frac{1}{2}$ inches inside diameter.
- For application rates of up to 2 gallons per acre, all main piping and fittings should be at least 1 inch inside diameter.
- For ULV applications, hoses to individual nozzles should be ⅛-inch inside diameter. Main line hoses and fittings should be at least ⅜ inches inside diameter.

Tees should be the size of the incoming pipe and fitted with reducers to couple the branch pipes. All joints should be made with the minimum of pockets that can trap materials. The number of bends should be kept to a minimum and all bends should be made with as large a radius as possible. All spray plumbing connections using hoses should be double-clamped or clamped and safety-wired to secure the connections.

Booms.   Booms need strong support. Placing the boom at the trailing edge of the wing reduces the drag resistance of the boom. They should be far enough back to avoid airstream interference with the control surfaces. Booms should be fitted with caps having openings the full size of the pipe to permit flushing and cleaning with a bottle brush when needed. Twin-boom or return line systems are used to eliminate air pockets at the boom tips and also to provide agitation in the boom for chemicals that settle out rapidly. Nozzle orientation to the direction of flight is important for droplet size control. If the boom cannot be rotated, use 45-degree elbows or swivel connectors to provide for nozzle adjustment. It has been proven that shortening booms to 57% of the wingspan will not narrow the swath, and will nearly completely stop any feeding of the wing tip vortice.

Nozzles*

Nozzle location.   The location of the outboard nozzle on each end of the boom is critical. Since wingtip vortices or main rotor vortices are used to develop the width of the pattern, the end nozzles must be inboard enough to prevent the vortex from trapping the fine droplets. This entrapment creates peaking in the pattern and drift. On fixed-wing aircraft, the propeller disturbance shifts the spray from the right to the left (as the pilot sees it). Nozzles need shifting to the right within the area of this disturbance to compensate for the uneven pattern. The choice of nozzles and the amount of shift cannot be determined without testing.

Nozzle selection.   A great deal of information is available from nozzle manufacturers for use by operators. Only general guides will be given here. Nozzles perform satisfactorily if they are operated at the manufacturer's recommendations of

*See additional information in section on Aerial Equipment Calibration in Chapter 12.

pressure and if flow is not restricted. Departures from these values must be made with care since output and droplet size both vary with pressure. Nozzle types and operating pressures should be selected to handle the material being applied and break it up into droplets of desired size.

Spraying Systems' Teejet® 8005 through 8010 are good choices, along with D6 and higher with *no* swirl plates (swirl plates tend to increse "fines"). For the ultimate in low drift on short booms, the ML® nozzle has proven very valuable. Two other nozzles also fit into this category. One is called the Dent LD® nozzle and the other is a low drift nozzle built by ACCU-FLO®. Some operators say that "next to GPS equipment, CP® nozzles and check valves are the hottest product on the market."

Water solutions of pesticides should be pumped with nozzle pressures of 35 to 45 psi to get manufacturer's recommended breakup. Suspensions, heavy emulsions, and slurries should be handled with nozzles having large openings to prevent clogging. Bifluid systems use special mixing chambers and tips.

Positive shutoff of nozzles is achieved by the use of:

- Diaphragm check valves
- Ball check valves
- "Suck back" connection on pump, working on boom or return line

These items will need regular attention if they are to perform properly. Aged or worn diaphragms or scarred ball seats need replacement. Any springs used to maintain pressure will require checking to see that they move freely. Seating faces should be smooth and free from cracks. Return lines need flushing to be sure that they are not clogged.

As a guide, the following suggestions are given to select the nozzle type for a treatment:

- Atomizing or hollow-cone-type spray gives the finest breakup.
- Flat spray nozzles have intermediate breakup.
- Solid cone nozzles have the coarsest breakup.
- Spinning nozzles tend to have a narrower range of droplet sizes than the preceding hydraulic nozzles.

Ultralow-Volume Systems

ULV spraying is a fairly recent development in aircraft spray work, using concentrate materials (no water added) in a spray made up of fine droplets. Some pesticides exhibit better action in their concentrated form. Less evaporation takes place since water is absent from the spray. The density of the droplets is slightly greater than the same sized droplets of water-based sprays, increasing the rate of fall. Application rates are reduced so more acres can be treated before reloading is needed. At present, however, only a few materials are registered for this use.

Plans are available for the modification of aircraft dilute spray systems to handle ULV. If the aircraft is used for different spray jobs during the season, the operator can install a small ULV system entirely separate from the dilute system. This system can be removed when the ULV applications are done, making cleanup much easier.

One system already marketed consists of a 5-gallon stainless steel pressure tank to hold the chemical. Pressure is obtained from a liquid carbon dioxide fire extinguisher hooked up to the tank with an air pressure regulator. An electric valve controls the flow from the tank, using a ½- or ¾-inch solenoid valve with a 12-volt DC coil. Half-inch hose leads from the valve to a tee, which center-feeds the ULV boom. This boom is made of hoses and tees feeding the nozzles. The ULV boom and nozzles are clamped to the regular boom for support. The solenoid valve can be wired to the aircraft electrical system or a dry cell battery in the cockpit and to a push-button switch taped to the control stick. The 5-gallon tank and fire extinguisher can be set in the regular spray tank, where it should stand in an upright position.

The following points should be observed: ULV spraying must create fine droplets to be effective. Flat-fan nozzles discharging 0.1 gpm or less are recommended operating at 40 to 55 psi. Spinning nozzles can also be used. Do not use fewer than four flat-fan nozzles to avoid gaps in the distribution pattern. On a helicopter, a single spinning nozzle may provide sufficient output if very low rates are needed (for example, mosquito control). Because this system produces fine droplets, the extreme outboard nozzles must be located away from the wingtips on fixed-wing aircraft. This avoids spray entrainment in the wingtip vortices. Use two-thirds of wingspan as a guide for the limit. Shift the central nozzles to the right to compensate for higher flight than usual (20 to 25 feet above the ground) in order to provide for a wider swath and greater uniformity.

When full-strength chemicals are used, the solvent carrier for the chemical has to be considered. The carrier for malathion will attack rubber and neoprene. The seals in the solenoid valve should be of Teflon® or Viton®. Hoses may have to be replaced at the end of the season if they are not resistant to the solvent. Diaphragms in the nozzle bodies should be checked frequently and replaced each season. Nozzle screens are important since plugging is easier with the smaller tips. Use 100-mesh screens.

In calculating conversions from gallons per acre to ounces per acre (liquid), use the factor 1 gallon = 128 ounces.

Granular Dispersal Equipment

Dusts, impregnated granulars, granular fertilizers, prilled fertilizers, and seeds all fit into this category. For the finer materials (smaller than 60 mesh) some agitation may be needed in the hopper to prevent bridging. Because most commercial granulars contain fines, the seals between the components of the equipment (the tank, to the metering gate, to the disseminator) require frequent inspection to make sure that no leaks occur. Seal strips that look mechanically tight on the ground will often leak under the influence of air pressures developed in flight.

The size, shape, and weight of the particles and the flowability of the material affect the swath width, the application rate, and the pattern. Common effective swath widths are 35 to 40 feet when applied at 10- to 15-foot altitude. Some disseminators at low application rates achieve satisfactory swath widths of 50 feet.

Fixed wing. The usual disseminator for fixed-wing aircraft is the ram-air or venturi type. The throat of the venturi is attached to the fuselage under the hopper outlet. It has a broad aft section to impart lateral motion to the material. Some ram-air spreaders can deliver up to 250 pounds per acre. Overloading the disseminator tends to peak the application pattern at the center and to reduce the swath width. Overloading should be avoided because it can lead to uneven flow, making the pattern erratic.

Rotary wing.   Two systems are being used. One system uses saddle tanks mounted next to the engine, each tank having its own disseminator system. The metering devices have to be linked to prevent one tank being emptied faster than the other. The second system uses a single hopper and disseminator, hanging on a cable below the rotary-wing aircraft. A single tank and disseminator avoids the problem of balancing the output of two metering devices. Also, being on a cable, the equipment is more remote from the aircraft, where corrosive granules create problems. The disseminators are either spinning plate "stingers" or some form of blower and ducting. The lower forward speed of the rotary aircraft prevents the use of the ram-air principle.

Ground Support Equipment for Aircraft

Efficient ground equipment is necessary for an efficient aerial operation. Time saved on the ground in loading aircraft safely and quickly may make the difference between profit and loss in the total operation. The supply equipment and ground crew will vary greatly from one operation to another depending on the needs of the specific aerial applicator.

Loading of solid materials.   Dusts, granules, and seeds are often put into the hopper directly from the container or sack through the large filling door of the hopper. This operation can be made easier and faster by loading a separate hopper of equal capacity on the ground before the airplane lands and hoisting this above the hopper door and dropping it into the hopper of the airplane. Other mechanical loaders can also be used for loading solid materials into aircraft, such as auger loaders.

Loading of liquid materials.   Supply tanks for transport of the spray carrier to the field of operations on a truck or trailer may range from 200- to 5000-gallon capacity. It is usually desirable to carry the diluent as well as the mixed chemical on the same tanker in order to save time going to and from a water source at each new mix. Smaller nurse tanks of 200 to 400 gallons are sometimes used for mixing of chemicals and supplying them directly to the aircraft in a large operation where several airplanes are to be loaded. The mixing tank may consist of two compartments, one for mixing the chemical with the carrier and the other for keeping the mixture ready for loading into the aircraft. Adequate agitation is essential for some mixtures.

Most aircraft in use are loaded through the bottom; the spray concentrate is pumped into the aircraft tank through a bottom-loading connection on the side of the aircraft, followed by the diluent. This closed-system method eliminates much of the handling of toxic chemicals and makes a much safer operation for the loading crew.

Operations

General.   When an aircraft has been calibrated, the air speed, spraying pressure or gate setting for granulars, height of flight, and the effective swath width are fixed. Applications must be made at the same settings. Ferrying height between the airstrip and the field should be done at a minimum of 500-foot height, loaded or empty. Avoid flying over farm buildings, feedlots, or residential areas both for noise and for possible leakage. Courtesy to your neighbor costs so little and pays real dividends.

Field flight patterns. With rectangular fields, the normal procedure is to fly back and forth across the field in parallel lines. Flight directions should be parallel to the long axis of the field (reducing the number of turns). Where crosswinds occur, treatment should start on the downwind side of the field (Figure 11–32) to save the pilot flying through the previous swath. Where this fits in with crop rows or orchard rows, the pilot can line up the aircraft with a crop row. If the area is too rugged or steep for these patterns, flight lines should follow along the contours of the slopes. Where spot areas are too steep for contour work (mountainous terrain), make all treatments downslope.

Swath marking. Swaths can be marked with flags set above the height of the crop to guide the pilot. This method is useful if the field is going to be treated several times during the season. Automatic flagging systems (Figure 11–33) are in common use. These devices, attached to the aircraft and controlled by the pilot, release weighted streamers. These streamers give the pilot a visible mark to help judge the next swath.

Perhaps the most remarkable modern development has been the acceptance of electronic guidance systems that precisely guide aircraft over parallel ground tracks.

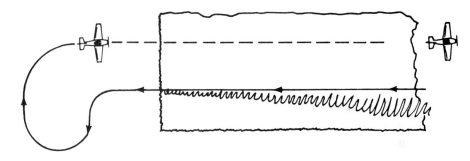

Figure 11–32   Start flight pattern on downwind side.

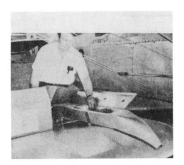

Figure 11–33   Automatic flagging device and its use.

The system's ability is derived from the global positioning system (GPS). Using software designed for aircraft dispensing operations, the plane picks up these signals allowing the pilot to select the swath width and flight pattern by use of cockpit controls. One very desirable example is the ability to guide the pilot and aircraft back to the exact spot along a swath line where the plane may have run out of material.

The GPS, a network of 18 satellites that rotate in a grid around the earth in a 12-hour period, is making the concept of site-specific pesticide application a reality. GPS enables an applicator to navigate accurately (to within a couple of feet) and consistently through a field time after time. Several companies offer aerial GPS guidance systems, which are not cheap. The SATLOC® system costs about $30,000, which translates to a cost of 15 to 35 cents an acre, depending on assumptions about how long the system will be used.

Automatic flagging costs about 15 cents an acre, and loading can be accomplished in about 2 minutes while a plane is being refueled. This lower cost is likely to ensure a continued market for paper streamers, which have been used for nearly 30 years.

The efficiency of the flying pattern can directly increase or decrease the flagging costs, as well as the overall cost of the job. Therefore, reducing flight time could appreciably reduce the overall cost of any particular application job.

The most commonly used flight pattern for spraying has been the "keyhole" method. But a newer, seemingly more efficient pattern, is also used. It is called the "racetrack" method because of its familiar shape. These are shown in Figure 11–34.

Utilizing the keyhole method allows the pilot to spray sequential strips across the field. But it requires more air time because a greater turn distance is required to realign the aircraft with the spray path. The racetrack method minimizes the turn arc to 180 degrees and alternately sprays two halves of the field. The actual flight time saved by utilizing the new racetrack method has been shown to approach 50% according to several studies. The conclusions drawn by the studies must be necessarily tempered by each individual's own flagging needs.

Turnaround.   At the end of each swath, the pilot should stop the disseminator and pull up out of the field before beginning the turn. The turn should be completed before dropping into the field again. The pilot should fly far enough beyond the field going out to turn to permit slight course corrections before dropping into the field again for the next swath, as shown in Figure 11–35.

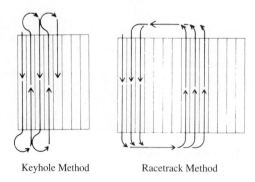

Keyhole Method   Racetrack Method

Figure 11–34   Keyhole and racetrack flight patterns.

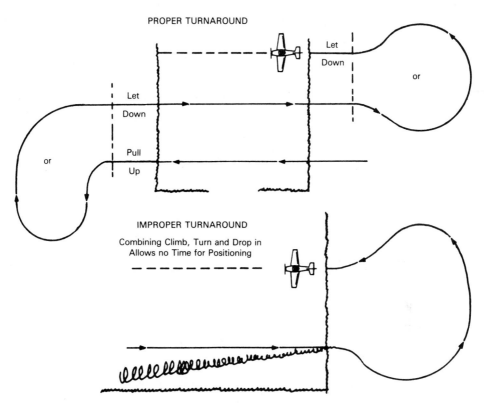

Figure 11–35 Proper and improper turnaround procedures.

SPRAY DRIFT MANAGEMENT

The following spray drift management requirements must be followed to avoid off-target movement from aerial applications.

- Do not apply within 150 feet by air or 100 feet by ground of an unprotected person or occupied building.
- All application equipment must be properly maintained and calibrated using appropriate carriers.
- The spray boom should be mounted on the aircraft so as to minimize drift caused by wingtip vortices. The boom length must not exceed three-fourths of the wing or rotor length (i.e., the distance of the outermost nozzles on the boom must not exceed three-fourths the length of the wingspan or rotor).
- Nozzles must always point backward and never be pointed downward more than 45 degrees.
- Do not apply at a height greater than 10 feet above the crop canopy unless a greater height is required for aircraft safety.
- Make applications when wind velocity favors on-target product deposition (approximately 3 to 10 mph). Do not apply when gusts or sustained winds exceed 15 mph.

12

Pesticide Equipment Calibration

CALIBRATION IMPORTANCE

Millions of dollars go into research to determine application rates of various pesticides on certain crops and animals. Proper rates must be applied to provide effective control. Accurate application is also essential to keep residues at acceptable levels. The application of the right pesticide at the right time and at the proper rate is important to prevent contamination of the environment. To get the correct rate, application equipment must be properly adjusted and operated. *The correct application rate is one factor that the operator can control.*

Application equipment is often not utilized to its maximum efficiency. This is partly because applicators often do not realize the importance of precise application and also because they do not have a clear understanding of the principles involved in applying pesticides. Even though pesticide application equipment is often the least expensive component of pest control, the proper nozzle selection, calibration, operation, and maintenance can pay big dividends.

The calibration of spray equipment is the most important step in good pest control. You may have carefully read the pesticide label, be wearing the appropriate protective clothing and equipment, and have the correct amount of pesticide measured and mixed in the spray tank, but if the spray equipment is not calibrated and applying the correct amount of pesticide per acre or other unit of measurement, then you have ruined the whole operation. Therefore, it is highly essential that any spray operation be preceded by correct and thorough calibration of the equipment to be used.

CALIBRATION METHODS

Hand Sprayers

Spray equipment with single nozzles or sometimes short booms with three or four nozzles are used for spot treatment and application of pesticides to smaller areas. Single high-pressure nozzles are also widely used for rights-of-way spray operations, ornamental spraying, and spot treatment of noxious weed patches. Accurate calibration of this type of equipment is very simple but seldom done. Pesticide applicators should pay strict attention to correct calibration of hand sprayers because overapplication of pesticides is a common occurrence with this type of equipment.

Compressed-air sprayers

Step 1. Mark out a square rod (16 feet by 16 feet).

Step 2. Determine the pressure and spray pattern needed to get good coverage of the area being treated.

Step 3. Fill the sprayer and pump up the sprayer to 30 or 40 pounds pressure if it has a pressure gauge, or count the number of strokes used to pump up the sprayer.

Step 4. Spray the square rod, walking at the same speed you plan to use in spraying the actual area to be treated. While spraying, time the operation in seconds needed to spray the area.

Step 5. Using a suitable measuring device, catch spray from the nozzle while spraying for the same amount of time it took to cover the one square rod area. Measure the amount of spray.

Step 6. Determine the amount applied per acre by using the following formula:

$$\text{gallons per acre} = \text{amount sprayed out (cups)} \times 10$$

There are 160 square rods per acre and 16 cups per gallon. By dividing 16 into 160, we get a constant of 10 that can be used in our formula.

Example: Assume that three cups of water are applied to 1 square rod. The application rate is $3 \times 10 = 30$ gallons per acre.

High-pressure handgun. Single-nozzle, high-pressure equipment is calibrated exactly the same way as compressed-air or knapsack sprayers. Accuracy is increased if the plot (1 square rod) is marked out on the species of plant to be sprayed. After determining the time required (in seconds) to spray this square rod plot, catch the spray from the nozzle for the same amount of time required to spray this square rod plot and measure. The rate of application is determined by the same formula used in step 6.

Boom Sprayers

The performance of any pesticide depends on the proper application of the correct amount of chemical. Most performance complaints about agricultural chemicals are

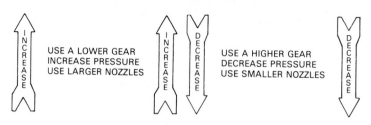

Figure 12–1 Variables that affect sprayer output.

directly related to errors in dosage or to improper application. The purpose of calibration is to ensure that your sprayer is applying the correct amount of material uniformly over a given area.

Three variables affect the amount of spray mixture applied to a unit area (such as an acre of a field or 1000 square feet of lawn): (1) the nozzle flow rate, (2) the ground speed of the sprayer, and (3) the effective sprayed width per nozzle. To calibrate and operate your sprayer properly, you must know how each of these variables affects sprayer output (see Figure 12–1).

Nozzle flow rate.   The flow rate through a nozzle varies with the size of the tip and the nozzle pressure. Installing a nozzle tip with a larger orifice or increasing the pressure will increase the flow rate. Nozzle flow rate varies in proportion to the square root of the pressure. Doubling the pressure will *not* double the flow rate. To double the flow rate, you must increase the pressure four times. For example, to double the flow rate of a nozzle from 0.28 gpm at 20 psi to 0.56 gpm, you must increase the pressure to 80 psi (4 × 20).

Pressure cannot be used to make major changes in application rate, but it can be used to correct minor changes because of nozzle wear. To obtain a uniform spray pattern and minimize drift hazard, you must keep the operating pressure within the recommended range for each nozzle type. Actual nozzle flow rates may not be the same as those shown in nozzle catalogs. This discrepancy may be the result of nozzle wear, inaccurate pressure measurement, pressure drop in booms and check valves, and flow variations due to the density and viscosity of the spray material, such as in fertilizer suspensions.

In practice, pressure is often measured several feet from the nozzle, and line losses of 5 to 10 psi between the gauge and the nozzle are not uncommon. Remember, if you use check valves to prevent nozzle drip, the pressure at the nozzle is 5 to 7 psi lower than the boom pressure indicated on the pressure gauge.

Density.   The weight per gallon of liquid will vary the discharge rate through the nozzle at a given pressure. The heavier a liquid, the slower it is discharged at the same pressure. Most nozzles are rated for water. When applying lighter materials (oils) or heavier materials (fertilizer and fumigants), adjustments must be made for the density of the material.

The density of the spray solution, especially when pesticides are applied along with fertilizers, affects the flow rate through the nozzle. High viscosity results in a reduction of nozzle flow rate in some suspension fertilizers. Tests have shown that the actual flow rate for a moderately viscous fertilizer suspension is 10% below the rate shown in nozzle catalogs. Presently, there is no good method for accurately predicting flow rate. Therefore, custom applicators must measure actual flow rates with the materials they will be applying. Also, flow meters monitoring the flow to the boom cannot detect variations in flow rate between individual nozzles.

Viscosity. The viscosity or thickness of a material will affect the flow rate. Sprayers are usually calibrated with water. If the viscosity of the spray material is considerably different than water, calibrate with the liquid that will be used in spraying. Most liquid pesticides have a viscosity very close to water and thus viscosity is usually not a factor.

Ground Speed. The spray application rate varies inversely with the ground speed. Doubling the ground speed of the sprayer reduces the gallons of spray applied per acre (gpa) by one-half. For example, a sprayer applying 20 gpa at 3 mph would apply 10 gpa if the speed were increased to 6 mph and the pressure remained constant.

Some low-pressure field sprayers are equipped with control systems that maintain a constant spray application rate (gpa) over a range of ground speeds. The pressure is changed to vary the nozzle flow rate according to changes in ground speeds. These systems require calibration at a set ground speed. In the field, speed changes must be limited to a range that maintains the nozzle pressure within its recommended range.

Sprayed Width per Nozzle. The effective width sprayed per nozzle also affects the spray application rate. *Doubling the effective sprayed width per nozzle decreases the gpa applied by one-half.* For example, if you are applying 40 gpa with flat-fan nozzles on 20-inch spacings and change to flooding nozzles with the same flow rate on 40-inch spacings, the application rate decreases from 40 to 20 gpa.

The gallons of spray applied per acre can be determined by using the following equation:

$$\text{gpa} = \text{gpm} \times 5940/\text{mph} \times \text{W} \qquad \text{(Equation 1)}$$

where gpm = output per nozzle in gallons per minute

mph = ground speed in miles per hour

W = effective sprayed width per nozzle in inches

5940 = a constant to convert gallons per minute, miles per hour, and inches to gallons per acre

There are many methods for calibrating low-pressure sprayers, but they all involve the use of the variables in Equation 1. Any technique for calibration that provides accurate and uniform application is acceptable. No single method is best for everyone.

The calibration method described next has three advantages. First, it allows you to complete most of the calibration before going to the field. Second, it provides a simple means for frequently adjusting the calibration to compensate for changes because of nozzle wear. Third, it can be used for broadcast, band, directed, and row crop spraying. This method requires a knowledge of nozzle types and sizes and the recommended operating pressure ranges for each type of nozzle used.

Selecting the proper nozzle tip. The size of the nozzle tip depends on the application rate (gpa), ground speed (mph), and effective sprayed width (W) that you plan to use. Some manufacturers advertise "gallons-per-acre" nozzles, but this rating is useful only for standard conditions (usually 30 psi, 4 mph, and 20-inch spacings). The gallons-per-acre rating is useless if any one of your conditions varies from the standard.

A more exact method for choosing the correct nozzle tip is to determine the gallons per minute (gpm) required for your conditions; then select nozzles that provide this flow rate when operated within the recommended pressure range. By following the five steps described next, you can select the nozzles required for each application well ahead of the spraying season.

Step 1. Select the spray application rate in gpa that you want to use. Pesticide labels recommend ranges for various types of equipment. The spray application rate is the gallons of carrier (water, fertilizer, and the like) and pesticide applied per treated acre.

Step 2. Select or measure an appropriate ground speed in miles per hour (mph) according to existing field conditions. Do not rely on speedometers as an accurate measure of speed. Slippage and variation in tire sizes can result in speedometer errors of 30% or more. If you do not know the actual ground speed, you can easily measure it.

If your sprayer does not have a speedometer, you must measure the speed at all the settings that you plan to use in the field. By measuring and recording the ground speed at several gear and throttle settings, you will not have to remeasure speed each time that you change settings.

To measure ground speed, lay out a known distance in the field to be sprayed or in a field with similar surface conditions. Suggested distances are 100 feet for speeds up to 5 mph, 200 feet for speeds from 5 to 10 mph, and at least 300 feet for speeds above 10 mph. At the engine throttle setting and gear that you plan to use during spraying with a loaded sprayer, determine the travel time between the measured stakes in each direction. Average these speeds and use the following equation to determine ground speed.

$$\text{speed (mph)} = \frac{\text{distance (feet)} \times 60}{\text{time (seconds)} \times 88}$$

where   1 mph = 88 feet in 60 seconds

Example: You measure a 200-foot course and discover that 22 seconds are required for the first pass and 24 seconds for the return pass.

$$\text{average time} = \frac{22 + 24}{2} = 23 \text{ seconds}$$

$$\text{mph} = \frac{200 \times 60}{23 \times 88} = \frac{12,000}{2024} = 5.9$$

Once you have decided on a particular speed, record the throttle setting and drive gear used.

Step 3.   Determine the effective sprayed width per nozzle (W) in inches.

- For broadcast spraying, W = the nozzle spacing
- For band spraying, W = the band width
- For row crop applications, such as spraying from drop pipes or directed spraying,

$$W = \frac{\text{row spacing (or band width)}}{\text{number of nozzles per row (or band)}}$$

Step 4.   Determine the flow rate required from each nozzle in gpm by using a nozzle catalog, tables, or the following equation:

$$gpm = \frac{gpa \times mph \times W}{5940} \qquad \text{(Equation 2)}$$

where   gpm = gallons per minute of output required from each nozzle
   gpa = gallons per acre from step 1
   mph = miles per hour from step 2
   W = inches sprayed per nozzle from step 3
   5940 = a constant to convert gallons per minute, miles per hour, and inches to gallons per acre

Step 5.   Select a nozzle that, when operated within the recommended pressure range, will give the flow rate determined in step 4 . You should obtain a catalog listing available nozzle tips. These catalogs may be obtained free of charge from equipment dealers or nozzle manufacturers. If you wish to use nozzles that you already have, return to step 2 and select a speed that allows you to operate within the recommended pressure range.

Example 1:   You want to broadcast an herbicide at 15 gpa (step 1) at a speed of 7 mph (step 2), using flooding nozzles spaced 40 inches apart on the boom (step 3). What nozzle tip should you select?

The required flow rate for each nozzle (step 4) is as follows:

$$gpm = \frac{gpa \times mph \times W}{5940}$$

$$= \frac{15 \times 7 \times 40}{5940} = \frac{4200}{5940} = 0.71$$

The nozzle that you select must have a flow rate of 0.71 gpm when operated within the recommended pressure range of 8 to 25 psi. Tables 12–1 and 12–2 show the gpm at various pressures for several Spraying Systems and Delavan nozzles. For example, the Spraying Systems TK5 and Delavan D5 nozzles (Table 12–1) have a rated output of 0.71 gpm at 20 psi (step 5). Either of these nozzles can be purchased for this application.

Example 2:   You want to spray an insecticide on corn plants growing in 30-inch rows, using two nozzles per row. The desired application rate is 15 gpa at 6 mph. Which disc-core, hollow-cone nozzle should you select?

Calibration Methods

TABLE 12–1 FLOODING FLAT-FAN NOZZLES

Manufacturer Delavan	Spraying systems	Liquid pressure (psi)	Capacity (gpm)
DI	TKI	10	0.10
		20	0.14
		30	0.17
		40	0.20
D2	TK2	10	0.20
		20	0.28
		30	0.35
		40	0.40
D3	TK3	10	0.30
		20	0.42
		30	0.52
		40	0.60
D4	TK4	10	0.40
		20	0.57
		30	0.69
		40	0.80
D5	TK5	10	0.50
		20	0.71
		30	0.87
		40	1.00
D7.5	TK7.5	10	0.75
		20	1.10
		30	1.30
		40	1.50

Because two nozzles spray each 30-inch row, W = 30/2 = 15 inches. The required flow rate for each nozzle is as follows:

$$\text{gpm} = \frac{\text{gpa} \times \text{mph} \times \text{W}}{5940} = \frac{15 \times 6 \times 15}{5940} = \frac{1350}{5940} = 0.23$$

Either a Delavan DC3-25 or a Spraying Systems D3-25 disc-core nozzle (Table 12–2) has a rated output of 0.23 gpm at 60 psi. This pressure is within the recommended pressure range of 40 to 80 psi.

Precalibration checking. After making sure that your sprayer is clean, install the selected nozzle tips, partially fill the tank with clean water, and operate the sprayer at a pressure within the recommended range. Place a container (e.g., a quart jar) under each nozzle. Check to see whether all the jars fill at about the same time. Replace any nozzle that has an output of 5% more or less than the average of all the nozzles, an obviously different fan angle, or a nonuniform appearance in spray pattern.

To obtain uniform coverage, you must consider the spray angle, spacing, and height of the nozzle. The height must be readjusted for uniform coverage with various

TABLE 12–2 HOLLOW-CONE NOZZLES (DISC AND CORE TYPE)

Manufacturer		Liquid pressure (psi)	Capacity (gpm)
Delavan	Spraying systems		
DC3-23	D3-23	40	0.12
		60	0.14
		80	0.16
		100	0.18
		150	0.21
DC2-25	D2-25	40	0.16
		60	0.19
		80	0.22
		100	0.25
		150	0.29
DC3-25	D3-25	40	0.19
		60	0.23
		80	0.26
		100	0.29
		150	0.35
DC3-45	D3-45	40	0.23
		60	0.28
		80	0.33
		100	0.36
		150	0.44
DC4-25	D4-25	40	0.29
		60	0.35
		80	0.40
		100	0.45
		150	0.54
DC5-45	D5-45	40	0.45
		60	0.55
		80	0.64
		100	0.71
		150	0.86

spray angles and nozzle spacings. Do *not* use nozzles with different spray angles on the same boom for broadcast spraying.

Worn or partially plugged nozzles produce nonuniform patterns. Misalignment of nozzle tips is a common cause of uneven coverage. The boom must be level at all times to maintain uniform coverage. Skips and uneven coverage will result if one end of the boom is allowed to droop. A practical method for determining the exact nozzle height that will produce the most uniform coverage is to spray on a warm surface such as a road and observe the drying rate. Adjust the height to eliminate excess streaking.

Calibrating your sprayer. Now that you have selected and installed the proper nozzle tips (steps 1 to 5), you are ready to complete the calibration of your sprayer. Check the calibration every few days during the season or when changing the pesticides being applied. New nozzles do not lessen the need to calibrate because some nozzles "wear in" and will increase their flow rate most rapidly during the first few hours of use. Once you have learned this method, you can check application rates quickly and easily.

Step 6. Determine the required flow rate for each nozzle in ounces per minute (opm). To convert gpm (step 4) to opm, use the following equation:

$$\text{opm} = \text{gpm} \times 128 \qquad (1 \text{ gal} = 128 \text{ oz}) \qquad \text{(Equation 3)}$$

From Example 1, the required nozzle flow rate = 0.71 gpm.

$$\text{opm} = 0.71 \times 128 = 91 \text{ opm}$$

From Example 2, the required nozzle flow rate = 0.23 gpm.

$$\text{opm} = 0.23 \times 128 = 29 \text{ opm}$$

An alternative method for determining the opm is to use Table 12–3. Locate the gpa selected in step 1 in the left column of the table. Follow this row across to the column indicating the ground speed that you determined in step 2. The value shown at the intersection of the row and column is the number of ounces per minute (opm) per inch of spray width. To determine the output required per nozzle, multiply this value by the inches of sprayed width per nozzle.

Example: From Example 1, gpa = 15, mph = 7, and W = 40 inches. From the table, the value for opm = 2.26. For a nozzle spacing of 40 inches, the required flow rate = 2.26 × 40 = 90.4 opm.

Step 7. Collect the output from one of the nozzles in a container marked in ounces. Adjust the pressure until the ounces per minute (opm) collected is the same as the amount that you determined in step 6. Check several other nozzles to determine if their outputs fall within 5% of the desired opm.

If it becomes impossible to obtain the desired output within the recommended range of operating pressures, select larger or smaller nozzle tips or a new ground speed and then recalibrate. It is important for spray nozzles to be operated within the recommended pressure range. (The range of operating pressures is for pressure at the nozzle tip. Line losses, nozzle check valves, and so on, may require the main pressure gauge at the boom or at the controls to read much higher.)

TABLE 12–3 OUNCES PER MINUTE PER INCH OF SPRAY WIDTH

Volume (gpa)	Speed (mph)										
	3	4	5	6	7	8	9	10	12	15	20
5	0.32	0.43	0.54	0.65	0.75	0.86	0.97	1.08	1.29	1.62	2.15
10	0.65	0.86	1.08	1.29	1.51	1.72	1.94	2.15	2.59	3.23	4.31
15	0.97	1.29	1.62	1.94	2.26	2.59	2.91	3.23	3.88	4.85	6.46
20	1.29	1.72	2.15	2.59	3.02	3.45	3.87	4.31	5.17	6.46	8.62
30	1.94	2.59	3.23	3.88	4.52	5.17	5.82	6.46	7.76	9.70	12.93
40	2.58	3.44	4.30	5.18	6.04	6.90	7.74	8.62	10.34	12.92	17.24

Step 8. Determine the amount of pesticide needed for each tankful or for the acreage to be sprayed (see calculations for mixing in Chapter 13). Add the pesticide to a partially filled tank of carrier (water, fertilizer, and the like); then add the carrier to the desired level with continuous agitation.

Step 9. Operate the sprayer in the field at the ground speed that you determined in step 2 and at the pressure that you determined in step 7. You will be spraying at the application rate that you selected in step 1. After spraying a known number of acres, check the liquid level in the tank to verify that the application rate is correct.

Step 10. Check the nozzle flow rate frequently. Adjust the pressure to compensate for small changes in nozzle output resulting from nozzle wear or variations in other spraying components. Replace the nozzle tips and recalibrate when the output has changed 10% or more from that of a new nozzle or when the pattern has become uneven.

Determining spray rate with installed nozzle tips. You may already have a set of nozzle tips in your sprayer, and you want to know the spray rate (gpa) when operating at a particular nozzle pressure and ground speed. There are many methods for measuring the spray rate, including spraying a known area such as an acre or fraction of an acre and measuring the gallons of spray required to spray the area.

In the method described next, the nozzle flow rate is measured, and then the gpa is calculated from Equation 1. Add water to the spray tank and make a precalibration check to be sure that all the sprayer components are working properly. Remember, the type, size, and fan angle of all the nozzle tips must be the same. The flow rate from each nozzle must be within 5% of the average flow rate from the other nozzles.

Step 1. Operate the sprayer at the desired operating pressure. Use a container marked in ounces to collect the output of a nozzle for a measured length of time, such as 1 minute. Check several other nozzles to determine the average opm of output from each nozzle.

Step 2. Convert opm of flow to gpm of flow by dividing the opm by 128 (the number of ounces in 1 gallon).

Step 3. Determine the sprayer ground speed (mph).

Step 4. Determine the sprayed width per nozzle (W) in inches.

- For broadcast spraying, W = the nozzle spacing
- For band spraying, W = the band width
- For row crop spraying, $W = \dfrac{\text{row spacing (or band width)}}{\text{number of nozzles per row (or band)}}$

Step 5. Calculate the sprayer output (gpa) using Equation 1.

$$\text{gpa} = \frac{\text{gpm} \times 5940}{\text{mph} \times W}$$

Example: The measured nozzle output is 54 opm, the measured ground speed is 6 mph, and the nozzle spacing (W) is 20 inches.

$$\text{gpm} = \frac{54}{128} = 0.42$$

$$\text{gpa} = \frac{0.42 \times 5940}{6 \times 20} = \frac{2495}{120} = 20.8$$

Gallons per acre of output (step 5) can be adjusted by changing the ground speed or nozzle pressure, and recalibrating. Changes in nozzle pressure should be used only to make small changes in output and must be maintained within the recommended pressure range. Considering all the variables, calibration must be done at the same speed used in a particular field, and speed should be held constant during application.

Failures and factors. Most complaints are usually the result of nonuniform application. Other causes of poor distribution include the following:

- Failure to make and keep a good suspension of wettable powders or liquid flowables and dry flowables in the tank
- Failure to get uniform coverage due to ground speed, spray angle, or differential nozzle output
- Adjusting pressures based on a faulty pressure gauge (two gauges in the system are desirable; this lets you know that one is bad if the readings are significantly different; you just have to figure out which gauge is bad)
- Failure to maintain uniform swath width; failure to adjust pressure to compensate for nozzle wear or to replace nozzles after their flow rate varies by more than 10% of the desired pressure
- Foaming of the spray mixture (usually as the tank nears empty)
- Failure to make frequent checks of the amount of material applied versus acreage covered

Swath marking is often difficult when applicators are using high-speed, volume flotation sprayers with wide booms. Overlaps and skips can easily occur as most operators are using visual observation. Sun glare, cross tillage, and visual perception of the applicator aggravate the problem. Errors tend to increase with applications made late in the day. The foam marker seems to be the most widely used system, although new systems using high-tech apparatus are becoming available.

Two other factors you can control for better pattern distribution include the following:

1. Since individual nozzle patterns are narrower when the sprayer is moving than when the sprayer is stationary, always make your final adjustment for proper overlap under field conditions.

2. Boom height is significant, so watch out for variations in height from bouncing, low tire pressure on one side or on one wheel, improper boom mounting, and bent booms. Ideally, keep the boom parallel to the spray surface. But if height is going to vary, it is better to be a little too high than a little too low. Also, try to

keep the boom from swinging fore and aft because coverage will decrease as the boom swings forward and increase as it swings back.

Granular Applicators

The application rate of granules is determined by the size of the metering opening, the speed of the agitator or rotor, the travel speed, the roughness of the surface to be treated, and the flowability of the granules. Granules flow at different rates, depending on the size, density, and type of granule, temperature, and humidity. Granules vary greatly in the size and density of the particles and in the carrier used to formulate the pesticide. For these reasons, a different orifice setting may be necessary for each pesticide applied. A different setting may even be required for the same pesticide formulated by two different manufacturers or for two batches of pesticide from the same manufacturer.

Except for the orifice setting, ground speed is the most significant variable affecting the application rate. A granular application must be calibrated at the same speed as that used during application, and the speed must be kept constant. Even though gravity-flow applicators use a rotating agitator that varies with the ground speed, the flow of granules through the orifice is not directly proportional to the speed. It is not uncommon to find a 50% variation in the application rate when changing the ground speed by only 1 or 2 miles per hour.

Temperature and humidity affect the moisture content of the granules, resulting in flow changes from day to day, and surface roughness can affect the application rate when changing from one location to another. For these reasons, it is important to calibrate frequently in order to maintain the proper application rate.

Individual units on row applicators should be calibrated independently. Even if all the applicators have the same setting, large differences in application rate may occur between the various row units on the same planter. The correct application rate can be maintained only by regularly checking calibration on all units. Calibration charts should be used only as a guideline for the initial setting of the application during calibration.

Determining rate. The first step in calibrating granular application equipment is to read the pesticide label carefully. The label indicates the application rate and where the granules should be placed. Recommendations vary according to the type of pesticide. Rates are usually expressed as pounds per broadcast area (1 acre or 1000 square feet) or as ounces per 1000 feet of row.

Because the same pesticides may be available in different concentrations of granules, written recommendations are sometimes given as pounds of active ingredient (a.i.) to apply per acre. First, read the label to determine the percent of a.i. in the granules; then convert the pounds of a.i. per acre to pounds of product per acre by using the following formula:

$$\text{lb product per acre} = 3 \text{ lb a.i. per acre} \times \frac{100\%}{\% \text{ a.i. in product}}$$

Example: The recommended rate is 3 pounds of a.i. per acre. You have purchased a 20% granular formulation. How many pounds of product should you apply per acre?

$$\text{lb per acre} = 3 \text{ lb a.i. per acre} \times \frac{100\%}{20\%} = \frac{300}{20} = 15$$

Broadcast: For all herbicides (banded and broadcast) and for broadcast insecticides and nematicides, apply the recommended pounds per broadcast acre. Remember, even though you do not treat the entire area when banding, you still apply the same rate of granules in the band as when broadcasting. Because less total product is applied to a field, however, you must drive over more than one acre to actually treat one acre of banded area.

Example: You want to apply a granular herbicide at 20 pounds per acre in a 14-inch band over 30-inch soybean rows. How many pounds should you purchase to treat 200 acres of soybeans?

$$200 \text{ acres} \times \frac{14\text{-inch band}}{30\text{-inch row}} = 93.3 \text{ acres treated}$$

$$93.3 \text{ acres} \times 20 \text{ lb per acre} = 1866 \text{ lb of granules needed}$$

In-the-row: Applications of insecticides and nematicides (banded or furrow treatments) are based on ounces per 1000 linear feet of row rather than on pounds per broadcast acre. These materials require a constant concentration of pesticide per linear foot of row, regardless of whether the pesticide is banded or placed in the furrow. Once you have calibrated a granular row applicator to apply the recommended ounces per 1000 feet of row, it will apply the correct rate of granules for any row spacing used. The same rate will be applied down the row, with the total amount applied per crop acre changing according to row spacing.

In addition to ounces per 1000 feet of row, manufacturers usually list the recommended pounds per acre for a standard row width, such as 40 inches, 36 inches, or 22 inches. For row spacings other than standard widths, the amount of granules required per acre to maintain a uniform concentration per foot of row depends on the row spacing used. The total feet of row in 1 acre increases as row spacing decreases.

For example, if a recommendation calls for 15 pounds per acre for 40-inch rows and you are planting 20-inch rows, you will need to apply 30 pounds per acre to maintain the same concentration of pesticide in the row. One acre contains 13,068 feet of 40-inch rows. At the rate of 15 pounds per acre, each 1000 feet of row would receive 18.3 ounces of granules. One acre planted in 20-inch rows contains 26,136 feet of rows. To maintain the same concentration of 18.3 ounces of granules per 1000 feet of row would require 30 pounds of granules applied per acre, rather than the 15 pounds per acre applied to the 40-inch rows.

Calibration techniques.   Most calibration techniques are based on determining the amount of granules dispensed when treating a known area. The following procedure is based on adjusting the orifice setting until you collect the required amount of granules while traveling a measured distance.

Step 1.   Determine the number of ounces required for application over a known distance.

Broadcast or band applications: Calculate the number of ounces required to be distributed over a measured course.

$$\text{ounces required} = \frac{\text{lb per acre} \times \text{area treated (square feet)}}{2722}$$

2722 = a constant arrived at by dividing the number of square feet in 1 acre (43,560) by the number of ounces in 1 pound (16)

The area treated is the area actually covered with granules and is the length of a measured course multiplied by the width of spread in feet. For broadcast applications, measure the effective swath width; for band applications, measure the band width.

Example 3: The recommended rate is 20 pounds per acre of herbicide granules in a 14-inch band (14 inches = 1.2 feet) over each row. A 1000-foot measured course is to be used. How many ounces should be collected when the applicator is operated over the 1000-foot measured course?

$$\text{area treated} = 1000 \times 1.2 \text{ feet} = 1200 \text{ square feet}$$

$$\text{ounces required} = \frac{20 \text{ lb per acre} \times 1200 \text{ square feet}}{2722} = 8.8$$

Example 4: Insecticide granules are to be applied broadcast to turf at the rate of 3 pounds per acre. The effective swath width is 60 inches (5 feet). A 500-foot distance is used for calibration. How many ounces should be collected?

$$\text{area treated} = 500 \times 5 \text{ feet} = 2500 \text{ square feet}$$

$$\text{ounces required} = \frac{3 \text{ lb per acre} \times 2500 \text{ square feet}}{2722} = 2.8$$

In-the-row applications: The number of ounces per 1000 feet of row is usually listed on the label. If the label lists the rate only in pounds per acre for a standard row spacing, you can convert to ounces per 1000 feet of row by using the following formula:

$$\text{ounces per 1000 feet of row} = \frac{\text{lb per acre} \times \text{std. row spacing (ft)} \times 1000 \text{ ft of row}}{2722} = 9.2$$

Step 2. Adjust the initial orifice setting on each applicator according to the equipment manufacturer's recommendation.

Step 3. Attach a container (plastic bag or jar) over the outlet of the applicator, and collect the granules while operating the applicator over the measured distance. Make the run in the field to be treated so that speed and traction conditions will be constant.

Step 4. Weigh the collected granules. Because the amount collected may be only a few ounces, the granules should be weighed on a postal scale, baby scale, or food scale. Volume measurements may be inaccurate because of nonuniform settling or segregation of the granules. If the amount collected is not equal to the ounces required (step 1), adjust the gauge setting and repeat the calibration until you have collected the required amount.

After you have calibrated the applicator to apply the proper amount of granules, make periodic field checks to verify that the application rate does not change from one day to the next or from one field to another. One simple method for verifying calibration is to place a strip of masking tape vertically on the inside of the application hopper, then fill the hopper in increments of 1 or 2 pounds. After adding each increment, shake the hopper to settle the material and mark the tape at the level of the chemical. The application rate can be verified throughout the season simply by reading the level of chemical before and after treating a known number of acres.

$$\text{pounds per acre} = \frac{\text{pounds used}}{\text{acres treated}}$$

Example: In making a field check, you discover that 12 pounds of granules were applied to 5 acres. What is the application rate?

$$\text{pounds per acre} = \frac{12 \text{ lb}}{5 \text{ acres}} = 2.4 \text{ lb per acre}$$

Even though the typical gravity-flow applicators use a rotating agitator where the speed varies with ground speed, the flow of granules is not proportional to speed. Researchers have found it is not uncommon to find a 50% variation in the application rate when changing ground speed by as little as 2 to 3 miles per hour. When applying granules, calibrate frequently—between fields, between loads, and between drivers.

Band Applicators

Preemergence spraying (spraying after the crop is planted but before it comes up) and preplant spraying (application before the crop is planted) are becoming widely used practices in cultivated crops. Many of the herbicides used are expensive, and one way to overcome the problem of cost is to apply the chemical as a band spray. Band spraying is the application of the herbicide in a band, usually about one-third to one-half as wide as the row spacing, immediately over the crop row, leaving the area between the crop rows unsprayed. In this way, only one-third to one-half as much material is used per cropped acre as when full coverage spraying is used, with a resultant saving in chemical cost. The area between the rows can be cultivated clean to reduce the weed infestation.

When application rates are recommended for weed control chemicals, such as 2 pounds per acre, this amount of active ingredient is to be applied to the area covered by spray. With full coverage spraying, the entire field would receive this amount of chemical, but with band spraying only the sprayed band receives chemical at this rate. Thus, if 14-inch bands were sprayed on 42-inch rows, for every acre of cropland treated only ⅓ acre would be sprayed. Therefore, if 2 pounds per acre of chemical were recommended, 2 pounds of active ingredient would be applied to each acre actually sprayed, but only ⅔ pound of chemical would be required to treat an acre of banded cropland.

The purpose of calibration is to determine the amount of spray applied to the band area. This figure is used to determine the amount of chemical to mix with carrier (water) in the tank. The concentration is figured exactly the same way as it would be if the spray were full coverage.

Calibration. As with full-coverage spraying, calibration is very important. The margin of selectivity or safety of preplant or preemergence herbicides on such crops as sugar beets, field beans, and corn is sometimes narrow, and accurate application is necessary.

Calibration can be done with various calibration jars on the market or by using various other methods. The following is the **refill method:**

1. Measure off a known distance, such as 300 or 400 feet.
2. Fill the sprayer tank with water to a known mark. Spray the measured area at the *same speed and pressure* that would be used in the field.
3. Refill the tank to the known mark, measuring carefully the amount of water used.
4. Calculate the gallons per acre (gpa) sprayed by the following formula:

$$\frac{43{,}560 \times \text{gallons used}}{\text{distance traveled (feet)} \times \text{band width (in feet)} \times \text{no. of bands}} = \text{gpa}$$

Example: If ½ gallon of water was sprayed on a 300-foot strip in two 14-inch bands, the acre rate would be

$$\frac{43{,}560 \times 0.5}{300 \times 1.2 \times 2} = 30 \text{ gpa sprayed}$$

Example: If the same amount of water was sprayed on two 6-inch bands, the acre rate would be

$$\frac{43{,}560 \times 0.5}{300 \times .5 \times 2} = 72.6 \text{ gpa sprayed}$$

The **nozzle method** described for a boom sprayer can also be used to calibrate a band sprayer. The formula is

$$\text{Spray rate (gpa)} = \frac{\text{nozzle output (opm)} \times 46.4}{\text{nozzle coverage (band width) (inches)} \times \text{speed (mph)}}$$

Step 1. Adjust the pressure to the amount that will be used in the field and collect the spray from each nozzle for 1 minute and measure. (If all nozzles deliver equal amounts as they should, only one nozzle needs to be measured.)

Step 2. Measure the band width (coverage) in inches.

Step 3. Substitute the values from steps 1 and 2 into the formula and calculate the gpa sprayed on the band.

Example: If the nozzle delivers 36 ounces of water in 1 minute at 30-pound pressure and the speed to be used is 3 mph, the spray rate on a 14-inch band will be

$$\frac{36 \times 46.4}{14 \times 3} = 39.8 \text{ gpa}$$

Using the same example, if the band width were reduced to 7 inches, the spray rate on the band would be

$$\frac{36 \times 46.4}{7 \times 3} = 79.5 \text{ gpa}$$

Changing the width of the band has a pronounced effect on the spray rate and thus on the concentration of chemical in the spray tank. It should be pointed out, however, that if the speed and row spacing remain the same in the two examples, the amount of water used per crop acre will also remain the same.

Air-Blast Sprayers

1. Observe carefully the nozzle arrangement on the machine as to location and sizes of nozzles.
2. Determine the rate of nozzle discharge at the normal operating pressure by consulting the service manual that accompanies the machine, or ask for help from your sprayer dealer or distributor.
3. Check the actual output per minute of the sprayer by timing the emptying of either a portion or all of a tank of water. If your spray tank is not marked or graduated, it may be necessary to measure known amounts of water into the tank before starting your calibration.
4. Make sure the speed of travel of the tractor and sprayer is correct. You may determine your speed in the following manner:
   - Determine the number of tree spaces per minute that you will pass at the speed you wish to travel using the following formula:

$$\text{tree spaces per minute} = \frac{\text{mph desired} \times 88}{\text{tree spacing in feet}}$$

Example: You have a tree spacing of 20 feet and you wish to go $1\frac{1}{2}$ mph:

$$\text{tree spaces} = \frac{1.5 \times 88}{20} = 6.6 \text{ tree spaces}$$

This means that your sprayer should cover 6.6 tree spaces in 1 minute to be traveling 1.5 mph. Check this by starting the sprayer and tractor exactly at one tree and then travel 6.6 trees and determine the time that was required. If you covered this distance in less than a minute, you should slow down; if you took more than a minute, you must speed up. *Repeat this* several times to make sure that you are accurate. Then mark the spot on your throttle and your gear setting. If the tractor has an automatic transmission, mark the speed setting on the selection lever.

5. Check your pump pressure and nozzles often to make sure that you are maintaining a good spray pattern.
6. Read carefully for any special instructions on handling of certain chemicals in air-blast sprayers.

7. Be sure to get your mixing ratios correct.
   - **Example:** Assume that you are going to apply 16 pounds of compound Z per acre and you are applying 600 gallons of water per acre. You have a sprayer with a 400-gallon tank. How much material do you use per tank? 400 gallons of water will do 400/600 of an acre or two-thirds of an acre. Consequently, two-thirds of 16 = $10\frac{2}{3}$ pounds of compound Z per 400 gallons.
   - **Example:** Assume that you are going to apply 16 pounds of compound Z per acre and you are applying 90 gallons of water per acre. Your sprayer has a 400-gallon tank. How much do you use per tank?

$$400 \text{ gallons of water will do } \frac{400}{90} = 4.4 \text{ acres}$$

$$4.4 \times 16 = 70.4 \text{ lb per 400-gallon tank}$$

Fumigation Applicators

Fumigants are metered from the system with orifices. These orifices serve the same purpose as spray nozzles except that they are placed in the line going to the injector. The orifice size is selected in the same manner as a nozzle. First determine the required flow rate with the gallons per hour formula. Then, using an orifice (spray nozzle) catalog or a flow formula, determine the size orifice needed.

Low-pressure fumigators. The low-volatile fumigator may be calibrated by collecting the material applied over a given area. The fumigant must remain as a liquid until it passes through the metering orifices. With the low-volatile fumigators, this is a problem only within the metering pump at high speeds. Even though the low-volatile fumigants are slow to vaporize, they will form vapor in the metering pump at high speeds due to the sudden vacuum created on the intake stroke. If this happens, the rate is drastically reduced.

There are two methods of collecting the material.

Method 1
1. Measure 100 feet in the field.
2. With the applicator equipment running at the desired speed and engaging the soil, collect the material applied in the 100-foot section from one or more soil tubes. (The entire applicator output may be collected.)
3. Determine the application width being collected. **Example:** If the material is collected from six tubes spaced 8 inches apart, the application width is 48 inches. If the material for one row is collected, the application width is the row spacing.
4. Compare the material collected with Table 12–4 to determine the rate.

Method 2
1. Measure 100 feet in the field.
2. While operating the equipment with the fumigant shut off, record the time required to travel the 100 feet.

TABLE 12–4 ROW FUMIGATION OR BAND SPRAYING RATE

	Quantity per 100-ft row									
Rate (gpa)	24-inch row		30-inch row		36-inch row		42-inch row		48-inch row	
	oz	cc	oz	cc	oz	cc	oz	cc	oz	cc
1	.6	17.4	.7	21.7	.9	26.1	1.0	30.4	1.2	34.8
3	1.7	52.1	2.2	65.0	2.6	78.2	3.1	91.2	3.5	104.3
5	2.9	86.9	3.7	108.4	4.4	130.3	5.1	152.1	5.9	173.8
7	4.1	121.6	5.1	151.7	6.2	182.5	7.2	212.9	8.2	235.1
9	5.3	156.4	6.1	195.1	7.9	234.6	9.3	273.7	10.6	312.8
12	7.1	208.5	8.8	260.1	10.6	312.8	12.3	365.0	14.1	417.1
15	8.8	260.7	11.0	325.2	13.2	391.0	15.4	456.2	17.6	521.3

3. Determine the application width for the orifice or orifices to be used. For row application, use the row width and collect all material for one row.

4. While the application equipment is not moving, adjust orifices and/or pressure to apply the quantity shown in Table 12–4 for the desired rate in the time required to move 100 feet through the field.

High-pressure fumigators. To calibrate the high-volatile applicators (methyl bromide), the container must be weighed before and after applying fumigant to a measured plot. The material cannot be collected because it would immediately volatilize.

Weigh the container of fumigant. Apply the fumigant to 100 linear feet of row for row crops or 1000 square feet of area for broadcast treatments. Reweigh the container to determine the amount of fumigant used. Multiply the amount (lb) used per 100 linear feet by:

145 for 30-inch width rows to give the pounds per acre

121 for 36-inch width rows to give the pounds per acre

104 for 42-inch width rows to give the pounds per acre

90 for 48-inch width rows to give the pounds per acre

For broadcast applications, multiply the amount used per 1000 square feet by 43.5 to determine the rate being applied per acre.

Remember that speed, pressure, and applicator injector spacing will affect the rate.

AERIAL EQUIPMENT

Fixed-wing aircraft and helicopters exhibit similar flight characteristics (wingtip vortex and main rotor vortex). Since the airflow patterns around and in the wake of each aircraft are sufficiently different, each type and series of aircraft needs testing. Changing the horsepower of the engine, the type of propeller, or wingtip shape will change the distribution pattern. Generalizations can be used to guide the operator on nozzle placement or granular disseminator adjustment. Pattern testing is needed to check the effect of each feature added to the aircraft.

Pattern tests should be made in calm air to avoid crosswind distortion. If wind is unavoidable, the tests should be made in a direction parallel to the wind. Testing should be carried out in winds less than 3 mph at all times. The best time for this is in the early morning before the sun heats up the ground, creating eddies and inversions. The tests must duplicate the use for which the application is required in terms of airspeed, height of flight, nozzle pressure or gate setting for granulars, nozzle angle and placement or disseminator adjustment, and so on. It is better to test with the same materials to be applied if at all possible. Substitute materials do not always act in quite the same manner as the chemicals. This is evident with granulars, where minor changes in the surface characteristics of the granules (shape, surface finish, fineness or grind, and so on) alter the discharge rate.

Spray Testing

The nozzle type and pressure should be selected for the material being used and the atomization required for the job. The application rate (gpa) will be set by the chemical being applied and crop being treated as listed in chemical recommendation handbooks or on the manufacturer's label. Because each aircraft exhibits a normal or effective swath width, this value should be used with Tables 12–5 and 12–6 to determine the acres per minute being treated.

Computation of acreage and materials. Use the following formula:

$$\text{acres covered} = \frac{\text{length of swath (miles)} \times \text{width of swath (feet)}}{8.25}$$

The number of acres in a swath of given width and length can be determined from Table 12–5.

Example: An aircraft with a 40-foot effective swath treats a strip 1 mile long. To find the number of acres, follow the 40-foot vertical column down until it intersects the 1-mile line. The answer to the nearest tenth is 4.8 acres. For swath widths other than those shown interpolate or use combinations of the figures shown. To determine the amount of chemical required, multiply the acres by the desired rate of application.

TABLE 12–5 ACREAGE DETERMINATION

Swath length (miles)	Swath width							
	30	35	40	45	50	75	100	200
¼	0.9	1.1	1.2	1.4	1.5	2.3	3.0	6.1
½	1.8	2.1	2.4	2.7	3.0	4.5	6.1	12.1
¾	2.7	3.2	3.6	4.1	4.6	6.8	9.1	18.2
1	3.6	4.2	4.8	5.5	6.1	9.1	12.1	24.2
2	7.2	8.4	9.8	10.9	12.1	18.2	24.2	48.5
3	10.8	12.6	14.5	16.4	18.2	27.3	36.4	72.7
4	14.4	16.8	19.4	21.8	24.2	36.4	48.5	97.0
5	18.0	21.0	24.2	27.3	30.3	45.5	60.6	121.1

TABLE 12–6 ACRES COVERED PER MINUTE

Speed (mph)	Swath width (ft)							
	30	35	40	45	50	75	100	200
40	2.4	2.8	3.2	3.6	4.0	6.0	8.0	16.0
50	3.0	3.5	4.0	4.5	5.0	7.5	10.0	20.0
60	3.6	4.2	4.8	5.4	6.0	9.0	12.0	24.0
70	4.2	4.9	5.6	6.3	7.0	10.5	14.0	28.0
80	4.8	5.6	6.4	7.2	8.0	12.0	16.0	32.0
90	5.4	6.3	7.2	8.1	9.0	13.5	18.0	36.0
100	6.0	7.0	8.0	9.0	10.0	15.0	20.0	40.0
110	6.6	7.7	8.8	9.9	11.0	16.5	22.0	44.0
120	7.2	8.4	9.6	10.8	12.0	18.0	24.0	48.0

Aircraft Calibration

Use the following formula:

$$\text{acres per minute} = \frac{2 \times \text{swath width} \times \text{mph}}{1000}$$

Table 12–6 shows the rate, in acres per minute, at which spray or dry material can be applied when swath width and speed of aircraft are known. For swath widths or aircraft speeds other than those shown, interpolate or use combinations of the figures shown. To find the rate of flow in gallons per minute or pounds per minute, multiply the acres per minute figure by the number of gallons or pounds per acre to be applied.

Example: A 100-mph aircraft has a 40-foot effective swath. Follow the vertical 40-foot column down until the figure opposite 100 mph is intersected. The aircraft would cover 8.0 acres per minute. If 1 gallon of spray is to be applied per acre, the aircraft should be calibrated to disperse liquid at the rate of 1 × 8.0 or 8.0 gpm. (If 10 pounds of dry material is to be applied per acre, the aircraft should be calibrated to disperse material at the rate of 10 × 8.0 or 80 pounds per minute.)

To determine gallons (or pounds) per acre discharged from the aircraft, divide the gallons (or pounds) per minute discharged by the acres per minute that the aircraft covers in a swath. Knowing the gpm required, the number of nozzles can be calculated based on the manufacturer's data for that type and pressure. The pressure and the airspeed are now fixed for the tests and the application.

Discharge calibration. Having installed the desired type, size, and number of nozzles, the output of the system should be checked to see that the correct discharge in gpm is taking place. If the pump can be run at operating speed with the aircraft stationary, nozzle discharge can be checked with a measuring container and a stopwatch. Boom pressure must remain constant. If this stationary test cannot be done, the aircraft should be filled with water to a suitable mark. The aircraft is flown and the spray system is run for a timed period (30, 60, 90, or 120 seconds). The aircraft is brought back to the same point used previously, and the amount of water used is determined by reading the tank scale(s) or refilling to the first mark using measuring devices.

Swath pattern tests. With the application rate now established, the swath pattern should be checked to see that the distribution across the swath is as uniform as possible. The best method of spray pattern testing consists of adding a tracer (dye, fluorescent material) to water in the tank(s) of the aircraft. The aircraft is then flown at the chosen airspeed and height and the spraying system is operated at the chosen pressure. One pass is made over a row of target plates or cards laid out at right angles to the direction of flight. The aircraft flies over the center of the target line 100 to 150 feet wide. The targets are collected and the spray deposit on each target is measured by the quantity of tracer. From the results, the distribution curve of the pattern can be determined. Corrections to the nozzle location can be made and the results checked by further testing.

The National Agricultural Aviation Association (NAAA), along with state and regional associations, has sponsored fly-ins featuring "Operation Safe," wherein aircraft are pattern tested with equipment attached to a computer for rapid analysis. This electronic testing has been well accepted. Several other similar techniques are being employed in different areas of the United States.

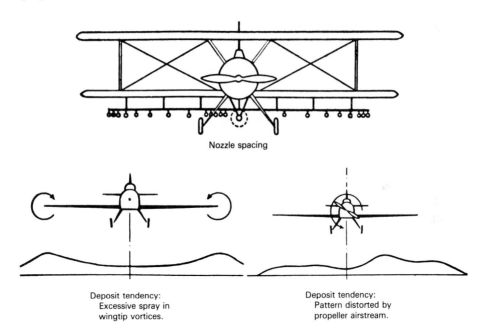

Nozzle spacing

Deposit tendency:
Excessive spray in
wingtip vortices.

Deposit tendency:
Pattern distorted by
propeller airstream.

Droplet size. Controlling droplet size is the key to success in applying liquid chemicals from aircraft. Droplet size is affected by the following factors:

- Shape of the nozzle cone (flat fan, whirljet, flood, and so on)
- Size of the tip
- Spraying pressure (20, 40, or 80 psi)
- Liquid or mixture of liquids being used (water, emulsion, chemical wetting agent, evaporation control agent, and so on)

- Angle of the nozzle with respect to the direction of flight (down, forward, back)
- Airspeed of the aircraft
- Density, viscosity, and surface tension of the liquid
- Evaporative conditions in the air between the point of release from aircraft and the point of impingement on ground

All commercial spraying nozzles tend to produce a range of droplet sizes. The hydraulic nozzles (flat fan, hollow cone, solid cone, floodjet, and so on) produce a broad range of sizes. The choice of these nozzles shifts the range as a whole, as well as the width of the range of sizes. The spinning nozzles, using rotating discs, screens, or brushes, produce a narrower range than the hydraulic nozzles. Overloading the spinning element will produce larger droplets and a wider range of droplets.

Pesticide work normally requires fine droplets to be effective, depending on the mode of operation of the chemical. This provides a large number of droplets. Herbicides (and systemic insecticides) require coarse droplets so that the plant will absorb the chemical. Fertilizer slurries can use very coarse droplets where they are normal NPK mixtures to be absorbed by the roots. Foliar fertilizers, however, should be applied like herbicides.

To increase droplet size

- lower spray pressure (not below 20 psi) and add nozzles to keep application rates up; or
- rotate the spray boom so that the nozzles discharge down and back with respect to the direction of flight; or
- change to larger tips of the same type, adjusting the number of nozzles being used; or
- use thickening agents in the spray; or
- stop spraying until cooler or calmer weather exists.

The control of drift is making the use of spray thickeners popular, reducing the number of fine droplets that create problems. These agents are often used with herbicides where sensitive crops are growing in the adjacent fields. Some sprays and spray mixtures demand fine droplets because they are phytotoxic to the crop when they are applied in coarse droplets (for example, cover sprays in orchards).

To reduce droplet size

- increase spray pressure (not above 60 psi); or
- rotate spray boom so that nozzles discharge down and forward with respect to the direction of flight; or
- change to smaller tips of the same type, adjusting the number of nozzles being used to keep the gallonage constant; or
- use thinning agents (wetting agents) in the spray.

It should be realized that all these changes affect the pattern and the rate of application with the exception of spray boom rotation. Rotation affects only the droplet

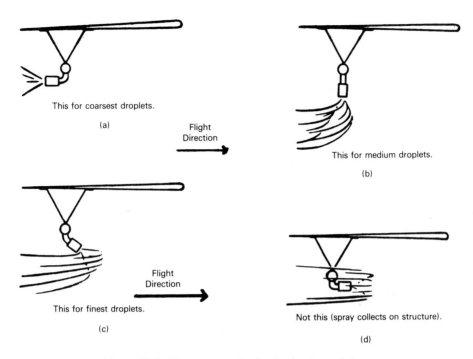

Figure 12–2 Nozzle orientation for droplet size variation.

size, with little or no effect on the pattern. The other changes will interact to affect the application unless you compensate for them.

Droplet size can be varied by changing the direction of the nozzle orifice in relation to the airstream. For example, when the same orifice and pump pressure are used, the coarsest droplets can be made by aiming the orifice straight back (Figure 12–2(a)); somewhat smaller droplets can be made by aiming the orifice straight downward (Figure 12–2(b)); the finest droplets can be made by aiming the orifice forward and downward at about 45 degrees (Figure 12–2(c)); aiming the nozzle more directly into the airstream will cause the spray to collect on the body of the nozzle or attachments and fall in large drops (Figure 12–2(d)). This wastes material.

Granular Materials Testing

Disseminators are sensitive to adjust and the differences between granular materials have a pronounced effect on the rate of delivery and the pattern. Some disseminators are restricted as to quantity or type of materials being handled. These limitations should be checked before testing.

Discharge calibration. Several runs should be made with the disseminating equipment installed to determine the quantity of material metered out for a given gate setting. If the disseminating equipment can be run with the aircraft on the ground, the material can be caught in large linen or paper bags and weighed. Ram-air disseminators require actual flight tests to get true discharge rates, since the air currents and the engine vibration in flight affect the metering gate discharge rate. After running the disseminator for a given time (30, 60, 90, 120 seconds), collected material is weighed. If flight tests are used, the quantity needed to refill the hopper is weighed. Where flight is needed to calibrate the system, use blank granulars (the granular carrier, without the pesticide, of the same type used to carry the chemical). Test for three gate settings to determine the gate setting that will give the required discharge in pounds per minute of granular material. Use the figures in Table 12–6 to calculate pounds per acre applied.

Important Considerations

Obstructions. Where obstructions occur (trees, power and telephone lines, or buildings) at the beginning or end of the swath, it is preferable to turn the equipment on late or shut off early. Then, when the field is completed, fly one or two swaths crossways (parallel to the obstruction) to finish out the field. Do not run the disseminator when dropping in or pulling out of the field; the pattern will be distorted. Obstructions inside the field should be treated in the same way. Skip the treatment as you avoid the obstruction, then, at the finish, come back and spot treat the skipped part, flying at right angles to the rest of the job.

Areas adjacent to buildings, residences, and livestock should be treated with extra care. Try to fly parallel to the property line, leaving a border of untreated crop, to avoid possible drift onto unwanted areas. Adjust pullout and drop-in paths and avoid making turns over houses. Use caution when fields include or are adjacent to waterways, canals, or reservoirs. Treat fields with care if sensitive crops are planted next to them. Be certain that beekeepers are warned if they have beehives near the field to be treated and you are applying chemicals harmful to the bees.

Ferrying. Ferrying height between airstrip and worksite should be at least 500 feet, whether loaded or empty. When possible, avoid flight over farm buildings, feedlots, or residential areas.

Speed. Calibration of the dispersal apparatus on the aircraft is actually rate of flow per minute. No device is at present available to aerial applicators to change flow rate *automatically* and proportionately as the speed of flight changes. Once the dispersal apparatus has been calibrated, the speed of flight should be kept *at the calibrated speed* as closely as possible during each swath run. Increasing the speed will result in too light a deposit; decreasing the speed will result in overapplication.

Height. Height of flight during application is usually governed by the form of chemical being applied. The amount of drift is increased as the height of application is increased. When deposit pattern and calibration have been set, the height of the application should be at the same height from which deposit pattern and calibration were made. Keep this height constant during each swath run to obtain uniform coverage. Whenever possible, application height should be increased up to a height equal to one wingspan (where this can be done without increasing drift hazard or decreasing penetration of chemical into foliage canopy). This increases safety of application and swath width.

13

Pesticide Calculations and Useful Formulas

The success of a spraying operation, whether you are spraying small areas or large fields, depends on accurate control of the application rate. After the equipment is accurately calibrated to apply the volume of spray desired, you must determine how much chemical to put in the tank to apply the correct dosage recommended.

In mixing a finished spray, it is most important to add the correct amount of pesticide to the tank. Too little may result in a poor job, while too much may result in injury to the treated surface, illegal residues on food crops, or unnecessary expense. Directions for mixing are given on the label, and only very simple calculations are necessary.

To determine how much chemical to put in the tank, you must know the capacity of the tank so that you can determine the number of acres that can be sprayed with one tankful of spray. This is found by dividing the capacity of the tank (in gallons) by the gallons applied per acre as determined when you calibrated the sprayer. For example, if the capacity of the tank is 200 gallons and the sprayer is applying 10 gallons per acre, you know that 200 gallons divided by 10 gallons per acre equals 20 acres that you can spray per tankful.

Most, but not all, recommendations are made on pounds of active ingredient to be applied per acre. To determine the amount of chemical to add to the tank, multiply the acres one tank will spray by the recommended rate per acre. For example, if the tank will spray 20 acres and the recommended rate is $1\frac{1}{2}$ pounds of active ingredient per acre, simply multiply 20 times $1\frac{1}{2}$ to arrive at 30 pounds of active ingredient per tank to spray 20 acres. The following are calculations and formulas for various pesticide mixes.

CALCULATIONS FOR MIXING

Liquid Formulations

The recommended application rate of the pesticide is listed on the label. The rate is usually indicated as pints, quarts, or gallons per acre for liquids. Sometimes the recommendation is given as pounds of active ingredient (lb a.i.) per acre rather than the amount of total product per acre. The active ingredient must be converted to actual product. Calculate as follows:

$$\frac{\text{no. of acres (area actually treated)} \times \text{rate per acre}}{\text{lb a.i. per gal concentrate}} = \text{no. gal concentrate needed}$$

Example: To treat 60 acres with a pesticide at a rate of 1.5 pounds per acre using a concentrate that contains 4.0 pounds per gallon will require 22.5 gallons of concentrate.

$$\frac{60 \times 1.5}{4} = 22.5 \text{ gal concentrate needed to treat 60-acre field}$$

Example (liquid formulation): A trifluralin recommendation calls for 1 pound of active ingredient (a.i.) per acre. You have purchased Treflan 4E® (4-pounds-per-gallon formulation). Your sprayer has a 300-gallon tank and is calibrated at 15 gallons per acre. How much Treflan® should you add to the spray tank?

Step 1. Determine the number of acres that you can spray with each tankful. Your sprayer has a 300-gallon tank and is calibrated for 15 gallons per acre.

$$\frac{\text{tank capacity (gal per tank)}}{\text{spray rate (gal per acre)}} = \frac{300}{15} = 20 \text{ acres sprayed with each tankful}$$

Step 2. Determine the amount of product needed per acre by dividing the recommended active ingredient per acre by the concentration of the formulation.

$$\frac{1 \text{ lb a.i. per acre}}{4 \text{ lb a.i. per gal}} = \frac{1}{4} \text{ gal per acre}$$

One-fourth gallon or 1 quart of product is needed for each "acre's worth" of water in the tank to apply 1 pound of active ingredient (a.i.) per acre.

Step 3. Determine the amount of pesticide to add to each tankful. With each tankful, you will cover 20 acres (step 1), and you want ¼ gallon (1 quart) of product per acre (step 2). Add 20 quarts (20 acres × 1 quart per acre = 20) of trifluralin to each tankful.

Example (adjuvant): The manufacturer may recommend that you add a small amount of an adjuvant (spreader-sticker, surfactant, and the like) in addition to the regular chemical. This recommendation is often given as "percent concentration."

If you use an adjuvant at a ½% concentration by volume, how much should you add to a 300-gallon tank?

Solution 1: 1% of 100 gallons = 1 gallon (100 × 0.01). You will need ½ gallon per 100 gallons, or 1½ gallons for 300 gallons (½ × 3 = 1½).
Solution 2: ½% = 0.005.

$$0.005 \times 300 \text{ gal} = 1.5 \text{ gal needed}$$

Dry Formulations

Wettable powders have active ingredients expressed as a percentage of the total weight and may vary from 25% to 80% active ingredient. To determine the total amount of material needed at a certain rate of active ingredient per acre as specified on the label, calculate as follows:

$$\frac{100 \times \text{no. acres} \times \text{rate per acre}}{\text{percent strength}} = \text{no. lb material needed}$$

Example: To treat 60 acres with a pesticide at a rate of 2 pounds per acre using a wettable powder that is 40% active ingredient will require 300 pounds of the wettable powder.

$$\frac{100 \times 60 \times 2}{40} = 300 \text{ pounds of the wettable powder to treat the 60-acre field}$$

Example: An atrazine recommendation calls for 2 pounds of active ingredient (a.i.) per acre. You have purchased AAtrex® (80% wettable powder). Your sprayer has a 400-gallon tank and is calibrated to apply 20 gallons per acre. How much AAtrex® should you add to the spray tank?

Step 1. Determine the number of acres you can spray with each tankful. Your sprayer has a 400-gallon tank and is calibrated to apply 20 gallons per acre.

$$\frac{\text{tank capacity (gal per tank)}}{\text{spray rate (gal per acre)}} = \frac{400}{20} = 20 \text{ acres sprayed per tank}$$

Step 2. Determine the pounds of pesticide product needed per acre. Because not all the atrazine in the bag is an active ingredient, you will have to add more than 2 pounds of the product to each "acre's worth" of water in your tank. How much more? The calculation is simple: divide the percentage of active ingredient (80) into the total (100).

$$2 \text{ lb a.i. per acre} \times \frac{100\%}{80\%} = 2 \times 1.25 = 2.5 \text{ lb product per acre}$$

You will need 2.5 pounds of product for each "acre's worth" of water in the tank to apply 2 pounds of active ingredient per acre.

Step 3. Determine the amount of pesticide to add to each tankful. With each tankful, you will cover 20 acres (step 1), and you want 2.5 pounds of product per acre (step 2). Add 50 pounds (20 acres × 2.5 pounds per acre = 50 pounds of product) of atrazine to each tankful.

Calculations for Mixing

Percentage Mixing

Sometimes you will find directions telling you to make a finished spray of a specific percentage, for instance a 1% spray. The pesticide may be formulated as a 57% emulsifiable concentrate. To make a 1% finished spray, add 1 part of pesticide to 56 parts of water, for example, 1 fluid ounce in 56 fluid ounces ($1\frac{3}{4}$ quarts) of water.

When mixing percentages, you should remember that 1 gallon of water weighs about 8.3 pounds and 100 gallons weigh about 830 pounds. Thus, to make a 1% mix of pesticide in 100 gallons of water, add 8.3 pounds of active ingredient of pesticide to 100 gallons of water. The formulas are as follows:

Formula for wettable powder percentage mixing. To figure the amount of wettable powder (WP) to add to get a given percentage of active ingredient (actual pesticide) in the tank, use the following formula:

$$\frac{\text{gal of spray wanted} \times (\% \text{ a.i. wanted}) \times 8.3 \text{ (lb per gal)}}{(\% \text{ a.i. in pesticide used})}$$

Example: How many pounds of an 80% wettable powder are needed to make 50 gallons of 3.5% spray for application by mist blower?

$$\frac{50 \text{ (gal wanted)} \times 3.5 \text{ (\% wanted)} \times 8.3 \text{ (lb per gal)}}{80 \text{ (\% a.i.)}} = \frac{1452.5}{80} = 18.1 \text{ lb of 80\% WP}$$

Formula for emulsifiable concentrate percentage mixing. To figure the amount of emulsifiable concentrate to add to get a given percentage of active ingredient (actual pesticide) in the tank, use the following formula:

$$\frac{\text{gal of spray wanted} \times (\% \text{ a.i. wanted}) \times 8.3 \text{ (lb per gal)}}{\text{lb a.i. per gal of concentrate} \times 100}$$

Example: How many gallons of 25% emulsifiable concentrate (2 pounds pesticide per gallon) are needed to make 100 gallons of 1% spray?

$$\frac{100 \text{ (gal wanted)} \times 1 \text{ (\% wanted)} \times 8.3 \text{ (lb per gal)}}{2 \text{ (lb a.i.)} \times 100} = \frac{830.0}{200} = 4.15 \text{ gal of 25\% EC}$$

Preparation of a cattle spray from a wettable powder. To obtain the quantity of pesticide needed to give a desired strength in the diluted spray, use the following formula:

$$\frac{\text{no. cattle} \times \text{gal spray per head} \times 8.3 \times \text{strength of spray desired}}{\text{strength of WP concentrate}} = \begin{array}{c} \text{lb WP} \\ \text{concentrate} \\ \text{needed} \end{array}$$

Example: One hundred and fifty head of cattle are to be sprayed with a 0.5% pesticide spray at the rate of 1 gallon of spray per animal; 50% pesticide WP is to be used.

$$\frac{150 \times 1 \times 8.3 \times 0.5}{50} = \begin{array}{l} \text{12.45 lb of 50\% pesticide needed to prepare 150 gal of} \\ \text{0.5\% pesticide finished spray} \end{array}$$

Preparation of a spray from wettable powder to treat an orchard.
Figure exactly as the preceding example except substitute trees for cattle. The number of gallons used per tree will vary depending on size of tree.

Oil sprays. Use the following formula to treat a certain number of trees with a certain percent of oil spray during the dormant season. *Note: Oil is figured on a volume basis.*

$$\frac{\text{no. trees} \times \text{gal finished spray per tree} \times \text{strength of spray desired}}{\text{strength of oil concentrate}}$$

Example: An orchardist wishes to spray 200 peach trees with a 3% dormant oil spray. The oil emulsion concentrate contains 85% oil. The trees require 3 gallons of spray for good coverage.

$$\frac{200 \times 3 \times 3}{85} = \begin{array}{l} \text{21.2 gal of 85\% oil emulsion concentrate required} \\ \text{to make 600 gal of 3\% finished oil spray} \end{array}$$

Dust Mixing

To figure the pounds of insecticide needed to mix a dust containing a given percent of active ingredient, use the following formula:

$$\frac{\text{\% a.i. wanted} \times \text{lb mixed dust wanted}}{\text{\% a.i. in insecticide used}}$$

Example: Five pounds of 3% dust are wanted. How much talc should be added to a 50% active ingredient insecticide powder to make the dust?

$$\frac{3 \times 5}{50} = 0.3 \text{ lb of 50\% insecticide}$$

Then add 4.7 pounds of talc to make the 5 pounds of 3% dust.

Granular Materials Calculation

These materials have active ingredients expressed as a percentage of the total weight and are usually manufactured in strengths of 5% to 15%. To determine how much formulation (commercial product) you need to apply to meet the recommended rate expressed in pounds of active ingredient per acre, the following formula is used:

$$\frac{\text{lb per acre desired}}{\text{\% a.i.}} = \text{lb of commercial product to use per acre}$$

Example: Forty acres are to be treated with a 15% granular pesticide at the rate of 1 pound per acre active ingredient. How many total pounds of commercial product are needed?

$$\frac{1}{0.15} \times 40 = 6.66 \text{ lb per acre} \times 40 \text{ acres} = 266.4 \text{ lb}$$

Square Feet Calculation and Mixing

Some labels will give mixing instructions in terms of amounts of active ingredient to be applied per 1000 square feet. This is often the case with lawn, golf course, and turfgrass treatments. The calculation for 1000 square feet areas can be carried out as follows:

Example: If an insecticide contains 4 pounds of active ingredient per gallon and the recommended rate of application is 0.5 pound of active ingredient per acre, how much would be needed (in ounces) to treat 20,000 square feet?

If 1 gallon of the insecticide contains 4 pounds of active ingredient then 1 gallon can be used to treat 8 acres at the rate of 0.5 pound of active ingredient per acre.

$$1 \text{ gal} = 4 \text{ lb a.i.}$$
$$\tfrac{1}{8} \text{ gal} = 0.5 \text{ lb a.i.}$$

The next step is to convert gallons to ounces. There are 128 ounces per gallon, so ⅛ gallon = 16 ounces. Sixteen ounces of the insecticide would contain 0.5 pound of active ingredient and would treat 1 acre.

Next, determine what percent of an acre is 20,000 square feet:

$$\frac{20,000}{43,560} = 46\%$$

Because 16 ounces of the insecticide will treat 1 acre, and the area to be treated (20,000 square feet) is 46% of an acre, then 46% of 16 ounces will treat the area at the recommended rate.

$$0.46 \ (46\%) \times 16 \text{ oz} = 7.36 \text{ oz}$$

So 7.36 ounces of the insecticide will treat 20,000 square feet at the rate of 1 pound of active ingredient per acre.

Example: The label on a liquid pesticide indicates a 25% concentration of the active ingredient. The recommended rate of application is 2 ounces of active ingredient per 1000 square feet. How much of the pesticide will it take to treat a 5000 square feet turf area?

Because the pesticide is only 25% active, this means that the rate of application for the pesticide must be 4 times (100%/25% = 4) the rate for the active ingredient.

$$2 \text{ oz a.i. per } 1000 \text{ ft}^2 \times 4 = 8 \text{ oz total pesticide per } 1000 \text{ ft}^2$$

The turf area is 5000 square feet, so 5 × 8 ounces = 40 ounces of pesticide required.

USEFUL FORMULAS AND TABLES

Figuring the Volume of Ponds

Before treating a pond, it is necessary to determine the pond volume in order to calculate the amount of a pesticide needed. The volume of a pond is based on the surface acreage plus the average depth of the water.

The surface acreage of ponds can be found as follows:

1. Compare one pond with another pond of the same shape and size for which the acreage is known.

2. If a pond is rectangular in shape, the surface acreage equals the length in feet times the width in feet divided by 43,560, that is:

$$\text{surface acres} = \frac{\text{length in ft} \times \text{width in ft}}{43,560}$$

Example: If the pond were 200 feet on each side and 100 feet wide at each end, the surface acreage would be

$$\frac{200 \times 100}{43,560} = \frac{20,000}{43,560} = 0.46$$

3. If a pond is circular in shape, measure the total distance (in feet) around the edge of the pond. Multiply this number by itself, and divide by 547,390.

$$\text{surface acres} = \frac{(\text{total ft of shoreline})^2}{547,390}$$

Example: A round pond has a total distance around the edge of 600 feet; the surface acreage is

$$\frac{600 \times 600}{547,390} = \frac{360,000}{547,390} = 0.66 \text{ acres}$$

Most farm ponds have uniformly sloping bottoms. Thus the average depth may be found by dividing the greatest depth of water by 2. The volume of the pond can then be determined using the following formula:

$$\text{vol in acre-feet} = \text{surface acreage} \times \tfrac{1}{2} \text{ maximum depth}$$

Example: A pond has a surface area of 0.5 acre, and the greatest depth is 10 feet. The volume of water in this pond is

$$0.5 \times 5 = 2.5 \text{ acre-feet}$$

Based on the volume of a pond, the following tables can be used to determine the correct amount of a pesticide that must be added to obtain control.

POUNDS OF PESTICIDE TO USE FOR OBTAINING THE DESIRED CONCENTRATION
OF PESTICIDE FOR PONDS OF VARIOUS VOLUMES

Pond volume		Concentration of pesticide desired, in parts per million							
Acre-feet	Gallons	1/10	1/4	1/2	3/4	1	5	10	40
0.1	32,585	0.03	0.07	0.14	0.2	0.3	1.4	2.7	10.9
0.2	65,170	0.05	0.14	0.27	0.4	0.5	2.7	5.4	21.7
0.3	97,755	0.08	0.20	0.41	0.6	0.8	4.1	8.1	32.6
0.4	130,340	0.11	0.27	0.54	0.8	1.1	5.4	10.9	43.4
0.5	162,925	0.14	0.34	0.68	1.0	1.4	6.8	13.6	54.3
0.6	195,510	0.16	0.41	0.82	1.2	1.6	8.2	16.3	65.2
0.7	228,095	0.19	0.48	0.95	1.4	1.9	9.5	19.0	76.0
0.8	260,680	0.22	0.54	1.09	1.6	2.2	10.9	21.7	86.9
0.9	293,265	0.24	0.61	1.22	1.8	2.4	12.2	24.4	97.8
1.0	325,850	0.27	0.68	1.36	2.0	2.7	13.6	27.2	108.6
2.0	651,700	0.54	1.36	2.72	4.1	5.4	27.2	54.3	217.3
3.0	977,550	0.81	2.04	4.08	6.1	8.1	40.8	81.5	325.9

DATA FOR CHANGING HUNDREDTHS OF POUNDS
TO OUNCES

Hundredths of pounds	Corresponding number of ounces
0.06	1
0.13	2
0.19	3
0.25	4
0.31	5
0.37	6
0.44	7
0.50	8
0.56	9
0.62	10
0.69	11
0.75	12
0.81	13
0.88	14
0.94	15
1.00	16

CONVERSION TABLES

LINEAR MEASURE

1 inch	= 2.54	centimeters
12 inches = 1 foot	= 0.3048	meter
3 feet = 1 yard	= 0.9144	meter
$5\frac{1}{2}$ yards or $16\frac{1}{2}$ feet = 1 rod	= 5.029	meters
1760 yards or 5280 feet = 1 (statute) mile	= 1609.3	meters

SQUARE MEASURE

1 square inch	= 6.452	square centimeters
144 square inches = 1 square foot	= 929	square centimeters
9 square feet = 1 square yard	= 0.8361	square meter
$30\frac{1}{4}$ square yards = 1 square rod	= 25.29	square meters
160 square rods or 4840 square yards or 43,560 square feet = 1 acre	= 0.4047	hectare
640 acres = 1 square mile	= 259	hectares or 2.59 square kilometers
16.5 feet = 1 rod		
272 square feet = 1 square rod		

AVOIRDUPOIS WEIGHT

437.5 grains = 1 ounce	= 28.3495	grams
16 ounces = 1 pound	= 453.59	grams
100 pounds = 1 hundredweight	= 45.36	kilograms
2000 pounds = 1 ton	= 907.18	kilograms

LIQUID MEASURE

3 teaspoonfuls = 1 tablespoonful	= 14.8	milliliters
2 tablespoonfuls = 1 fluid ounce	= 29.6	milliliters
16 tablespoonfuls = 1 cup	= 237	milliliters
8 fluid ounces = 1 cup		
2 cups = 1 pint (16 fluid ounces)	= 473.167	milliliters
2 pints = 1 quart	= 0.946	liters
4 quarts = 1 gallon (8.34 pounds water)	= 3.785	liters

APPLICATION FACTORS

1 cup per square rod	=	10 gallons per acre
1 pint per square rod	=	20 gallons per acre
1 quart per square rod	=	40 gallons per acre
1 gallon per square rod	=	160 gallons per acre

CONVENIENT CONVERSION FACTORS

Multiply	By	To get
Acres	43,560	Square feet
Acres	4,840	Square yards
Bushels	2,150.42	Cubic inches
Bushels	4	Pecks
Bushels	64	Pints
Bushels	32	Quarts
Centimeters	0.3937	Inches
Centimeters	0.01	Meters
Centimeters	10	Millimeters
Cubic feet	1,728	Cubic inches
Cubic feet	0.03704	Cubic yards
Cubic feet	7.4805	Gallons
Cubic feet	59.84	Pints (liquid)
Cubic feet	29.92	Quarts (liquid)
Cubic inches	16.39	Cubic centimeters
Cubic meters	1,000,000	Cubic centimeters
Cubic meters	35.31	Cubic feet
Cubic meters	61,023	Cubic inches
Cubic meters	1,308	Cubic yards
Cubic meters	264.2	Gallons
Cubic meters	2,113	Pints (liquid)
Cubic meters	1,057	Quarts (liquid)
Cubic yards	27	Cubic feet
Cubic yards	46,656	Cubic inches
Cubic yards	0.7646	Cubic meters
Cubic yards	202	Gallons
Cubic yards	1,616	Pints (liquid)
Cubic yards	807.9	Quarts (liquid)
Feet	30.48	Centimeters
Feet	12	Inches
Feet	0.3048	Meters
Feet	⅓ or 0.33333	Yards
Miles	5,280	Feet
Miles	320	Rods
Miles	1,760	Yards
Miles per hour	88	Feet per minute
Miles per hour	1.467	Feet per second
Miles per minute	88	Feet per second
Miles per minute	60	Miles per hour
Ounces (dry)	437.5	Grains
Ounces (dry)	28.3495	Grams
Ounces (dry)	0.0625	Pounds
Ounces (liquid)	1.805	Cubic inches
Ounces (liquid)	0.0078125	Gallons
Ounces (liquid)	29.573	Milliliters (cubic centimeters)
Ounces (liquid)	0.0625	Pints (liquid)
Ounces (liquid)	.03125	Quarts (liquid)
Parts per million	0.0584	Grains per U.S. gallon
Parts per million	0.001	Grams per liter
Parts per million	8.345	Pounds per million gallon
Pecks	0.25	Bushels
Pecks	537.605	Cubic inches
Pecks	16	Pints (dry)
Pecks	8	Quarts (dry)
Pints (dry)	0.015625	Bushels
Pints (dry)	33.6003	Cubic inches
Pints (dry)	0.0625	Pecks
Pints (dry)	0.5	Quarts (dry)
Pints (liquid)	28.875	Cubic inches
Pints (liquid)	0.125	Gallons
Pints (liquid)	0.4732	Liters
Pints (liquid)	16	Ounces (liquid)
Pints (liquid)	0.5	Quarts (liquid)

Multiply	by	to obtain
Feet per minute	0.01667	Feet per second
Feet per minute	0.01136	Miles per hour
Gallons	3,785	Cubic centimeter
Gallons	0.1337	Cubic feet
Gallons	231	Cubic inches
Gallons	128	Ounces (liquid)
Gallons	8	Pints (liquid)
Gallons	4	Quarts (liquid)
Gallons of water	8.3453	Pounds of water
Grains	0.0648	Grams
Grams	15.43	Grains
Grams	0.001	Kilograms
Grams	1,000	Milligrams
Grams	0.0353	Ounces
Grams per liter	1,000	Parts per million
Inches	2.54	Centimeters
Inches	0.08333	Feet
Inches	0.02778	Yards
Kilograms	1,000	Grams
Kilograms	2.205	Pound
Kilometers	3,281	Feet
Kilometers	1,000	Meters
Kilometers	0.6214	Miles
Kilometers	1,094	Yards
Liters	1,000	Cubic centimeters
Liters	0.0353	Cubic feet
Liters	61.02	Cubic inches
Liters	0.001	Cubic meters
Liters	0.2642	Gallons
Liters	2.113	Pints (liquid)
Liters	1.057	Quarts (liquid)
Meters	100	Centimeters
Meters	3.281	Feet
Meters	39.37	Inches
Meters	0.001	Kilometers
Meters	1,000	Millimeters
Meters	1.094	Yards
Pounds	7,000	Grains
Pounds	453.5924	Grams
Pounds	16	Ounces
Pounds	0.0005	Tons
Pounds of water	0.01602	Cubic feet
Pounds of water	27.68	Cubic inches
Pounds of water	0.1198	Gallons
Quarts (dry)	0.03125	Bushels
Quarts (dry)	67.20	Cubic inches
Quarts (dry)	0.125	Pecks
Quarts (dry)	2	Pints (dry)
Quarts (liquid)	57.75	Cubic inches
Quarts (liquid)	0.25	Gallons
Quarts (liquid)	0.9463	Liters
Quarts (liquid)	32	Ounces (liquid)
Quarts (liquid)	2	Pints (liquid)
Rods	16.5	Feet
Square feet	144	Square inches
Square feet	0.11111	Square yards
Square inches	0.00694	Square feet
Square miles	640	Acres
Square miles	27,878,400	Square feet
Square miles	3,097,600	Square yards
Square yards	0.0002066	Acres
Square yards	9	Square feet
Square yards	1,296	Square inches
Temperature (°C) + 17.98	1.8	Temperature, °F
Temperature (°F) − 32	5/9 or 0.5555	Temperature, °C
Ton	907.1849	Kilograms
Ton	32,000	Ounces
Ton	2,000	Pounds
Yards	3	Feet
Yards	36	Inches
Yards	0.9144	Meters
Yards	0.000568	Miles

FIGURING CAPACITY OF SPRAYER TANKS

The capacity of tanks of hand or power sprayers in gallons can be calculated as follows:

Cylindrical tanks (circular cross section): Multiply length in inches by square of diameter in inches; multiply the product by 0.0034.

Tanks with elliptical cross section: Multiply length in inches by short diameter in inches by long diameter in inches; multiply the product by 0.0034.

Rectangular tanks (square or oblong cross section): Multiply length by width by depth, all in inches; multiply product by 0.004329.

PESTICIDE DILUTION TABLE (AMOUNT OF PESTICIDE FORMULATION FOR EACH *ONE GALLON* OF WATER)

Pesticide Formulation	Percentage of actual chemical wanted								
	0.0313%	0.0625%	0.125%	0.25%	0.5%	1.0%	2.0%	3.0%	5.0%
15% WP	2½ t	5 t	10 t	7 T	1 C	2 C	4 C	6 C	10 C
25% WP	1½ t	3 t	6 t	12 t	8 C	1 C	2 C	3 C	5 C
40% WP	1 t	2 t	4 t	8 t	5 T	10 T	1¼ C	2 C	3¼ C
50% WP	¾ t	1½ t	3 t	6 t	4 T	8 T	1 C	1½ C	2½ C
75% WP	¼ t	1 t	2 t	4 t	8 t	5 t	10 t	1 C	2 C
10%–12% EC 1 lb actual/gal	2 t	4 t	8 t	16 t	10 t	⅔ pt	1⅓ pt	1 qt	3½ qt
15%–20% EC 1.5 lb actual/gal	1½ t	3 t	6 t	12 t	7½ t	½ t	1 pt	1½ pt	2 pt
25% EC 2 lb actual/gal	1 t	2 t	4 t	8 t	5 T	10 T	⅔ pt	1 pt	1¾ pt
33%–35% EC 3 lb actual/gal	¾ t	1½ t	3 t	6 t	4 T	8 T	½ pt	¾ pt	1⅓ pt
40%–50% EC 4 lb actual/gal	½ t	1 t	2 t	4 t	8 t	5 T	10 T	½ pt	⅘ pt
57% EC 5 lb actual/gal	7/16 t	⅞ t	1¾ t	3½ t	7 t	4¼ T	9 T	14 t	1½ C
60%–65% EC 6 lb actual/gal	⅜ t	¾ t	½ T	1 T	2 T	4 T	8 T	12 T	½ C
70%–75% EC 8 lb actual/gal	¼ t	½ t	1 t	2 t	4 t	8 t	5 T	7½ T	13 T

EC = emulsifiable concentrate; gal = gallon; lb = pound; pt = pint; C = cup; T = tablespoon; t = teaspoon; WP = wettable powder; 3 level teaspoonfuls = 1 level tablespoonful; 2 tablespoonfuls = 1 fluid ounce; 8 fluid ounces or 16 tablespoons = 1 cupful; 2 cupfuls = 1 pint; 2 pints = 1 quart or 32 fluid ounces; 4 quarts = 1 gallon or 128 fluid ounces.

DILUTION TABLE FOR PESTICIDES GIVEN IN POUNDS PER ACRE

Formulation	Amount of actual chemical recommended per acre								
	⅛ lb	¼ lb	½ lb	¾ lb	1 lb	$1\frac{1}{2}$ lb	2 lb	$2\frac{1}{2}$ lb	3 lb
	Amount of formulation needed to obtain the above amounts of actual chemical								
10%–12% EC (contains 1 lb chemical per gal)	1 pt	1 qt	2 qt	3 qt	1 gal	$1\frac{1}{2}$ gal	2 gal	$2\frac{1}{2}$ gal	3 gal
15%–20% EC (contains $1\frac{1}{4}$ lb chemical per gal)	⅓ qt	⅔ qt	$1\frac{1}{3}$ qt	2 qt	$2\frac{2}{3}$ qt	1 gal	$1\frac{1}{3}$ gal	$1\frac{2}{3}$ gal	2 gal
25% EC (contains 2 lb chemical per gal)	½ pt	1 pt	1 qt	3 pt	2 qt	3 qt	1 gal	5 qt	$1\frac{1}{2}$ gal
40%–50% EC (contains 4 lb chemical per gal)	¼ pt	½ pt	1 pt	$1\frac{1}{2}$ pt	1 qt	3 pt	2 qt	5 qt	3 qt
60%–65% EC (contains 6 lb chemical per gal)	⅙ pt	⅓ pt	⅔ pt	1 pt	$1\frac{1}{3}$ pt	1 qt	$2\frac{2}{3}$ pt	$3\frac{1}{3}$ pt	2 qt
70%–75% EC (contains 8 lb chemical per gal)	⅛ pt	¼ pt	½ pt	¾ pt	1 pt	$1\frac{1}{2}$ pt	1 qt	$2\frac{1}{2}$ pt	3 pt
25% WP	½ lb	1 lb	2 lb	3 lb	4 lb	6 lb	8 lb	10 lb	12 lb
40% WP	5 oz	10 oz	$1\frac{1}{4}$ lb	$1\frac{7}{8}$ lb	$2\frac{1}{2}$ lb	$3\frac{3}{4}$ lb	5 lb	$6\frac{1}{4}$ lb	$7\frac{1}{2}$ lb
50% WP	¼ lb	½ lb	1 lb	$1\frac{1}{2}$ lb	2 lb	3 lb	4 lb	5 lb	6 lb
75% WP	⅙ lb	⅓ lb	⅔ lb	1 lb	$1\frac{1}{3}$ lb	2 lb	$2\frac{2}{3}$ lb	$3\frac{1}{3}$ lb	4 lb
80% WP	$2\frac{1}{2}$ oz	5 oz	⅝ lb	¹⁵⁄₁₆ lb	$1\frac{1}{4}$ lb	$1\frac{7}{8}$ lb	$2\frac{1}{2}$ lb	$3\frac{1}{8}$ lb	$3\frac{3}{4}$ lb
1% Dust	$12\frac{1}{2}$ lb	25 lb	50 lb	75 lb	100 lb	150 lb	200 lb	250 lb	300 lb
5% Dust	$2\frac{1}{2}$ lb	5 lb	10 lb	15 lb	20 lb	30 lb	40 lb	50 lb	60 lb
0% Dust	$1\frac{1}{4}$ lb	$2\frac{1}{2}$ lb	5 lb	$7\frac{1}{2}$ lb	10 lb	15 lb	20 lb	25 lb	30 lb

CONVERSION TABLES FOR LIQUID FORMULATIONS

Per 1000 square feet

Concentration of active ingredient in formulation

(cc or tablespoon (T) of formulation per 1000 square feet)

Rate desired lbs/A	1 cc	1 T	2 cc	2 T	2.5 cc	2.5 T	3 cc	3 T	4 cc	4 T	5 cc	5 T	6 cc	6 T
1	87	$(5\frac{3}{4})$	43	(3)	35	$(2\frac{2}{3})$	29	(2)	22	$(1\frac{1}{2})$	17	$(1\frac{1}{4})$	14	(1)
2	173	$(11\frac{1}{2})$	87	$(5\frac{3}{4})$	69	$(4\frac{2}{3})$	58	$(3\frac{3}{4})$	43	(3)	35	$(2\frac{1}{3})$	29	(2)
3	260	$(18\frac{1}{3})$	130	$(8\frac{2}{3})$	104	(7)	87	$(5\frac{3}{4})$	65	$(4\frac{1}{3})$	52	$(3\frac{1}{2})$	43	(3)
4	348	$(23\frac{1}{4})$	174	$(11\frac{2}{3})$	139	$(9\frac{1}{4})$	116	$(7\frac{3}{4})$	87	$(5\frac{3}{4})$	70	$(4\frac{2}{3})$	58	$(3\frac{3}{4})$
5	434	(29)	217	$(14\frac{1}{2})$	174	$(11\frac{1}{3})$	145	$(9\frac{2}{3})$	109*	$(7\frac{1}{4})$	87	$(5\frac{3}{4})$	72	$(4\frac{3}{4})$
6	521	$(34\frac{3}{4})$	260	$(17\frac{1}{3})$	208	$(13\frac{3}{4})$	174	$(11\frac{2}{3})$	130	$(8\frac{2}{3})$	104	(7)	87	$(5\frac{3}{4})$
7	608	$(40\frac{1}{2})$	304	$(20\frac{1}{4})$	243	$(16\frac{1}{4})$	203	$(13\frac{1}{2})$	152	(10)	122	(8)	101	$(6\frac{3}{4})$
8	694	$(46\frac{1}{4})$	347	(23)	278	$(18\frac{1}{2})$	231	$(15\frac{1}{2})$	174	$(11\frac{2}{3})$	139	$(9\frac{1}{4})$	116	$(7\frac{3}{4})$
9	781	(52)	390	(26)	312	$(20\frac{3}{4})$	260	$(17\frac{1}{3})$	195	(13)	156	$(10\frac{1}{2})$	130	$(8\frac{1}{2})$
10	867	$(57\frac{3}{4})$	433	$(28\frac{3}{4})$	347	(23)	289	$(19\frac{1}{4})$	217	$(14\frac{1}{2})$	173	$(11\frac{1}{2})$	144	$(9\frac{2}{3})$

*Example: To spray a 1000 sq ft area at the rate of 5 lb/A active ingredient using a formulation containing 4 lb/A active ingredient, use 109 cc or $7\frac{1}{4}$ tablespoons of the 4 lb/gal formulation in the amount of carrier your application equipment is applying per unit area (1000 sq ft).

	Per 1000 square feet												
	Concentration of active ingredient in formulation												
Rate desired lb/A	100%	90%	80%	75%	70%	60%	50%	40%	30%	25%	20%	10%	5%
	(Grams of formulation per 1000 square feet)												
1	10	12	13	14	15	17	21	26	35	42	52	104	208
2	21	23	26	28	30	35	42	52	69	83	104	208	417
3	31	35	39	42	45	52	63	78	104	125	156	312	625
4	42	46	52	56	60	69	83*	104	139	167	208	417	833
5	52	58	65	69	74	87	104	130	174	208	260	521	1040
6	63	69	78	83	89	104	125	156	208	250	312	625	1250
7	73	81	91	97	104	121	146	182	243	292	364	729	1460
8	83	93	104	111	119	139	167	208	278	333	417	833	1670
9	94	104	117	125	134	156	187	234	312	375	469	937	1870
10	104	116	130	139	149	174	208	260	347	417	521	1040	2080

*Example: To treat a 1000 sq ft area at the rate of 4 lb/A active ingredient using a formulation containing 50% active ingredient, use 83 grams of the 50% formulation in the amount of carrier your application equipment is applying per unit area (1000 sq ft).

SPRAY CONCENTRATION CONVERSION CHART

Ounces per 100 gallons	ppm	% Solution	Grams per 100 liters
⅔	50	.005	5
1	75	.0075	7.5
$1\frac{1}{3}$	100	.01	10
2 (⅛ lb)	150	.015	15
$2\frac{2}{3}$	200	.02	20
$3\frac{1}{3}$	250	.025	25
4 (¼ lb)	300	.03	30
$5\frac{1}{3}$	400	.04	40
$6\frac{2}{3}$	500	.05	50
8 (½ lb)	600	.06	60
$9\frac{1}{3}$	700	.07	70
$10\frac{2}{3}$	800	.08	80
12 (¾ lb)	900	.09	90
$13\frac{1}{3}$	1000	0.10	100
16 (1 lb)	1200	0.12	120
20 ($1\frac{1}{4}$ lb)	1500	0.15	150
24 ($1\frac{1}{2}$ lb)	1800	0.18	180

A DIPSTICK GAUGE FOR 55-GALLON DRUMS*

You will need a 3-foot, flat stick. A length of 1×2, or an old yardstick, works fine. Mark the correct gallonage figure directly on the stick for each inch. Half inches can be estimated. Use the appropriate listing below, depending on whether your drum is mounted vertically or sideways.

Horizontal drum		Upright drum	Horizontal drum		Upright drum
Inches	Gallons	Inches	Inches	Gallons	Inches
22	55	33	$10\frac{7}{8}$	27	$16\frac{1}{4}$
21	54	$32\frac{1}{2}$	$10\frac{1}{2}$	26	$15\frac{1}{2}$
$20\frac{1}{4}$	53	$31\frac{3}{4}$	$10\frac{1}{4}$	25	15
$19\frac{3}{4}$	52	$31\frac{1}{4}$	10	24	$14\frac{1}{2}$
$19\frac{1}{4}$	51	$30\frac{1}{2}$	$9\frac{3}{4}$	23	$13\frac{3}{4}$
$18\frac{3}{4}$	50	30	$9\frac{1}{4}$	22	$13\frac{1}{4}$
$18\frac{1}{4}$	49	$29\frac{1}{2}$	9	21	$12\frac{1}{2}$
$17\frac{3}{4}$	48	$28\frac{3}{4}$	$8\frac{3}{4}$	20	12
$17\frac{1}{2}$	47	$28\frac{1}{4}$	$8\frac{1}{4}$	19	$11\frac{1}{2}$
17	46	$27\frac{1}{2}$	8	18	$10\frac{3}{4}$
$16\frac{3}{4}$	45	27	$7\frac{3}{4}$	17	$10\frac{1}{4}$
$16\frac{1}{2}$	44	$26\frac{1}{2}$	$7\frac{1}{4}$	16	$9\frac{1}{2}$
16	43	$25\frac{3}{4}$	7	15	9
$15\frac{3}{4}$	42	$25\frac{1}{4}$	$6\frac{3}{4}$	14	$8\frac{1}{2}$
$15\frac{1}{4}$	41	$24\frac{1}{2}$	$6\frac{1}{4}$	13	$7\frac{3}{4}$
15	40	24	6	12	$7\frac{1}{4}$
$14\frac{3}{4}$	39	$23\frac{1}{2}$	$5\frac{3}{4}$	11	$6\frac{1}{2}$
$14\frac{1}{4}$	38	$22\frac{3}{4}$	$5\frac{1}{2}$	10	6
14	37	$22\frac{1}{4}$	5	9	$5\frac{1}{2}$
$13\frac{3}{4}$	36	$21\frac{1}{2}$	$4\frac{1}{2}$	8	$4\frac{3}{4}$
$13\frac{1}{4}$	35	21	4	7	$4\frac{1}{4}$
13	34	$20\frac{1}{2}$	$3\frac{3}{4}$	6	$3\frac{1}{2}$
$12\frac{3}{4}$	33	$19\frac{3}{4}$	$3\frac{1}{4}$	5	3
$12\frac{1}{4}$	32	$19\frac{1}{4}$	3	4	$2\frac{1}{2}$
12	31	$18\frac{1}{2}$	$2\frac{1}{2}$	3	$1\frac{3}{4}$
$11\frac{3}{4}$	30	18	$1\frac{3}{4}$	2	$1\frac{1}{4}$
$11\frac{1}{2}$	29	$17\frac{1}{2}$	$1\frac{1}{4}$	1	$\frac{1}{2}$
$11\frac{1}{8}$	28	$16\frac{3}{4}$	0	0	0

*Most drums are 22 inches in diameter, 33 inches tall, and weigh, when full (55 gallons), 500 pounds.

CONVERSION TABLE FOR GRANULAR RATES

If % of active ingredients in granules is:	And recommended rate of active ingredients is:	Then apply this amount of granules:
5%	½ lb/acre	10 lb/acre
5%	1 lb/acre	20 lb/acre
5%	1½ lb/acre	30 lb/acre
5%	2 lb/acre	40 lb/acre
5%	3 lb/acre	60 lb/acre
5%	4 lb/acre	80 lb/acre
10%	½ lb/acre	5 lb/acre
10%	1 lb/acre	10 lb/acre
10%	1½ T/acre	15 lb/acre
10%	2 lb/acre	20 lb/acre
10%	3 lb/acre	30 lb/acre
10%	4 lb/acre	40 lb/acre
20%	½ lb/acre	2½ lb/acre
20%	1 lb/acre	5 lb/acre
20%	1½ T/acre	7½ lb/acre
20%	2 lb/acre	10 lb/acre
20%	3 T/acre	15 lb/acre
20%	4 lb/acre	20 lb/acre
25%	½ lb/acre	2 lb/acre
25%	1 lb/acre	4 lb/acre
25%	1½ lb/acre	6 lb/acre
25%	2 lb/acre	8 lb/acre
25%	3 lb/acre	12 lb/acre
25%	4 lb/acre	16 lb/acre

WIDTH OF AREA COVERED TO ACRES PER MILE TRAVELED

Width of strip (feet)	Acres/Mile
6	.72
10	1.21
12	1.45
16	1.93
18	2.18
20	2.42
25	3.02
30	3.63
50	6.04
75	9.06
100	12.1
150	18.14
200	24.2
300	36.3

AIDS IN CONVERTING DOSAGE, VOLUME, RATES, AND AMOUNTS (APPROXIMATE)

Wettable powders

Rates		Approximate amount for less than 1 acre			
Amt per acre	Area treated per oz	½ acre	¼ acre	4000 sq ft	1000 sq ft
1 lb	2722 sq ft	$5\frac{1}{3}$ oz	4 oz	1.4 oz	.37 oz
2 lb	1360 sq ft	11 oz	½ lb	2.9 oz	¾ oz
3 lb	907 sq ft	1 lb	¾ lb	$4\frac{1}{3}$ oz	1.1 oz
4 lb	681 sq ft	$1\frac{1}{3}$ lb	1 lb	$5\frac{3}{4}$ oz	$1\frac{1}{2}$ oz
5 lb	545 sq ft	$1\frac{2}{3}$ lb	$1\frac{1}{4}$ lb	$7\frac{1}{5}$ oz	$1\frac{4}{5}$ oz
6 lb	454 sq ft	2 lb	$1\frac{1}{2}$ lb	$8\frac{2}{3}$ oz	$2\frac{1}{3}$ oz
7 lb	380 sq ft	$2\frac{1}{3}$ lb	$1\frac{3}{4}$ lb	10 oz	$2\frac{1}{2}$ oz
8 lb	316 sq ft	$2\frac{2}{3}$ lb	2 lb	$12\frac{2}{5}$ oz	$3\frac{1}{5}$ oz
9 lb	303 sq ft	3 lb	$2\frac{1}{4}$ lb	12.9 oz	$3\frac{1}{3}$ oz
10 lb	272 sq ft	$3\frac{1}{3}$ lb	$2\frac{1}{2}$ lb	$14\frac{2}{5}$ oz	$3\frac{3}{4}$ oz
11 lb	250 sq ft	$3\frac{2}{3}$ lb	$2\frac{3}{4}$ lb	$15\frac{4}{5}$ oz	4 oz
12 lb	222 sq ft	4 lb	3 lb	$1\frac{1}{8}$ lb	$4\frac{1}{2}$ oz
13 lb	209 sq ft	$4\frac{1}{3}$ lb	$3\frac{1}{4}$ lb	$1\frac{1}{5}$ lb	$4\frac{4}{5}$ oz
14 lb	194 sq ft	$4\frac{2}{3}$ lb	$3\frac{1}{2}$ lb	$1\frac{1}{4}$ lb	$5\frac{1}{5}$ oz
15 lb	180 sq ft	5 lb	$3\frac{3}{4}$ lb	$1\frac{1}{3}$ lb	$5\frac{2}{3}$ oz
16 lb	170 sq ft	$5\frac{1}{3}$ lb	4 lb	1 lb 7 oz	6 oz

Determining small quantities of spray materials

Liquids		Powders			
Amt per 100 gal	Amt per 1 gal	Approx. Dilutions	Amt per 100 gal	Amt per 1 gal	Approx. dilutions
¼ pt	¼ t	1-3200	½ lb	¾ t	1-1600
1 pt	1 t	1-8000	⅝ lb	1 t	1-1400
$1\frac{1}{2}$ pt	$1\frac{1}{2}$ t	1-550	1 lb	$1\frac{1}{2}$ t	1-800
1 qt	2 t	1-400	$1\frac{1}{4}$ lb	2 t	1-700
3 pt	1 T	1-233	$1\frac{1}{2}$ lb	$2\frac{1}{2}$ t	1-600
2 qt	4 t	1-200	2 lb	1 T	1-400
3 qt	2 T	1-150	$2\frac{1}{2}$ lb	4 t	1-350
1 gal	8 t	1-100	3 lb	$4\frac{1}{2}$ t	1-266
2 gal	5 T	1-50	4 lb	2 T	1-200
3 gal	½ C	1-33	5 lb	8 t	1-160
4 gal	⅔ C	1-25	8 lb	4 T	1-100
5 gal	1 C	1-20	10 lb	5 T	1-80
11 gal	⅞ pt	1-9	16 lb	8 T	1-50

Liquids					
Rates		**Approximate amount for less than 1 acre**			
Amt per acre	Area treated per gal	⅓ acre	½ acre	4000 sq ft	1000 sq ft
⅛ pt (12 t)	64 acres	4 t	3 t	1 t	¼ t
⅙ pt (16 t)	48 acres	5½ t	4 t	1.5 t	⅜ t
¼ pt (8 T)	32 acres	8 t	2 T	2.3 t	½ t
⅓ pt (32 t)	24 acres	11 t	8 t	2⅘ t	¾ t
⅖ pt (38 t)	20 acres	13 t	9½ t	3⅓ t	1 t
½ pt (16 T)	16 acres	16 t	4 T	4⅓ t	1⅛ t
¾ pt (24 T)	11 acres	8 T	6 T	6½ t	1¾ t
1 pt (32 T)	8 acres	⅔ C	8 T	2⅘ t	2¼ t
1⅓ pt (43 T)	6 acres	14 T	11 T	3⅘ t	1 T
⅕ gal (51 T)	5 acres	17 T	13 T	4½ T	4 t
1 qt (64 T)	4 acres	⅔ pt	½ pt	5¾ T	1½ T
⅓ gal (85 T)	3 acres	⅞ pt	¾ pt	⅛ pt	2 T
2 qt (128 T)	2 acres	⅔ qt	1 pt	⅜ pt	3 T
3 qt (190 T)	1⅓ acres	1 qt	¾ qt	½ pt	4½ T
1 gal (256 T)	1 acre	1⅓ qt	1 qt	⅘ pt	6 T

Useful Conversion Tables

APPLICATION EQUIVALENTS

1 oz per sq ft	= 2722.5 lb/acre
1 oz per sq yd	= 302.5 lb/acre
1 oz per 100 sq ft	= 27.2 lb/acre
1 lb per 100 sq ft	= 435 lb/acre
1 lb per 1000 sq ft	= 43 lb/acre
1 lb per acre	= ½ oz/1000 sq ft
5 gal per acre	= 1 pt/1000 sq ft
100 gal per acre	= 2.5 gal/1000 sq ft

AREA AND CUBIC MEASURES

1 acre	= 43,560 sq ft	= 160 sq rd	
1 acre	= 1 mile 8 ft wide or 2 miles 4 ft wide		
1 sq rd	= 272 sq ft	= 30¼ sq yd	
3⅔ sq rd	= 1000 sq ft		
1 sq yd	= 1296 sq in	= 9 sq ft	
1 sq ft	= 144 sq in		
1 cu ft	= 1728 cu in	= 7.48 gal	
1 cu yd	= 27 cu ft	= 202 gal	

LIQUID MEASURES

1 teaspoon (t) = 5 cubic centimeters (cc) = 60 drops
1 tablespoon (T) = 3 t = ½ fl oz = 15 ml
2 T = 1 fl oz
1 cup = ½ pt = 16 T = 48 t
1 pt = 32 T = 16 fl oz = 28.8 in^3
1 gal = 4 qt = 231 in^3 = 8.33 lb water

WEIGHTS

1 oz = 2 T liquid = 28⅓ grams
1 lb = 16 oz = 454 grams = 0.453 kilograms
1 gram (g) = 0.03527 oz
1 milligram (mg) = 1/1000 g
1 kilogram (kg or kilo) = 2.2046 lb
1 gal water = 8.345 lb

Tree spacing (ft)	10	12	16	20	25	30	35	40
Trees/acre	435.2	302.5	170	108.9	69.7	48.4	35.6	27.2
Trees/minute at:								
1 mph	8.8	7.3	5.5	4.4	3.5	2.9	2.5	2.2
$1\frac{1}{2}$ mph	13.2	11.0	8.2	6.6	5.3	4.4	3.8	3.3
2 mph	17.6	14.6	11.0	8.8	7.0	5.9	5.0	4.4
$2\frac{1}{2}$ mph	22.0	18.3	13.7	11.0	8.8	7.3	6.3	5.5
3 mph	26.4	22.0	16.5	13.2	10.6	8.8	7.5	6.6

FIGURING THE COST OF CHEMICALS

Using the information given, which pesticide is a better buy?
Example:

- *Pesticide A:* $1.45/lb; rate: 8 oz/1000 ft^2
- *Pesticide B:* $8.90/lb; rate: 1 oz/1000 ft^2

Because there are 16 ounces in a pound, pesticide A costs $1.45/16 = $0.09/ounce and pesticide B costs $8.90/16 = $0.56/ounce. Pesticide A is cheaper per ounce, but the rate of application is eight times that of pesticide B. It therefore will cost 8 × $0.09 or $0.72 per 1000 square feet to use pesticide A compared to $0.56 per 1000 square feet to use pesticide B. Pesticide B is the better buy.

Example:

- *Pesticide A:* $8.00/gal (4 lb/gal material); rate: 2 lb/acre
- *Pesticide B:* $10.60/gal (4 lb/gal material); rate: 3 lb/acre

Pesticide A would cost $4.00 per acre for application since each pound costs $2.00 (8/4) and the rate is pounds per acre; so 2 × 2 = $4.00. Pesticide B would cost $7.95 per acre for application since each pound costs $2.65 (10.60/4) and the rate is 3 pounds per acre; so 3 × $2.65 = $7.95. Pesticide A is the better buy.

AREA MEASUREMENTS

To determine how much pesticide you will need to do a job, you must measure the area to be treated. If the area is a rectangle, circle, or triangle, simple formulas may be used. Determining the area of an irregularly-shaped site is more difficult. The following examples will help you in computing the area of both regularly- and irregularly-shaped surfaces.

Regularly-shaped Areas

Rectangles. The area of a rectangle is found by multiplying the length (L) by the width (W).

$$\text{area} = \text{length} \times \text{width}$$

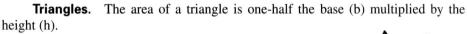

Example: $40 \times 125 = 5000$ square feet

Circles. The area of a circle is the radius (one-half the diameter) squared and then multiplied by 3.14.

$$\text{area} = 3.14 \times \text{radius squared}$$

Example: $35 \times 35 \times 3.14 = 3846.5$ square feet

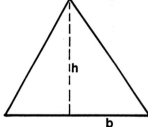

Triangles. The area of a triangle is one-half the base (b) multiplied by the height (h).

$$\text{area} = \frac{b \times h}{2}$$

Example: b = 55 ft
h = 53 ft
$55 \times 53 \div 2 = 1457.5 \text{ ft}^2$

Irregularly-shaped Areas

Irregularly-shaped areas often can be reduced to a combination of rectangles, circles, and triangles. Calculate the area of each and add them together to obtain the total area.

Example: b = 25 ft
h = 25 ft
L = 30 ft
L_1 = 33 ft
W_1 = 31 ft
area = $(b \times h \div 2) + (L \times W) + (L_1 \times W_1)$
$25 \times 25 \div 2 = 312.5$; $30 \times 42 = 1260$;
$31 \times 33 = 1023$
$= 312.5 + 1260 + 1023 = 2595 \text{ ft}^2$

Another way is to establish a line down the middle of the property for the length, and then measure from side to side at several points along this line. Areas with very irregular shapes require more side to side measurements. The average of the side measurements can be used as the width. The area is then calculated as a rectangle.

Example: ab = 45 ft
c = 22 ft
d = 21 ft
e = 15 ft
f = 17 ft
g = 22 ft

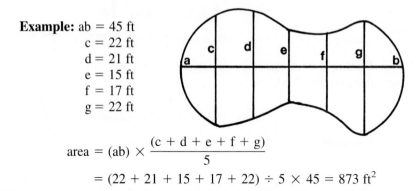

$$\text{area} = (ab) \times \frac{(c + d + e + f + g)}{5}$$

$$= (22 + 21 + 15 + 17 + 22) \div 5 \times 45 = 873 \text{ ft}^2$$

A third method is to convert the area into a circle. From a center point, measure distance to the edge of the area in 10 to 20 increments. Average these measurements to find the average radius. Then calculate the area, using the formula for a circle.

Example:

$$\text{radius} = \frac{(a + b + c + d + e + f + g + h + i + j + k + l)}{12}$$

area = 3.14 × radius squared (10 + 12 + 16 + 15 + 11 + 12 +
10 + 9 + 13 + 12 + 13 + 16) ÷ 12 = 12.42 (radius)
= 12.42 × 12.42 × 3.14 = 484.1 ft^2

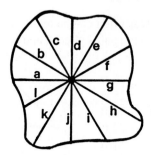

14

Pesticide Transportation, Storage, Decontamination, and Disposal

SAFE TRANSPORT OF PESTICIDES

When transporting toxic chemicals, it is possible for accidents to happen. It is the responsibility of the person transporting the material to take preventive measures to reduce the likelihood of anyone being hurt by accidental contact with a poison. Ignorance of the nature of the materials being transported should not be a contributing factor to an accident.

The federal government has applied a preemption of states' rights to govern their *intrastate* operations. All states will be adopting the United States Department of Transportation (USDOT) regulations. These regulations will be enforced by the federal government, state patrol, and state departments of transportation. Several states will maintain some exemptions for certain industries such as agriculture. The key to these regulations is one question: "Are you involved in commerce?" It is advisable to check with your state and local regulatory agencies.

Several precautions and regulatory requirements must be followed to provide safe transportation of pesticides. USDOT has stepped up their roadside checks for commercial motor vehicles that transport hazardous materials. The following are brief items of concern for transporting hazardous materials safely in commerce.

Transporting Vehicles

The Federal Highway Administration definition of a *commercial motor vehicle* is one that has a gross vehicle weight rating of greater than 10,000 pounds or a vehicle that is *transporting hazardous materials in a quantity requiring placarding*. This means that a pickup, a ½-ton straight truck with a stake bed, or a dry van could be considered a commercial motor vehicle if it is carrying pesticides in a quantity requiring placarding.

Pesticides should never be placed in the cab of a truck, and it is not advisable to transport them in automobiles. They should be properly loaded in the cargo area of the truck and special attention should be given to securing the load, as well as providing protection from the weather. The shipper or dealer will have more information regarding load securement when you pick up your pesticides.

Hazardous Materials

USDOT regulates the transportation of hazardous materials. It is the responsibility of the manufacturer to properly classify, mark, and label the packaged pesticide materials. The shipper of hazardous materials must meet the regulations of USDOT to prepare the proper shipping paper that communicates the items that make up the "bill of lading," giving their proper shipping descriptions and proper placarding requirements. The shipping description must meet the proper sequence specified by USDOT: the proper shipping name, hazard class, the four-digit UN code, the packing group, and technical names for not otherwise specified (NOS) entries. For example, a 50-pound bag of Dursban® may have the following shipping description: "RQ Environmentally Hazardous Substance, Solid, NOS, 9, UN3077, Pg III (contains chlorpyrifos)."

If you pick up hazardous materials from the shipper or dealer, they will offer you a proper bill of lading and the required placards for your vehicle to meet the requirements of USDOT. The person who operates the commercial vehicle picking up the hazardous materials, must meet the requirements of a properly qualified driver in most states. This means that the driver must have a Commercial Driver License (CDL). Persons with a CDL must have additional training every 2 years as outlined by the USDOT.

Container Requirements

USDOT requires the manufacturer to meet their requirements for proper packaging. USDOT has adopted the United Nations guidelines for packaging. This packaging is known as Performance Oriented Packaging (POP). Using these guidelines, USDOT allows the manufacturers to design their packaging to meet their own needs as long as they pass the required testing guidelines.

The packaging guidelines are divided into two groups: nonbulk and bulk packaging. Nonbulk packaging has a maximum capacity of 119 gallons (450 liters) for liquids or 882 pounds (400 kg) for solids. The Dursban® mentioned earlier was a 50-pound bag, so this qualifies as a nonbulk package. Bulk packaging is any container for liquids or solids that exceeds the nonbulk guidelines. These would be such items as portable tanks, intermediate plastic bulk containers, and intermodal containers. They can also be tankers (3500 gallons or more) such as liquid fertilizers pulled by tractors.

Placarding Requirements

Hazardous materials are classified into two different groups. The first group must always be placarded. Two common pesticides used by applicators are hazard class 2.3, "Poison Gas" (methyl bromide) and class 4.3, "Dangerous When Wet" (aluminum phosphide).

Figure 14–1   Appropriate placards for transporting certain pesticides that are classified as hazardous materials.

The second group includes such pesticides as class 3, "Flammable" (parathion); class 4.1, "Flammable Solid" (sulfur); class 5.1, "Oxidizer" (fertilizer); class 6.1, "Poison" (organophosphorus pesticides); class 8, "Corrosive" (paraquat); and class 9 "Environmentally Hazardous Substance" (Dursban®). If you exceed 1000 pounds of these materials, you must placard the vehicle. Placards must be placed on the front, back, and sides of the trucks. Samples of some of the appropriate placards are depicted in Figure 14–1.

These are only a few examples of groups 1 and 2. Consult USDOT regulations for the complete lists.

Checklist for Shippers of Hazardous Materials

In order to determine and follow the appropriate regulations for the pesticides or other hazardous materials you may be carrying, you should:

- Check with your supplier. Your supplier can prepare proper shipping bills of lading (BOLs) with the proper shipping descriptions.
- Check with the local DOT agency for information that may be needed beyond the federal requirements.
- Develop manifests, using the supplier's BOL as a guide, that best suit your needs.
- Make sure service containers are properly sealed and marked.
- Make sure all pesticides are properly USDOT marked for transportation and have appropriate labels intact.
- Have appropriate phone numbers in the vehicle in case of an emergency.
- Secure the containers to avoid a potential spill.
- Be aware of the temperature; too hot or too cold can create problems in shipment of hazardous materials.
- Load the materials in such a way as to avoid cross contamination.
- Never load hazardous materials along with foodstuffs or feed.
- Never allow children to ride in the area where the hazardous materials are secured.
- If there is a spill, follow the decontamination guidelines addressed later in this chapter.

Safe Transport of Pesticides

PESTICIDE STORAGE

Proper storage of pesticides will protect the health and well-being of people, help protect against environmental contamination, and protect the chemical shelf life. A number of fairly rigid conditions are required for the storage of pesticides intended for agricultural and industrial uses. The following are the major items with which everyone should be completely familiar.

- Storage in a separate building is preferable. A separate, isolated storage room can be used in some instances.
- The storage area should be well ventilated and have a source of heat for pesticides that cannot tolerate cold temperatures. When chemicals are subjected to high temperatures, they may expand, causing drum heads to bulge and leak. High temperatures have also been noted to reduce the effectiveness of emulsifiers, speed up container erosion, and, in some instances, cause the pesticide to deteriorate. Products should be stored in heated rooms with caution because excess heat may cause emission of ignitable fumes from solvents or cause emission of toxic fumes. High temperatures are unnecessary. Low temperatures also cause problems in pesticide storage. Some compounds will freeze and upon expansion rupture metal or glass containers. Other compounds under low temperatures may settle or crystallize out of solution. Refer to the pesticide label for storage temperature recommendations. See also "Cold Weather Handling of Liquid Chemical Products," Appendix H.
- Store all pesticides in their original, labeled containers; never store in food, feed, or beverage containers.
- Keep lids of containers tight and tops of bags closed when containers are not being used.
- Liquid containers should be stored on pallets to avoid rusting of metal containers. Dusts and powders will tend to cake when wet or subjected to extremely high humidity. They too should be stored on pallets to prevent moisture absorption.
- Check all containers frequently for leaks or tears to avoid contamination.
- Try to avoid storing unnecessarily large quantities of pesticides by keeping good records of previous requirements and making good estimates of future needs. When large amounts of pesticides are being stored, records should be kept on the kind and quantity. Make an inventory of all pesticides available in storage, and mark containers with date of purchase. On some products, the EPA mandates a stop-sale date and will not allow sale after a specified date unless the product bears an up-to-date label. You could also inadvertently be subjected to federal or state hazardous waste system requirements by storing suspended products past their legal use date, so do not store more than is necessary, and sell or use pesticides based on the first in, first out concept.
- Dispose of pesticides showing the following signs of deterioration:

Formulation	General signs of deterioration
Emulsifiable concentrates	Milky coloration does not occur when the addition of water and sludge is present or any separation of components is evident.
Oil sprays	Milky coloration does not occur by adding water.
Wettable powders	Lumping occurs and the powder will not suspend in water.
Dusts	Excessive lumping.
Granulars	Excessive lumping.
Aerosols	Generally effective until the opening of the aerosol dispenser becomes obstructed.

- Do not store pesticides in the same room with food, feed, or water.
- Do not store herbicides, especially the hormone types, in the same room with other pesticides or fertilizers. Cross contamination may occur.
- Post warning signs on the outside of *all* walls, doors, and windows of the pesticide storage facility where they will be readily seen by anyone attempting to enter. Keep the facility locked when not in use (Figure 14–2). If the storage or operational facility involves the use of migrant workers, all warning signs should be in dual language.
- The storage facility should be constructed of fire-resistant material and should include portable fire extinguishers and a sprinkler system if possible.
- Do not store clothes, respirators, lunches, cigarettes, or drinks with pesticides. They may pick up poisonous fumes or dusts or soak up spilled poison.
- Plenty of soap and water should be available in the storage area. Seconds count when washing poisons from your skin. For chemical locations that handle corrosive materials, OSHA requires an emergency shower and an eye wash station. It is strongly recommended that any chemical storage building have these facilities even if no corrosive materials are handled there.
- If possible, try to avoid storing pesticides from one year to the next. The storage life of various pesticides is considerably different. Some pesticides belonging to the chlorinated hydrocarbon group can be stored for a number of years with little

Figure 14–2 Keep the storage area locked.

or no chemical change. Other pesticides, however, such as organic phosphorous compounds, tend to have a rather short storage life. Climatic conditions such as high temperatures, high humidity, and sunlight may cause a chemical breakdown or degradation of certain pesticides, especially when they are formulated as wettable powders or dusts. Compounds that are subject to degradation will have labels that provide instructions to prevent it. Sometimes, when pesticides have been stored for long periods, there is doubt as to their effectiveness. In this case, try small amounts of the pesticide according to label directions. If the treatment is satisfactory, use the pesticide up as promptly as possible.

Note: There are several sources for construction plans for storage buildings, loading areas, and chemical rinse pads. Most Cooperative Extension offices at state universities can provide information through their Agricultural Engineering Departments. A booklet entitled The "How To's" of Agricultural Chemical Storage is available from the Midwest Agricultural Chemicals Association, P.O. Box 2125 Northside Station, Sioux City, Iowa 51104-0125. Telephone: (712) 277-7380.

Minimizing Pesticide Fire and Explosion Hazards

Pesticides with low flash points (140° F or less) are dangerous in storage. Pesticide formulations in pressurized containers that have flash points at or below 20° F must have the following on the label: "Extremely flammable. Contents under pressure. Keep away from fire, sparks, and heated surfaces. Do not puncture or incinerate container. Exposure to temperatures above 130° F may cause bursting." Those products with flash points above 20° F but not over 80° F must be labeled with the following statement: "Flammable. Contents under pressure. Keep away from heat, sparks, and open flame. Do not puncture or incinerate container. Exposure to temperatures above 130° F may cause bursting." All other pressurized containers must bear this statement: "Contents under pressure. Do not use or store near heat or open flame. Do not puncture or incinerate container. Exposure to temperatures above 130° F may cause bursting."

Nonpressurized containers at or below 20° F must have the following statement on the label: "Extremely flammable. Keep away from fire, sparks, and heated surfaces." Labels on those above 20° F and not over 80° F must say "Flammable. Keep away from heat and open flame." On those above 80° F and not over 150° F, the label must say "Do not use or store near heat or open flame."

When pesticides are purchased, check the label for flashpoint warnings and make sure that they are used and stored according to the directions to prevent the hazard of fires and explosions. In the case of 5-gallon containers and larger, federal law requires a 4-inch sticker be placed on the container if the product is poison, flammable, corrosive, and so on. Normally, case lots of 1-gallon and $1\frac{1}{2}$-gallon containers will have these stickers on the shipping carton. Smaller units will not have this designation.

Pesticides containing oils or aromatic petroleum distillates are the ones most likely to have such warnings on the label. Certain dry formulations also present fire and explosion hazards. Sodium chlorate is well known for its potential to ignite when in contact with organic matter, sulfur, sulfides, phosphorus, powdered metals, strong acids, or ammonium salts. Whenever a container of sodium chlorate is opened, the entire contents should be used. A container partially full should never be stored. Certain dusts or powders, particularly those that are very fine, such as sulfur, may ignite as easily as gases or vapors. The following suggestions should help to reduce fire hazards:

- Keep the storage area locked at all times when not in actual use to prevent the possibility of fires being accidentally set.
- Do not store glass containers in sunlight where they might concentrate heat rays and start fires.
- Store combustible materials away from steam lines and heating devices.
- Sheet rock or other fireproof materials should be used to line the storage area.
- Major environmental problems may result should large storage areas equipped with sprinkler systems discharge in case of a fire. Systems in place should be adapted to a selective head system so that the total system does not discharge in case of fire. In addition, the facility property should be evaluated to be sure contaminated water will remain on site. Prior to installation of a sprinkler system, consider the ramification of its use in a pesticide warehouse.
- Be familiar with fire alarms and fire exits, and be sure to have proper fire fighting equipment.
- Storage areas should be located as far away as possible from other buildings and populated areas.
- Local fire personnel should be notified of the contents of the storage facility. It may save their lives and the lives of others in the event of a fire. Federal law (Superfund Amendments and Reauthorization Act, 1986) requires that pesticide facilities notify state and county regulatory agencies (commissions) of the location of pesticide storage facilities. The nearest local fire department must also be notified. An emergency fire disaster plan should be developed in cooperation with the local emergency response agencies for the protection of personnel and property.

The following suggestions should be helpful to firefighters in the event of a fire in a pesticide storage area:

- If a fire occurs in a pesticide storage facility, someone thoroughly familiar with the hazards of resulting smoke, fumes, splashes, or other possible types of contamination should be on hand to warn firefighters or anyone in the vicinity.
- Firefighters should wear the proper protective clothing and self-contained breathing apparatus (SCBA).
- Firefighters should attempt to stay upwind of the fire while fighting it.
- If necessary, residents downwind of the fire should be evacuated.
- Firefighters should avoid dragging hoses through pesticide-contaminated water.
- Firefighters should assume that all equipment used to fight such fires is contaminated and hazardous until decontaminated.
- Firefighters should avoid using heavy hose streams if possible, because the force of the stream spreads contamination and will cause dusts to become airborne and present a possible explosion hazard, as well as the toxic hazard.
- Firefighters should be aware that overheated containers may erupt at any time and should keep a safe distance from the fire.

- The runoff of water from a fire involving pesticides should be diked to prevent it from entering sewers or streams. Pesticide storage facilities should plug all internal building drains.
- Firefighters should wash and change clothing immediately after the fire.
- All clothing, boots, and other equipment should be thoroughly washed.
- Firefighters should prevent curious people from entering burned-out areas by erecting "Toxic Chemical" signs or barriers until cleanup is completed.
- Rubble and surrounding areas should be checked for evidence of contamination.
- If water is used, use as little water as possible. If possible, allow fire to burn out; however, this issue should be discussed in a prefire plan with the local fire department. If large quantities of foam are available, use foam in place of water.
- Facility managers should keep a copy of a monthly inventory and safety data sheets at their residence as they will be needed in case of fire.
- A prefire or contingency plan is extremely desirable for any pesticide facility. A review of the plan with fire and police departments is strongly advised. A major fire may cost hundreds of thousands of dollars to clean up, so preplan and think prevention.
- Contaminated rubble, ground, water, and the like, will be considered to be federally-regulated hazardous waste and must be disposed of at a regulated facility.
- Request a visit twice a year by the local fire department.

PESTICIDE DECONTAMINATION

The subject of decontamination concerns application and protective equipment, personnel, and areas involved in spills, fires, and highway accidents.

Decontamination of Spray Equipment

Most pesticides can be washed out of sprayers. Dicamba (Banvel®), 2,4-D, 2,4-DB, and MCPP are more difficult to wash out, however, and many crops are very sensitive to these herbicides. So it is best to have a separate sprayer to apply these herbicides. Certain crops are very sensitive to sulfonylurea herbicides (Classic®, Canopy®, Gemini®, Glean®, Harmony®), but these can be easily washed out of sprayers if the proper procedure is used.

Before applying a pesticide with a sprayer that was previously used for some other pesticide, *always* wash out the sprayer thoroughly. It is best to wash out the sprayer immediately after use. Some pesticide labels give instructions on how to properly clean that pesticide out of the sprayer. If the label does not contain this information, it does *not* imply that residues of that particular pesticide in a sprayer will not harm other crops.

Use the following steps to clean your sprayer system.

For pesticides except Dicamba, 2,4-D, 2,4-DB, MCPP, and sulfonylureas

1. Hose down the outside surfaces of the tank and other components.
2. Fill the tank half full of water and flush it by operating the sprayer until all the rinsewater is drained.

3. Fill the tank with water while adding 1 pound of *household detergent* per 100 gallons of water. Operate the pump to circulate this detergent solution for 10 minutes.

4. Flush this solution from the tank through the boom and nozzles.

5. Remove nozzles and strainers and flush the tank and system with two tanks of water. Wash off the strainers. Replace nozzles and strainers.

For Dicamba (Banvel®), 2,4-D, 2,4-DB, and MCPP

1. Hose down the outside surfaces of the tank and other components.

2. Fill the tank half full of water and flush by operating the sprayer until all rinse-water is drained.

3. Fill the tank with water and add 1 quart of *household ammonia* per 25 gallons of water.

4. Operate the pump to circulate this ammonia solution for 15 to 20 minutes. Discharge a small amount of solution through the nozzles. Let the solution stand in the tank overnight (or at least for several hours).

5. Flush the ammonia solution from the tank through the boom and nozzles.

6. Remove the nozzles and screens and flush the tank and spray system with two tanks of water. Wash off strainers. Replace nozzles and strainers.

7. Consult the individual product labels for information regarding rinsate disposal.

For sulfonylurea herbicides (Accent®, Ally®, Canopy®, Classic®, Escort®, Express®, Finesse®, Gemini®, Glean FC®, Harmony Extra®, Lorox Plus®, Pinnacle®, Preview®, Telar®, and others in the future). Specific spray tank cleanout procedures will vary with different sulfonylurea herbicides. For most of these products, the use of *household ammonia* is recommended.

Chlorine bleach (Clorox®) also decomposes most sulfonylureas, forming herbicidally inactive by-products. However, the sodium hypochlorite in Clorox® is not as efficient as other cleaners at penetrating and removing the pesticide deposits that can form in spray equipment. Specific spray tank cleanout procedures for each sulfonylurea herbicide are available from dealers.

Several commercial tank cleaners also do a good job. These include Chem-Tank Cleaner®, manufactured by Farmbelt Chemicals, Inc.; Incide-Out®, a trademark of Precision Laboratories, Inc.; Nutra-Sol®, compounded for Thomas G. Kilfoil Co., Inc.; Protank Cleaner®, manufactured for Cenex/Land O'Lakes Agronomy Co.; Tank-Aid®, manufactured for Cornbelt Chemical Co.; and Tank and Equipment Cleaner®, manufactured by Loveland Industries, Inc.

There are two types of tank cleaner activity. Some highly alkaline decontaminants simply neutralize and inactivate acidic residues. On the other hand, detergent-type materials physically remove residues. Some commercial cleaners have both types of activity. Tank and Equipment Cleaner® works both ways according to Bob Reeves, Technical Service Manager for Loveland Industries, Inc. Always read and follow label directions for sprayer cleanout and rinsate disposal.

Caution: Do not use chlorine bleach with ammonia. Also, all traces of liquid fertilizer (especially those containing ammonia, ammonium nitrate, or ammonium sulfate) must be rinsed from the sprayer with clean water before adding bleach. Failure to heed this caution will result in release of chlorine gas, which can cause eye, nose, throat, and lung irritation.

- Be sure to perform cleaning procedures in an area where you will not contaminate the environment or your water source. Do not clean equipment in an enclosed area.
- Always dispose of equipment washwater as you would excess pesticides.
- Keep children and pets away from the cleaning operation.

To be sure you have a clean sprayer, you can spray a few sensitive plants with "clean" water added to the "clean" tank. Wait at least 2 hours (longer if possible) and check plants for symptoms of herbicide injury before applying a newly mixed chemical to the crop.

Environmental issues are not at risk nor is there a major regulatory concern if proper decontamination procedures are followed for small spray units. However, in the case of commercial spray units, proper disposal is regulated by federal and/or state law. Not all pesticides are regulated, so efforts need to be made to determine the regulated products plus the regulated rinsate. Proper preplanning can go a long way toward keeping you in regulatory compliance.

- Use a separate spray tank for regulated pesticides.
- Clean out in a fashion so that decontamination is not needed; thus rinsate residue need not be generated. This can be done by using large amounts of water, with rinsate being applied to the spray site.
- Preplan the spray schedule so that cross contamination is not a problem; thus decontamination is not needed.
- Rinsate can be maintained in holding tanks until the next job requiring the use of the same active ingredient.

Are Washwaters Hazardous?

Concerns have been expressed regarding the washwaters resulting from the washing of exteriors of pesticide equipment. In a letter dated July 22, 1985, from John H. Skinner, Director, Office of Solid Waste, United States Environmental Protection Agency, the following statement was made:

> "Airplane washing rinsewater is not hazardous via mixture rule. . . The Agency does not believe that the pesticide residue left on the aircraft is a discarded commercial chemical product. The residue does not qualify as material discarded or intended to be discarded."

> "Consequently, we are withdrawing our previous interpretation that airplane washing rinsewater is a hazardous waste via the mixture rule."

An additional letter, dated May 30, 1986, from Marcia E. Williams, Director, Office of Solid Waste, U.S. EPA, states:

"Since the Agency sees no difference between washwaters from aerial versus ground application equipment, it is logical that the interpretation issued in July 1985 should also extend to the washwaters from ground equipment."

"Consequently, this rinsewater would not be considered a hazardous waste under the mixture rule and would only be considered hazardous if the rinsewater exhibited one of the characteristics of a hazardous waste identified in Subpart C of Part 261."

The items identified in Subpart C include ignitability, corrosivity, reactivity, or the characteristic of toxicity if, using the Toxicity Characteristic Leaching Procedure test method, the extract from a representative sample of the waste contains any of 40 contaminants at levels equal to or greater than the respective values shown in a table in Part 261 as referenced in the preceding paragraph.

It should be noted that, although there is a form of an "exemption" under the hazardous waste law, as cited in the letters discussed, this is for *collected rinsate*. This exemption does not apply for rinsate that falls on the ground, which would then be classified as a spill. To properly make use of this exemption, rinsate needs to be collected and used or disposed of in the proper manner.

Decontaminating Protective Equipment

Decontaminating respiratory devices is accomplished simply by discarding the filter pads and cartridges and washing the respirator with soap and water after each use. After washing, rinse the facepiece to remove all traces of soap. Dry the respirator with a clean cloth and place the facepiece in a well-ventilated area to dry. When you are ready to use the respirator, insert new filters and cartridges. Do not wash or abuse the cartridges; they should be replaced after the length of use specified by the manufacturer.

Rubber boots and gloves should also be washed inside and out with soap and water and rinsed thoroughly daily or more often if contamination has occurred or is suspected. Rubber or plastic protective pants, coats, and headgear should also be washed, rinsed, and dried in a method similar to that for respirators. The Worker Protection Standard (refer to Chapter 6) requires that workers and handlers be trained in the use and care of personal protective equipment.

Decontamination of Clothing

Clothing worn by pesticide applicators should be laundered daily. The problem of how to launder pesticide-contaminated clothing is sometimes difficult and is determined on the basis of the toxicity of the chemical and the type of formulation. Clothing is easily contaminated by pesticides. Once contaminated, it is difficult to remove all the pesticide through home laundering procedures.

Clothing contaminated with highly toxic and concentrated pesticides must be handled most carefully, as these pesticides are easily absorbed through the skin. If the clothes have been completely saturated with concentrated pesticides, they should be discarded. Dispose of them as you would other contaminated materials or unwanted pesticides. Clothing contaminated by moderately toxic pesticides does not warrant such drastic measures. Hazards are less pronounced in handling clothing exposed to low-toxicity pesticides. However, the ease of pesticide removal through laundering does *not* depend on toxicity level; it depends on the formulation of the pesticide. For example, 2,4-D amine is easily removed through laundering because it is soluble in water, while 2,4-D ester is much more difficult to remove through laundering.

Disposable clothing helps limit contamination of clothes because the disposable garments add an extra layer of protection. This is especially important when you are in direct contact with pesticides, such as when mixing and loading pesticides for application.

Laundering recommendations. Wash contaminated clothing separately from the family wash. Research has shown that pesticide residues are transferred from contaminated clothing to other clothing when they are laundered together. It is vital that the person doing the wash know when pesticides have been used so that all clothing can be properly laundered. If a granular pesticide has been used, shake clothing outdoors and be sure to empty pockets and cuffs.

Prerinsing contaminated clothing before washing will help remove pesticide particles from the fabric. Prerinsing can be done by:

- Presoaking in a suitable container prior to washing.
- Spraying or hosing garments outdoors.

Prerinsing is especially effective in dislodging the particles from clothing when a wettable powder pesticide formulation has been used. Clothing worn while using slightly toxic pesticides may be effectively laundered in one machine washing. It is strongly recommended that multiple washings be used on clothing contaminated with concentrated pesticides to draw out excess residues. Always wear rubber gloves or use a stick when handling highly contaminated clothing to prevent pesticide absorption into the body.

Washing in hot water removes more pesticide from the clothing than washing in other water temperatures. Remember, the hotter, the better. Avoid cold-water washing. Although cold-water washing might save energy, it is relatively ineffective in removing pesticides from clothing.

Laundry detergents, whether phosphate, carbonate, or heavy-duty liquids, are similarly effective in removing pesticides from fabric. However, research has shown that heavy-duty liquid detergents are more effective than other detergents in removing emulsifiable concentrate pesticide formulations. Emulsifiable concentrate formulations are oil-based, and heavy-duty liquid detergents are known for oil-removing ability. Granular detergents have been found to be effective in removing water-soluble pesticides.

Laundry additives such as bleach or ammonia do not contribute to removing pesticide residues. Either of these additives may be used, if desired, but caution must be used. *Bleach should never be added to or mixed with ammonia* because they react together to form a fatal chloride gas. Be careful; do not mix ammonia and bleach.

If several garments have been contaminated, wash only one or two garments in a single load. Wash garments contaminated by the same pesticide together. Launder using a full water level to allow the water to thoroughly flush the fabric.

During seasons when pesticides are being used daily, clothing exposed to pesticides should be laundered daily. This is especially true with highly toxic or concentrated pesticides. It is much easier to remove pesticides from clothing by daily laundering than attempting to remove residues that have accumulated over a period of time.

Pesticide carryover to subsequent laundry loads is possible because the washing machine is likely to retain residues, which are then released in following laundry loads. It is important to rinse the washing machine with an "empty load" using hot water and the same detergent, machine settings, and cycles used for laundering the contaminated clothing.

Line drying is recommended for these items. Although heat from an automatic dryer might create additional chemical breakdown of pesticide residues, many pesticides break down when exposed to sunlight. This also eliminates the possibility of residues collecting in the dryer.

Decontamination of Spray Personnel

Decontamination of personnel, especially after a splash or spill of a concentrated chemical, whether on clothing or skin or in the eyes, should be handled quickly and efficiently. Speed is particularly important when highly toxic chemicals are involved, but it is only slightly less urgent with the less toxic compounds. Keep in mind the factors that contribute to poisoning: the inherent toxicity of the chemical, its concentration, its physical form, the actual portion of the body exposed, and the length of time exposed. If, for example, you have spilled a highly toxic, concentrated liquid formulation on your skin or clothing, you must get the clothing off immediately and wash thoroughly with soap and water. *Speed is essential.* Don't forget to wash your hair and under your fingernails.

Alcohol is an excellent decontaminating agent, particularly if the contaminated area is limited. It has been found that 30 minutes after a test application of parathion to the skin, vigorous scrubbing with soap and water will remove 80% or more of the material, and alcohol will remove most of the remainder. After 5 hours, however, 40% *cannot* be washed off with soap and water and 10% will remain even after scrubbing with alcohol.

It is worth noting that studies presented to the American Chemical Society in 1984 showed that the scrotal, jaw, forehead, and scalp areas of the body absorb pesticides most rapidly. Damaged skin (sunburn) absorbs four to ten times faster than normal skin.

The Worker Protection Standard requires that handlers (applicators) have decontamination facilities available. However, if facilities are not available, the nearest irrigation water, pond, or practically any source of water that is not contaminated with pesticides will serve the purpose in an emergency. Following emergency decontamination, call your physician about the accident, as he or she may want to check for possible aftereffects (see Chapter 8).

Decontamination of Pesticides Spilled During Transportation

In the case of a significant pesticide incident, whether on-site or off-site, help is available through CHEMTREC (Chemical Transportation Emergency Center). CHEMTREC is sponsored by the Chemical Manufacturers Association based in Washington, D.C. CHEMTREC operates 24 hours a day with a toll-free, nationwide telephone number. Refer to Appendix B for the telephone number. CHEMTREC is not a contact for general chemical information, but is a source for assistance in chemical emergency incidents. It is designed to provide immediate data on how to handle these emergencies for those trained to do so.

The National Pesticides Telecommunications Network (NPTN) is a 24-hour service with a toll-free telephone number. Refer to Appendix B for the telephone number. The NPTN provides information about pesticides to the medical, veterinary, and professional communities, as well as to the general public. Originally a service for physicians, the NPTN has expanded service to the public by providing information on pesticide products, basic safety practices, health, and environmental effects.

Spills of any type of chemical, hazardous or nonhazardous, should be taken seriously and action taken immediately, regardless of the size of the spill. The immediate goal of all cleanup procedures is to contain the spill. Under no circumstances should an individual wash a spill off the road or property. To do so could put you in violation of federal statutes and entail costly cleanup measures.

The following cleanup procedures will vary depending on the extent of the spill and whether outside agencies are involved.

- Isolate the contaminated area and confine entry to those persons who are properly protected.
- If there is a visible spread of a *large* quantity of hazardous material, you must consider recommending to local authorities to evacuate all residents in the path of spread and maintain a close check on wind directions and conditions until the hazard abates. In most spill situations, evacuation is not necessary.
- Prompt medical attention is necessary for persons known to have been exposed or suspected of having been exposed. Ambulance and hospital personnel must be informed that they will be dealing with chemical-contaminated individuals and should wear protective clothing, such as rubber gloves and aprons. As soon as the active ingredient of the chemical is determined, the hospital should be informed immediately.
- Federal and state law requires that spills of certain products in specified amounts (reportable quantities) be reported. The amount that triggers a notification may vary in some states. Failure to report will likely trigger a fine. This reporting is regardless of whether the products end up in water or not.
- If the leak or spill involves a vehicle and it is possible to move the vehicle without extending the exposure or contributing to the occurrence of secondary accidents, the vehicle should be moved to a cleanup area.
- Proper personnel protective equipment must be worn at all times. Do not enter a closed trailer involving a spill without proper respiratory protection. Volatile materials or class B poisons should not be handled unless a self-contained breathing unit is used. Proper ventilation of the trailer may change the need for the type of respiratory protection needed. Remember, many pesticides will penetrate through clothes, through the skin, and into the blood system. Rubber boots and gloves are absolutely necessary. The proper form of respiratory equipment should also be worn. If leather articles, such as shoes, should become contaminated, they must be disposed of with the waste pesticide. They cannot be decontaminated and should never be worn again.

- The leak or spill should be confined to the smallest area possible, utilizing natural terrain or diking, or covering with a plastic tarp.
- Washing down with hose streams or large amounts of water should be avoided or kept to a minimum and only after as much as possible of the spill has been picked up. Runoff water should not be permitted to enter bodies of water, such as rivers or lakes, or to flow indiscriminately into storm sewers.
- Contaminated ground areas should be dug out a depth of 3 to 4 inches and covered with a layer of lime followed by a top layer of clean earth.
- All contaminated soil and waste pesticide is regulated under federal law by RCRA and/or state law. After waste is containerized and the spill emergency is abated, contact state authorities for proper disposal information.
- In the case of fairly small spills, two commonly available materials, household bleach and hydrated lime, will neutralize many of the pesticides used in agriculture. Sprinkle or spray the spill with hypochlorite (household bleach) and water, adding 1 gallon of water to each 1 gallon of the hypochlorite. Spread hydrated lime over the entire area and let stand at least one hour.

Important: Hydrated lime and the hypochlorites will not neutralize all pesticides that are being used agriculturally, but they are suggested at this time as a practical approach to help reduce the hazards of small pesticide spills.

Recognize that several federal and state laws come into play as a result of a pesticide spill. Regulatory agency involvement will depend on the product involved, amount spilled, potential or actual environmental damage, water contamination, and proper disposal of contaminated material. Abate the emergency; then check into regulatory issues before proceeding further.

Deactivating Residual Herbicides with Activated Charcoal

Activated charcoal is generally effective against organic chemicals and *not* against inorganic chemicals. Almost all herbicides used today are in the organic category. Combination herbicides (those containing more than one active chemical) may contain both organic and inorganic ingredients, so charcoal may be effective against only some of the ingredients. Fortunately, most inorganic ingredients of herbicides are relatively short-lived, compared to the organic ones. However, their presence should be considered in any deactivation program.

How much herbicide is present. You must determine, insofar as possible, the amount of herbicide you are dealing with. Herbicide formulations vary in the amount of active chemical they contain; and since it is only the active ingredients you are concerned with, you must determine the relative amount of active ingredients involved. Table 14–1 will help you make the determination. Estimating the amount of active ingredients present will usually be accurate enough for your purpose. On small areas where the amount of charcoal needed is relatively small, be sure to use plenty.

Another, more accurate way to determine the amount of chemical you are dealing with, is to have a chemical analysis run on a soil sample taken from the area. The sample should be taken from the top 3 inches of soil and deeper if the chemical has

TABLE 14–1 AMOUNT OF ACTIVE INGREDIENT IN VARIOUS FORMULATIONS

Formulation	Unit	Pounds a.i. per unit
25G	1 lb	0.25
25E	1 gal	2.0
80W	1 lb	0.8
4L	1 qt	1.0
4G	1 lb	.04
5PS	10 lb	0.5
8P	10 lb	0.8
50WP	1 gal	0.5
4E	1 gal	1.0

TABLE 14–2 CHARCOAL NEEDED TO DEACTIVATE A CHEMICAL

Pounds a.i.* present per acre	Pounds a.i.* present per 1000 ft^2	Pounds charcoal needed per 1000 ft^2
Up to 1.0	Up to 0.02	2.3– 3.5
2.0	0.04	4.6– 6.9
3.0	0.06	6.9–10.3
5.0	0.11	11.5–17.2
10.0	0.22	22.9–34.4
20.0	0.45	45.0–67.5

*Active ingredient.

been applied some time before, if considerable rainfall has occurred since application, or if the soil is especially coarse-textured and porous. If the analysis from a 3-inch sample shows 1 part per million (ppm) of chemical present, this is equivalent to 1 pound of chemical per acre in the top 3 inches of soil. One ppm in a 6-inch sample is equivalent to 2 pounds of chemical per acre in the top 6 inches of soil.

1000-square-foot basis. Assuming the equivalent of 1 pound of active chemical per acre (0.02 lb/1000 square feet) you will need from 2.3 to 3.5 pounds of activated charcoal per 1000 square feet (from Table 14–2). Use at least 2.3 to 3.5 pounds of activated charcoal per 1000 square feet even though the amount of active chemical might call for less. If you find more than the equivalent of 1 pound of active chemical per acre present, increase the rate of charcoal as recommended in Table 14–2, which will help you determine the amount of chemical you must deactivate on a unit basis.

How to apply charcoal. If you use dry, powdered charcoal, spread it evenly over the affected area. Some charcoal is treated so it can be added to water to form a slurry and applied as a spray. Incorporate the charcoal 4 to 6 inches deep with a rototiller or similar implement. This assures getting the charcoal intermixed with the herbicide in the soil.

After incorporation, water the area thoroughly every day for 3 to 4 days before reseeding or planting. If possible, wait a few more days because, under some conditions, deactivation is slowed. If working with large areas, test seed a small plot to check the effectiveness of the treatment. If the seedlings die, water for 3 or 4 more days before seeding again. It may be necessary to add more charcoal, but usually the extra time will be sufficient.

PESTICIDE AND CONTAINER DISPOSAL

Proper pesticide waste and container disposal is extremely important. It is the pesticide user's responsibility to see that unused chemicals and empty containers are disposed of properly. Improper disposal of pesticide wastes and containers over the past few years has resulted in incidents involving animal poisonings and environmental contamination.

Unfortunately, there are no easy or perfect means to dispose of excess pesticides and empty containers. State regulations should be reviewed, as some states have specific regulations on containers. Pesticide wastes may range in type from materials left over from excess spray mixtures, to accidental spillage, to pesticides left after a warehouse fire. Pesticide containers vary from aerosol cans or paper bags to 1- and 5-gallon cans on up to the 55-gallon drums, which are sometimes hard to dispose of.

Supposedly empty pesticide containers are never really empty as several ounces of pesticide can generally be found in discarded containers. Many times the concentrations of the active ingredient in liquid formulations will tend to increase in toxicity as the solvents evaporate from the discarded container.

The American Crop Protection Association has suggested a procedure to use when rinsing containers prior to disposal that reduces their hazard potential. The procedure involves the triple rinse method, as described in the following steps:

1. When the container is being emptied, allow it to drain in a vertical position for 30 seconds.
2. Water or other diluting materials being used in a spray program can be used to rinse the container as follows: use 1 quart for each rinse of a 1-gallon can or jug; 1 gallon for each 5-gallon can; and 5 gallons for either 30- or 55-gallon drums.
3. Best results are obtained when the container is rinsed three times, allowing 30 seconds for draining after each rinse. Each rinse should be drained into the spray tank before filling it to the desired level.
4. Crush or break the container and dispose of it in an approved manner. Do not reuse for any other purpose.

Review Figure 14–3 to memorize the proper steps in the rinse and drain procedure. It makes good sense to empty and triple rinse pesticide containers because it will ensure that you are getting all usable product out of the container, as well as ensuring that there will be no residue left to contaminate the environment or injure someone. Figure 14–4 attests to the fact that you are saving money when you use the triple rinse system.

Pesticide and Container Disposal

Follow this Rinse and Drain Procedure for Pesticide Containers

 Empty container into spray tank. Then drain in vertical position for 30 seconds.

  Add a measured amount (of rinsewater or other diluent), so container is ¼ or ⅛ full. For example, 1 quart in a 1-gallon container.

Rinse container thoroughly, pour into tank, and drain 30 sec. Repeat three times. Add enough fluid to bring tank up to level.

Crush pesticide container immediately. Sell as scrap for recycling or take to the landfill. Do not reuse.

Figure 14–3 Triple rinse procedure.

EMPTY AND RINSE

 Pesticide Containers thoroughly

 "Empty" Containers thoroughly

Laboratory tests show that even a good effort to empty a drum leaves about 6 ounces of pesticide in a 5-gallon container and 32 ounces in a 55-gallon drum.

Rinse with several quarts of water—once, twice, or THREE times and pour the rinse-water into a sprayer. You save money with each of the rinses.

Your in-the-can loss with pesticides costing $20 and $30 per gallon is shown below.

The value of rinsing a 5-gallon "empty" container is illustrated below.

Empty to Save Money

AMOUNT OF RESIDUE	$ LOSS AT:	
	$20/GAL.	$30/GAL.
6½ OZ.	$1.00	$1.50
32 OZ.	$5.00	$7.50

AN EASY WAY TO CUT COSTS!

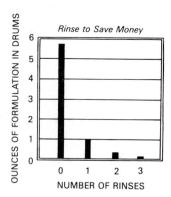

Rinse to Save Money

Figure 14–4 Empty and rinse containers to save money.

Note: Not only is it smart to triple rinse based on economics, but it is also a requirement found on most pesticide labels. The federal hazardous waste law (RCRA) requires that empty containers that once contained "acute hazardous material" must be triple rinsed or else the "empty" container becomes a regulated waste.

Surplus or unwanted pesticide disposal can be kept at a minimum with some advance planning. The safest means of disposal is to use the pesticide exactly as the label directs for the purpose for which it is intended. To do so requires some advance planning to determine the kinds and amounts needed for the job; such planning means you will use all the spray, dust, or granule prepared for the application.

Warning: Never pour pesticides down the drain or flush them down the toilet!

Present methods of pesticide container and waste disposal include burial, incineration, degradation, deep well injection, and in some instances permanent storage. All pesticides and containers should be kept in storage until used or disposed of properly. The several methods just mentioned for pesticide waste and container disposal are feasible under certain conditions. It is the user's responsibility to select the most feasible method. Every possible alternative should be considered in relation to the problem, giving consideration to the environment.

The choice or selection of the disposal method (refer to Figure 14–5) depends on several factors such as pesticide type; container type; facilities available for disposal; nearness to communities, streams, and crops; and any other geological or environmental considerations. The failure to consider any of these factors may result in more complex problems in the future.

The Resource Conservation and Recovery Act (RCRA) governs larger accumulations of pesticides, rinsings collected from aircraft washing aprons, disposal of empty containers, and so on. RCRA gives an exemption to the farmer, but all other users may be directly affected depending on the pesticide involved, amount of waste generated, and other factors. Many states also have hazardous waste disposal laws that the pesticide applicator must know about and comply with.

In 1976 Congress passed the RCRA, which directed the EPA to develop and implement a program to protect human health and the environment from improper hazardous waste management practices. The program is designed to control the management of hazardous waste from its generation to its ultimate disposal, from cradle to grave. The EPA first focused on large companies, which generate the greatest portion of hazardous waste. Business establishments producing less than 2200 pounds of hazardous waste in a calendar month (known as small-quantity generators) were exempted from most of the hazardous waste management regulations published by the EPA in 1980. In recent years, however, public attention has been focused on the potential for environmental and health problems that may result from mismanaging even small quantities of hazardous waste. For example, small amounts of hazardous waste dumped on the land may seep into the earth and contaminate underground water that supplies drinking water wells.

In November 1984, the Hazardous and Solid Waste Amendments to RCRA were signed into law. With these amendments, Congress directed the EPA to establish new requirements that would bring small-quantity generators who generate between 220 and

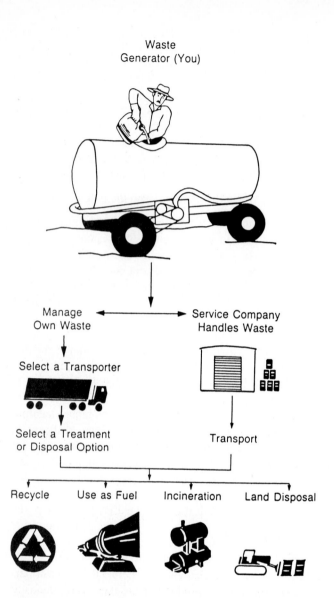

Waste
Generator (You)

Manage
Own Waste
→
Service Company
Handles Waste

Select a Transporter

Select a Treatment
or Disposal Option

Transport

Recycle   Use as Fuel   Incineration   Land Disposal

As a hazardous waste generator, you normally will go through the decision process shown here. You can choose the disposal option and (1) hire the transporter or a disposer who will take care of transportation or (2) you can use a service firm. The service firm can combine your waste with those of other generators to reduce transportation costs to a disposal site.

Figure 14–5   What can be done with hazardous wastes?

2200 pounds of hazardous waste in a calendar month into the hazardous waste regulatory system. The EPA issued final regulations for these generators on March 24, 1986. In the case of certain highly toxic products, generation of over 2 pounds in a month will trigger permit requirements.

Because the rules cover only *waste,* pesticides are not regulated under RCRA until they become *waste.* This can occur in several ways:

- *Rinsewater* used to clean pesticide application equipment and to triple rinse product containers, unless it is used as the diluent in another load
- *Empty containers,* unless cleaned according to label instructions or hazardous waste rules
- *Unusable or unidentifiable pesticide materials,* including those for which registration has been canceled
- *Contaminated material,* such as soil or other material contaminated by pesticide spills
- *Maintenance and repair wastes,* including parts washing solvents, paint and thinner wastes, epoxies and adhesives, certain caustics, acids or alkalines, and certain batteries.

TIPS FOR MINIMIZING WASTE GENERATION

- Don't use more of a hazardous product, such as a pesticide, than you need to do the job.
- Use all the hazardous product that you mix or that is in the container, and make sure that containers are completely empty and properly rinsed and prepared for disposal.
- Avoid spills or leaks of hazardous products. If you have a cleanup problem, the materials used to contain and clean up the hazardous product also become hazardous materials.
- Do not mix nonhazardous wastes with hazardous wastes. This just makes your disposal problem worse, may make recycling impossible, or may make disposal costs more expensive.

For help, call the EPA RCRA Superfund Hotline listed in Appendix B. Details of some of the state laws, as well as interpretations of RCRA, are complicated and difficult to understand in some cases, but it is the responsibility of the pesticide applicator to make sure he or she knows what the law is and to obey all provisions.

15

Integrated Pest Management

THE CONCEPT AND PRINCIPLE OF IPM

Integrated pest management (IPM) is a relatively new approach to an old problem: how to ensure crop* protection and maintain appearance and quality through controlling pest populations while minimizing effects on humans and the environment. IPM attempts to make the most efficient use of the strategies available to control pest problems by taking action to prevent problems, suppress damage levels, and use pesticides only where and when needed. Rather than seeking to eradicate all pests entirely, IPM strives to prevent their development or to suppress their population numbers below levels that would be economically damaging.

- *Integrated* means that a broad, interdisciplinary approach is taken using scientific principles of crop protection in order to fuse into a single system a variety of methods and tactics.
- *Pest* includes insects, mites, nematodes, plant pathogens, weeds, and vertebrates that adversely affect crop quality and usefulness.
- *Management* refers to the attempt to control pest populations in a planned, systematic way by keeping their numbers or damage within acceptable levels.

THE EVOLUTION OF IPM

For the last 35 to 40 years, pest control has been achieved almost exclusively by use of pesticides. There are several reasons for this:

*The term *crop* as used in this chapter means all forms of crops, including field crops, turfgrass, golf courses, parks and other green areas, landscape and ornamental plantings, gardens, and pastures.

392

- Pesticides provide the easiest, quickest way of reducing pest populations.
- The broad-spectrum nature of early pesticides protected crops from a variety of pest species, and a single application was often sufficient for control.
- Development of crop technology and application equipment made it economically feasible to apply pesticides; that is, cost-return benefits for pesticides were maximized.

Although chemicals have been extremely beneficial in crop protection, an almost total dependence on synthetic pesticides has resulted in unintended and unforeseen problems. Some of these problems are:

- *Environmental contamination.* Pesticides have aroused concern because of possible adverse effects on humans and other life in the environment.
- *Pest resistance.* Several pests have developed resistance to commonly used pesticides. This resistance has rendered pesticides less effective against certain pests and shortened the useful life of some pesticides. Pest resistance generally leads to an increase in the amount of pesticides applied, a search for newer, more effective replacement chemicals, more sensible use of pesticides, or a search for alternatives to pesticide usage.
- *Misuse of pesticides.* Because pesticides were relatively inexpensive and easy to apply, growers often resorted to a higher number of pesticide applications than might have been necessary in order to protect their crops. Not only was this economically unsound, it tended to increase other problems associated with pesticides.
- *Secondary pest outbreaks.* Complete control of one pest by means of pesticides often led to secondary population outbreaks. In other words, eliminating one pest would upset the ecological balance, and another organism, which was previously not harmful, would emerge as a pest.
- *Death of nontarget organisms.* The broad-spectrum pesticides killed not only pests, but also their natural enemies. For example, lady bird beetles and their larvae are natural enemies of other insects and their larvae. Pesticides often kill the beneficial insects as well as the harmful insects.
- *Pest resurgence.* Using pesticides sometimes led to a resurgence of the original pest population, which in turn called for the use of more and more pesticides. This often occurred because the use of pesticides would upset the ecological balance by eliminating both pests and nontarget organisms (such as natural enemies). Thus, any pests that survived or reinvaded would have an excellent opportunity for increasing their numbers, even to a level higher than before pesticide application, because neither their natural enemies nor competitors would be present.

Because pesticide use is being more highly regulated and more restrictions are in place due to possible adverse effects on humans and the environment, and since some pesticides are becoming less effective for the variety of reasons just listed, a more comprehensive, ecologically-based approach to crop protection is often called for. To be sure, pesticides will continue to play an important role in an IPM program.

The primary difference, however, is that these products will be used selectively and judiciously. The new approach, IPM, seeks to decrease the dependence on pesticides as the exclusive tool for pest control. IPM attempts to meet the present needs of modern society and crop management. Thus, IPM strives to (1) protect the health and welfare of growers, workers, and society as a whole by reducing pesticide entry into the environment, and (2) control pests in a more effective, economical, and ecologically stable manner.

THE GOALS OF IPM

The four goals of an IPM program are:

1. *Improved control.* IPM will provide more effective pest control to maintain and sometimes improve crop quality. For example, by implementing alternatives to strict dependence on pesticides, IPM makes use of a balanced approach relying, for example, on cultural practices, natural enemies (parasites or pathogens), and host plant resistance, as well as chemicals. By reducing the use of pesticides, IPM emphasizes biological control and the conservation of natural enemies already occurring in the environment.

2. *Pesticide management.* IPM will supply a more efficient and sensible approach to pesticides, thus increasing their effectiveness and useful life span and decreasing possible adverse effects.

3. *Economical crop protection.* IPM will control pest populations more economically. For example, simply by treating crops *as needed,* instead of merely by the calendar, IPM can often reduce protection costs by reducing the amount of pesticide used and the number of applications.

4. *Reduction of potential hazards.* IPM will better safeguard human health and the environment from possible harmful side effects associated with pesticides.

WHO BENEFITS FROM IPM?

- *Farmers, gardeners, and turfgrass and parks managers* benefit from decreased management costs, a more balanced and effective means to control pests, and reduced risks of crop injury or harm to human and animal health.
- *Workers* benefit by reduced exposure to pesticides in the work area.
- *Users of parks and turfgrass areas* benefit by enjoying the area with minimal or no pesticide residue.
- *Fish and wildlife* benefit by less exposure to pesticides.
- *The water, air, and soil* will remain less contaminated from misuse of pesticides.
- *Society as a whole* will ultimately reap benefits from an environment less contaminated by potentially harmful chemicals.

AN APPROACH TO EFFECTIVE IPM

Effective IPM consists of four basic principles:

1. *Exclusion* seeks to prevent pests from entering the crops in the first place, thus stopping problems before they arise.
2. *Suppression* refers to the attempt to suppress pests below the level at which they would be economically damaging.
3. *Eradication* strives to eliminate entirely certain pests whose presence, however minimal, cannot be tolerated.
4. *Plant resistance* stresses the effort to develop healthy, vigorous strains of crops that will be resistant to certain pests.

It is interesting that these are the same principles used in plant disease control. To carry out these four basic principles, the following steps are often taken.

1. *The identification of key pests and beneficial organisms* is a necessary first step. In addition, biological, physical, and environmental factors that affect these organisms must be ascertained.
2. *Preventive cultural practices* are selected to minimize pest population development. These practices include soil preparation procedures, use of resistant plants, specified planting dates, and the like.
3. *Pest populations must be monitored* by trained personnel who routinely check crop areas and keep adequate records.
4. *A prediction of loss and risks* involved is made by setting an economic threshold. Pests are controlled only when the pest population threatens acceptable levels of crop quality. The level at which the pest population or its damage endangers crop quality is often called the economic threshold. The economic threshold is set by predicting potential losses and risks at a given population density. This estimation takes into account weather data, state of crop development, risk/benefit, costs, and kinds of control available.
5. *An action decision must be made.* In some cases pesticide application will be necessary to reduce the threat to the crop, while in other cases a decision will be made to wait and rely on closer monitoring.
6. *Evaluation and follow-up* must occur throughout all stages in order to make corrections, assess levels of success, and project future possibilities for improvement.

IMPLEMENTATION OF IPM

To be effective, IPM must make use of the following tools:

- *Resistant crop varieties* are bred and selected when available in order to protect against key pests.
- *Natural enemies* are used to regulate the pest population whenever possible.

- *Pheromone (sex lure) traps* are used to lure and destroy male insects, thus helping the monitoring procedures. Pheromone traps have control potential and have been used to keep a population within acceptable levels.
- *Avoidance* of peak pest populations can sometimes be brought about by a change in planting times or management practices.
- *Pesticides.* Some pesticides are applied preventively, for example, herbicides, fungicides, and nematicides. The proper choice of herbicides can best be made utilizing a "weed map" drawn up at the end of the season. At the beginning of the next season, this map will be helpful in determining which preplant or preemergence herbicide will be most effective for the specific weeds in a certain crop area. In an effective IPM program, pesticides are applied on a prescription basis tailored to the particular pest and chosen so as to have minimum impact on humans and the environment. They are applied only when a pest population has been diagnosed as large enough to threaten acceptable levels of crop quality. Pesticides are usually chosen only after all feasible alternatives have been considered.
- *Improved pesticide application* can be achieved through reliance on accurate pressure, timing, agitation, and the like, and by keeping equipment up to date and in excellent shape.
- *Preventive measures* such as soil fumigation for nematodes and assurance of good soil fertility help to provide a healthy, vigorous crop.
- *Other assorted cultural practices* can be used to influence pest populations.

Pest control is important to community health and welfare, farms, gardens, urban lawns, and public green areas. A more judicious, better informed use of all existing control measures, including IPM in a systems context, promises to reduce environmental hazards and yield significant benefits for everyone.

Glossary for Pesticide Users

A

Abrasion   The process of wearing away by rubbing or grinding. Also, a scrape, scratch, or sore that breaks the skin.

Abrasive   Something that grinds down or wears away an object. For example, wettable powders are abrasive to pumps and nozzles.

Abscission   The formation of a layer of cells that results in fruit, leaf, or stem drop from a plant.

Absorption   Movement of a pesticide from the surface into a plant, animal, or the soil. For instance, in animals, absorption may take place through the skin, breathing organs, stomach, or intestines; in plants, through leaves, stems, or roots.

Absorptive clay   A special type of clay powder that can take up chemicals and hold them. It is sometimes used to clean up pesticide spills.

Acaricide   A pesticide used to control mites and ticks. Same as miticide.

Accelerate   Increase in rate of speed.

Acceptable daily intake (ADI)   The maximum dose of a substance that is anticipated to be without lifetime risk to humans when taken daily.

Accepted method   A commonly used way of doing something.

Accident   An unexpected, undesirable event, caused by the use or presence of a pesticide, that adversely affects man or the environment.

Accordance   Agreement. Example: To follow directions on the pesticide label.

Accumulate   To build up, add to, store, or pile up.

Accumulative pesticides   Chemicals that tend to build up in animals or the environment.

Acid   A compound whose hydrogen atom is replaceable by positive ions or radicals to form salts. Usually very dangerous in concentrated form.

Acid equivalent   The theoretical yield of parent acid from an ester or salt such as esters of 2,4-D or the amine salt of 2,4-D.

Acre   43,560 square feet. An area of land about 210 feet long by 210 feet wide.

Activated charcoal   Very finely ground, high quality charcoal that absorbs liquids and gases very easily.

Activator   A material added to a pesticide to increase, either directly or indirectly, its toxicity.

Active ingredient   The chemical or chemicals in a product responsible for the desired effects, which are capable, in themselves, of preventing, destroying, repelling, or mitigating insects, fungi, rodents, weeds, or other pests.

Actual dosage   The amount of active ingredient (not formulated product) that is applied to an area or other target.

Acute dermal toxicity   Poisoning from a single dose of a chemical absorbed through the skin.

Acute inhalation toxicity   How poisonous a single dose (or exposure) of a pesticide is when breathed into the lungs.

Acute oral toxicity   Poisoning from a single dose of a chemical taken by mouth.

Acute poisoning   Poisoning that occurs after a single dose (or exposure) of a pesticide.

Acute toxicity   How poisonous a pesticide is to an animal or person after a single dose (or exposure).

Additive   Any material added to a pesticide (not necessarily a wetting agent or a surfactant).

Adherence   The property of a substance to adhere or stick to a given surface.

Adhesive   A substance that will cause a spray material to stick to the sprayed surface, often referred to as a sticking agent.

Adjacent   Next to. Neighboring.

Adjuvant   A chemical or agent added to a pesticide mixture that helps the active ingredient do a better job. Examples: wetting agent, spreader, adhesive, emulsifying agent, penetrant.

Adsorption   The process by which materials are held or bound to the surface in such a manner that the chemical is only slowly available. Clay and high organic soils tend to adsorb pesticides in many instances.

Adult   A full-grown, sexually mature insect, mite, nematode, or other animal. This stage is often the one to migrate and begin new colonies or infestations.

Adulterated   When strength or purity of a pesticide falls below the standard claimed or if any substance has been added, or any part subtracted.

Adulticide   An insecticide that is toxic to the adult stage of insects.

Adverse effect   An undesirable result such as an illness or pollution.

Aerate   To bring in contact with more air. To loosen or stir up the soil mechanically or use additives such as peat. To let in or pull in fresh air after fumigation.

Aerial application   Treatment applied with the use of an airplane or helicopter.

Aerobic   Living or functioning in air or free oxygen. The opposite of anaerobic.

Aerosol spray   A fine spray produced by pressurized gas that leaves very small droplets of pesticide suspended in the air.

Agitate To keep a pesticide chemical mixed up. To keep it from separating or settling in the spray tank.

Agitator A paddle, air, or hydraulic action to keep a pesticide chemical mixed in the spray or loading tank.

Agricultural commodity Any plant or part of a plant, animal, or animal product that is to be bought or sold.

Air-blast sprayer A machine that can deliver high and low volumes of spray. The arrangement of the sprayer and its operation cause spray to be carried at a distance from the sprayer by air for deposition on the objects being treated. It is used for orchards, shade trees, vegetables, and fly control.

Alcohol Although in ordinary conversation ethyl alcohol is generally referred to merely as alcohol, that term is applied to a long series of hydroxy organic compounds beginning with the one-carbon compound methanol, or methyl poison. Both methyl alcohol and ethyl alcohol are common solvents, frequently used in formulating pesticidal mixtures.

Algae Nonvascular chlorophyll-containing plants, usually aquatic.

Algicide A chemical intended for the control of algae, especially in water that is stored or is being used industrially.

Aliphatic Chemically, those compounds that possess open-chain molecular structures. Generally are considered less toxic to plants than aromatic compounds.

Alkali The opposite of an acid. It is usually dangerous in concentrated form.

Alkaloid Naturally occurring nitrogenous materials appearing in some plants that are used in preparing the botanically derived insecticides.

Alternative One of two or more choices. Another way to do something.

Amide A compound derived from carboxylic acids by replacing the hydroxyl of the COOH by the amino group NH_2.

Amine An organic compound containing hydrogen derived from ammonia by replacing one or more hydrogen atoms by as many hydrocarbon radicals.

Amine salt A formulation that includes an acid that has been neutralized by an amine, a basic compound.

Amphibians Animals of the class Vertebrata that are intermediate between fish and reptiles. They are cold-blooded and have moist skin without scales, feathers, or hair. Examples are frogs, toads, and salamanders.

Amphoteric An amphoteric compound has the capacity of behaving either as an acid or base. An amphoteric surface-active compound is capable of anionic or cationic behavior depending on whether it is in an acidic or basic system.

Anaerobic Living or functioning in the absence of air or free oxygen.

Analog A compound that is very similar in both structure and formula to another compound.

Animal All vertebrate and invertebrate species.

Animal sign Evidence of an animal's presence in an area.

Anionic An ion having a negative charge is an anion. When the surface-active portion of a surfactant molecule possesses a negative charge it is termed an *anionic surface-active agent*. Contrast with *cationic*.

Anionic surfactant   A surface-active additive to a pesticide having a negative surface charge. The anionics perform better in cold, soft water. Most wetting agents are of this class.

Annual   A plant that grows from a seed; produces flowers, fruit, or seed the same year; then dies.

Antagonism   Opposing action of different chemicals such that the action of one or more is impaired, or the total effect is less than that of each component used separately. (Opposite of *synergism*.)

Antibiotics   Substances that are "against life." They are chemical substances produced by certain living cells such as bacteria, yeasts, and molds. They are antagonistic or damaging to other living cells such as disease-producing bacteria. Antibiotics may kill living cells or prevent them from growing and multiplying. Penicillin is an example of an antibiotic that damages certain bacteria that cause disease in man.

Anticoagulant   A chemical used in a bait to destroy rodents. It destroys the walls of the small blood vessels, and keeps the blood from clotting. As a result, the animal bleeds to death.

Antidote   A practical immediate treatment including first aid in case of poisoning; a remedy used to counteract the effects of a poison.

Antidrift agent   A chemical used to reduce spray drift during the actual spraying operation by various physical factors.

Antifeeding compound   A compound that will prevent the feeding of pests on a treated material without necessarily killing or repelling them. It is not a repellent.

Antifoaming agent   An adjuvant that reduces foam in spray mixtures.

Antioxidant   A substance capable of chemically protecting other substances against oxidation or spoilage.

Antisiphoning device   A device attached to a filling hose to prevent water in the spray tank from draining into the water source.

Antitranspirant   A chemical applied directly to a plant that reduces the rate of transpiration or water loss by the plant.

Aphicide   A compound used to control aphids or plant lice.

Apiary   A place where colonies of bees are intentionally kept for the benefit of their honey-producing and pollinating activities.

Apiculture   Pertaining to the care and culture of bees.

Application   The placing of a pesticide on a plant, animal, building, or soil; or its release into the air or water to prevent damage or destroy pests.

Application rate   The amount of pesticide formulation applied to a given area.

Applicator   A person or piece of equipment that applies pesticides to destroy pests or prevent damage by them.

Apply uniformly   To spread or distribute a pesticide evenly.

Aquatic plants or weeds   Plants or weeds that grow in water. The plants may float on the surface, grow up from the bottom of the body of water (emergent), or grow under the surface of the water (submergent).

Aqueous   Indicating the presence of water in a solution. A solution of a chemical in water.

Aromatics   Oil or oil-like materials that kill plant and animal tissue similar to burning. Chemically, a compound having a closed ring structure.

ARS   Agricultural Research Service of the U.S. Department of Agriculture.

Arsenicals   One of the most important groups of early insecticides, comprised principally of the arsenates and arsenites. The killing power of these materials is directly related to the percentage of metallic arsenic contained, and in addition the metal component of the material (for example, lead in lead arsenate) may also have some toxicity.

Arsenate   An arsenical compound with a hydrocarbon group connected to the arsenic atom; hence, an organic arsenical.

Artificial respiration   First aid given to a person who has stopped breathing.

Aseptic   Free of disease-causing organisms.

Asexual   Reproduction not involving union of two nuclei.

Association of American Pesticide Control Officials, Inc.   This association (AAPCO) is composed of officials charged by law with the active execution of the laws regulating sale of economic poisons.

Atomize   To break up a liquid into fine droplets by passing it through an apparatus under pressure.

Atropine sulfate or **atropine**   An antidote used by doctors to treat people or animals poisoned by organic phosphate and carbamate pesticides in an attempt to save their lives.

Attractants   Substances or devices capable of attracting insects or other pests to areas where they can be trapped or killed.

Attractive nuisance   A legal term for any object that might attract children or other persons to it and then might injure or hurt them as a result. Examples are sprayers, empty pesticide containers, or a bottle or food container filled with pesticides.

Auxin   A generic term for compounds characterized by their capacity to induce elongation in shoot cells. They resemble indole-3-acetic acid in physiological action.

Avicide   A substance to control pest birds. Generally not designed to kill but to repel or to so affect a few individuals that others are frightened away.

Avoid prolonged contact   Do not get exposed to a pesticide or breathe in the vapor (gas) for very long.

B

Back-siphoning   Fluid or spray material siphoning from the sprayer back to the original source.

Bacteria   Organisms (germs), some of which cause diseases in plants or animals. They are too small to be seen without a microscope.

Bactericide   Any chemical used to kill bacteria.

Bait   An edible material that is attractive to the pest. Normally contains a pesticide unless used as a prebait.

Bait shyness   The tendency for rodents, birds, or other pests to avoid a poisoned bait.

Balance of nature   The developing, evolving, and diversified life and state of adjustment (balance) all organisms have reached in relation to the environment. Nature is in a constant state of change and this constant evolutionary process is nature's way of trying to create a balance, which is never reached.

Band application   An application to a continuous restricted area such as in or along a crop row rather than over the entire field area.

Barrier application   The use of pesticide or another agent to stop pests from entering a container, area, field, or building.

Basal application   A treatment applied to the stems or trunks of plants at and just above the ground line.

Base   Alkali; opposite of acid.

Bed   1. A narrow flat-topped ridge on which crops are grown with a furrow on each side for drainage of excess water. 2. An area in which seedlings or sprouts are grown before transplanting.

Bed-up   To build up beds or ridges with a tillage implement.

Beetle   A hard-shelled insect. Examples: lady beetles, June bugs.

Beneficial   Useful or helpful to humans. A lady beetle is beneficial because it feeds on other insects that damage plants.

Benzenoid chemical   One having the benzene ring form of molecular structure (see also *Aromatics*).

Biennial   A plant that completes its life cycle in two years. The first year it produces leaves and stores food. The second year it produces fruits and seeds.

Bioactivity   Pertains to the property of affecting life.

Bioassay   The qualitative or quantitative determination of a substance by the systematic measurement of the response of living organisms as compared to measurement of the response to a standard or standard series.

Biocide   A chemical that has a wide range of toxic properties, usually to members of both the plant and animal kingdoms.

Bioconcentration   The process of a pesticide becoming concentrated in plants or animals. When a chemical increases in concentration at each succeeding link in the food chain.

Biological control   Control of pests by means of predators, parasites, and disease-producing organisms.

Biological insecticide   A biological agent such as *Bacillus thuringiensis,* which kills insects like a chemical insecticide and then rapidly dissipates in the environment.

Biotic   Relating to life.

Bipyridyliums   A group of synthetic organic pesticides, which includes the herbicide paraquat.

Bird repellent   A substance that drives away birds or discourages them from roosting.

Blanket application   Application of a chemical over an entire area.

Blight   A general term that may include spotting, discoloration, sudden wilting, or death of leaves, fruit, flowers, stems, or the entire plant.

Boom A section of pipe (or tubing) that connects several nozzles so that a pesticide can be applied over a wider area.

Boom sprayer An apparatus used for applying liquid pesticide. Consists of a tank, laterally extended pipe (boom) supporting and carrying the liquid spray nozzles, and a pump to produce the desired pressure on the liquid as it flows to the nozzles. The entire apparatus may be mounted on a tractor or a truck or supported by its own wheels.

Boot stage When the seed head of a grass begins to emerge from the sheath.

Botanical name A scientific name made up of the genus and species. It is universal and more reliable than common names.

Botanical pesticide A pesticide produced by and extracted from plants. Examples are nicotine, pyrethrum, strychnine, and rotenone.

Bract A specialized leaf or leaflike part usually at the base of a flower.

Brand The name, number, trademark, or designation of a pesticide or device made by the manufacturer, distributor, importer, or vendor. Each pesticide differing in the ingredient statement, analysis, manufacturer or distributor name, number, or trademark is considered as a distinct and separate brand.

Broadcast application An application over an entire area.

Broadleaf species Botanically, those plans classified as a dicotyledoneae; morphologically, those having broad, rounded, or flattened leaves as opposed to the narrow bladelike leaves of the grasses, sedges, rushes, and onions.

Broad-spectrum pesticide One that controls a wide range of pests when applied correctly.

Brood The offspring, all of approximately the same age, rising from a single species of organism.

Brush Economically useless woody plants, including a range in size from small shrubs to large trees; woody weeds.

Brush control Control of woody plants.

Bucket pump The simplest hydraulic sprayer is the bucket pump. It is a plunger pump adapted by a clamp or otherwise for use in an open pail and delivering the liquid through a spray nozzle at the end of a hose.

Bud spray A pesticide application to trees any time between the time the first color appears in the ends until the buds begin to open.

Buffer An adjuvant that allows mixing of pesticides of different acidity or alkalinity.

Bug bomb An aerosol can containing insecticides.

Buildup Accumulation of a pesticide in soil, animals, or in the food chain.

C

Calculate To figure out by mathematical process by working with numbers.

Calibrate To measure or figure out how much pesticide will be applied by the equipment to the target in a given amount of time.

Calibrated Checked, measured. Knowing how much of a pesticide chemical is being applied by each nozzle or opening of a sprayer, duster, or granular applicator to a given area, plant, or animal.

Cambium   The actively growing plant layer, beneath the bark, that produces all secondary tissues and results in growth in diameter.

Canceled   A pesticide use that is no longer registered as a legal use by the EPA. (*Note:* A cancellation order is less severe than a suspension order.)

Canister   A metal or plastic container filled with absorbent material that filters fumes and vapors from the air before they are breathed in by an applicator.

Canker   A definite, localized, dead, often sunken or cracked area on a stem, twig, limb, or trunk, surrounded by living tissues. Cankers may girdle affected parts resulting in a dieback starting in the tip.

Carbamates   A group of chemicals that are salts or esters of carbamic acid. Includes insecticides, herbicides, and fungicides.

Carbon dioxide   Heavy colorless gas that does not support combustion.

Carcinogen   A substance or agent capable of producing cancer.

Carcinogenic   The term used to describe the cancer-producing property of a substance or agent.

Carrier   1. The liquid or solid that is used to dilute the active ingredient in manufacturing a pesticide formulation. Examples: talc, petroleum solvents. 2. The material used to carry the pesticide to the target. Example: water in a hydraulic sprayer, air in a mist blower.

Cartridge   The part of a respirator that adsorbs fumes and vapors from the air before the applicator breathes them in.

Catalyst   A substance that speeds up the rate of a chemical reaction but is not itself used up in the reaction.

Caterpillar   The wormlike stage of moths and butterflies. Usually feeds on plants.

Cationic   An ion having a positive charge is a cation. When the surface-active portion of a surfactant molecule possesses a positive charge it is termed a *cationic surfactant.*

Causal organism   The organism that produces a given disease.

Caustic   A chemical that will burn if it gets on the skin.

Caution   A warning to the user of pesticide chemicals. Used on labels of pesticide containers having slightly toxic pesticides in toxicity category III as defined by the Federal Insecticide, Fungicide, and Rodenticide Act.

Cell   The structural and functional unit of all plant and animal life. The living organism may have from one cell (bacteria) to billions (a large tree).

Centigrade (C)   A thermometer scale in which water freezes at 0°C and boils at 100°C. To change to degrees Fahrenheit, multiply degrees centigrade by nine-fifths and add 32. Celsius (C) is the preferred term approved by the Ninth General Conference on Weights and Measures in 1948.

Certification   The recognition by a certifying agency that a person is competent and thus authorized to use or supervise the use of restricted use pesticides (FIFRA amended).

Certified applicator   Any individual who is certified to use or supervise the use of any restricted use pesticide covered by his or her certification as defined by the EPA.

cfs   Cubic feet per second of flow, as water in a stream.

Chemical   Often used in this book to mean pesticide chemical.

Chemical control   Using pesticides to control pests.

Chemically inactive   Will not easily react with any other chemical or object.

Chemical name   One that indicates the chemical composition and/or chemical structure of the compound being discussed.

Chemical reaction   When two or more substances are combined and, as a result, undergo a complete change to make new substances or materials.

Chemosterilant   A chemical compound capable of producing reproductive sterilization.

Chemotherapy   The treatment of a desired plant or animal with chemicals to destroy or control a pathogen without seriously harming the plant or animal.

Chlorates   Herbicides and defoliants. They act as contact poisons, are translocated, and may be adsorbed from the soil to kill both plant roots and tops. Chlorates cause chlorosis of leaves and a starch depletion in stems and roots when applied in less than lethal doses.

Chlorinated hydrocarbon   A chemical compound containing chlorine, carbon, and hydrogen. Aldrin, chlordane, DDT, dieldrin, heptachlor, lindane, methoxychlor, and toxaphene are chlorinated hydrocarbons.

Chlorosis   The yellowing of a plant's normally green tissue because of a partial failure of the chlorophyll to develop.

Cholinesterase   A body enzyme necessary for proper nerve function that is destroyed or damaged by organic phosphates or carbamates taken into the body by any path of entry.

Cholinesterase inhibitor   Any organophosphate, carbamate, or other pesticide chemical that can interrupt the action of enzymes that inactivate the acetylcholine associated with the nervous system.

Chronic effect   A slow and long-continued effect.

Chronic poisoning   Resulting from small repeated doses over long periods of exposure to a chemical.

Chronic toxicity   Ability to cause injury or death from a series of small doses applied (or received) over a long period.

-cide   Suffix meaning "to kill."

Circulate   To move completely through something in a path that returns to the starting point.

Classification   The process of assigning pesticides into groups according to common characteristics.

Climatic   Having to do with weather conditions, sunshine, temperature, rainfall, moisture in the air, and so on.

Clods   Hard lumps of dirt formed by plowing or cultivating soil.

Cluster break   The separation of pear or apple flower buds from each other in the cluster before blooming.

Codistillation   In the process of distillation, vapors are driven off by heat and then condensed. In codistillation, an additional material is carried off with the vapor.

Coloration   Certain white or colorless pesticides must be treated with a coloring agent that will produce a uniformly colored product not subject to change in color beyond the minimum requirements during ordinary conditions of marketing, storage, and use. Seed treatment materials are examples of pesticides that have been colored to indicate upon visual inspection what seed has been treated.

Combustible   Able to catch fire and burn.

Commercial applicator   A certified applicator (whether or not he or she is a private applicator with respect to some uses) who uses or supervises the use of any pesticide that is classified for restricted use for any purpose or on any property other than as provided by the definition of "private applicator."

Common exposure route   A likely way (oral, dermal, respiratory) by which a pesticide may reach and/or enter an organism.

Common name   Official common names of pesticides such as *atrazine, carbaryl, captan,* and *warfarin,* which when present must appear in the active ingredients list on container labels. Other unofficial common names given to a plant or animal. It is possible for two living things to have the same common name in different places. In addition, the same living thing may have several common names.

Compatibility   The ability of two or more substances to mix without objectionable changes in their physical or chemical properties or without reducing the effectiveness of any individual component.

Compatible   Two compounds are said to be compatible when they can be mixed without affecting each other's properties.

Competent   Properly qualified to perform functions associated with pesticide application, the degree of capability required being directly related to the nature of the activity and the associated responsibility.

Complete metamorphosis   The process of insect development involving egg, larva, pupa, and adult stages.

Compressed-air sprayer   Sprayer of usually 1- to 3-gallon capacity with extension rod, equipped with air pump to develop pressure, often with shoulder strap for carrying. Not suitable for spraying at heights over 6 to 8 feet.

Concentrate   Opposite of *dilute.* A liquid or dry formulation containing a high percentage of toxicant to save shipping and storage charges and yet be of convenient strength and composition for dilution.

Concentration   The amount of a substance contained in a unit volume, for example, 4 pounds per gallon; or the percentage of the substance, as in a 50% wettable powder.

Condemnation   The removal of a crop or product that does not meet the legal standards for tolerance on foods and is therefore not to be sold.

Conditions favorable to infection   The situation a disease organism must have to attack and grow on or in a plant or animal. Disease, injury, or destruction may follow. Some of the factors that can be important are temperature, humidity, and light intensity.

Confined areas   Rooms, buildings, and greenhouses with limited or inadequate ventilation.

Conidium (plural, conidia) A spore formed asexually, usually at the tip or side of a hypha (for example, the conidia of the apple scab fungus; causes secondary infection).

Conifers Trees and shrubs with needlelike leaves. Examples: pine, cedar, larch, spruce, and hemlock.

Contact herbicide A compound that kills primarily by contact with plant tissue rather than as a result of translocation. Only that portion of a plant contacted is directly affected.

Contact poison A pesticide that kills when it touches or is touched by a pest.

Contaminate To alter or render a material unfit for a specified use, by the introduction of a foreign substance (a chemical).

Contract An agreement with someone to do a job or perform a service.

Control To prevent from doing damage. To reduce or keep down the number of pests so that little disease, damage, or injury occurs to a crop or property.

Conventional Usual. Everyday. Customary.

Copper Compounds of cooper from one of the most useful groups of fungicides, and various forms are applied to plants, cordae, fabric, and leather and are used as algicides, seed disinfectants, and wood preservatives. Copper materials have also been used as insecticides and repellents for certain insects.

Corm The fleshy, bulblike base of a stem, as of a crocus.

Corrosion The effect of being worn down or eaten away.

Corrosive Having the power to eat away slowly. Some pesticides eat or wear away rubber hoses, nozzles, and other parts of spray machinery.

Cotyledon The first leaf, or pair of leaves, of the embryo of seed plants.

Coupling agent A solvent that has the ability to solubilize or increase the amount of solubility of one material in another.

Coverage The amount of spread of a pesticide over a surface.

Creeping perennial A perennial plant that reproduces by both seed and asexual means (rhizomes, stolons, and so on).

Critical areas Places where pests are most likely to be troublesome.

Crook stage That stage of plant growth as it emerges from the soil, for example, bean seedlings that have broken through the soil but before the stem becomes erect.

Crop A plant growing where it is desired.

Crop tolerance The degree or the ability of the crop to be treated with a chemical but not injured.

Cross contamination When one pesticide gets into or mixes with another pesticide accidentally. This usually occurs in a pesticide container or in a poorly cleaned sprayer.

Cross-resistant When a pest population that has become resistant to one pesticide also becomes resistant to other chemically related pesticides.

Crown The point where stem and root join in a seed plant. This term is also used to describe the foliage and branches of trees.

Crucifers   Plants belonging to the mustard family including mustard, cabbage, turnips, radish, and the like.

Cube   The root of a tropical plant (*Lonchocarpus* spp.) valued as a source of rotenone. Obtained commercially mostly from Peru. (Pronounced koo-bay.)

Cucurbits   Plants belonging to the gourd family, including pumpkins, cucumbers, squash, and the like.

Culm   The jointed stem of a grass, which is usually hollow except at the nodes or joints.

Cultural control   Control measures including modifications of the planting, growing, cultivating, and harvesting of crops aimed at prevention of pest damage rather than destruction of an existing infestation.

Cumulative effect   The result of some poisons that build up or are stored in the body so that small amounts eaten or contacted over a period of time can sicken or kill an animal or person.

Cumulative pesticides   Chemicals that tend to accumulate or build up in the tissues of animals or in the environment (soil, water). Examples of chemicals that accumulate are anticoagulant rodenticides, mercury compounds, and thallium sulfate.

Curative   Having the power to heal or cure.

Curative pesticide   A pesticide that can inhibit or eradicate a disease-causing organism after it has become established in the plant or animal.

Current   Now in use.

Custom applicator   Any person who for hire, by contract or otherwise, applies by aerial, ground, or any hand or mechanical equipment, pesticides to any waters, lands, plants, farm structures, or animals.

Cuticle   A nonliving outer layer covering all or part of an organism.

Cutin   Waxy, fatty material that forms the cuticle, or waxy layer, covering plant surfaces such as leaves.

Cut-surface application   Treatments made to frills or girdles that have been made through the bark into the wood of a tree.

Cyclodienes   Compounds with a ring structure such as aldrin, chlordane, heptachlor, and so on.

D

Damaging   Harmful. Injurious.

Damping off   Rotting of seeds in the soil, or the sudden wilting and death of seedlings anytime after germination. Usually caused by fungi that attack the seed or stem of the plant.

Danger   Risk, hazard.

Days to harvest   The minimum number of days permitted by law between the final pesticide application and the harvest of crops. (Same as *preharvest interval*.)

Days to slaughter   The minimum number of days, established by law, between the final pesticide application and the date an animal is to be slaughtered.

Debris   Trash or unwanted plant parts or remains.

Deciduous plants   Those plants that are perennial in habit but lose their leaves during the winter.

Decontaminate   To make safe by removing any pesticide from equipment or other surfaces as directed on a pesticide label or by an agricultural authority.

Deflocculating agent   A substance that prevents rapid precipitation (settling out) of solids in the liquid in the spray tank.

Defoliant   A preparation intended for causing leaves to drop from crop plants such as cotton, soybeans, or tomatoes, usually to facilitate harvest.

Defoliate   To lose or become stripped of leaves.

Degradability   The ability of a chemical to decompose or break down into less complex compounds or elements.

Degradation   Breakdown of a complex chemical by the action of microbes, water, air, sunlight, or other agents.

Degrade   Decomposure or break down.

Degree of exposure   The amount or extent to which a person has been in contact with a toxic pesticide.

Delayed action   With some herbicidal chemicals delayed response is expected. Considerable time may elapse before maximum effects can be observed. Usually treated plants stop developing soon after treatment, then gradually die.

Delayed dormant spray   A spray applied to deciduous trees during the period from the first swelling of the buds until the first color starts to show.

Deleterious   Harmful. Injurious.

Density   The size of a population within a given unit of space.

Deoxygenation   Depletion of oxygen.

Depleted   Exhausted. Reduced.

Deposit   The amount of pesticide laid down immediately following an application.

Dermal   Of or pertaining to the skin.

Dermal toxicity   The passage of pesticides into the body through the unbroken skin. Most pesticides pass through unbroken skin to some extent. Many are not adsorbed readily unless in certain solvents. However, many, including most organophosphates, will pass through unbroken skin as technical material and in all formulations. Dermal exposure results from spillage onto skin and clothing (including gloves and other protective equipment), from drift, or from damaged or improperly maintained equipment. It is important to be aware of hazards resulting from the dermal toxicity of the materials you use. This is the greatest hazard to people handling pesticides.

***Derris* species**   Formerly the most important plant sources of rotenone-containing roots. Grown principally in Malaya and East Indies. U.S. industry now depends on *Lonchocarpus* species from South America, mostly Peru.

Desiccant   A compound that promotes drying or removal of moisture from plant tissues. (See *defoliant* for difference in these two terms.)

Desiccation   Dehydration or the removal of tissue moisture by chemical or physical action. Drying chemicals that promote desiccation are *desiccants*. They are used primarily for preharvest drying of actively growing plant tissues when seed or other plant parts are developed but only partially mature, or for drying of plants that normally do not shed their leaves, such as rice, corn, small grains, and cereals.

Detergent   A chemical (not soap) having the ability to remove soil and grime. Detergents can be used as surfactants in some pesticide sprays.

Deteriorate   To break down or wear away. To decay or grow worse.

Detoxify   To make harmless.

Diagnosis   Identification of the nature and cause of a problem.

Dicot (dicotyledon)   A plant that has two seed leaves or cotyledons; the broadleaf plants.

Dieback   Death of branches or shoots of plants from the tips back toward the trunk or stem.

Diluent   A material, liquid or solid, serving to dilute the technical toxicant to field strength for adequate plant coverage, maximum effectiveness, and economy.

Dilute   To make less concentrated. To mix a pesticide chemical with water, oil, or other material before it can be safely and correctly used.

Dilution rate   The amount of diluent that must be added to a unit of a pesticide to obtain the desired dosage.

Dinitros   A common designation for dinitrophenol contact pesticides.

Dip tank   A large tank used for dunking animals in a pesticide to protect against or destroy ectoparasites.

Dip treatment   The application of a liquid chemical to an organism by momentarily immersing it, wholly or partially under the surface of the liquid, so as to coat the organism with the chemical.

Direct supervision   Unless otherwise prescribed by its labeling, a restricted use pesticide must be applied by a certified applicator or under a certified applicator's direct supervision.

Directed application   An application to a restricted area such as a row, bed, or at the base of plants.

Discing   Breaking up the top few inches of the soil with a disc. This destroys many weeds and levels the soil for planting or treating with a fumigant or other pesticide.

Disease   A condition in which any part of a living organism is abnormal as the result of an infectious or noninfectious agent.

Diseased   Unhealthy or abnormal. A plant or animal suffering from an infection by a fungus, bacteria, nematode, virus, other pest, or poor growing conditions that disturb normal activity. As opposed to an insect injuring a plant or animal by chewing or sucking.

Disinfectant   Similar to bactericide. Chemicals that kill or inhibit growth of bacteria.

Disinfestant   An agent that kills or inactivates organisms present on the surface of the plant or plant part in the immediate environment. In the case of seeds (seed disinfestant) or soils (soil disinfestant).

Disperse   To spread out or scatter. To separate and move apart.

Dispersing agent   A material that reduces the cohesive attraction between like particles. Dispersing and suspending agents are added during the preparation of wettable powders to facilitate wetting and suspension of the active ingredient.

Disposal The process of discarding or throwing away unused spray material, surplus pesticides, and pesticide containers.

Dissipate To get rid of by scattering or spreading out. Gases and vapors dissipate through the air.

Dissolve Usually refers to getting solids into suspension and/or solution that is necessary for uniform application results.

Distribution The amount and way in which a pesticide chemical is spread when applied to a plant, animal, or other surface. Also, part of the United States or the world in which an insect, fungus, or other pest is found.

Domestic animal Tame animal used for human's benefit. Example: cow, sheep, horse.

Dormant Inactive. Not growing. In the case of plants, it is after the leaves fall or growth stops and before the buds open in the spring. Also, the period when seeds fail to sprout due to internal controls.

Dormant spray A spray applied when plants are in a dormant condition. The temperature should be as high as 40° to 45° F for application.

Dose, dosage A dose is a measured quantity, as of medicine, taken at one time or in one period of time. Dosage, therefore, is the amount of medicine in a dose. Used in connection with pesticides, it refers in general to rate of application.

Dose ratio Ratio between successively increasing doses.

Douse Drench. Soak.

Downwind Direction toward which the wind is blowing.

Drench Saturation (thorough soaking) of the soil with a pesticide, or an oral treatment of an animal.

Drench treatment The application of a liquid chemical to an area until the area is completely soaked.

Drift Movement of spray or dust material by wind or air currents outside the intended area, usually as fine droplets, during or shortly after application.

Droplet size The diameter size of the droplet, usually measured in micrometers or microns.

Dry flowable Pesticide formulated into dust-free, highly concentrated, water-dispersible granules.

Duckfoot cultivator A field cultivator equipped with small sweep shovels.

Dust A dry mixture consisting of the pesticide and some inert carrier such as clays, talc, attapulgite clays, walnut shell, calcium carbonate, and others as carriers or diluents to facilitate application.

E

Early cover spray A pesticide chemical applied to fruit soon after the petals fall.

Ecology The science concerned with the interrelationships of organisms and their environments.

Economic injury level Lowest pest population density that will cause economic damage.

Economic poison Any substance or mixture of substances intended for preventing, destroying, repelling, or mitigating any insects, rodents, nematodes, fungi, or weeds or any other form of life declared to be a pest and any substance or mixture of substances intended for use as a plant regulator, defoliant, or desiccant.

Economic threshold   That point of pest infestation where application of a control measure would prevent an increasing pest population from reaching the economic injury level and would return more money than the cost of the control procedure.

Ecosystem   A community of life and its environment functioning as a unit in nature.

Ectoparasites   Plants and animals that live and feed on the outside of an animal or plant. Most are annoying, cause injury, or carry disease organisms. Examples: lice, fleas, some fungi.

Edible   Safe to eat. A food.

Eelworms   See *nematode*.

Efficacy   Capacity for serving to produce effects; effectiveness.

Electrostatic charging   Providing a spray or dust particle with an electric charge as it leaves the nozzle.

Emergence   The action of a young plant breaking through the surface of the soil or of an insect coming out of an egg or pupa.

Emersed aquatic   A plant rooted below the water surface. However, the main plant parts are above the surface.

Emetic   A substance used to make humans or animals vomit.

Emulsifiable concentrate   A concentrated pesticide formulation usually consisting of a pesticide dissolved in an organic solvent and a surface-active agent that permits dispersion of the total formulation in water.

Emulsifier   A chemical that helps one liquid form tiny droplets and thus remain mixed in another liquid. It is used to form a stable mixture between two liquids that usually would not mix. Example: oil in water.

Emulsify   To make into an emulsion. When small drops of one liquid are finely dispersed (distributed) in another liquid, an emulsion is formed. The drops are held in suspension by an emulsifying agent, which surrounds each drop and makes a coating around it.

Emulsifying agent   A material that facilitates the suspending of one liquid in another; for example, oil dispersed in water.

Emulsion   A mixture in which one liquid is suspended as minute globules in another liquid; for example, oil in water.

Encapsulation   A method of formulating pesticides, in which the active ingredient is encased in a material (often polyvinyl) resulting in sustained pesticidal release and decreased hazard. Also, a method of disposal of pesticides and pesticide containers by sealing them in sturdy, waterproof, chemical-proof containers which are then sealed in thick plastic, steel, or concrete to resist damage of breakage so that contents cannot get out.

Encephalitis   Inflammation of the brain. Virus-caused encephalitis can affect humans and horses. These viruses are transmitted by mosquitoes.

Endangered species   A group of organisms on the brink of extinction.

Endemic   Referring to a disease that is regularly present in a given region or area but does not necessarily cause significant loss. Such a disease may become *epidemic*.

Endemic disease   A disease peculiar to a particular locality.

Endotoxin   Any of a group of toxic substances found in certain disease-producing bacteria and liberated by the disintegration of the bacterial cell.

Entomology   A branch of science that studies insects.

Environment   Water, air, land, all plants, humans, and other animals living therein, and the interrelationships that exist among them.

Environmental manipulation   Skillful or artful management or use of the environment in such a way as to assist people in pest control.

Enzyme   A natural substance that regulates the rate of a reaction but which itself remains chemically unchanged.

EPA   Environmental Protection Agency. Responsible for the protection of the environment in the United States.

EPA establishment number   A number assigned to each pesticide production plant by the EPA. The number indicates the plant at which the pesticide product was produced and must appear on all labels of that plant.

EPA registration number   A number assigned to a pesticide product by the EPA when the product is registered by the manufacturer or its designated agent. The number must appear on all labels for a particular product.

Epidemic   A sudden widespread increase in the incidence of a disease or organism.

Epidermis   The outer cellular tissue of an animal or plant.

Epinasty   Increased growth on the upper surface of a plant organ or part (especially of leaves) causing it to bend downward.

Epiphyte   A plant that grows upon another plant (or on a building or telegraph wire), which it uses as a mechanical support but not as a food source. Some (such as Spanish moss) may harm the plants they grow on, however, by excluding light or smothering them. Most orchids are epiphytes.

Eradicant   A chemical used to eliminate a pest from a plant or a place in the environment.

Eradication   The complete elimination of weeds, insects, disease organisms, or other pests from an area.

Erode   To wear away.

Escape   An organism in a treated area that either missed treatment or failed to respond to treatment in the same manner as other treated organisms.

Essential   Necessary. A must.

Ester   A compound formed by the union of an organic acid and an organic base (an alcohol). An example is 2,4-D and isooctyl alcohol to form the isooctyl ester of 2,4-D.

Ethane   A colorless gas present in the gases flowing from gas wells and in refinery off-gases. One of the major raw materials for producing ethylene. (Ethane, propane, and butane comprise LPG.)

Ethers   Organic compounds in which two hydrocarbon radicals are joined through an atom of oxygen.

Ethylene   An olefin; a basic chemical. The most widely used petrochemical building block in the world. A colorless, sweet-smelling gas made principally from ethane and propane in the United States and from naphtha in other countries.

Ethylene oxide   Used for making chemical derivatives such as glycol ethers, ethanolamines, and synthetic detergents.

Evaporate   To form a gas and disappear into the air; to vaporize.

Evaporation   The process of a solid or liquid turning into a gas.

Evergreen   Plants that retain their functional leaves throughout the year.

Exclusion   Control of disease by preventing its introduction (for example, by quarantines) into disease-free areas.

Exemption   An exception to a policy, rule, regulation, law, or standard.

Exoskeleton   The segmented, external skeleton of an insect; the insect's "skin."

Exposed surfaces   Surfaces that can be successfully attacked by a pest. Surfaces that need to be protected by a pesticide chemical or other agent.

Exposure   When contact occurs with a pesticide through skin (dermal), mouth (oral), lungs (inhalation/respiratory), or eyes.

Exposure period   The length of time something has been under attack by a pest. Also, the length of time a pest or nontarget organism is in contact with a pesticide chemical.

Extender   Anything added to a pesticide to extend its activity or effectiveness.

Exudate   Matter diffused from within a cell.

F

Face shield   A transparent piece of protective equipment used by a pesticide applicator to protect his or her face from exposure to pesticides.

Facilitate   Make it easier to do or to understand something.

Facultative   Incidental, not necessarily compelled to live under one type of environment.

Fahrenheit (F)   A thermometer scale that marks the freezing point of water at 32° F and the boiling point at 212° F.

Fatal   Deadly. Able to cause death.

FDA   Food and Drug Administration, U.S. Department of Health, Education, and Welfare.

Feeder roots   The hairlike roots of a plant that take up water and most of the food materials needed by a plant.

FEPCA   Federal Environmental Pesticide Control Act of 1972. Greatly revised FIFRA and mandated for the newly created U.S. Environmental Protection Agency (EPA) to direct its efforts toward protecting the environment from the unreasonable adverse effects of pesticides.

Fetotoxicity   The ability to cause toxic effects to a developing embryo or fetus.

Field margin   Edge of a field.

FIFRA   Federal Insecticide, Fungicide, and Rodenticide Act (Amended).

Filler   A diluent in a powdered form.

Filter   To screen out the unwanted material; clean by straining out the undesirable parts, or a piece of equipment for doing this.

Final treatment   Last pesticide chemical application before harvesting a crop or slaughtering an animal.

Finite tolerance   The maximum amount of pesticide that can legally remain on a food or feed crop at harvest after the pesticide has been directly applied to the crop.

First aid   The first effort to help a victim while medical help is on the way.

Flaccid   Plant tissues becoming limp.

Flag stage   Stage of growth in cereals and other grasses at which the sheath and leaf have been produced from which the head will emerge.

Flexible   Easy to bend.

Flowable pesticide   Very finely ground solid material that is suspended in a liquid and usually contains a high concentration or large amount of the active ingredient and must be mixed with water when applied.

Fluid   In a liquid form.

Fluorine compounds   This group includes a number of fluorine salts, many of which are highly toxic to warm-blooded animals. Sodium fluoride was the first to be used as an insecticide being applied to cockroaches, poultry lice, and ants.

Fly   An insect, usually with one pair of wings.

Foaming agent   A material that causes a pesticide mixture to form a thick foam. It is used to reduce drift.

Fog   Particles between 0.1 and 50 micrometers (μ) in diameter that make a fine mist.

Fog application   The application of a pesticide as a fine mist for the control of pests.

Fogger   An aerosol generator. A piece of pesticide equipment that breaks some pesticides into very fine droplets (aerosols or smokes) and blows or dries the "fog" onto the target area.

Foliage   The leaves, needles, and blades of plants.

Foliar sprays   Droplets of a pesticide applied to leaves, needles, and blades of plants.

Food chain   Phrase that describes how all living organisms are linked together and depend on each other for food; that is, plant eaters, plant and meat eaters, and meat eaters.

Formula   A brief way of writing a complicated idea by using abbreviations and symbols.

Formulation   The pesticide product containing the active ingredient, the carrier, and other additives required to make it ready for sale.

Frill   One or a series of overlapping cuts made through bark into the sapwood of unwanted trees or brush into which herbicides are applied.

Fruit set   The number of apples or other fruit that begin to grow after the petals fall.

Fuel oils   Petroleum fractions used for burning are used also as solvents. Distillate fuel oil has been recommended for application by itself against mosquito larvae.

Full coverage spray   This term on a label signifies that the total volume of spray to be applied will thoroughly cover the crop being treated to the point of runoff or drip.

Fuller's earth   A silicate mineral used as an absorbent base or dust diluent. It is flowable and possesses excellent grindability, acting as a grinding aid.

Fumes   A smoke, vapor, or gas.

Fumigant   The AAPCO has adopted this definition: "A substance or mixture of substances that produce gas, vapor, fume, or smoke intended to destroy insects, bacteria or rodents." Fumigants may be volatile liquids and solids as well as substances already gaseous. They may be used to disinfest the interiors of buildings, objects, and materials that can be enclosed so as to retain the fumigant, and the soil where crops are valuable enough to warrant the treatment.

Fungi   Small, often microscopic plants without chlorophyll (green coloring). Fungi produce tiny threadlike growths. They grow from seedlike spores. Some fungi can infect and cause disease in plants or animals; other fungi can attack and destroy nonliving things.

Fungicide   A chemical used to kill fungi; a compound used to destroy or inhibit fungi (usually plant diseases).

Fungistat   A material to prevent growth or multiplication of fungi without necessarily destroying the organisms after the latter have gained a foothold.

G

Gall   A growth, lump, or swelling of plant tissue caused by mites, insects, bacteria, nematodes, viruses, fungi, or chemicals.

Gamma irradiation   The use of rays from a radioactive source.

Gas mask   A device that filters out chemicals in the spray, dust, or gas form from air breathed by the wearer. A full-face gas mask must be worn to protect from gases; it should be equipped with adequate canisters of absorbent materials (or with oxygen supply). Simple respirators protect from spray and dust without covering the eyes, but not from poisonous gases.

General use pesticide   A pesticide that can be purchased and used by the general public without undue hazard to the applicator and the environment as long as the instructions on the label are followed carefully (see *restricted use pesticide*).

Germicide   A chemical or agent that kills microorganisms such as bacteria or prevents them from causing disease.

Germination   Process of germinating or beginning of vegetative growth. Often refers to the beginning of growth from a seed.

Girdling   Practice of completely removing a band of bark all the way around a woody stem.

gpa (G.P.A.)   Gallons per acre.

gpm   Gallons per minute. A measure of liquid moved by a pump.

Gram   A metric weight measurement equal to 1/1000 kilogram; approximately 28.5 grams equal 1 ounce.

Granary   A storage area for threshed grain.

Granular pesticide   A pesticide chemical mixed with or coating small pellets or a sandlike material. They are applied with seeders, spreaders, or special equipment. Granular pesticides are often used to control or destroy soil pests.

Granule   A type of formulation in which the active ingredient is mixed with, adsorbed, absorbed, or pressed on an inert carrier forming a small pellet.

Grassy weeds   Weeds belonging to the grass family, characterized by round jointed stems, sheathing leaves that are parallel veined, and flowers borne in spikelets.

Groundwater   Waterways located beneath the soil surface from which wells get their water.

Growing season   In general the number of days from the last killing frost in the spring until the first killing frost in the fall. This varies according to the resistance of each crop to freezing temperatures.

Growth regulator   An organic substance effective in minute amounts for controlling or modifying plant processes; organic compounds, other than nutrients, which in small amounts promote, inhibit, or otherwise modify any physiological process in plants.

Growth stages of cereals   1. Tillering stage: when a plant produces additional shoots from a single crown. 2. Jointing stage: when the internodes of the stem are elongating and the nodes can be felt by pinching the stem. 3. Boot stage: when leaf sheath swells due to the growth of developing spike or panicle. 4. Heading stage: when seed head is emerging from the sheath.

Grubs   The larvae of certain beetles, wasps, bees, and ants.

H

Habitat   Physical place where an organism lives.

Hand duster   In a hand duster a plunger expels a blast of dust-laden air. The dust chamber may be at the end of the plunger tube itself, or an enlargement at the end, or it may be located below the plunger tube.

Hand sprayer   Small, portable, pesticide sprayers that can be carried and operated by a person.

Hard water   Water with minerals such as calcium, iron, and magnesium dissolved in it. Some pesticides added to hard water will curdle or settle out.

Harvest aid   A material used to remove the leaves from cotton plants, kill potato vines, and in any other way facilitate machine harvesting of a crop. (See *defoliant, desiccant.*)

Harvest date   A day a crop is removed from its site of growth, as from a tree, bush, or vine, or cut as in the case of alfalfa. (*Note:* "Removed from" does not refer to when a crop is removed from the field, but rather to when the crop is picked, cut, dug up, and the like.)

Harvest interval   Period between last application of pesticide to a crop and the allowable time for harvest.

Hazard   The probability that injury will result from use of a substance in a proposed quantity and manner. The sum of the toxicity plus the exposure to a pesticide.

Hazardous   Dangerous, risky. Pesticide chemicals that may cause injury or death if not used as directed on the label.

Hectare   In the metric system, a land measure equal to 100 ares, or 10,000 square meters. One hectare is equivalent to 2.471 acres.

Herbaceous plant   A vascular plant that remains soft or succulent and does not develop woody tissue.

Herbicide   A pesticide used for killing or inhibiting plant growth. A weed or grass killer.

High-pressure sprayer   Same as *hydraulic sprayer.*

Highly toxic   Substances are considered highly toxic by law (1) if the LD_{50} of a single oral dose is 50 milligrams or less per kilogram of body weight (see *LD_{50}*); (2) if LC_{50} of toxicity by inhalation is 2000 mcg or less of dust or mist per liter of air or 200 ppm or less by volume of a gas or vapor when administered by continuous inhalation for one hour to both male and female rats or to other rodent or nonrodent species if it is reasonably foreseeable that such concentrations will be encountered by people; or (3) if LD_{50} of toxicity by skin absorption is 200 milligrams or less per kilogram of body weight when administered by continuous contact for 24 hours with the bare skin of rabbits or other rodent or nonrodent species as specified.

Highly volatile   A liquid that quickly forms a gas or vapor (evaporates) at room temperature.

High-volume sprays   Spray applications of more than 50 gallons per acre.

Homolog   A series of compounds that express similarities of structure but regular differences of formulae.

Hormone   A naturally occurring substance in plants or animals that controls growth or other physiological processes. It is used with reference to certain synthetic chemicals that regulate or affect growth activity.

hp   Horsepower.

Host   Any plant or animal on or in which another lives for nourishment, development, or protection.

Humidity   Refers to the dampness or amount of moisture in the air.

Hydrated   The addition of water (chemically combined).

Hydraulic   Pertaining to water or other liquids.

Hydraulic agitator   A device that keeps the tank mix from settling out by means of water flow under pressure.

Hydraulic sprayer   A machine that applies pesticides by using water at high pressure and volume to deliver the pesticide to the target. Same as *high-pressure sprayer.*

Hydrocarbon   A chemical whose molecules contain only carbon and hydrogen atoms. The simplest hydrocarbon is methane (natural gas), whose molecules each contain one atom of carbon and four atoms of hydrogen. Crude oil is largely a mixture of hydrocarbons.

Hydrogen-ion concentration   A measure of the acidity. The hydrogen-ion concentration is expressed in terms of the pH of the solution. For example, a pH of 7 is neutral, from 1 to 7 is acid, and from 7 to 14 is alkaline.

Hydrolysis   The splitting of a substance into the smaller units of which it is composed by the addition of water elements.

Hydrophilic   A substance or system that attracts or is attracted to water is hydrophilic in nature.

Hydrophobic   A substance or system that repels or is repelled by water is hydrophobic in nature.

Hygroscopic   Substances capable of absorbing water from the atmosphere under normal conditions of temperature, pressure, and humidity, are called hygroscopic substances.

Hypertrophy   Abnormal growth and hypoplasia; underdevelopment; stunting.

Hypha (plural, hyphae)   One of the threadlike elements of the mycelium; a tubular filament.

Hypo   A prefix denoting a deficiency, lack, or less than the normal or desirable amount.

Hypo-hatchet   An instrument used to inject a pesticide directly into the woody part of a plant.

I

IGR   Insect growth regulator.

Illegal residue   Residue that is in excess of a preestablished government-enforced safe level.

Imminent hazard   A situation that exists when the continued use of a pesticide would likely result in unreasonable, harmful effects on the environment.

Immune   A state of not being affected by disease or poison; exempt from or protected against.

Impermeable   Not capable of being penetrated. *Semipermeable* means permeable to some substances but not to others.

Impervious   Hard to penetrate or soak through. The condition of soil when it will not soak up water.

Inactive   Not involved in the pesticide action; not reacting chemically with anything.

Incinerator   A special high-heat furnace or burner that reduces everything to a nontoxic ash or vapor. Used for disposing of some highly toxic and moderately to slightly toxic pesticides.

Incompatible   Not capable of being mixed or used together; in the case of pesticides, the effectiveness of one or more is reduced, or they cause injury to plants or animals.

Incorporate into soil   The mixing of a pesticide into the soil by mechanical means.

Induce vomiting   To make a person or animal throw up.

Inert ingredient   An inactive ingredient. An ingredient in a formulation which has no pesticidal action.

Infection   The development and establishment of a pathogen (for example, a bacterium) in its host that will produce a disease.

Infectious disease   A disease caused by a pathogen that multiplies and can be transmitted from plant to plant.

Infest   To be present in number (for example, insects, mites, nematodes, bacteria, fungi) on the surface matter or in the soil. Do not confuse with *infect* (infection), which applies only to living, diseased plants or animals.

Infestation   Pests that are found in an area or location where they are not wanted.

In-furrow   An application to or in a furrow in which a crop is planted.

Ingested   Taken into the digestive system.

Ingredients   The simplest constituents of the economic poison that can reasonably be determined and reported.

Ingredient statement   The part of the label on a pesticide container that gives the name and amount of each pesticide chemical and the amount of inactive material in the mixture.

Inhalation   To take air into the lungs; to breathe in.

Inhalation toxicity   Poisoning through the respiratory system.

Inhibitor   A chemical, usually of the regulator type, that prevents or suppresses growth or other physiological processes in plants. 2-amino- 1, 2, 4-triazole (ATA) is an example of a growth inhibitor.

Inject   To force a pesticide chemical into a plant, animal, building or other enclosure, or the soil.

Injurious   Harmful. Can or will cause damage.

In lieu of   Instead of. In place of.

Inoculation   The introduction of an infective agent (the inoculum) into or on living tissues producing a specific disease.

Inorganic compounds   Those compounds lacking carbon.

Insect   Any of the numerous small invertebrate animals generally having segmented bodies and for the most part belonging to the class *Insecta,* comprising six-legged, usually winged forms.

Insect growth regulator   A synthetic, organic pesticide that mimics insect hormonal action so that the exposed insect cannot complete its normal development cycle, and dies without becoming an adult.

Insecticide   A substance or mixture of substances intended to prevent or destroy any insects that may be present in any environment.

Instar   The form of an insect between molts, numbered to designate the various periods; for example, the first instar is a stage between the egg and the first molt.

Integrated pest management (IPM)   A system in which two or more methods are used to control a pest. These methods may include cultural practices, natural enemies, and selective pesticides.

Intermittently   Starting and stopping from time to time.

Internally   Inside. See *take internally.*

Interval   Period of time. The time between two applications. The distance between two points.

Intradermal   Within the skin.

Inversion   When temperature increases with elevation from the ground, an inversion condition exists.

Invert emulsion   One in which oil is the continuous phase and water is suspended in it. Makes a mixture about the consistency of mayonnaise.

Ion   An electrically charged atom or group of atoms. Ions may be either positively (cations) or negatively (anions) charged.

Ionic surfactant   One that ionizes or dissociates in water.

IPM   Integrated pest management.

Irritating   Annoying. Making a person or animal uncomfortable by burning, stinging, tickling, making the eyes water, and the like.

Isomers   Two or more chemical compounds having the same structure but different properties.

J

Jet (hydraulic) agitator   A device that keeps a tank mix from settling *out of suspension* by means of water flowing under pressure (see *hydraulic agitator*).

Jointing stage   The elongation of the areas between the joints or nodes of a grass or cereal stem to elevate and expose the young seed head.

Jointly liable   When two or more persons or companies share legal responsibility for negligence.

Jurisdiction   The extent or range of judicial or other authority.

Juvenile hormone   A specific group of complex organic substances that, as hormones, regulate the development of larval characteristics in insects. These are internally secreted by the endocrine glands in insects; however, synthetic chemical analogs are being developed as insecticides.

K

kg   The abbreviation for kilogram.

Kilogram   A metric measurement of weight equivalent to 1000 grams or approximately 2.2 pounds.

Knapsack duster   A duster carried on the back. It is operated by a bellows on top of a cylindrical dust container, the bellows being actuated by a hand lever at the side of the operator.

Knapsack sprayer   A sprayer that can be strapped on the back and used to apply liquid pesticide chemicals. The attached hose has a nozzle at the tip that can be aimed at the spot to be treated.

L

Label   All written, printed, or graphic matter on or attached to the economic poison or the immediate container thereof, and the outside container or wrapper to the retail package of the economic poison.

Labeling   All information and other written, printed, or graphic matter upon the economic poison or any of its accompanying containers or wrappers to which reference is made on the label or in supplemental literature accompanying the economic poison.

Labile   Easily destroyed.

Lactating animal Any animal that is producing milk.

Lactation The period during which an animal produces and secretes milk.

Larva The wormlike or grublike immature or growing stage of an insect. It hatches from an egg and later goes into a resting stage called the pupa. The larva looks very different from the adult. Many insects cause most or all of their damage as larvae. Example: caterpillar.

Larvicide An insecticide used to kill larvae of insects.

Latent Dormant.

Lateral movement Chemical movement in a plant or in the soil to the side or horizontal movement in the roots or soil layers.

Lay-by treatments Applications of pesticides after the last cultivation.

LC_{50} A means of expressing the toxicity of a compound present in air as a dust, mist, gas, or vapor. It is generally expressed as micrograms per liter as a dust or mist but in the case of a gas or vapor as ppm. The LC_{50} is the statistical estimate of the dosage necessary to kill 50% of a very large population of the test species, through toxicity on inhalation under stated conditions by law, the concentration that is expected to cause death in 50% of the test animals so treated.

LD_{50} A common method of expressing the toxicity of a compound. It is generally expressed as milligrams of the chemical per kilogram of body weight of the test animal (mg/kg). An LD_{50} is a statistical estimate of the dosage necessary to kill 50% of a very large population of the test species under stated conditions (for example, single oral dose of aqueous solution), or by law, the dose that is expected to cause death within 14 days in 50% of the test animals so treated. If a compound has an LD_{50} of 10 mg/kg it is more toxic than one with an LD_{50} of 100 mg/kg.

Leaching Movement of a substance downward or out of the soil in or with water as the result of water movement.

Legal residue Residue that is within safe levels according to the regulations.

Legume One of a family of plants called *Legumenosae*. Examples: peas, beans, alfalfa, and soybeans.

LEL Lowest effect level. In a series of dose levels tested, it is the lowest level at which an effect is observed in the species tested.

Lesion A small diseased or abnormal area on a plant or an animal. It can be a spot, scab, canker, or blister, often caused by fungi, bacteria, viruses, or nematodes.

Lethal Fatal or deadly.

Liability Legal responsibility for actions performed.

Licensing The issuing of a certificate of permission from a constituted authority to carry on a business, profession, or service.

Life cycle The complete succession of developmental stages in the life of an organism.

Limitation Restriction. The most that is allowed.

Linear feet Running feet. Example: The length of a field is measured in linear feet.

Liter A unit of volume in the metric system equal to a little more than a quart.

Localized Limited to a given area or part. Something that occurs in a small area or on a certain part of a plant or animal.

Lodged Flattened or bent over and tangled. Refers to hay, grain, corn, and other plants that fall over if a disease, insects, or a storm damages them.

Low concentrate solution A solution that contains a low concentration or a small amount of active ingredient in a highly refined oil. The solutions are usually purchased as stock sprays meant for use in aerosol generators.

Low-pressure boom sprayer A machine that can deliver low to moderate volumes of pesticide at pressures of 30 to 60 psi. The sprayers are most often used for field and forage crops, pastures, and rights-of-way.

Low volatile A liquid or solid that does not evaporate quickly at normal temperatures.

Low-volatile ester An ester with a high molecular weight and a low vapor pressure.

Low-volume spray A spray application of 0.5 to 5.0 gallons per acre.

M

Macro A combining form meaning *large*.

Macroscopic Visible to the naked eye without the aid of a microscope.

Maggot Larval stage of flies.

Mammals Warm-blooded animals that nourish their young with milk; their skin is more or less covered with hair.

Mandibles The first pair of jaws in insects, stout and toothlike in chewing insects.

Marine Of the ocean or sea; having to do with plants and animals that live in, on, or around an ocean or sea.

Mass median diameter (MMD) The diameter at which half of the spray mass is in droplets of larger diameter and half is in smaller droplets. It is basically equal to volume mean diameter (VMD).

Material A substance; often used to mean a pesticide, pesticide formulation, pesticide chemical, active ingredient, or additive ingredient.

Maximum dosage The largest amount of a pesticide that is safe to use without resulting in excess residues or damage to whatever is being protected.

Mechanical agitation The stirring, paddling, or swirling action of a device that keeps a pesticide and any additives thoroughly mixed in the spray tank.

Mechanical control Using people, machines, or tools to physically remove pests like pruning, burning, or bulldozing.

Medium-volume spray A spay application of 5.0 to 50 gallons per acre.

MESA Mining Enforcement and Safety Administration. All respirators designed for use with pesticides are jointly approved by MESA and NIOSH.

Mesh (screen) Standard screens are used to separate solid particles into size ranges. The mesh is stated in number of openings to each linear inch. The finest screen practical in this work is the 325-mesh, which has openings 44 micrometers in diameter, 1 micrometer being equivalent to 0.001 mm. This screen has over 10,500 openings per square inch. Fine dusting sulfur preferably has 95% of the particles passing a 325-mesh screen. A common range for granular formulations is the 15/30 range. Particles small enough to pass a 60-mesh screen are considered dusts.

Metabolite   A compound derived in the case of a pesticide by chemical action upon the pesticide within a living organism (plant, insect, higher animal). The action varies (oxidation, reduction, and so on) and the metabolite may be either more toxic or less toxic than before. The same derivative may in some cases develop upon exposure of the pesticide outside a living organism.

Metamorphosis   The series of changes through which an insect passes in its growth from the egg through the larvae and pupa to the adult; complete when the pupa is inactive and does not feed; incomplete when there is no pupa or when the pupa is active and feeds.

Meteorological conditions   Those factors dealing with the atmosphere and the weather in a given locality, for example, temperature, humidity, wind velocity, and temperature lapse rate.

Metric system   A system of measurement used by most of the world. Because of its international use and scientific application, the United States has stated its intention to gradually adopt the metric system. The units of the metric system are meters (for length), grams (for weight), and liters (for volume).

mg   The abbreviation for milligram; 1/1,000 of a gram.

mg/kg   Used to express the amount of pesticide in milligrams per kilogram of animal body weight to produce a desired effect. 1,000,000 milligrams = 1 kilogram = 2.2 pounds.

Micro   A combining form meaning *small.*

Microbial insecticide   Bacteria or other tiny plants or animals used to prevent damage by or to destroy insects. Example: milky spore disease of Japanese beetle.

Microfauna   Microscopic forms of animal life.

Microflora   Microscopic forms of plant life.

Microfoil boom   A boom with nozzles that spray a mist of relatively large particles, thus reducing drift.

Microgram   A metric weight measurement equal to 1/1,000,000 of a gram; approximately 28,500,000 micrograms equal 1 ounce.

Micrometer   A unit of length equal to 1/25,400 of an inch of 1/1,000,000 of a meter. Often called a micron.

Micron   A unit to measure the diameter of spray droplets. Approximately 25,000 μ (microns) equal an inch.

Microorganism   An microscopic animal or vegetable organism.

Microscopic   Visible under the microscope.

Migrating   Moving from place to place.

Mildew   A plant disease characterized by a thin, whitish coating of mycelial growth and spores on the surface of infected plant parts.

Milligram   A metric weight measurement equal to 1/1,000 of a gram; approximately 28,500 mg equals 1 ounce.

Mineral spirits   Similar to kerosene. It can be used as a solvent for some pesticides.

Minimize   To reduce to the smallest possible amount. Example: to minimize spray drift or hazard to the environment.

Miscible   Able to be mixed.

Miscible liquids   Two or more liquids capable of being mixed and which will remain mixed under normal conditions.

Misdiagnose   To make a mistake in deciding what pest has caused the problem.

Mist application   A liquid application having droplets with a volume median diameter of less than 100 micrometers.

Mist blower   Spray equipment in which hydraulic atomization of the liquid at the nozzle is aided by an air blast past the source of spray.

Mite   Tiny eight-legged animal with a body divided into two parts. It has no antennae (feelers). During the nymphal stage it has six legs.

Miticide   A chemical used to control mites (acarids).

Mitotic poison   A chemical that disrupts cell division and resultant growth.

MLD   Minimum lethal dose; the smallest of several doses that kills one of a group of test animals.

Mode of action   Manner in which herbicides kill or prevent weed or plant growth.

Mold   A fungus-caused growth often found in damp or decaying areas or on living things.

Molluscicide   A compound used to control slugs and snails that are intermediate hosts of parasites of medical importance to humans.

Mollusks   Any of a large family of invertebrate animals, including snails and slugs.

Monitoring   To check and keep track of all aspects of a pesticide application, including mixing chemicals, spray pressure, wind speed, pest reduction records, and so forth.

Monitoring system   A regular system of keeping track of and checking up on whether or not pesticides are escaping into the environment.

Monocot (monocotyledon)   A seed plant having a single cotyledon or leaf; includes grasses, corn, lilies, orchids, and palms.

Monoculture   Growing some single crop and not using the land for any other purpose.

Monophagous   Limited to a single kind of food, as in the case of the boll weevil, which restricts its feeding to the cotton plant.

Morbidity   The relative incidence of disease or the state of wasting away.

Mortality   Death rate.

Mosaic   A virus-type plant infection showing patchwork of discolored areas on a leaf.

mph   Miles per hour.

Mulch   A layer of wood chips, dry leaves, straw, hay, plastic strips, or other material placed on the soil around plants to hold moisture in the ground, keep weeds from growing, soak up rain, reduce soil temperatures, or keep fruits and vegetables from touching the ground.

Multipurpose spray or dust   One that controls a wide range of pests.

Mutagenic   Capable of producing genetic change.

Mutation   A new type of organism produced as a result of a heritable change.

Mycelium (plural mycelia)   Mass of hyphae constituting the body of a fungus.

Mycoplasma A bacteria-like organism that lacks a rigid cell wall and is variable in shape. Mycoplasmas cause certain plant diseases of the yellows type.

Mycoplasma-like organisms Organisms recently discovered to be the case of many plant diseases formerly attributed to viruses; organisms smaller than bacteria and larger than viruses.

N

Naphtha A hydrocarbon fraction of crude oil. When crude is distilled, it is separated into naphtha, kerosene, gas oil, fuel oil, lube oil, and residues, with each fraction itself a mixture of many different chemicals. Naphtha is especially valuable as a raw material for both chemicals and gasoline. European ethylene plants, for example, use naphtha as their primary feedstock.

Narrowleaf species Botanically, those plants classified as monocotyledonae; morphologically, those plants having narrow leaves and parallel veins. Examples: grasses, sedges, rushes, and onions. Compare to *broadleaf species.*

Natural control A control of undesirable pests by natural forces, usually predators and parasites, but may be by pathogens or physical means.

Natural enemies The predators and parasites in the environment that attack pest species.

Necrosis Localized death of living tissue, for example, death of a certain area of a leaf or of a certain area of an organ.

Necrotic A term used to describe tissues exhibiting varying degrees of dead areas or spots.

Negligence Failure to do a job or duty; an act or state of neglectfulness.

Negligible residue A tolerance that is set for a food or feed crop that contains a very small amount of pesticide at harvest as a result of indirect contact with a chemical.

Nematicide A material, often a soil fumigant, used to control nematodes infesting roots or crop plants.

Nematode A member of a large group (phylum Nematoda), also known as threadworms and roundworms. Some larger kinds of internal parasites of humans and other animals. Nematodes, sometimes called eelworms, injurious to crop plants, are slender, free-living, microscopic, wormlike organisms in the soil.

Neoprene A synthetic rubber often used to make gloves and boots that offer protection against most pesticides.

Nervous system All nerve cells and tissues in animals, including the brain, spinal cord, ganglia, nerves, and nerve centers.

Neurotoxicity The ability to cause poisonous effects to the nervous system (brain, spinal cord, and nerves).

Neutral soil A soil neither acid nor alkaline with a pH of 7.0.

Neutralize To destroy the effect of or to counteract the properties of something.

NIOSH National Institute for Occupational Safety and Health. All respirators designated for use with pesticides are jointly approved by NIOSH and MESA.

Nitrophenol A synthetic organic pesticide that contains carbon, hydrogen, nitrogen, and oxygen.

NOEL No Observable Effect Level. In a series of dose levels tested, it is the highest level at which no effect is observed; that is, it is safe in the species tested.

No-residue As the term applies to pesticides, the act of registration of an economic poison on the basis of the absence of a residue at time of harvest on the raw agricultural product when the economic poison is used as directed.

No-till Planting crop seeds directly into stubble or sod with no more soil disturbance than is necessary to get the seed into the soil.

Nonaccumulative Does not build up or store in an organism or in the environment.

Noncorrosive Opposite of corrosive.

Noninfectious disease A disease that is caused by unfavorable growing conditions and cannot be transmitted from plant to plant.

Nonionic Chemically inert. A surfactant that does not ionize is classed as nonionic, in contrast to anionic and cationic compounds. Many emulsifiers used in pesticide formulations are nonionic.

Nonpersistent Only lasts for a few weeks or less. Example: A pesticide may disappear because it is broken down by light or microorganisms, or it may evaporate.

Nonselective A chemical that is generally toxic to plants or animals without regard to species. A nonselective herbicide may kill or harm all plants.

Nontarget Any area, plant, animal, or other organism at which a pesticide application is not aimed, but which may accidentally be hit by the chemical.

Nontoxic Not poisonous.

Nonvolatile A pesticide chemical that does not evaporate (turn to a gas or a vapor) at normal temperatures and pressures.

Noxious weed A weed arbitrarily defined by law as being especially undesirable, troublesome, and difficult to control. Definition of the term *noxious weed* will vary according to legal interpretations.

Nozzles Devices that control drop size, rate, uniformity, thoroughness, and safety of a pesticide application. The nozzle type determines the ground pattern and safety of a pesticide application. Examples: flat fan, even flat fan, cone, flooding, offset, atomizing, broadcast, and solid stream nozzles.

Number median diameter (NMD) The diameter at which half the number of spray droplets are larger and half are smaller.

Nymph The early stage in the development of insects that have no larva stage. It is the stage between egg and adult during which growth occurs in such insects as cockroaches, grasshoppers, aphids, and termites.

O

Obligate Compelled to live under only one type of environment.

Off-target An area outside the target to which the application is applied; not drift.

Oils Usually refers to aromatic paraffinic oils used as diluents in formulating products as carriers of pesticides or for direct use.

Olefins Reactive, hydrogen-deficient hydrocarbons characterized by one or more double bonds per molecule. Ethylene and propylene, widely used building block basic chemicals, are olefins.

Oligophagous   Restricted to a few kinds of food. For instance, the common cabbage worm feeds on plants related to the cabbage, such as turnips, mustard, and other plants of the crucifer family.

Oncogenic   The property to produce tumors (not necessarily cancerous) in living tissues.

Operating speed   The constant rate at which a pesticide sprayer moves during application; usually measured in miles per hour or feet per minute.

Oral   Through the mouth.

Oral toxicity   Ability of a pesticide chemical to sicken or kill an animal or human if eaten or swallowed.

Organic compounds   A large group of chemical compounds that contain carbon.

Organic matter   Plant and animal debris or remains found in the soil in all stages of decay. The major elements in organic matter are oxygen, hydrogen, and carbon.

Organic phosphorous insecticide   A synthetic compound derived from phosphoric acid. Organic phosphorous insecticides are primarily contact killers with relatively short-lived effects. They are decomposed by water, pH extremes, high temperature, and microorganisms. Examples are malathion, parathion, diazinon, phorate, TEPP, dimethoate, and fenthion.

Organism   Any living thing; plant, animal, fungus, bacteria, insect, and so on.

Organochlorine   Same as *chlorinated hydrocarbon.*

Orifice   The opening or hole in a nozzle through which liquid material is forced out and broken up into a spray.

Original container   The package (can, bag, or bottle) in which a company sells a pesticide chemical. A package with a label telling what the pesticide is and how to use it correctly and safely.

Ornamentals   Plants used to add beauty to homes, lawns, and gardens. They include trees, shrubs, and small colorful plants.

Osmosis   The transfer of materials that takes place through a semipermeable membrane that separates two solutions, or between a solvent and a solution that tends to equalize their concentrations. The walls of living cells are semipermeable membranes and much of the activity of the cells depends on osmosis.

Overlap   To extend over and cover part of the previous swath.

Over-the-top   A pesticide application over the top of a growing plant.

Ovicides   Pesticides or agents used to destroy insect, mite, or nematode eggs.

Oviposition   Egg laying by insects.

Oxidation   To combine with oxygen.

Oximes   Chemical compounds characterized by the general formula $R - CH = NOH$.

P

Parasite   A plant of animal that harms another living plant or animal (the host) by living and feeding on or in it. Some of our worst pests are parasites that cause disease or injury to animals and plants grown by humans or to humans themselves.

Parasitic insect   An insect that lives in or on the body of another insect.

Pathogen   Any microorganism that can cause disease. Most pathogens are parasites but there are a few exceptions.

Pathology   Branch of science of the origin, nature, and course of diseases.

Pellet   A dry formulation of pesticide mixed with other components in discrete particles, usually larger than 10 cubic millimeters.

Penetrant   Wetting agents that enhance the ability of a liquid to enter into pores of a substrate, to penetrate the surface. Also termed *penetrating agents.*

Penetration   The ability to get through; the process of entering.

Percent by weight   A percentage that expresses the active ingredient weight as a part to the total weigh of the formulation. Example: one pound of active ingredient added to and mixed with three pounds of inert materials results in a formulation that is 25% pesticide by weight.

Percent concentration   The weight or volume of a given compound in the final mixture expressed as a percentage.

Perennial   A plant that normally lives for more than two years. Trees and shrubs are perennial plants. Some perennials die back to the roots each winter but new shoots grow again in the spring.

Persist   To stay. To remain.

Persistent pesticide   A pesticide chemical (or the metabolites) that remains active in the environment more than one growing season. These compounds often accumulate and are stored in animal and plant tissues.

Pest   Any pesticide that injures humans, their property, or their environment or that annoys them; any insect, mite, rodent, nematode, fungus, weed, or other plant or animal (except those microorganisms in or on living humans or animals) that is injurious to the health of humans, animals, or plants or to the environment.

Pesticide (economic poison)   As defined under the Federal Insecticide, Fungicide, and Rodenticide Act, economic poison (pesticide) "means any substance or mixture of substances intended for preventing, destroying, repelling, or mitigating any insects, rodents, nematodes, fungi, or weeds, or any other forms of life declared to be pests; and any substance or mixture of substances intended for use as a plant regulator, defoliant, or desiccant."

Pesticide tolerance   The amount of pesticide residue that may legally remain in or on a food crop. Federal residue tolerances are established by the Environmental Protection Agency.

Pest management   A management system that uses all suitable techniques and methods in a compatible manner to maintain pest populations at levels below those that cause economic injury.

Petrochemical   Generally, a chemical derived from crude oil or natural gas. The term embraces "basic chemicals" and "derivatives," but is difficult to use to precisely designate specific chemical materials. Some of today's petrochemicals, for example, may someday be profitably derived from coal, oil shale, or manufactured gas.

Petroleum distillate   Kerosene.

Petroleum oils   Pesticides used to control insects in plants; they are refined from crude oil.

Petroleum products   Anything that contains gasoline, kerosene, oil, or similar products.

pH value   The degree of acidity or alkalinity. The pH scale of 0 to 14 expresses intensity of acidity or alkalinity in the same manner that degrees in the thermometer express intensity of heat. The pH value of 7.0, halfway between 0 and 14, is neither acid nor alkaline. pH values below 7.0 indicate acidity with its intensity increasing as the numbers decrease. Conversely, pH values above 7.0 indicate alkalinity with its intensity increasing as the numbers increase.

Phenoxy   Referring to a chemical class of herbicides including 2,4-D.

Pheromones   Chemicals produced by insects and other animals to communicate with and influence the behavior of other animals of the same species. (*Note:* Most pheromones used today in pest management are synthetic and are used to monitor insect populations.)

Phloem   The living tissue in plants that functions primarily to transport metabolic compounds from the site of synthesis to the site of use.

Photodegradation   The process of breaking down a substance through reaction to light.

Photosynthesis   The manufacture of simple sugars by green plants utilizing light as the energy source; a process by which carbohydrates are formed in the chlorophyll-containing tissues of plants exposed to light.

Physical properties   Examples are solubility, volatility, inflammability, and the state of being solid, liquid, or gas.

Phytoplankton   Microscopic plant life living suspended in water.

Phytotoxic   Injurious or lethal to plants.

Phytotoxicity   The degree to which a chemical or other agent is toxic to plant life. This may be specific to particular kinds or types of plants.

Piscicide   A pesticide used to control fish.

Plant-derived pesticide   Same as botanical insecticides.

Plant diseases   Harmful conditions or sicknesses that negatively affect plant life; fungi, bacteria, and viruses most often cause plant diseases.

Plant growth regulator   A substance that alters the growth of plants. The term does not include substances intended solely for use as plant nutrients or fertilizers.

Point of drip or runoff   When a spray is applied until it starts to run or drip off the ends of the leaves and down the stems of plants or off the hair or feathers of animals.

Poison   A chemical causing a deleterious effect when absorbed by a living organism (biocide).

Poison control center   An agency, generally a hospital, that has current information as to the proper first aid techniques and antidotes for all poisoning emergencies.

Poisonous bait   A chemical that causes a deleterious effect when absorbed or eaten by a living organism.

Pollinators   Bees, flies, and other insects that visit flowers and carry pollen from flower to flower in order for many plants to produce fruit, vegetables, buds, and seeds.

Pollutant   A harmful chemical or waste material discharged into the water, soil, or atmosphere. An agent that makes something dirty or impure.

Pollute   To add an unwanted material (often a pesticide) that may do harm or damage. To render unsafe or unclean or impure by carelessness or misuse.

Polyphagous   Feeding on a wide range of food species, not necessarily related; for instance, the corn earworm, which, because of its damage to crops other than corn, is also called the tomato fruitworm and the bollworm.

Port of entry   Place where foreign goods (plants, animals, crops) enter the United States.

Postemergence   After the appearance of a specified weed or crop.

Postplant incorporated   Applied and incorporated into the soil after the planted crop emerges.

Potable   Water suitable for drinking.

Potency   The strength of something. Example: how deadly a poison is.

Potentiation   The joint action of two pesticides to bring about an effect greater than the sum of their individual effects.

Pour-on   A pesticide that is poured along the midline of the backs of livestock.

ppb   Parts per billion. A way of expressing amounts of chemicals in foods, plants, animals, and so on. One part per billion equals 1 lb in 500,000 tons.

ppm   Parts per million. A way of expressing amounts of chemicals in foods, plants, animals, and so on. One part per million equals 1 lb in 500 tons.

ppt   Parts per trillion. A way of expressing amounts of chemicals in foods, plants, animals, and so on. One part per trillion equals 1 lb in 500,000,000 tons.

Practical knowledge   The possession of pertinent facts and comprehension together with the ability to use them in dealing with specific problems and situations.

Precautions   Warnings. Safeguards.

Precipitate   To settle out. A solid substance that forms in a liquid and sinks to the bottom of the container.

Precipitation   The amount of rain, snow, sleet, or hail that falls.

Predacide   A pesticide used to control vertebrate predator pests.

Predator   An insect or other animal that attacks, feeds on, and destroys other insects or animals. Predators help to reduce the number of pests that cause disease, damage, and destruction.

Predispose to chemical injury   To increase the chances of damage or harm by careless handling and incorrect use of pesticide chemicals. One must read the label and follow directions carefully to use the pesticide chemicals safely.

Preemergence   Prior to emergence of the specified weed or crop.

Preharvest   The time just prior to the picking, cutting, or digging up of a crop.

Preharvest interval   See *days to harvest.*

Preplant application   Application of a pesticide prior to planting a crop.

Preplant soil incorporated (PPI)   Applied and tilled into the soil before seeding or transplanting.

Pressure   The amount of force on a certain area. The pressure of a liquid pesticide forced out of a nozzle to form a spray is measured in pounds per square inch.

Prevailing   The predominant or general occurrence or use.

Private applicator   A certified applicator who uses or supervises the use of any pesticide that is classified for restricted use for purposes of producing any agricultural commodity on property owned or rented by the applicator or his or her employer or (if applied without compensation other than trading of personal services between producers of agricultural commodities) on the property of another person.

Product   A term used to describe a pesticide as it is sold. It usually contains the pesticide chemical plus a solvent and additives.

Prolonged exposure   Contact with a pesticide chemical or its residue for a long time.

Promulgated   To put into operation (such as a law or rule), or to make known by declaration.

Propellant   Agent in self-pressurized pesticide products that produces the force required to dispense the active ingredient from the container.

Proper dosage   The right amount, according to label instructions.

Properties   The characteristics or traits that describe a pesticide.

Proprietary chemical   A chemical made and marketed by a person or company having the exclusive right to manufacture and sell it.

Propylene   A colorless gas usually produced as a by-product of ethylene, manufactured from propane, butane, or naphtha. An important ingredient for making isopropanol, butanol, propylene oxide, polypropylene, glycerin, acrylonitrile, and phenol.

Protectant   A chemical applied to the plant or animal surface in advance of the pest (or pathogen) to prevent infection or injury by the pest.

Protective equipment   Clothing or any other materials or devices that shield against unintended exposure to pesticides.

Protopam chloride (2-PAM)   An antidote for organophosphate poisoning, *but not for carbamate poisoning.*

Protozoa   One-celled animals or a colony of similar cells.

psi   Pounds per square inch.

Pubescent   Hairy. It affects ease of wetting of foliage; also retention of spray on foliage.

Pupa   The pupa is the resting stage of many insects. The stage between the larva (caterpillar or maggot) and the adult (butterfly or fly). Some caterpillars spin a silk cocoon before they change to a pupa inside.

Putrefaction   The decomposition of proteins by microorganisms under anaerobic conditions, resulting in the production of incompletely oxidized compounds, some of which are foul smelling.

Pyrethrin   Either of two liquid esters derived from chrysanthemums, the active ingredient of pyrethrum.

Pyrethroids   Synthetic compounds produced to duplicate the biological activity of the active principles of pyrethrum. There are at least 26 compounds that have been synthesized to date.

Pyrethrum   An insecticide made from the dried chrysanthemum flower heads.

Pyrophoric   Tendency for a material to ignite spontaneously.

Q

Quarantine Regulation forbidding sale or shipment of plants or plant parts, usually to prevent disease, insect, nematode, or weed invasion of an area.

R

Radioactive Giving off atomic energy in the form of radiation, such as in alpha, beta, or gamma rays.

Radioisotopes One of a broad class of elements capable of becoming radioactive and giving off atomic energy. Some radioisotopes occur naturally, others are produced artificially. The word is synonymous with radioactive elements and includes tracer elements.

Rate Rate refers to the amount of active ingredient or acid equivalent of a pesticide applied per unit area (such as 1 acre). Rate is preferred to the occasionally used terms *dosage* and *application.*

Recirculating sprayer A sprayer system with the nozzle aimed at a catchment device that recovers and recirculates herbicide that does not hit plants or weeds that pass between nozzles and the catchment device.

Recommendation Suggestion or advice from a county agent, extension specialist, or other agricultural authority.

Reentry interval The length of time between the pesticide applications and when workers can safely go back into an area without protective clothing.

Registered pesticides Pesticides approved by the U.S. Environmental Protection Agency for use as stated on the label of the container.

Registration Approval the EPA or a state agency for the use of a pesticide as specified on the label.

Regulated pest A specific organism considered by a state or federal agency to a be a pest requiring regulatory restrictions, regulations, or control procedures in order to protect the host, humans, and the environment.

Regulatory officials People who work with the federal or state government and enforce rules, regulations, and laws.

Reinfestation The return of insects or other pests after they left or were destroyed.

Relative humidity The amount of moisture in the air compared to the total amount that the air could hold at that temperature.

Remote Very slight or faint chance; far apart or distant.

Repeated contact or inhalation To touch or breathe in a pesticide several times over a period of time.

Repellent A compound that is annoying to a certain animal or other organism, causing it to avoid the area in which it is placed.

Reptiles Animals of the class Vertebrate that are cold-blooded and possess scaly skin. Examples: snakes, turtles, and lizards.

Residual pesticide A pesticide chemical that can destroy pests or prevent them from causing disease, damage, or destruction for more than a few hours after it is applied.

Residue The amount of chemical that remains on the harvested crop.

Resistance   The ability of an organism to suppress or retard the injurious effects of a pesticide.

Resistant species   One that is difficult to kill with a particular herbicide.

Respiration   Breathing.

Respirator   A face mask used to filter out poisonous gases and dust particles from the air so that a person can breathe and work safely. A person using the most poisonous pesticide chemicals must use a respirator as directed on the pesticide label.

Respiratory toxicity   Intake of pesticides through air passages into the lungs. Pesticides may reach the lungs as a vapor or as extremely fine droplets or particles. Most compounds are more toxic by this route. Safety equipment is *essential* to protect the pesticide handler or applicator from the hazards of respiratory exposure *whenever it is recommended on the label.*

Restricted use pesticide   A pesticide that is available for purchase and use only by certified pesticide applicators. This growth of pesticides is not available for use by the general public because of the very high toxicities and/or environmental hazards associated with these materials.

Restrictions   Limitations.

Revocation   To nullify or withdraw. To revoke a license.

Rhizome   Underground rootlike stem that produces roots and leafy shoots. Examples: the white underground parts of johnsongrass and horse nettle, the black parts of Russian knapweed.

Risk   The probability that a substance will produce harm under specified conditions.

Rodent   All animals of the order Rodentia, such as rats, mice, gophers, woodchucks, and squirrels.

Rodenticide   A substance or mixture of substances intended to prevent, destroy, repel, or mitigate rodents.

RPAR   Rebuttable Presumption Against Registration. An EPA process whose objectives are to identify pesticide chemicals that present "unreasonable adverse effects on the environment."

rpm   Revolutions per minute.

Runaway pest   Any pest organisms that enter a new territory where they have no natural enemies, and therefore reproduce with little interference, resulting in a large population that can overrun an area.

Runoff   The sprayed liquid that does not remain on the plant. See *point of drip or runoff.*

Russetting   Rough, brownish markings on leaves, fruit, or tubers. This can be caused by some insects, fungi, pesticide chemicals, or possibly weather.

S

Safener   A material added to a pesticide to eliminate or reduce phytotoxic effects to certain plant species.

Safety   The practical certainty that injury will not result from the proper use of a pesticide or implement.

Salt   One of a class of compounds formed when the hydrogen atom of an acid radical is replaced by a metal or metal-like radical. The most common salt is sodium chloride, the sodium salt of hydrochloric acid. Other metal or metal-like salts in food may include phosphorus, calcium, potassium, sodium, magnesium, sulfur, manganese, iron, cobalt, zinc, and other metals. They may be present as chlorides, sulfates, phosphates, lactates, citrates, or in combination with proteins as in calcium caseinate.

Sanitation   Cleaning up. Keeping gardens, fields, animals, or buildings clean to reduce the number of insects and other pest and disease problems. This is one way to cut down on the use of pesticides.

Saprophyte   An organism living upon dead or decaying organic matter.

Scientific name   The one name used throughout the world by scientists for each animal and plant. These names are made up of two words based on the Greek and Latin languages and are called the genus and species.

Seed bed   Land prepared and used to plant seeds.

Seed protectant   A chemical applied to seed before planting to prevent disease and insect attack on seeds and new seedlings.

Seed treatment   The application of a pesticide chemical to seeds before planting in order to protect them from injury or destruction by insects, fungi, and other soil pests.

Segment   A ring or subdivision of the body or of an appendage between areas of flexibility associated with muscle attachment.

Seizure   To take or impound a crop of animal if it contains more than the allowable pesticide residue. Also the onset of a fit or convulsions.

Selective pesticide   A chemical that is more toxic to some species (plant, insect, animal, microorganisms) than to others.

Semipermeable   Can be penetrated or entered only by some substances.

Sensitive areas   Places where pesticides could cause great harm if not used with special care and caution. Examples: houses, barns, parks, ponds, streams, and hospitals.

Sensitive crops   Crops that are easily injured by pesticide chemicals; even slight drift could cause great damage.

Sensitivity   Susceptible to effects of toxicant at low dosage; not capable of withstanding effects; for example, many broadleaved plants are sensitive to 2,4-D.

Sex attractants   See *pheromones.*

Shock   The severe reaction of the human body to a serious injury, which can result in death if not treated (even if the actual injury was not a fatal one).

Short-term pesticide   A pesticide that breaks down almost immediately after application into nontoxic by-products.

Side-dressing   To put fertilizer or a pesticide in granular form on or in the ground near plants after they have started to grow.

Signal words   Words that must appear on pesticide labels to denote the relative toxicity of the product. The signal words are "DANGER-POISON" (for highly toxic), "WARNING" (for moderately toxic), and "CAUTION" (for low-order toxicity). The symbol of the skull and crossbones must appear on the label of highly toxic pesticides along with the words "DANGER-POISON."

Sign   Some evidence of exposure to a dangerous pesticide; an outward signal of a disease or poisoning in a plant of animal, including people. Compare to *symptom*.

Signs of poisoning   Warnings or symptoms of having breathed in, touched, eaten, or drunk a dangerous pesticide that could cause injury or death.

Silvicide   A term applied to herbicides used to control undesirable brush and trees, as in wooded areas.

Site   An area, location, building, structure, plant, animal, or other organism to be treated with a pesticide to protect it from, or to reach and control, the target pest.

Slimicide   A chemical used to prevent slimy growths, as in wood pulping processes for manufacture of paper and paperboard.

Slurry   A thick suspension of a pesticide made from wettable powder and water.

Smoke   Particles of a pesticide chemical between 0.001 and 0.1 micrometer in diameter. The particles are released into the air by burning. See *micron*.

Soft water   Water with few minerals or other chemicals dissolved in it. Soap makes suds easily in soft water.

Soil application   Application of a chemical to the soil rather than to vegetation.

Soil drench   To soak or wet the surface of the ground with a pesticide chemical. Generally, fairly large volumes of the pesticide preparation are needed to saturate the soil to any depth.

Soil fumigant   A pesticide that will evaporate quickly. When added to the soil the gas formed kills pests in the soil. Usually a tarpaulin, plastic sheet, or layer of water is used to trap the gas in the soil until it does the job.

Soil incorporation   Mechanical mixing of the pesticide with the soil.

Soil injection   Mechanical placement of the pesticide beneath the soil surface with a minimum of mixing or stirring. Common method of applying liquids that change into gases.

Soil sterilant   A chemical that prevents the growth of plants, microorganisms, and the like, when present in soil. Soil sterilization may be temporary or relatively permanent, depending on the nature of the chemical being applied.

Solubility   The ability of a chemical to dissolve in another chemical or solvent; a measure of the amount of substance that will dissolve in a given amount of another substance.

Soluble   Will dissolve in a liquid.

Soluble powder   A powder formulation that dissolves and forms a solution in water.

Solution   A preparation made by dissolving a solid, liquid, or gaseous substance into another substance (usually a liquid) without a chemical change taking place. Example: sugar in water.

Solvent   A liquid that will dissolve a substance forming a true solution (liquid in molecular dispersion).

Space bomb   An aerosol spray that can be used in rooms or buildings. A container having a pesticide plus a chemical under pressure that forces the pesticide out as a spray or mist.

Space spray   A pesticide forced out of an aerosol container or sprayer as tiny droplets that fill the air in a room or building and destroy insects and other pests.

Species   A group of living organisms that are very nearly alike, are called by the same common name, and can interbreed successfully.

Specific category   Any one of the pesticide use categories designated by FIFRA or by state regulations; a special area (such as agricultural plant pest control or regulatory pest control) requiring state certification for use of restricted use pesticides by commercial applicators.

Specific gravity   Density. The ratio of the mass (weight) of a material to the mass of an equal volume of water at a specified temperature such as 20° C.

Spermatotoxicity   The ability to cause adverse changes in sperm, such as lowering the viable sperm count.

Spiders   Small animals closely related to insects; they have eight jointed legs, two body regions, no antennae (feelers), and no wings. Spiders are often grouped with mites and ticks.

Spillage   The leaking, running over, or dripping of any chemical. It should be cleaned up immediately for safety.

Spore   An inactive form of a microorganism that is resistant to destruction and capable of becoming active again.

Spot treatment   A treatment directed at specific plants or areas rather than a general application.

Spray   A mixture of a pesticide with water or other liquid applied in tiny droplets.

Spray concentrate   A liquid formulation of pesticide that is diluted with another liquid (usually water or oil) before using.

Spray deposit   The amount of pesticide chemical that remains on a spray surface after the droplets have dried.

Spray drift   The movement of airborne spray particles from the intended area of application.

Spray height   The distance between the target being treated and the spray boom.

Spray volume   The amount of spray liquid applied to a unit receiving treatment, expressed in volume per unit treated. (For an area treatment, gallons/acre or liters/hectare; for space treatment, milliliters/cubic meter or ounces/1,000 cubic feet; for individual units, milliliters/plant, milliliters/animal, gallons/tree).

Spreader   A device for spreading granular materials.

Spreader-sticker   A chemical added to a pesticide mixture to make the droplets of the spray spread out and stick better to the animal, plant, or other treated surface.

Stage   Any definite period in the development of an insect, for example, egg stage, caterpillar stage; or in the development of a plant, for example, seedling stage, tillering stage, milk stage.

Stage of development   A defined period of growth; usually refers to an insect. An insect goes through many changes in its growth, from an egg to an adult; each such change is a stage of development.

Standard   The measure of knowledge and ability that must be demonstrated as a requirement for certification.

Sterilize   Treat with a chemical or other agent to kill every living thing in a certain area.

Sticker   A material added to a pesticide to improve its adherence to plants or other surfaces.

Stolon   Aboveground runners or slender stems that develop roots, shoots, and new plants at the tip of nodes, as in the strawberry plant or Bermuda grass.

Stomach poison   Pesticide that must be eaten by an insect or other animal in order to kill the animal.

Structural pests   Pests that attack and destroy buildings and other structures, clothing, stored food, and manufactured and processed goods. Examples: termites, cockroaches, clothes moths, rats, dry rot fungi.

Stump treatment   Herbicides applied to cut stems or stumps to halt sprouting.

Stylet   A small, stiff instrument used by insects to pierce plant or animal tissue for the purpose of feeding.

Subacute toxicity   Results produced in test animals of various species by long-term exposure to repeated doses or concentrations of a substance.

Submersed plant   An aquatic plant that grows with all or most of its vegetative tissue below the water surface.

Suction hose   A hose through which water is pulled from a pond or stream or spray from the spray tank to the pump.

Summer annuals   Plants that germinate in the spring, make most of their growth in the summer, and die in the fall after flowering and seeding.

Supplement   Same as *adjuvants*. Substance added to a pesticide to improve its physical or chemical properties. May be a sticker, spreader, wetting agent, safener, and so on, but usually not a diluent.

Suppress   To keep from building up in numbers. To stop a pest from causing injury or destruction.

Surface-acting agent   A substance that reduces the interfacial tension of two boundary lines. Most pesticide adjuvants may be considered surface-active agents. Also known as *surfactants*.

Surface spray   A pesticide spray that is applied in order to completely cover the outside of the object to be protected.

Surface tension   Due to molecular forces at the surface, a drop of liquid forms an apparent membrane that causes it to ball up rather than to spread as a film.

Surface water   Rivers, lakes, ponds, streams, and the like, that are located aboveground.

Surfactant   A material that reduces surface tension between two unlike materials such as oil and water. A spreader or wetting agent.

Susceptibility   The degree to which an organism is affected by a pesticide at a particular level of exposure.

Susceptible   Capable of being injured, diseased, or poisoned by a pesticide; not immune.

Susceptible species   A plant or animal that is affected at rates listed on the label.

Suspended   A pesticide use that is no longer legal and the remaining stocks cannot be used. (*Note:* This order is more severe than *canceled.*) Also, describes particles that are dispersed (or held) in a liquid.

Suspension   A system consisting of very finely divided solid particles dispersed in a liquid or gas, but not dissolved in it.

Swath   The width of a treated area when a ground rig or spray plane makes one trip across a field.

Symbiosis   The living together in intimate association of two diverse types of organisms.

Symptom   Warning that something is wrong. Any indication of disease or poisoning in a plant or animal. This information is used to figure out what insect, fungus, other pest, or pesticide is causing the disease, damage, or destruction.

Syndrome   Symptoms characterizing a particular abnormality.

Synergism   When the effect of two or more pesticides applied together is greater than the sum of the individual pesticides applied separately.

Synergist   A chemical that when mixed with a pesticide increases its toxicity. The synergist may or may not have pesticidal properties of its own.

Synthesis   A coming together of two or more substances to form a new material.

Systemic pesticide   A chemical that is absorbed and translocated throughout the plant or animal making it toxic to pests. For example, a systemic insecticide can be applied to the soil, enter the roots of the plant, travel to the leaves, and kill insects feeding on the leaves.

T

Take internally   To eat or swallow.

Tank mix   The mixture of two or more compatible pesticides in a spray tank in order to apply them simultaneously.

Taproot   The main root of some plants. It acts as an anchor and reaches water deep in the soil.

Target   The plants, animals, structures, areas, or pests to be treated with a pesticide application.

Technical material   The pesticide as it is manufactured by a chemical company. It is then formulated with other materials to make it usable in wettable powder, dust, liquid, or other form.

Temperature inversion   When air is coolest at ground level, warms with an increase in elevation, and then gets cooler again at further elevation. This normally occurs in the early morning and late evening.

Temporary tolerance   A tolerance established on an agricultural commodity by the EPA to permit a pesticide manufacturer or its agent time, usually one year, to collect additional residue data to support a petition for a permanent tolerance. In essence, this is an experimental tolerance.

Tenacity   Adherence. The resistance of pesticide deposit to weathering as measured by retention.

Tenacity index   A ratio of the residual deposit over the initial deposit when subjected to prescribed washing techniques.

Teratogenicity   The ability to cause birth defects.

Terminal growth   The tips or ends of a growing plant.

Terminate   Finish. Stop.

Test animals   Laboratory animals, usually rats, fish, birds, mice, or rabbits, used to determine the toxicity and hazards of different pesticides.

Therapeutants   Remedies for disease, drugs.

Thermal   Of or pertaining to heat.

Thickener   A type of adjuvant added to a liquid formulation to increase its viscosity; an adjuvant that reduces drift by increasing droplet size.

Thorax   The second or intermediate region of the insect body bearing the true legs and wings, made up of three rings, named in order prothorax, mesothorax, and metathorax.

Thorough coverage   Application of spray or dust where all parts of the plant or area treated are covered.

Threatened species   A group of organisms likely to become endangered in the foreseeable future.

Ticks   Tiny animals closely related to insects; they have eight jointed legs, two body regions, no antennae (feelers), and no wings. They are bloodsucking organisms and are often found on dogs, cows, or wild animals. Ticks are often grouped with mites and spiders.

Tillering stage   The development of side shoots from the base of a single stemmed grass or cereal plant.

Time interval   The period of time that is required between the last application of a pesticide and harvesting in order to ensure that the legal residue tolerance will not be exceeded.

Tolerance   1. Capacity to withstand pesticide treatment without adverse effects on normal growth and function. 2. The maximum residue concentration legally allowed for a specific pesticide, its metabolites, or breakdown products in or on a particular raw agricultural product, processed food, or feed item. Expressed as parts per million (ppm).

Tolerant   Same meaning as *resistant*. For example, grass is tolerant of 2,4-D to the extent that this herbicide can be used selectively to control broadleafed weeds without killing the grass.

Topical application   Implies application to the top or to the upper surface of the plant; thus applied from above.

Toxic   Poisonous; injurious to animals and plants through contact or systemic action.

Toxicant   An agent capable of being toxic; a poison.

Toxicity   The natural capacity of a substance to produce injury. Toxicity is measured by oral, dermal, and inhalation studies on test animals.

Toxin   A natural poison produced by a plant or animal.

Tracer element   A radioactive element used in biological and other research to trace the fate of a substance or follow stages in a chemical reaction, such as the metabolic pathway or a nutrient or growth formation in plants or animals. Radioactive elements that have proved useful for tracer work in nutrition research are carbon 14, calcium 45, cobalt 60, strontium 90, and phosphorus 32.

Trademark   A word, letter, device, or symbol, used in connection with merchandise and pointing distinctly to the origin or ownership of the article to which it is applied.

Trade name (trademark name, proprietary name)   Name given a pesticide or pesticide product by its manufacturer or formulator, distinguishing it as being produced or sold exclusively by that company.

Translocated pesticide   One that is moved within the plant or animal from the site of entry. Systemic pesticides are translocated.

Transport   To carry from one place to another, usually in a truck or trailer.

Treated area   A building, field, forest, garden, or other place where a pesticide has been applied.

Trial use   This notation indicates that this material is worthy of trial use on a portion (perhaps 10%) of the grower's acreage. Results may be variable, but the recommendations are the best available based on the present knowledge.

Triazine   A chemical class of herbicides.

Trivial name   A name long commonplace and used everywhere; example, nicotine.

Tuber   An underground stem used for storage of reserve food, for example, Irish potato.

Turbidity   Suspended matter in water preventing light penetration.

Turbulence   Instability or erratic airflow consisting of horizontal and vertical eddies.

U

Ultrahazardous   A job or activity that is very dangerous.

Ultralow volume (ULV)   Pesticide formulation for application at a rate no greater than 0.5 gallon per acre.

Ultralow-volume spray   A spray application of 0.05 to 0.5 gallon per acre.

Unauthorized persons   People who have no right doing something because they have not been told or trained to do it.

Uncontaminated   Does not contain hazardous pesticide residues, filth, or other undesirable material.

Under the direct supervision of   The act or process whereby application of a pesticide is made by a competent person acting under the instruction and control of a certified applicator who is responsible for the actions of that person and who is available if and when needed, even though such certified applicator is not physically present at the time and place the pesticide is applied.

Underground water   Water and waterways that are below the soil surface; this is where wells get their water.

Uniform coverage   The application of a pesticide chemical evenly over a whole area, plant, or animal.

Uninformed persons   People who are not trained to use and handle pesticides safely.

Unintentionally   Did not mean to do, done accidentally.

U.S. Bureau of Mines   An agency of the U.S. government that tests respirators and gas masks to find ones that can be used with highly poisonous pesticide chemicals.

USDA   U.S. Department of Agriculture.

V

Vapor   Gas, steam, mist, fog, or fume.

Vapordrift   The movement of chemical vapors from the area of application. Some herbicides when applied at normal rates and normal temperatures have vapor pressures that change into vapor form, which may seriously injure susceptible plants away from the application site. (*Note:* Vapor injury and injury from spray drift are often difficult to distinguish.)

Vaporize   To evaporate; to form a gas and disappear into the air.

Vapor pressure   The property that causes a chemical compound to evaporate. The lower the vapor pressure, the more volatile the compound.

Vector   An insect of other animal that carries a disease organism from one host to another. Example: aphids, leafhoppers, and nematodes can transfer viruses from plant to plant.

Vermin   Pests; usually rats, mice, or insects.

Vertebrate   An animal with a bony spinal column. Examples: mammals, fish, birds, snakes, frogs, toads.

Viability   Being alive. Example: A seed is viable if it is capable of sprouting or germinating.

Victim   Someone who is injured, poisoned, or hurt in any way.

Virus   A submicroscopic pathogen that requires living cells for growth and is capable of causing disease in plants or animals. Plant viruses are often spread by insects.

Viscosity   A property of liquids that determines whether they flow readily or resist flow. Viscosity of liquids usually increases with a decrease in temperature.

Volatile   A compound is said to be volatile when it evaporates (changes from a liquid to a gas) at ordinary temperatures on exposure to air.

Volatility   The ability of a solid or liquid to evaporate quickly at ordinary temperatures when exposed to air.

Volume   The amount, mass, or bulk.

Vomitus   What comes up when you throw up; matter that is vomited.

W

Waiting period   See *time interval.*

Warning   Beware. Used on labels of pesticide containers having moderately toxic pesticides as defined by the Federal Insecticide, Fungicide, and Rodenticide Act.

Warranty clause   A statement on the label limiting the liability of the chemical company.

Water solubility   Capable of being homogeneously mixed with water.

Weathering   The wearing away of pesticides from the surfaces they were applied to because of rain, snow, ice, and heat.

Weed   A plant that is undesirable due to certain characteristics or its presence in certain areas. A plant growing in a place where it is not wanted.

Weed control   The process of inhibiting weed growth and limiting weed infestations so that crops can be grown profitably or other operations can be conducted efficiently.

Weed eradication   The elimination of all live plants, plant parts, and seeds of a weed infestation from an area.

Weed suppression   The process of retarding weed growth.

Wettable powder   A finely ground, dry powder formulation that can be readily suspended in water for application in spray equipment.

Wetting agent   A compound that reduces surface tension and causes a liquid to contact plant surfaces more thoroughly.

Whorl   The point where leaves or other plant parts unfold or are formed in a circular pattern. Example: a corn plant before the tassel grows out.

Wide range   Pesticides' ability to kill several different kinds of pests.

Wildlife   All living things that are not human or domesticated, nor pests, as used here, including birds, mammals, and aquatic life.

Wilting   Drooping of leaves.

Wind shear   In aerial pesticide applications, the effect that wind leaving a wing edge has on spray droplets, causing the droplets to split; a wind across the face of a nozzle directed downward at an angle to the airstream. Also applies to any situation when an airstream is used to atomize liquids.

Winter annual   Plants that germinate in the fall and complete their life cycle by early summer.

Woody plants   Perennials with a tough, thick stem or trunk covered with bark.

X

Xerophyte   A plant adapted for growth under dry conditions.

Xylem   The nonliving tissue that functions primarily to conduct water and mineral nutrients from the roots of plants.

Xylene   Any of three isomeric hydrocarbons of the benzene series used as solvents.

Y

Yellows   One of the various plant diseases, such as *aster yellows,* whose prominent symptom is a loss of green pigment in the leaves.

Z

Zero tolerance   By law, no detectable amount of the pesticide may remain on the raw agricultural commodity when it is offered for shipment. Zero tolerances are no longer allowed.

Zooplankton   Microscopic animal life living suspended in water.

Appendixes

APPENDIX A: SELECTED REFERENCES

AGRIOS, GEORGE N. 1988. *Plant Pathology.* 3rd Ed. Academic Press. 803 p.

ALTMAN, JACK. 1993. *Pesticide Interactions in Crop Production.* CRC Press.

ANONYMOUS. 1960. *Index of Plant Diseases in the United States.* Agriculture Handbook No. 165. ARS USDA. Govt. Printing Office. Washington, DC.

ANONYMOUS. 1970. *Selected Weeds of the United States.* Agriculture Handbook No. 366. ARS USDA. Govt. Printing Office. Washington, DC.

ANONYMOUS. 1987. *Agricultural Chemicals in Groundwater: Proposed Pesticide Strategy.* Office of Pesticides and Toxic Substances. EPA. Washington, DC.

ANONYMOUS. 1988. *Applying Pesticides.* Am. Assoc. of Voc. Materials.

ASHTON, F. M. and A. S. CRAFTS. 1981. *Mode of Action of Herbicides.* New York: John Wiley and Sons, Inc. 525 p.

ASHTON, F. M. and THOMAS J. MONACO. 1991. *Weed Science: Principles and Practices.* 3rd Ed. John Wiley, New York, NY. 466 p.

BODE, L. E. et al. (Eds.) 1990. *Pesticide Formulation and Application Systems.* Special Technical Series No. 1078. ASTM. 260 p.

BODE, L. E. et al. (Eds.) 1992. *Pesticide Formulation and Application Systems.* Special Technical Series No. 1112. ASTM. 310 p.

BRIGGS, SHIRLEY A. 1992. *Basic Guide to Pesticides: Their Characteristics.*

BROWN, C. L. and W. K. HOCK (Eds.) 1988. *Pesticide Education Manual.* The Pennsylvania State University. University Park, PA.

CURTIS, C. R. (Ed.) 1988. *Agricultural Benefits Derived from Pesticide Use: A Study of the Assessment Process.* Dept. of Plant Path. The Ohio State University. Columbus, OH.

ENVIRON CORPORATION. 1988. *Elements of Toxicology and Chemical Risk Assessment.* Revised Ed. 1000 Potomac St. NW, Washington, DC.

MARER, PATRICK J. (Ed.) 1991. *Residential, Industrial and Institutional Pest Control.* University of California. 232 p.

MARER, P. J., M. L. FLINT, and M. W. STIMMANN. 1988. *The Safe and Effective Use of Pesticides.* Pub. 3324. University of California. Oakland, CA. 387 p.

McWHORTOR, C. G. and M. R. GEBHARDT (Eds.) 1987. *Methods of Applying Herbicides.* Monograph 4. Weed Science Society of America. Champaign, IL.

MEISTER, R. T. (Ed.) 1995. *Farm Chemicals Handbook.* Meister Publishing Company. Willoughby, OH.

METCALF, R. L. 1992. *Destructive and Useful Insects.* 5th Ed. New York: McGraw-Hill Book Co. 1087 p.

PEDIGO, LARRY P. 1989. *Entomology and Pest Management.* MacMillan, New York, NY. 904 p.

PORTER, K. S. and M. W. STIMMANN. 1988. *Protecting Groundwater: A Guide for the Pesticide User.* NYS Water Resources Institute. Cornell University. Ithaca, NY.

ROMOSER, WILLIAM and JOHN G. STOFFOLANO. 1994. *The Science of Entomology.* 3rd Ed. William C. Brown Pubs. Dubuque, IA.

SHURTLEFF, M. C., T. W. FERMANIAN, and R. RANDELL. 1987. *Controlling Turfgrass Pests.* Prentice Hall, Inc. Englewood Cliffs, NJ.

TIMM, R. M. (Ed.) 1987. *Prevention and Control of Wildlife Damage.* Cooperative Extension Service. University of Nebraska. Lincoln, NE.

THOMSON, W. T. 1993. *Agricultural Chemicals: Herbicides.* Thomson Pubs. 330 p.

WARE, G. W. 1986. *Fundamentals of Pesticides—A Self-Instruction Guide.* Thomson Publications. P.O. Box 9335. Fresno, CA 93791. 304 p.

WATSCHKE, THOMAS L. 1994. *Managing Turfgrass Pests.* Lewis Pubs. Boca Raton, FL. 400 p.

WEED SCIENCE SOCIETY OF AMERICA. 1994 *Herbicide Handbook.* 7th Ed. Champaign, IL. 352 p.

WILSON, J. H., JR. 1987. *Private Pesticide Applicator Recertification Manual.* Agricultural Extension Service. North Carolina State University. Raleigh, NC.

APPENDIX B: PESTICIDE INFORMATION TELEPHONE NUMBERS

The following telephone numbers should be helpful to you in obtaining information concerning pesticides. In addition to those listed, remember that each state land grant university has a Cooperative Extension Service Pesticide Coordinator who can provide pesticide information.

1-202-296-1585	American Crop Protection Association (ACPA)
1-800-262-8200	Chemicals Referral Center
1-800-424-9300	CHEMTREC Emergency Hotline. Twenty-four-hour emergency hotline with help dealing with major spills that may contaminate water or endanger public health.
1-800-262-7937	Disposal of Hazardous Pesticides
1-800-535-0202	EPA Community Right-to-Know Hotline
1-800-858-7378	EPA National Pesticide Telecommunications Network (NPTN). Call between 9:30 A.M. and 7:30 P.M. EST.
1-800-424-8802	EPA National Response Center
1-800-426-4791	EPA Safe Drinking Water Hotline
1-800-424-9346	RCRA Superfund Hotline

APPENDIX C: RESTRICTED USE PESTICIDES

Common name	Trade name	Pesticide type	Formulations restricted	Use pattern
Acetamide	Guardsman	Herbicide	Dimethenamide 25.0%, atrazine 28.8%	For weed control in field corn, seed corn, and popcorn/or forage
Acetochlor	Harness Plus, Surpass, Doubleplay	Herbicide	Emulsifiable concentrate	Field corn, popcorn, forage/fodder corn
Acrolein	Magnacide H and B, Aqualin	Herbicide	All formulations	All uses
Alachlor	Lasso, Lasso II, Lasso Micro-Tech, Partner, Cannon, Bullet, Bronco, Freedom, Ala-Scept, others	Herbicide	All formulations	All uses
Aldicarb	Temik	Nematicide, Insecticide	As sole active ingredient and in combination with other actives. All granular formulations.	All uses

Common name	Trade name	Pesticide type	Formulations restricted	Use pattern
Aluminum phosphide	Detia, Phoskill, L-Fume, Fumitoxin, Tri-Tox, Phostoxin, Quick-Phos, Gastoxin, Quik-Fume, Detta Fumex, others	Fumigant	As sole active ingredient	All uses
Amitraz	Mitac, Taktic	Insecticide, Miticide	All formulations	Pears
Amitrole	Amizol, Amitrol T	Herbicide	All formulations	All uses except homeowner
Arsenic acid	Desiccant L-10, Hy-Yield H-10, Poly Brand Desiccant, CCA Type C, Chemonite Part A	Herbicide, wood preservative	All formulations except brush-on	All desiccant uses, all wood preservative uses
Arsenic pentoxide	Osmose K-33, Chromated copper Arsenate, CCA, Wolmanac Concentrate	Wood preservative	All formulations	Wood preservative uses
Atrazine	Atrazine, AAtrex, Bicep, Extrazine, Bullet, Altratol, Atraol, Laddok, Sutazine, Marksman, Gladezine, others	Herbicide	All manufacturing and end use formulations	Agricultural and/or industrial herbicide. Home use exempted from restrictions.
Avermectin	Zephyr, Agri-Mek	Insecticide	Emulsifiable concentrate	Cotton, citrus
Avitrol	Avitrol	Bird control	All formulations	All uses
Azinphos-methyl	Guthion, Beetle Buster, Ketokil No. 52, others	Insecticide	All liquids with a concentration greater than 13.5%. All other formulations on a case by case basis.	All uses
Bendiocarb	Turcam	Insecticide	Granular and wettable powder	Turf
Biphenthrin	Capture 2	Insecticide	Emulsifiable concentrate	Cotton
Bis(tributylin) oxide	Interlux Micron, Iinterswift BKA007, Super Sea Jacket, Sigmaplane 7284, Navicote 2000, AF-Seaflo Z-100	Biocide	Solution—ready to use	Antifouling paint

(continued)

Common name	Trade name	Pesticide type	Formulations restricted	Use pattern
Carbofuran	Furadan	Nematicide, insecticide	All formulations, except pellets/tablets	All uses
Chlorophacinone	Rozol Tracking Powder, Rozol Blue Tracking Powder	Rodenticide	Tracking powder, dust and ready-to-use formulations 0.2%	Inside buildings
Chloropicrin	Timberfume,Chlor-O-Pic, Tri-Con, Brom-O-Gas, Terr-O-Gas, Pic-Brom, Bro-Mean, Pic-Chlor, Dowfume, others	Fumigant, fungicide, rodenticide	All formulations greater than 2% and all formulations for rodent control	All uses
Chlorothalonil	Consyst, Dacobre, Echo 75	Fungicide	Water-dispersible granules	Cranberries, strawberries, almonds, walnuts, crab apples, pears, quinces, apricots, cherries, and nectarines
Chlorpyrifos	Lorsban 4E-SG	Insecticide	Emulsifiable concentrate	Wheat
Chromic acid	CCA (Chromated Copper Arsenate), Osmose K-33, others	Wood preservative	All formulations except brush-on	All wood preservative uses
Clofentezine	Apollo SC	Miticide	Apollo SC	All uses
Coal tar	60/40 Creosote Coal/Tar Solution	Wood preservative	Solution—ready to use	Wood preserving compounds
Coal tar creosote	Osmoplastic-D, Creosote Oil, Creosote Coal Tar, Smoplastic-F	Wood preservative	All formulations	Wood preservative uses
Coumaphos	CO-RAL	Insecticide	Flowable concentrate	Indoor
Creosote oil	Original Carbolineum	Wood preservative	All formulations	Wood preservative
Cyanazine	Bladex, Extrazine, Cycle	Herbicide	All formulations	All uses
Cyfluthrin	Baythroid 2	Insecticide	25% emulsifiable concentrate	Agricultural
Cyhalothrin	Karate C50	Insecticide	Emulsifiable concentrate	Cotton
Cypermethrin	Ammo	Insecticide	All formulations	All agricultural crop uses
Diazinon	Diazinon	Insecticide	14G, AG500, 50W, 4EC, 4AG	All uses

Common name	Trade name	Pesticide type	Formulations restricted	Use pattern
Dichloropropene	Telone Soil Fumigant, Tri-Form, Pic Clor, Brom 70/30	Fumigant	All formulations (94% liquid concentrate is the only formulation)	All uses
Diclofop methyl	Hoelon 3 EC or 3 EW, Brestan H 47.5	Herbicide	All formulations	All uses
Dicrotophos	Penetrex, Chiles' Go-Better, Mauget Inject-A-Cide B	Insecticide	All liquid formulations 8% and greater	All uses
Diflubenzuron	Dimilin	Insecticide	Wettable powders	All uses
Dioxathion	Cooper Del-Tox, Delnav	Insecticide, miticide	All concentrate solutions or emulsifiable concentrates greater than 30%. All solutions 3% and greater for domestic uses.	All uses
Disulfoton	Di-Syston, Root-X, Dot-Son Brand Stand-Aid, Rigo Insyst-D, Terraclor Super-X	Insecticide	All ECs 65% and greater. All ECs and concentrate solutions 21% and greater with fensulfothion 43% and greater. All ECs 32% and greater in combination with 32% fensulfothion and greater.	All uses. Commercial seed treatment (nonaqueous solution 95% and greater).
Dodemorph	Milban	Fungicide	All formulations	All uses
Endrin	Endrin 1.6 EC	Insecticide	9.4% liquid	Bird perch use
EPTC	Doubleplay	Insecticide	Emulsifiable concentrate	Grapes, citrus, grapefruit, lemons, oranges, tangerines, almonds, walnuts, tomatoes, potatoes, sweet potato
Ethion	Ethion 8	Insecticide	Emulsifiable concentrate	All uses
Ethoprop	Mocap, Holden	Insecticide	Emulsifiable concentrates 40% and greater (aquatic). All granular and fertilizer formulations.	Aquatic uses. All uses

(continued)

Common name	Trade name	Pesticide type	Formulations restricted	Use pattern
Fenamiphos	Nemacur	Nematicide	Emulsifiable concentrates 35% and greater	All uses
Fenitrothion	Sumithion	Insecticide	Emulsifiable concentrate, 93% soluble concentrate/liquid	Only forestry uses
Fenpropathrin	Danitol	Insecticide, miticide	2.4 EC spray	Cotton
Fenthion	Mosquitocide 700, Rid-a-Bird, BX-1, BX-2, Baytex	Insecticide	Emulsifiable concentrate	Mosquitocide
Fenvalerate	Asana XL, Fury 1.5	Insecticide	Emulsifiable concentrates (30%)	Outdoor uses
Fonofos	Dyfonate	Insecticide	Emulsifiable concentrates 44% and greater. Granulars 20% and greater.	All uses
Isazofos	Triumph 4E	Insecticide	All formulations	All uses
Isofenphos	Pryfon 6	Insecticide	65% liquid formulation	Termiticide
Lambdacyhalothrin	Karate, Scimitar	Insecticide	All formulations	All uses
Lindane	Lindane, Borer Spray, many trade names		All formulations for various uses	Avocados, pecans, livestock sprays, forestry, Christmas trees, ornamentals, structural treatment, dog dusts/shampoos
Magnesium phosphide	Magtoxin, Phostoxin, Fumi-Cel Plate, Magnaphos	Insecticide, fumigant	All formulations	All uses
Methamidophos	Monitor 4	Insecticide	Liquid formulation 40% and greater, dust formulations 2.5% and greater	All uses
Methidathion	Supracide	Insecticide	All formulations	All uses except nursery stock, safflower, and sunflower
Methiocarb	Mesurol, Slug & Snail Bait, Slug'm	Acaricide, insecticide, bird repellent	All formulations	Outdoor commercial and ag uses

Common name	Trade name	Pesticide type	Formulations restricted	Use pattern
Methomyl	Lannate, Methomyl 5G, Lannabait	Insecticide	As sole active ingredient in 1% to 2.5% baits (except 1% fly bait). All concentrated solution formulations and 90% wettable powder formulations (not in water soluble bags).	Nondomestic outdoor ag crops, ornamentals, and turf; all other registered uses
Methyl bromide	Metho-O-Gas, Terr-O-Gas, Brom-O-Gas, Bro-Mean, Pic-Brom, Metabrom, Tri-Con, Tri-Brom, others	Fumigant	All formulations	All uses
Methyl isothiocyanate	Degussa methyl isothiocyanate	Wood preservative	Solution—ready to use	Fungicide for wood, wood preservative
Mevinphos	Phosdrin, Duraphos, others	Insecticide	Emulsifiable concentrates, 2% dust	None
Nicolsamide	Bayluscide	Molluscicide, larvicide	70% WP and greater	All uses
Nicotine	Nicotine, Nico-Fume, others	Insecticide, fumigant	Liquid and dry formulations 14% and above for indoor use, all formulations to cranberries	Indoor (greenhouse) applications to cranberries
Nitrogen, liquid	Liquid nitrogen	Insecticide	Solution—ready to use	Termiticide
Osamyl	Vydate	Nematicide, insecticide	Liquid formulations, granular on a case by case basis	All uses
Oxydemeton methyl	Metasystox-R, Dylox/MSR, Inject-A-Cide, Harpoon	Insecticide	All products	All uses
Paraquat	Paraquat, Gramozone, Prelude, Surefire	Herbicide	All formulations and concentrates except for certain mixtures. See label.	All uses
Parathion, ethyl	Parathion, Phoskil, Parawet, Durathion, Dithion, Thionspray No. 84, others	Insecticide	All formulations	All uses

(continued)

Common name	Trade name	Pesticide type	Formulations restricted	Use pattern
Parathion, methyl	Methyl parathion, Penncap-M, Terrazole 5%, Dithon 63, Mal Methyl, Ketokil No. 52, Seis-Ties 6-3, Metaspray, Paraspray 6-3, others	Insecticide	All formulations	All uses
Pentachlorophenol	Penta, PCP, Pol-NU, Oz-88, Pentacon, Osmo-plastic, Forepen-50, Dura-Treat, Penwar, others	Wood preservative	All formulations	Wood preservative uses
Pentachlorophenol, Sodium S	Mitrol G-ST, Dura Treat	Wood preservative	All formulations	Wood preservative uses
Permethrin	Pounce, Ambush, Ketokil, Biomist	Insecticide	All formulations	Ag crop uses (broadcast spray)
Phorate	Thimet, Rampart, Phorate, Holdem, Milo Bait, others	Insecticide	Liquid formulations 65% and greater. All granular formulations on rice.	All uses
Phosphamidon	Phosphamidon 8	Insecticide	Liquid formulations 75% and greater, dust formulations 1.5% and greater	All uses
Picloram	Tordon 101/22K/K, Access, Grazon	Herbicide	All formulations and concentrations except Tordon 101R and Tordon RTU	All uses
Piperonyl butoxide	Pybuthrin	Insecticide	Emulsifiable concentrate	Small fruits, certain berries
Profenofos	Curacron	Insecticide	Emulsifiable concentrate 59.4%	Cotton
Pronamide	Kerb	Herbicide	All 50% wettable powders	All uses
Propanoic acid	Bugle	Herbicide	Emulsifiable concentrate	Wheat, cotton, rice, clover, alfalfa, wheatgrass, sideoats grama, little bluestem, edible chrysanthemum
Propetamphos	Safrotin, Zoecon	Insecticide	Emulsifiable concentrates 50%	Indoor domestic use

Common name	Trade name	Pesticide type	Formulations restricted	Use pattern
Resmethrin	Kill-Ko-Permgard, Ind-Sol, Vex, Oblique, Bonide	Insecticide	All formulations	Mosquito abatement and pest control treatments at non-agricultural sites
Rotenone	Rotenone, Synpren, Pren-fish, Fish-Tox, Chem Fish, NUSYN	Fish control	2.5 and 5.0 EC, 5.0% and 20.0% wettable powder	Fish kill—lakes, ponds, and streams immediately above lakes and ponds
Simazine	Lilly Miller 4G, Printrex, Simazat	Herbicide	Emulsifiable concentrate	Berries (cane, black, blue, logan, cran., rasp., straw.), grapes
Sodium cyanide	M-44, Cyanide	Rodenticide	All capsules and bait formulations	All uses
Sodium fluoracetate	Compound 1080 Livestock Pro-tection Collar	Rodenticide	All solutions and dry baits	All uses
Sodium hydroxide	Augus Hot Rod	Herbicide	Ready to use	Control tree roots in sewage systems
Sodium methyl-dithiocarbamate	Metam, Vapam, Metam-Sodium	Fumigant	32.7% anhydrous	Soil fumigant—control soilborne pests to orna-mentals, food and fiber crops
Starlicide	Gull Toxicant 98%	Bird repellent	98% concentrates	Bird repellent
Strychnine	Strychnine, Gopher Bait, Gopher Getter, Gopher-Rodent Killer, others	Rodenticide	Dry baits, pellets, and powder formulations. See label for specifics.	Various uses
Sulfotepp	Dithio Insecticidal Smoke	Insecticide	Sprays and smoke generators	All uses
Sulfuric acid	Sulfuric acid	Desiccant	Solution—ready to use	Desiccant for potato vines
Sulfuryl fluoride	Sulfuryl Fluoride Fumigant, Vikane	Fumigant	All formulations	All uses
Tefluthrin	Force	Insecticide	Granular product	Corn grown for seed
Terbufos	Counter	Insecticide	Granular formula-tions 15% and greater	All uses
Tergitrol	Compound PA-14	Bird control	Solution—ready to use	Single dose poison, avian control, limited to USDA approved and supervised situations
TFM	Sea Lamprey Larvicide, TFM Bar	Biocide	Impregnated material	Aquatic pest control

(continued)

Appendix C: Restricted Use Pesticides

Common name	Trade name	Pesticide type	Formulations restricted	Use pattern
Toxaphene	Toxaphene Methyl Parathion	Insecticide	All formulations	All uses
Tralomethrin	Scout	Insecticide	All formulations	All ag crop uses
Tributyltin fluoride	Polyflo, KL-990, Amercoat Biocide 635, Pro-Line 1077, Sea Hawk Biotin, Vin Clad Super Vinge	Biocide	Solution—ready to use	Antifouling paint
Tributyltin methacrylate	Interlux Micron, Interswift BKA007, Inter-smooth Hisol, M&T Polyflo, Amercoat, Biocop, AF-SeafloZ-100, Classic Yacht 625, Hempel's	Biocide	All formulations	Antifouling paint
Triphenyltin hydroxide	Super Tin Man, Du-Ter, Supertin, Brestan H/R, Photon	Fungicide	All formulations	All uses
Zinc phosphide	Ridall-Zinc, Rodent Field/Ag Bait, ZP Tracking Powder, Zp Rodent Bait, Zincphos, others	Rodenticide	All dry formulations 60% and greater; all bait formulations; all dry formulations 10% and greater	All uses; nondomestic outdoor uses (other than 1–2% formulation in/around bldg.); domestic uses

Note: Restricted use pesticides list is intended solely to assist applicators in recognizing products that may be classified for such use. The official list of restricted use pesticides is published by the U.S. Environmental Protection Agency (EPA) and is subject to periodic change. For a current list of restricted use pesticides, consult the Code of Federal Regulations or contact the nearest EPA office.

CHEMICALS DELETED FROM THE RESTRICTED USE LIST BECAUSE OF PRODUCT CANCELLATION

Common name	Trade name	Pesticide type
Acrylonitrile	Acritet 34-66	Fumigant, insecticide
Allyl alcohol	Allyl Alcohol Weed Seed Killer	Herbicide
Alpha-chlorohydrin	Epibloc	Rodenticide
Brodifacoum	Talon G	Rodenticide
Butylate	Sutazine	Herbicide
Cadmium chloride	Caddy	Fungicide
Calcium cyanide	Calcium Cyanide A-Dust or G-Fumigant	Insecticide
Carbon tetrachloride	Dowfume 75, Vulcan Formula 72	Fumigant
Chlordimeform	Galecron, Fundal	Insecticide, miticide
Chlorfenvinphos	Residual Surface Spray and Larvicide, Coopona Poultry Premise Larvicide	Insecticide
Chlorobenzilate	Acaraben, Benz-o-chlor, Benzilan, others	Insecticide
Creosote	Coal tar creosote (nonpressure), BL Coal Tar Creosote	Wood preservative
Cycloheximide	Acti-Aid Fumigant	Insecticide
DBCP (dibromochloropropane)	Nematocide EM or Solution	Fumigant
Demeton	Systox 2, Systox 6, Demox, Stemite	Insecticide
Diallate	Avadex	Herbicide
EPN	EPN, Barricade, Ketokil, Veto, others	Insecticide, miticide
Ethylene dibromide	TRI-X Garment Fumigant, Infuco Dibrome	Fumigant
Fensulfothion	Dasanit, BIG-D Granules	Insecticide
Flucythrinate	Pay Off, AASTAR	Insecticide
Fluoroacetamide	Fluoroacetamide/1081	Rodenticide
Hydrocyanic acid	Aero HCN	Fumigant
Monocrotophos	Azodrin, DPHMC 5, Chiles' Go-Better	Insecticide
Phosacetim	Gophacide, Gopher-Trol, others	Rodenticide
Phosalone	Zolone	Insecticide
Potassium pentachlorophenate	Permatox 180 or 182	Wood preservative
Sodium arsenate	Sodium arsenate, Osmosalts	Wood preservative
Sodium dichromate	Wolman Salts CCA-Type A, B, and C, Wood preservative CCA-Type C	Wood preservative
Sodium pyroarsenate	Wolman Salts CCA-Type B	Wood preservative
TEPP	Miller Kilmite-40	Insecticide

APPENDIX D: UNITED STATES AND CANADIAN PESTICIDE CONTROL OFFICES

United States Federal Pesticide Control Offices

ENVIRONMENTAL PROTECTION AGENCY

Office of Pesticide Programs
Environmental Protection Agency
401 M Street, S.W.
Washington, DC 20460
Phone: 703-557-7090

EPA REGIONAL PESTICIDE OFFICES

Region 1 (CT, MA, ME, NH, RI, VT)
John F. Kennedy Federal Bldg.
U.S. EPA
Boston, MA 02203
Phone: 617-565-3420

Region 2 (NJ, NY, Puerto Rico,
   Virgin Islands)
26 Federal Plaza
New York, NY 10278
Phone: 212-264-2525

Region 3 (DC, DE, MD, PA, VA, WV)
841 Chestnut St.
Philadelphia, PA 19107
Phone: 215-597-9800

Region 4 (AL, FL, GA, KY, MS, NC,
   SC, TN)
345 Courtland St., N.E.
Atlanta, GA 30365
Phone: 404-347-4727

Region 5 (IL, IN, MI, MN, OH, WI)
77 W. Jackson Blvd.
Chicago, IL 60604
Phone: 312-353-2000

Region 6 (AR, LA, NM, OK, TX)
1445 Ross Ave.
12th Floor, Suite 1200
Dallas, TX 75202
Phone: 214-665-6444

Region 7 (IA, KS, MO, NE)
726 Minnesota Ave.
Kansas City, KS 66101
913-551-7000

Region 8 (CO, MT, ND, SD, UT, WY)
999 18th St., Ste. 500
Denver, CO 80202
Phone: 303-293-1603

Region 9 (American Samoa, AZ, CA,
   Guam, HI, NV)
75 Hawthorne St.
San Francisco, CA 94105
Phone: 415-744-1305

Region 10 (AK, ID, OR, WA)
1200 6th Ave.
Seattle, WA 98101
Phone: 206-553-1200

State Pesticide Control Offices

Alabama
Ag Chemistry/Plant Industry Div.
AL Dept. of Ag & Industry
P.O. Box 3336-1445 Federal Dr.
Montgomery, AL 36193
Phone: 205-242-2656

Alaska
AK Dept. of Environmental
Conservation
500 S. Alaska St.
Palmer, AK 99645
Phone: 907-745-3236

Arizona
AZ Dept. of Agriculture
Environmental Services Div.
P.O. Box 234
Phoenix, AZ 85001
Phone: 602-407-2900

Arkansas
Div. of Feeds, Fertilizer & Pesticides
AR State Plant Board
P.O. Box 1069
Little Rock, AR 72203
Phone: 501-225-1598

California
CA/EPA Dept. of Pesticide
Regulations
1020 N. St., Rm. 100
Sacramento, CA 95814-5604
Phone: 916-445-4000

Colorado
Div. of Plant Industry
CO Dept. of Ag.
700 Kipling St., #400
Lakewood, CO 80215-5894
Phone: 303-239-4140

Connecticut
Pesticide, PCP, UST, & Marine
Terminals
Dept. of Environmental Protection
State Office Bldg.
Hartford, CT 06106
Phone: 203-566-8476

Delaware
DE Dept. of Ag.
Div. of Resource Management
2320 S. DuPont Hwy.
Dover, DE 19901-5515
Phone: 302-739-4811

Florida
Agricultural Environmental Services
FL Dept. of Ag & Consumer
Services
3125 Conner Blvd.
Tallahassee, FL 32399-1650
Phone: 904-488-3731

Georgia
Entomology & Pesticide Division
GA Dept. of Ag
14 MLK Jr. Dr., SW
Atlanta, GA 30334
Phone: 404-656-3641

Hawaii
Plant Industry Div.
HI Dept. of Ag
711 Keeaumoku St.
Honolulu, HI 96814
Phone: 808-973-9401

Idaho
Div. of Agrichemical Technology
ID Dept. of Ag
P.O. Box 790
Boise, ID 83701-0790
Phone: 208-334-3550

Illinois
Bureau of Environmental Programs
IL Dept. of Ag
State Fairgrounds
P.O. Box 19281
Springfield, IL 62794-9281
Phone: 217-785-2427

Indiana
Office of Indiana State Chemists
1154 Biochemistry Bldg.
Purdue University
West Lafayette, IN 47907-1154
Phone: 317-494-1492

Iowa
IA Dept. of Ag
Henry Wallace Bldg.
Des Moines, IA 50319
Phone: 515-281-8590

Kansas
KS State Board of Ag
Division of Plant Health
901 S. Kansas Ave. 7th Floor
Topeka, KS 66612-1272
Phone: 913-296-2263

Kentucky
KY Dept. of Ag
100 Fair Oak
Frankfort, KY 40601
Phone: 502-564-7274

Louisiana
LA Dept. of Ag & Forestry
P.O. Box 3596
Baton Rouge, LA 70821-3596
Phone: 504-925-3763

Maine
ME Dept. of AG
State House Station 28
Augusta, ME 04333
Phone: 207-287-2731

Maryland
Office of Plant Ind. & Pest Mgmt.
MD Dept. of Ag
50 Harry S. Truman Pkwy.
Annapolis, MD 21401
Phone: 410-841-5870

Massachusetts
MA Dept. of Food & Ag
100 Cambridge St., 21st Floor
Boston, MA 02202
Phone: 617-727-3020

Michigan
Pesticide & Plant Mgmt. Div.
MI Dept. of Ag
P.O. Box 30017
Lansing, MI 48909
Phone: 517-335-0880

Minnesota
MN Dept. of Ag
90 West Plato Blvd.
St. Paul, MN 55107
Phone: 612-297-2530

Mississippi
Div. of Plant Industry
MS State Chemical Laboratory
P.O. Box 5207
Mississippi State, MS 39762
Phone: 601-325-3390

Missouri
Bureau of Pesticide Control
MO Dept. of Ag
P.O. Box 630
Jefferson City, MO 65102-0630
Phone: 314-751-2462

Montana
MT Dept. of Ag
Agricultural Science Div.
P.O. Box 200201
Helena, MT 59620
Phone: 406-444-2944

Nebraska
Bureau of Plant Industry
NE Dept. of Ag
P.O. Box 94756
301 Centennial Mall
Lincoln, NE 68509
Phone: 402-471-2394

Nevada
Dept. of Business & Industry
NV Dept. of Ag
350 Capitol Hill Ave.
P.O. Box 11100
Reno, NV 89510
Phone: 702-688-1180

New Hampshire
NH Dept. of Ag
P.O. Box 2042
Concord, NH 03302-2042
Phone: 603-271-3550

New Jersey
NJ Dept. of Environmental
   Protection
CN-411
380 Scotch Rd.
West Trenton, NJ 08625
Phone: 609-530-4122

New Mexico
Div. of Ag & Environmental
   Services
NM Dept. of Ag
P.O. Box 30005, Dept. 3150
Las Cruces, NM 88003-3150
Phone: 505-646-3208

New York
NY Dept. of Environmental
   Conservation
Bureau of Pesticide Regulation
50 Wolf Rd., Room 440
Albany, NY 12233-7254
Phone: 518-457-7482

North Carolina
Food & Drug Protection Div.
NC Dept. of Ag
P.O. Box 27647
Raleigh, NC 27611-7647
Phone: 919-733-3556

North Dakota
ND Dept. of Agriculture
600 E. Blvd. 6th Floor
Bismarck, ND 58502
Phone: 701-328-4756

Ohio
Pesticide Regulation
Div. of Plant Industry
8995 E. Main St.
Reynoldsburg, OH 43608-3399
Phone: 614-866-6361

Oklahoma
Div. of Plant Industry
OK Dept. of Ag
2800 N. Lincoln Blvd.
Oklahoma City, OK 73105-4298
Phone: 405-521-3864

Oregon
Plant Div.
OR Dept. of Ag
635 Capitol St., N.E.
Salem, OR 97310
Phone: 503-986-4635

Pennsylvania
Div. of Agronomic Services
PA Dept. of Ag
2301 N. Cameron St.
Harrisburg, PA 17110-9408
Phone: 717-787-4843

Rhode Island
Div. of Ag & Marketing
RI Dept. of Environmental
   Management
22 Hayes St.
Providence, RI 02908
Phone: 401-277-2781

South Carolina
Regulatory & Public Service
   Programs Div.
212 Barre Hall
Box 340390
Clemson University
Clemson, SC 29634-0390
Phone: 803-656-3005

South Dakota
Feed, Fertilizer & Pesticide Program
Div. of Regulatory Services
SD Dept. of Ag
Foss Bldg.
523 E. Capitol
Pierre, SD 57501-3188
Phone: 605-773-3724

Tennessee
Plant Industry Div.
TN Dept. of Ag
P.O. Box 40627, Melrose Station
Nashville, TN 37204
Phone: 615-360-0130

Texas
Pesticide Programs
TX Dept. of Ag
P.O. Box 12847
Austin, TX 78711
Phone: 512-463-7476

Utah
Div. of Plant Industry
UT Dept. of Ag
350 N. Redwood Rd.
Salt Lake City, UT 84116
Phone: 801-538-7188

Vermont
Plant Industry, Laboratory &
    Consumer Assurance
VT Dept. of Ag
116 State St.
Montpelier, VT 05602
Phone: 802-828-2431

Virginia
VA Dept. of Ag & Consumer Services
P.O. Box 1163
Richmond, VA 23209
Phone: 804-371-6558

Washington
Registration
P.O. Box 42589
Olympia, WA 98504
Phone: 360-902-2026

West Virginia
WV Dept. of Ag
Plant Industries Div.
Capital Complex—Guthrie Center
Charleston, WV 25305
Phone: 304-558-2228

Wisconsin
Ag Resource Mgt. Div.
WI Dept. of Ag, Trade &
    Consumer Protection
801 W. Badger Road, P.O. Box 8911
Madison, WI 53705
Phone: 608-266-7129

Wyoming
Technical Services Section
Div. of Standards & Consumer
    Services
WY Dept. of Ag
2219 Carey Ave.
Cheyenne, WY 82002
Phone: 307-777-6590

Canada
Pest Management Regulatory
    Agency
Health Canada
59 Camelot Drive
Nepean, Ontario, Canada K1A OY9
Phone: 613-952-5330

Puerto Rico
Analysis & Registration of Ag
    Materials
PR Dept. of Ag
P.O. Box 10163
Santuce, PR 00908
Phone: 809-721-2120

Virgin Islands
Environmental Protection
8000 Nisky Center, Ste. 45
St. Thomas, VI 00802
Phone: 809-774-3320

Addresses of Canadian Provincial Regulatory

Alberta Environmental Protection
Chemicals Assessment and
  Management Division
9820-106 Street, 5th Floor
Edmonton, Alberta
T5K 2J6

British Columbia Environment
Pollution Prevention & Pesticide
  Mngt. Branch
777 Broughton Street–4th Floor
Victoria, B.C.
V8V 1X4

**Government of the Northwest
  Territories**
Department of Renewable Resources
Pollution Control Division
P.O. Box 1320
Yellowknife, NWT
X1A 2L9

Manitoba Environment
Environmental Management,
  Pesticides Approval
139 Tuxedo Ave.
Winnipeg, Manitoba
R3N 0H6

New Brunswick Environment
Hazardous Materials/Pest
  Management Unit
Box 6000-364 Argyle Street
Fredericton, N.B.
E3B 5H1

**Newfoundland Department
  of Environment**
Pesticides Control Section
Box 8700
St John's, Newfoundland
A1B 4J6

**Nova Scotia Department
  of the Environment**
Box 2107
Halifax, N.S.
B3J 3B7

Prince Edward Island Agriculture
Potato Services and Horticulture
Box 306
Kensington, P.E.I.
C0B 1M0

Quebec Ministere de l´Environment
Direction du Milieu Agricole
  et du Contrôle des Pesticides
2360 Chemin Saint-Foy
Ste-Foy, Quebec
G1V 4H2

Saskatchewan Agriculture and Food
Room 323-3085 Albert Street
Regina, Saskatchewan
S4S 0B1

APPENDIX E: TRADE NAME/COMMON NAME CROSS-REFERENCE

Pesticides are most frequently listed by common names in publications. It is sometimes difficult, however, to recognize a particular pesticide by the common name if an individual is familiar with it only by its trade name. This cross-reference will be helpful in looking for certain pesticides listed in the toxicity tables in Appendix F.

Trade name (listed alphabetically)	Common name	Trade name (listed alphabetically)	Common name
A-Rest	ancymidol	Arsonate	MSMA
Aaterra	etridiazole	Asana	esfenvalerate
AAtrex	atrazine	Assert	imazamethabenz
Abate	temophos	Assure II	quizalofop-p-ethyl
Accelerate	endothall	Asulox	asulam
Accent	nicosulfuron	Avadex BW	triallate
Access	picloram + triclopyr	Avenge	difenzoquat
Accord	glyphosate	Avid	abamectin
Acetellic	pirimiphos-methyl	Avitrol	aminopyridine
Acrobe	*Bacillus thuringiensis berliner*	Azodrin	monocrotophos
		Azofene	phosalone
Adios	carbaryl	B-Nine	daminozide
Agri-Mek	abamectin	Bactec Bernan	*Bacillus thuringiensis berliner*
Agri-Mycin 17	streptomycin		
Alanap-L	naptalam	Balan	benefin
Alfa	sulfur	Banvel	dicamba
Aliette	fosetyl-al	BareSpot Ureabor	Bromacil + sodium-chlorate + sodium metaborate
Ally	metsulfuron-methyl		
Altosid	methoprene		
Amber	triasulfuron	Barricade	prodiamine
Ambush	permethrin	Basagran	bentazon
Amdro	hydramethylnon	Basic Copper 53	copper sulfate
Amiben	chloramben	Basicap	copper sulfate
Amitrol T	amitrole	Bayfidan	triadimenol
Ammo	cypermethrin	Baygon	propoxur
Ansar	MSMA	Bayleton	triadimefon
Ansar 8100	DSMA	Baytax	fenthion
Apam	metam-sodium	Baythroid	cyfluthrin
Apex	methoprene	Baytan	triadimenol
Apollo	clofentezine	Beacon	primisulfuron
Apron	metalaxyl	Belmark	fenvalerate
Aquathol	endothall	Benlate	benomyl
Arbotect	thiabendazole	Bensumec	bensulide
Argold	cinmethylin	Betamix	desmedipham + phenmedipham
Arrosolo	molinate + propanil		
Arsenal	imazapyr	Betanal	phenmedipham
		Betanex	desmedipham

Trade name (listed alphabetically)	Common name	Trade name (listed alphabetically)	Common name
Biarbinex	heptachlor	Carbamate	ferbam
Bicep	atrazine + metolachlor	Carzol	formetanate hydrochloride
Bidrin	dicrotophos		
Biobit HP	*Bacillus thuringiensis*	Casoron	dichlobenil
Biocot FC	*Bacillus thuringiensis berliner*	Cerone	ethephon
		Champ	copper hydroxide
Bladex	cyanazine	Chiptox	MCPA
Blazer	acifluorfen	Chlor-O-Pic	chloropicrin
Blue Shield	copper hydroxide	Chopper	imazapyr
Bolstar	sulprofos	Clarity	dicamba
Bomyl	bomyl	Classic	chlorimuron ethyl
Bonzi	paclobutrazol	Co-Rax	warfarin
Boot Hill	bromadiolone	Cobra	lactofen
Botec	captan + DCNA	COC WP	copper oxychloride
Botran	dicloran	COCS	copper oxychloride sulfate
Brace	isazophos		
Bravo	chlorothalonil	Comite	propargite
Bravo S	sulfur + chlorothalonil	Command	clomazone
Broadstrike	flumetsulam	Commence	clomazine + trifluralin
Broadstrike Plus	clopyralid + flumetsulam	Concep II	oxabetrinil
		Concep III	fluxofenim
Brom-O-Gas	methyl bromide + chloropicrin	Concert	chlorimuron + thifensulfuron
Brom-O-Sol	methyl bromide + chloropicrin	Conclude G	sethoxydim
		Confront	clopyralid + triclopyr
Bromone	bromadiolone	Contour	atrazine + imazethapyr
Bronate	bromoxynil + MCPA	Copper Power	copper sulfate
Bronco	alachlor + glyphosate	Copper Z	copper sulfate
Broot	trimethacarb	Copper-Count N	copper ammonium carbonate
Buckle	triallate + trifluralin		
Buctril	bromoxynil	Coppercide	copper hydroxide
Bueno	MSMA	Copro	copper oxychloride sulfate
Bugle	fenoxaprop-P-ethyl		
Busan 1020	metam-sodium	Cotoran	fluometuron
Busan 30A	TCMBT	Cotton-Pro	prometryn
Butyrac	2,4-DB	Counter	terbufos
Caid	chlorophacinone	Cov-R-Tox	warfarin
Calar	cama	Coxysul	copper oxychloride sulfate
Caliber	simazine		
Calo-gran	mercurous chloride + mercuric chloride	Croak	flumetsulam + MSMA
		Cropstar	alachlor
Canopy	chlorimuron-ethyl + metribuzin	Crossbow	2,4-D + triclopyr
		CS-56	copper oxychloride sulfate
Caparol	prometryn		
Capture	bifenthrin		*(continued)*

Trade name (listed alphabetically)	Common name	Trade name (listed alphabetically)	Common name
Cuproxat	copper oxysulfate	Dual	metolachlor
Curacron	profenofos	Dyfonate	fonofos
Curtail	clopyralid + 2,4-D	Dylox	trichlorfon
Cutlass	*Bacillus thuringiensis*	Eclipse	fenoxycarb
Cybolt	flucythrinate	Edge	fonofos + pebulate
Cyclone	paraquat	ELF	diazinon
Cycocel	chlormequat chloride	Embark	mefluidide
Cymbush	cypermethrin	Endurance	prodiamine
Cythion	malathion	Envert 171	2,4-D + dichlorprop
Cytrolane	mephosfolan	Eptam	EPTC
D-D 92	dichloropropene	Eradicane	EPTC
D-z-n Diazinon	diazinon	Escort	metsulfuron-methyl
Dabantic	tetrachlorvinphos	Ethion	ethion
Daconate	MSMA	Ethrel	ethephon
Daconil 2787	chlorothalonil	Evik	ametryn
Dacthal	DCPA	Express	tribenuron methyl
Dakota	fenoxaprop-P-ethyl + MCPA	Extrazine II	atrazine + cyanazine
		Facet	quinclorac
Danitol	fenpropathrin	Fallow Master	dicamba + glyphosate
Deadline	metaldehyde	Far-Go	triallate
Debantic	tetrachlorvinphos	Ficam	bendiocarb
Decco	chlorpropham	Finesse	chlorsulfuron + metsulfuron
DEF 6	butifos		
Defol	sodium chlorate	Folex	tribufos
Demon	cypermethrin	Folicur	tebuconazole
Denarin	triforine	Force	tefluthrin
Desiccant H-10	arsenic acid	Fore	manganese ethylene bisdith iocarbamate
Devrinol	napropamide		
Di-Syston	disulfoton	Freedom	alachlor + trifluralin
Dibrom	naled	Frontier	dimethenamid
Dimecron	phosphamidon	Fruit Fix	1-naphthalenacetic acid
Dimethoate	dimethoate	Fruitone N	1-naphthalenacetic acid
Dimethyl-T	DSMA	Fumi-Cel	magnesium phosphide
Dimilin	diflubenzuron	Fumi-Strip	magnesium phosphide
DiPel	*Bacillus thuringiensis*	Funginex	triforine
Diphacin	diphacinone	Fungo 50	thiophanate-methyl
Dipterex	trichlorfon	Furadan	carbofuran
Dithane	mancozeb	Fury	cypermethrin
Diumate	MSMA	Fusilade 2000	fluazifop-P-butyl
Dividend	difenoconazzole	Fusion	fenoxaprop-ethyl + fluazifop-P
Domain	thiophanate-methyl		
Drexel DSMA	DSMA	Galaxy	acifluorfen + bentazon
Drexel Sulfur	sulfur	Gemini	chlorimuron-ethyl + linuron
Dropp	thiadiazuron		

Trade name (listed alphabetically)	Common name	Trade name (listed alphabetically)	Common name
Ginstar	diuron + thidiazuron	Legion	naled
Glean	chlorsulfuron	Lentagran	pyridate
Goal	oxyflurofen	Lexone	metribuzin
Golden Dew	sulfur	Linex	linuron
Golden Leaf Tobacco Spray	endosulfan	Lock-on	chlorpyrifos
		Logic	fenoxycarb
Gramoxone	paraquat	Londax	bensulfuron methyl
Grazon	picloram	Lorox	linuron
Guardsman	atrazine + dimethenamid	Lorox Plus	chlorimuron-ethyl + linuron
Guthion	azinphos-methyl	Lorsban	chlorpyrifos
Harmony Extra	thifensulfuron + tribenuron	M-Peril	*Bacillus thuringiensis*
		M-Trak	*Bacillus thuringiensis*
Harness	acetochlor	Maintain	chlorflurenol
Harvade	dimethipin	Maki	bromadiolone
Havoc	brodifacoum	Manzate	mancozeb
Herbicide 273	endothall	Marksman	atrazine + dicamba
Hi-Yield	arsenic acid	Marlate	methoxychlor
Hoelon	diclofop-methyl	Mavrik	fluvalinate
Horizon	fenoxaprop-ethyl	Mecomec	mecoprop
Hydrothol	endothall	Melorex	dodine
Hyvar	bromacil	Mepichlor	mepiquat chloride
Ignite	glufosinate-ammonium	Mertect	thiabendazole
Imidan	phosmet	Mesurol	methiocarb
Javelin	*Bacillus thuringiensis*	Metam 426	metam-sodium
Karate	lambdacyhalothrin	Metasystox-R	oxydemeton-methyl
Karmex	diuron	Meth-O-Gas	methyl bromide
Kelthane	dicofol	Methar 30	DSMA
Kerb	pronamide	Meturon	fluometuron
KM	sodium chlorate	Micro-Tech	alachlor
Knox Out	diazinon	Microthiol Special	sulfur
Koban	etridiazole	Milo-Pro	propazine
Kocide	copper hydroxide	Mitak	amitraz
Komeen	copper-ethylenediamine	Mocap	ethoprop
KOP 300	copper sulfate	Mocap Plus	disulfoton + ethoprop
Krenite	fosamine ammonium	Monitor	methamidophos
Krovar	bromacil + diuron	Montar	cacodylic acid
Kryocide	cryolite	Morestan	oxythioquinox
Laddok	atrazine + bentazon	Mustang	cypermethrin
Landmaster BW	glyphosate + 2,4-D	MVP	*Bacillus thuringiensis*
Lannate	methomyl	Mylone	dazomet
Lariat	alachlor + atrazine	NAA 800	1-naphthaleneacetic acid
Larvin	thiodicarb	Nemacur	fenamiphos
Lasso	alachlor		*(continued)*

Trade name (listed alphabetically)	Common name	Trade name (listed alphabetically)	Common name
Nimrod	burpimate	Prolate	phosmet
Nortron	ethofumisate	Promalin	benzyladenine and gibberellins A4, A7
Nova	mycobutanil		
Noxfish	rotenone	Promar	diphacinone
Nu-Flow	chloroneb	Prompt	atrazine + bentazon
Nustar	flusilazole	ProVide	gibberellins A4, A7
Oftanol	isofenphos	Prowl	pendimethalin
Omite	propargite	Proxol	trichlorfon
Option II	fenoxaprop-P-ethyl	Punch	flusilazole
Orbit	propiconazole	Pursuit	imazethapyr
Ordram	molinate	Pyramin	pyrazon
Ornalin	vinclozolin	Pyramin	chloridazon
Orthene	acephate	Pyrenone	pyrethrum
Oust	sulfometuron methyl	Pyrocide	pyrethrum
Ovasyn	amitraz	Rabon	tetrachlorvinphos
Pansoil	etridiazole	Rad-E-Cate	sodim cacodylate
Partner	alachlor	Rally	mycobutanil
Parzate C	zineb	Ramik	diphacinone
Payload	acephate	Rampage	cholecalciferol
Penncap-M	methyl parathion	Ramrod	propachlor
Penncozeb	mancozeb	Ranger	glyphosate
Pentac	dienochlor	Raptor	*Bacillus thuringiensis*
Phaser	endosulfan	RAX	warfarin
Phosdrin	mevinphos	Reach	chlorothalonil + triadimefon
Phostoxin	aluminum phosphide		
Phytar	cacodylic acid	Reflex	fomesafen
Pinnacle	thifensulfuron methyl	Reldan	chlorpyrifos-methyl
Pipron	piperalin	Release	gibberellic acid
Pivalyn	pindone	Resolve CP	imazapyr
Pix	mepiquat chloride	Resolve CP-A	imazethapyr
Plantvax	oxycarboxin	Resolve CP-B	dicamba
Poast	sethoxydim	Resource	flumiclorac pentyl ester
Polyram	metiram	Reward	diquat
Pounce	permethrin	Rhodocide	ethion
Pramitol	prometon	Rhomene	MCPA
Pre-San	bensulide	Rhonox	MCPA
Prefar	bensulide	Ridall-Zinc	zinc phosphide
Preview	chlorimuron-ethyl + metribuzin	Ridomil	metalaxyl
		Ridomil MZ58	metalaxyl + EBDC
Prime+	flumetralin	Ridomil PC	metalaxyl + PCNB
Princep	simazine	Ro-Neet	cycloate
Prism	clethodim	Rodenticide AG	zinc phosphide
Procure	triflumizole	Rodeo	glyphosate
ProGibb	gibberellic acid	Rodex	warfarin

Trade name (listed alphabetically)	Common name	Trade name (listed alphabetically)	Common name
Ronilan	vinclozolin	Sutan+	butylate
Ronstar	oxadiazon	Sutazine+	atrazine + butylate
Roundup	glyphosate	Syllit	dodine
Rovral	iprodione	Synchrony	chlorimuron + thifensulfuron
Roxion	dimethoate		
Royal MH-30	maleic hydrazide	Synthrin	resmethrin
Royal Slo Gro	maleic hydrazide	Talon	brodifacoum
Rozol	chlorophacinone	Tandem	tridiphane
RTU PCNB	quintozene	Tandex	karbutilate
Rubigan	fenarimol	TBZ	thiabendazole
RyzUp	gibberellic acid	Tecto	thiabendazole
Salute	metribuzin + trifluralin	Telar	chlorsulfuron
Savey	hexythiazox	Telone	dichloropropene
Scepter O.T.	acifluorfen + imazaquin	Temik	aldicarb
		Terr-O-Gas	methyl bromide + chloropicrin
Scout X-TRA	tralomethrin		
Select	clethodim	Terraclor	quintozene
Sencor	metribuzin	Terraclor Plus	disulfoton + PCNB
Sevin	carbaryl	Terraclor Super	disulfoton + etridiazole + PCNB
Signal	sulfur		
Sinbar	terbacil	Terraneb SP	chloroneb
Slam	carbaryl	Terrazole	etridiazole
Soil-Mend	lime-sulfur	Tersan 1991	benomyl
Solicam	norflurazon	Thibenzole	thiabendazole
Sonalan	ethalfluralin	Thimet	phorate
Sonar	fluridone	Thiodan	endosulfan
Spike	tebuthiuron	Thiolux	sulfur
Spotrete	thiram	Thistrol	MCPB
Sprout Nip	chlorpropham	Thuricide	*Bacillus thuringiensis*
Squadron	imazaquin + pendimethalin	Tillam	pebulate
		Tiller	2,4-D + fenoxaprop-P + MCPA
Stam	propanil		
Stampede	propanil		
Starfire	paraquat	Tilt	propiconazole
Stinger	clopyralid	Tomcat	diphacinone
Storite	thiabendazole	Topsin E	thiophanate
Strel	propanil	Topsin M	thiophanate-methyl
Sulferix	lime-sulfur	Tordon 22K	picloram
Super Tin	triphenyltin hydroxide	Tordon RTU	2,4-D + picloram
Super-sul	sulfur	Tornado	fluazifop-P + fomesafen
Supracide	methidathion		
Surflan	oryzaline	Torpedo	sethoxydim
Surpass	acetochlor	Touchdown	sulfosate
Surpass 100	acetochlor + atrazine		*(continued)*

Trade name (listed alphabetically)	Common name	Trade name (listed alphabetically)	Common name
Tough	pyridate	Vapona	DDVP
Tox-Hid	warfarin	Vault	*Bacillus thuringiensis*
Treflan	trifluraln	VectoBac	*Bacillus thuringiensis berliner*
Tri-4	trifluralin		
TRI-CON	methyl bromide + chloropicrin	Vectrim	resmethrin
		Velpar	hexazinone
Tri-Cut	mefluidide	Vendex	fenbutatin-oxide
Tri-Scept	imazaquin + trifluralin	Venturol	dodine
Trigard	cyromazine	Vernam	vernolate
Trilin	trifluralin	Vydate L	oxamyl
Triumph	isazophos	Vikane	sulfuryl fluoride
Truban	etridiazole	Vitavax	carboxin
Tupersan	siduron	Vorlan	vinclozolin
Turbo	metolachlor + metribuzin	Weatherblok	brodifacoum
		Weedmaster	2,4-D + dicamba
Turcam	bendiocarb	Weedone 170	2,4-D + dichlorprop
Turfcide	quintozene	Weedone CB	2,4-D + dichlorprop
2 Plus 2	2,3-D + mecoprop	Whip	fenoxaprop-ethyl
Typhoon	fluazifop-P + fomesafen	XenTari	*Bacillus thuringiensis*
		Zephyr	abamectin
Uniflow	sulfur	Ziram 76	ziram
Ureabor	borax + monuron	Zolone	phosalone
Vapam	metam-sodium	Zorial	norlurazon

APPENDIX F: TOXICITY CLASSIFICATION AND SIGNAL WORDS FOR PESTICIDES

TABLE 1 TOXICITY CLASSIFICATION AND SIGNAL WORDS FOR INSECTICIDES AND ACARICIDES

Common name	Trade name	Producer	Toxicity class	Signal word
abamectin	Agri-Mek Avid	Merck	IV	Caution
	Zephyr		III (dry)	Caution
acephate	Orthene Payload	Valent	III	Caution
aldicarb	Temik	Rhone-Poulenc	I	Danger
amitraz	Mitak Ovasyn	AgrEvo	II	Warning
azinphos-methyl	Guthion	Bayer	I	Danger–Poison
Bacillus thuringiensis (subspecies *kurstaki*)	DiPel Thuricide Vault Javelin Biobit HP MVP M-Peril Cutlass	Abbott Labs Sandoz DuPont Mycogen Ecogen	III	Caution
Bacillus thuringiensis (var. *tenebrionis*)	M-Trak	Mycogen	III	Caution
Bacillus thuringiensis berliner (subspecies *aizawai*) Lepidop-teran active toxin	ZenTari	Abbott	III	Caution
Bacillus thuringiensis berliner (subspecies *israelensis*)	VectoBac Acrobe	Abbott American Cyanamid	III	Caution
Bacillus thuringiensis berliner (subspecies *kurstaki*)	Biocot FC	Uniroyal	III	Caution
Bacillus thuringiensis berliner (var. *morrisoni*)	Bactec Bernan	Bactec	III	Caution
Bacillus thuringiensis (subspecies *kurstaki*) strain BMP 123	Raptor	American Cyanamid	III	Caution
bendiocarb	Ficam Turcam	AgrEvo	II	Warning
bifenthrin	Capture	FMC	II	Warning
bomyl	Bomyl	HACCO	I (tech) II (bait)	Danger–Poison Warning
carbaryl	Sevin Adios Slam	Rhone-Poulenc BASF	I	Poison

(continued)

TABLE 1 *(Continued)*

Common name	Trade name	Producer	Toxicity class	Signal word
carbofuran	Furadan	FMC	I (flowable) II (granule)	Danger Warning
chlordane	Many	Some formulations temporarily halted	II	Warning
chlorobenzilate		Ciba	III	Caution
chlorpyrifos	Lorsban Lock-on	Dow Elanco	II	Warning
chlorpyrifos-methyl	Reldan	Dow Elanco Gustafson	III	Caution
clofentezine	Apollo	AgrEvo	III	Caution
cryolite	Kryocide	ELF Atochem	III	Caution
cyfluthrin	Baythroid	Bayer	I, II	Danger (eyes) Warning
cypermethrin	Mustang Ammo Fury Cymbush Demon	FMC Zeneca	III	Caution
cyromazine	Trigard	Ciba	III	Caution
DDT	Many	Cancelled in U.S.	III	Caution
DDVP	Vapona	Fermenta	I	Danger–Poison
diazinon	D-z-n Diazinon Knox Out ELF	Ciba Atochem	II or III (depends on formulation)	Warning or Caution
dicofol	Kelthane	Rohm & Haas	II or III (depends on formulation)	Warning or Caution
dicrotophos	Bidrin	Amvac	I	Danger–Poison
dienochlor	Pentac	Sandoz	II	Warning (eyes)
diflubenzuron	Dimilin	Uniroyal	III	Caution
dimethoate	Roxion Dimethoate	American Cyanamid Helena	II	Warning
disulfoton	Di-Syston	Bayer	I	Danger–Poison
endosulfan	Phaser Thiodan Golden Leaf Tobacco Spray	AgroEvo FMC	I	Danger–Poison
esfenvalerate	Asana	DuPont	II	Danger
ethion	Ethion Rhodocide	FMC Rhone-Poulenc	I (EC) II (WP)	Danger–Poison Warning
ethoprop	Mocap	Rhone-Poulenc	I (depends on II formulation)	Danger Warning
fenbutatin-oxide (hexakis)	Vendex	DuPont	I	Danger
fenitrothion			II	Warning

TABLE 1 *(Continued)*

Common name	Trade name	Producer	Toxicity class	Signal word
fenoxycarb	Eclipse Logic	Ciba	IV	Caution
fenpropathrin	Danitol	Valent	I	Danger (eyes)
fenthion	Baytax	Bayer	II	Warning
fenvalerat	Belmark	American Cyanamid	II	Warning
flucythrinate	Cybolt	American Cyanamid	I	Danger
fluvalinate	Mavrik	Sandoz	I	Warning
fonofos	Dyfonate	ZENECA	I or II (depends on formulation)	Danger Warning
formetanate hydrochloride	Carzol	AgrEvo	I	Danger–Poison
heptachlor	Biarbinex	Some sale halted	II	Warning
hexythiazox	Savey	Gowan	III	Caution
hydrmethylnon	Amdro	American Cyanamid	III	Caution
isazophos	Brace Triumph	Ciba	II	Warning
isofenphos	Oftanol	Bayer	I	Danger
lambdacyhalothrin	Karate	ZENECA	II	Warning
lead arsenate		Used outside U.S.	I	Danger
lindane	Many	Drexel	II	Warning
malathion	Cythion	Helena	III	Caution
mephosfolan	Cytrolane	American Cyanamid	I	Danger
metaldehyde	Deadline	Valent	III	Caution
methamidophos	Monitor	Bayer, Valent	I	Danger–Poison
methidathion	Supracide	Ciba	I	Danger
methiocarb	Mesurol	Bayer	II	Warning
methomyl	Lannate	DuPont	I	Danger–Poison
methoprene	Altosid Apex	Sandoz	IV	Caution
methoxychlor	Marlate	Kincaid Enterprises	IV	Caution
methyl parathion	Penncap-M Methyl parathion	ELF Atochem Helena	I	Danger–Poison
mevinphos	Phosdrin	American Cyanamid	I	Danger
monocrotophos	Azodrin	American Cyanamid	I	Danger–Poison
naled	Dibrom Legion	Valent	I	Danger
nicotine			I	Danger
oxamyl	Vydate L	DuPont	I	Danger–Poison

(continued)

Appendix F: Toxicity Classification and Signal Words for Pesticides **471**

TABLE 1 *(Continued)*

Common name	Trade name	Producer	Toxicity class	Signal word
oxydemeton-methyl	Metasystox-R	Gowan	I	Danger
oxythioquinox	Morestan	Bayer	I, II or III (depends on formulation)	Danger, Warning, Caution
parathion	Many	Several	I	Danger–Poison
permethrin	Ambush Ounce	ZENECA FMC	II or III (depends on formulation)	Warning Caution
phorate	Thimet	American Cyanamid	I	Danger–Poison
phosalone	Zolone Azofene	Rhone-Poulenc	II	Warning
phosmet	Imidan Prolate	Gowan	II	Warning
phosphamidon	Dimecron	Ciba	I	Danger–Poison
pirimiphos-methyl	Acetellic	ZENECA	II	Warning
profenofos	Curacron	Ciba	II	Warning
propargite	Omite Comite	Uniroyal	I	Danger
propoxur	Baygon	Bayer	III	Caution
pyrethrum	Pyrenone Pyrocide	Roussel Uclaf MGK	III	Caution
resmethrin	Synthrin Vectrim	Roussel Uclaf	III	Caution
rotenone	Noxfire Noxfish	Roussel Uclaf	I (depends on III formulation)	Danger Caution
sulprofos	Bolstar	Bayer	II	Warning
tefluthrin	Force	ZENECA	II	Warning
temophos	Abate	American Cyanamid	III	Caution
terbufos	Counter	American Cyanamid	I	Danger–Poison
tetrachlorvinphos	Rabon Debantic	Fermenta	III	Caution
thiodicarb	Larvin	Rhone-Poulenc	II	Warning
toxaphene			II	Warning
tralomethrin	Scout X-TRA	AgrEvo	I	Danger
trichlorfon	Dylox Dipterex Proxol	Bayer AgrEvo	II	Warning
trimethacarb	Broot	Drexel	III	Caution

TABLE 2 TOXICITY CLASSIFICATION AND SIGNAL WORDS FOR HERBICIDES AND PLANT GROWTH REGULATORS

Common name	Trade name	Producer	Toxicity class	Signal word
acetochlor	Harness	Monsanto	I	Danger (eyes)
	Surpass	ZENECA	II	Warning
acifluorfen	Blazer	BASF	I	Danger (eyes)
alachlor	Lasso	Monsanto	I	Danger
	Cropstar			
	Micro-Tech			
	Partner			
alachlor + atrazine	lariat	Monsanto	II	Warning
alachlor + glyphosate	Bronco	Monsanto	I	Danger
alachlor + trifluralin	Freedom	Monsanto	III	Caution
ametryn	Evik	Ciba	III	Caution
amitrole	Amitrol T	Rohne-Poulenc	III	Caution
ancymidol	A-Rest	Lilly (Elanco)	III	Caution
arsenic acid	Hi-Yield Desiccant H-10	Pennwalt	I	Danger
asulam	Asulox	Rhone-Poulenc	IV	Caution
atrazine	AAtrex	Ciba	III	Caution
atrazine + bentazon	Laddok	BASF	III	Caution
	Laddok 600			
	Prompt			
atrazine + bromoxynil	Buctril + Atrazine	Rhone-Poulenc	III	Caution
atrazine + butylate	Sutazine+	ZENECA	II	Warning
atrazine + cyanazine	Extrazine II	DuPont	IV	Caution
atrazine + dicamba	Marksman	Sandoz	III	Caution
atrazine + dimethenamid	Guardsman	Sandoz	III	Caution
atrazine + imazethapyr	Contour	American Cyanamid		Caution
atrazine + metolachlor	Bicep	Ciba	III	Caution
benefin	Balan	DowElanco	IV (depends on formulation)	Caution
bensulfuron methyl	Londax	DuPont	IV	Caution
bensulide	Pre-San	PBI/Gordon	III	Caution
	Bensumec			
	Prefar	Gowan		
bentazon	Basagran	BASF	III	Caution
benzyladenine and gibberellins A4, A7	Promalin	Abbott	II	Warning
borax		Kerr-McGee	III	Caution
borax + monuron	Ureabor	J. R. Simplot	I	Danger

(continued)

TABLE 2 *(Continued)*

Common name	Trade name	Producer	Toxicity class	Signal word
bromacil	Hyvar	DuPont	II (liquid) III (dry)	Warning Caution
bromacil + diuron	Krovar	DuPont	III	Caution
bromacil + sodium chlorate + sodium metaborate	BareSpot Ureabor	Simplot	I	Danger
bromoxynil	Buctril	Rhone-Poulenc	II	Warning
bromoxynil + atrazine	Buctril + Atrazine	Rhone-Poulenc	II	Warning
bromoxynil + MCPA	Bronate	Rhone-Poulenc	II	Warning
butifos	DEF 6	Bayer	I	Danger
butylate	Sutan⁺	ZENECA	III	Caution
cacodylic acid	Montar Phytar	Monterey	III	Caution
cama	Calar	Drexel	III	Caution
calcium arsenate			I	Danger–Poison
chloramben	Amiben	Rhone-Poulenc	IV	Caution
chlorflurenol	Maintain CF 125 Maintain A	Uniroyal	III	Caution
chloridazon	Pyramin	BASF	III	Caution
chlorimuron ethyl	Classic	DuPont	III	Caution
chlorimuron-ethyl + linuron	Gemini Lorox Plus	DuPont	III	Caution
chlorimuron-ethyl + metribuzin	Canopy Preview	DuPont	III	Caution
chlorimuron + thifensulfuron	Concert Synchrony	DuPont	III	Caution
chlormequat chloride	Cycocel	American Cyanamid	III	Warning
chlorpropham	Decco Sprout Nip	ELF Atochem Platte	III	Caution
chlorsulfuron	Glean Telar	DuPont	IV	Caution
chlorsulfuron + metsulfuron	Finesse	DuPont	IV	Caution
cinmethylin	Argold	American Cyanamid	III	Caution
clethodim	Select Prism	Valent	II	Warning
clomazone	Command	FMC	II	Warning
clopyralid	Stinger	DowElanco	I III	Danger (eyes) Caution
clopyralid + 2,4-D	Curtail	DowElanco	I	Danger
clopyralid + flumetsulam	Broadstrike Plus	DowElanco	I	Danger

TABLE 2 *(Continued)*

Common name	Trade name	Producer	Toxicity class	Signal word
clopyralid + triclopyr	Confront	DowElanco	III	Caution
clomazine + trifluralin	Commence	FMC	III	Warning
copper-ethylenedi-amine complex	Komeen	Griffin	III	Caution
cyanazine	Bladex	DuPont	II	Warning
cycloate	Ro-Neet	ZENECA	III	Caution
2,4-D	Many	Many	I	Danger (eyes)
			III	Caution
2,4-D + dicamba	Weedmaster	Sandoz	IV	Caution
2,4-D + dichlorprop	Envery 171 Weedone 170 Weedone CB	Rhone-Poulenc	II	Warning
2,4-D + fenoxaprop-P + MCPA	Tiller	AgrEvo	III	Caution
2,4-D + mecoprop	2 Plus 2	ISK Biosciences	III	Caution
2,4-D + picloram	Tordon RTU	DowElanco	III	Caution
2,4-D + triclopyr	Crossbow	DowElanco	III	Caution
2,4-DB	Butyrac	Rhone-Poulenc	III	Caution
dalapon			II	Warning
daminozide	B-Nine	Uniroyal	III	Caution
dazomet	Mylone	UBC Bio-chemicals	III	Caution
DCPA	Dacthal	ISK Biosciences	IV	Caution
desmedipham	Betanex	AgrEvo	III	Caution
desmedipham + phenmedipham	Betamix	AgrEvo	II	Warning
dicamba	Banvel Banvel SGF Resolve CP-B	Sandoz American Cyanamid	II	Warning (eye)
dicamba + glyphosate	Fallow Master	Monsanto	I	Danger (eyes)
dichlobenil	Casoron	Uniroyal	III	Caution
diclofop-methyl	Hoelon	AgrEvo	III	Caution
3,6-dichloro-o-anisic acid	Clarity	Sandoz	III	Caution
difenzoquat	Avenge	American Cyanamid	II	Danger
dimethenamid	Frontier	Sandoz	II	Warning
dimethipin	Harvade	Uniroyal	III	Caution
diphenamide			III	Caution
diquat	Rewa	ZENECA	II	Warning

(continued)

Appendix F: Toxicity Classification and Signal Words for Pesticides   **475**

TABLE 2 *(Continued)*

Common name	Trade name	Producer	Toxicity class	Signal word
diuron	Karmex	DuPont	III	Caution
diruon + thidiazuron	Ginstar	AgrEvo	IV	Caution
DSMA	Methar 30 Drexel DSMA Ansar 8100 Dimethyl-T DSMA Liquid	W. A. Cleary Drexel ISK Biosciences	III	Caution
endothall	Endothal Herbicide 273 Accelerate Aquathol Hydrothol	ELF Atochem	II	Warning
EPTC	Eptam Eradicane	ZENECA	III	Caution
ethalfluralin	Sonalan	DowElanco	II	Warning
ethephon	Ethrel Cerone	Rhone-Poulenc	I, (depending II on concen- tration)	Danger
ethofumisate	Nortron	AgrEvo	I	Warning
fenoxaprop-ethyl	Whip Horizon	AgrEvo	III	Danger
fenoxaprop-P-ethyl	Bugle Option II	AgrEvo	II	Caution
flumetsulam	Broadstrike	DowElanco	III	Warning
flumetsulam + clopyralid	Broadstrike Plus	DowElanco	II	Caution
flumetsulam + trifluralin	Broadstrike + Treflan	DowElanco	I	Warning (eyes)
flumetsulam + metolachlor	Broadstrike + Dual	DowElanco	I	Danger (eyes)
fluxofenim	Concep III	Ciba	II	Danger (eyes)
fluazifop-butyl	Fusilade	ZENECA	III	Warning (eyes)
fluazifop-p-butyl + fomesafen	Tornado Typhoon	ZENECA	II	Caution
flumetralin	Prime+	Ciba	IV	Warning
flumiclorac pentyl ester	Resource	Valent	I	Caution
fluometuron	Cotoran Meturon	Ciba Griffin	II	Danger
fluometuron	Croak	Drexel	II	Warning
fluridone	Sonar	DowElanco	IV	Warning (eyes)
fomesafen	Reflex	ZENECA	II	Caution
fosamine ammonium	Krenite	DuPont	III	Warning (eyes, skin)
gibberellic acid	ProGibb Release RyzUp	Abbott	III	Warning (eyes)

TABLE 2 *(Continued)*

Common name	Trade name	Producer	Toxicity class	Signal word
gibberellins A4, A7	ProVide	Abbott	III	Caution
glufosinate-ammonium	Ignite	AgrEvo	III	Caution
glyhposate	Ranger	Monsanto	I	Danger (eyes)
	Roundup		II	Warning (eyes)
	Rodeo			
	Accord			
glyphosate + 2,4-D	Landmaster BW	Monsanto	I	Danger (eyes)
hexazinone	Velpar	DuPont	I	Danger (eyes)
imazamethabenz	Assert	American Cyanamid	III	Caution
imazapyr	Arsenal	American Cyanamid	IV	Caution
	Chopper			
	Resolve CP			
imazaquin	Scepter	American Cyanamid	III	Caution
imazaquin + pendimethalin	Squadron	American Cyanamid	I	Danger (eyes)
imazaquin + trifluralin	Tri-Scept	American Cyanamid	I	Warning (corrosive)
imazethapyr	Pursuit	American Cyanamid	III	Caution
	Resolve CP-A			
imazethapyr + pendimethalin	Pursuit Plus	American Cyanamid	III	Caution
karbutilate	Tandex	FMC	III	Caution
lactofen	Cobra	Valent	I	Danger (eyes)
linuron	Lorox	DuPont	III	Caution
	Linex	Griffin		
maleic hydrazide (MH)	Royal MH-30	Uniroyal	IV	Caution
	Royal Slo Gro			
MCPA	Rhomene	Rhone-Poulenc	I	Danger
	Rhonox			
	Chiptox			
MCPB	Thistrol	Rhone-Poulenc	III	Caution
mecoprop	Mecomec	PBI/Gordon	III	Caution
	Cleary's MCPP	W. A. Cleary		
mefluidide	Embark	PBI/Gordon	III	Caution
	Tri-Cut			
mepiquat chloride	Pix	BASF	II	Warning
	Mepichlor	Micro Flo		
metam-sodium	Vapam	ZENECA	I	Danger
metolachlor	Dual	Ciba	III	Caution
	Dual II			
metolachlor + metribuzin	Turbo	Bayer	III	Caution
metribuzin	Sencor	Bayer	III	Caution
	Lexone	DuPont		

(continued)

Appendix F: Toxicity Classification and Signal Words for Pesticides **477**

TABLE 2 *(Continued)*

Common name	Trade name	Producer	Toxicity class	Signal word
metribuzin + trifluralin	Salute	Bayer	III	Caution
metsulfuron-methyl	Escort Ally	DuPont	IV	Caution
molinate	Ordram	ZENECA	II (liquid) IV (granule)	Warning Caution
molinate + propanil	Arrosolo	ZENECA	II	Warning
monobor-chlorate	BareSpot	J. R. Simplot	I	Danger
MSMA	Ansar Daconate Bueno Arsonate Super Arsonate Diumate Drexel MSMA	Fermenta Drexel	III	Caution
napropamide	Devrinol	ZENECA	III	Caution
1-naphthalenacetic acid (NAA)	Fruitone N Fruit Fix NAA 800	Amvac	III	Caution
naptalam	Alanap-L	Uniroyal	II	Warning
nicosulfuron	Accent	DuPont	IV	Caution
norflurazon	solicam Zorial	Sandoz	IV	Caution
oryzaline	Surflan	DowElanco	IV	Caution
oxadiazon	Ronstar	Rhone-Poulenc	I (liquid) IV (dry)	Danger Warning
oxabetrinil	Concep II	Ciba	IV	Caution
oxyflurofen	Goal	Rohm & Haas	II	Warning
paclobutrazol	Bonzi	Sandoz	III	Caution
paraquat	Gramoxone Cyclone Starfire	ZENECA	I	Danger–Poison
pebulate	Tillam	ZENECA	III	Caution
pendimethalin	Prowl	American Cyanamid	III	Warning
phenmedipham	Betanal	AgrEvo	II	Warning
picloram	Tordon 22K Grazon	DowElanco	II (liquid)	Warning
picloram + triclopyr	Access	DowElanco	III	Caution
primisulfuron	Beacon	Ciba	IV	Caution
prodiamine	Barricade Endurance	Sandoz	IV	Caution
prometon	Pramitol	Ciba	I	Danger (corrosive)
prometryn	Caparol Cotton-Pro	Ciba Griffin	II (liquid) III (dry)	Warning Caution
pronamide	Kerb	Rohm & Haas	III	Caution

TABLE 2 *(Continued)*

Common name	Trade name	Producer	Toxicity class	Signal word
propachlor	Ramrod	Monsanto	I	Danger (skin)
propachlor + atrazine	Ramrod and Atrazine	Monsanto	II	Warning (skin)
propanil	Stam Stampede Strel	Rohm & Haas	II	Warning
propazine	Milo-Pro	Griffin	IV	Caution
pyrazon	Pyramin	BASF	III	Caution
pyridate	Tough Lentagran	Cedar Gowan	III	Caution
quinclorac	Facet	BASF	III	Caution
quizalofop-p-ethyl	Assure II	DuPont	III	Caution
sethoxydim	Poast	BASF	IV	Caution
siduron	Tupersan	DuPont	IV	Caution
simazine	Princep Caliber	Ciba	III	Caution
sodium cacodylate	Rad-E-Cate	Drexel	I	Danger
sodium chlorate	Defol KM	Drexel Kerr-McGee	IV	Caution
sulfometuron methyl	Oust	DuPont	III	Caution
Sulfosate (glyphosate-trimesium)	Touchdown	ZENECA	III	Caution
2,4,5-T	Many	Several	III	Caution
tebuthiuron	Spike	DowElanco	IV	Caution
terbacil	Sinbar	DuPont	III	Caution
terbutryn	Terbutrex	Cornbelt Marshal Thomas	III	Caution
thidiazuron	Dropp	AgrEvo	IV	Caution
thifensulfuron methyl	Pinnacle	DuPont	III (dry)	Caution
triallate	Avadex BW Far-Go	Monsanto	IV	Caution
triasulfuron	Amber	Ciba	IV	Caution
tribenuron methyl	Express	DuPont	II	Warning
tribufos	Folex	Rhone-Poulenc	I	Danger
tryclopyr	Garlon 3A Garlon 4	DowElanco	III	Caution
tridiphane	Tandem	DowElanco	II	Warning
trifluralin	Treflan Trilin Tri-4	DowElanco Griffin American Cyanamid	I, II, or III	Danger, Warning, or Caution
trisilate + trifluralin	Buckle	Monsanto	III	Caution
vernolate	Vernam	Drexel	III	Caution

Appendix F: Toxicity Classification and Signal Words for Pesticides **479**

TABLE 3 TOXICITY CLASSIFICATION AND SIGNAL WORDS FOR FUNGICIDES
AND PLANT DISEASE CHEMICALS

Common name	Trade name	Producer	Toxicity class	Signal word
abenomyl	Benlate Tersan 1991	DuPont	IV	Caution
buprimate	Nimrod	ICI Agrochemicals	III	Caution
captan	Captan	Many	I	Danger (eyes)
captan + DCNA	Botec	Gowan	I	Danger (eyes)
carboxin	Vitavax	Uniroyal	III	Caution
chloroneb	Terraneb SP Chloroneb 65W Nu-Flow	Kincaid	IV	Caution
chlorothalonil	Bravo Daconil 2787	Wilbur-Ellis	I (depends on II formulation)	Danger Warning
copper ammonium carbonate	Copper-Count N	Mineral Research	III	Caution
copper hydroxide	Kocide Champ Coppercide	Griffin Agtrol Old Bridge Chemicals	I	Danger (eyes)
copper oxychloride	COC WP	Cuproquim	II	Warning
copper oxychloride sulfate	Copro Coxysul CS-56 COCS	Agtrol Platte Chemical	III	Caution
copper oxysulfate	Cuproxat	Agrolinz		
copper sulfate	Basicap Copper Power KOP 300 Basic Copper 53 Copper Z	Griffin Agtrol Drexel Cuproquim Helena	II	Warning (eyes)
dicloran (DCNA)	Botran Botran 30C	Gowan Gustafson	IV	Caution
difenoconazzole	Dividend	Ciba	III	Caution
disulfoton	Disyston	Bayer Ag	I	Danger–Poison
dodine	Venturol Melorex Syllit	American Cyanamid Rhone-Poulenc UAP/Platte	I	Danger
etridiazole	Truban Koban Aaterra Pansoil Terrazole	Grace-Sierra Uniroyal	I (depends on II formulation) III	Danger Warning Caution
etridiazole + PCNB	Terraclor Super X	Uniroyal	I (depends on II formulation) III	Danger Warning Caution
fenarimol	Rubigan	DowElanco	II (depends on III formulation)	Warning Caution

TABLE 3 *(Continued)*

Common name	Trade name	Producer	Toxicity class	Signal word
ferbam	Carbamate Ferbam	UBC Chemicals	IV	Caution
flusilazole	Nustar Punch	DuPont	III	Caution
fosetyl-al	Aliette	Rhone-Poulenc	III	Caution
iprodione	Rovral	Rhone-Poulenc	IV	Caution
lime-sulfur	BSP Lime-Sulfur Soil-Mend Sulferix	Best Sulfur Prod.	I	Danger
manganese ethylene bisdithiocarbamate	Fore	Rohm & Haas	IV	Caution
mancozeb	Dithane Manzate Penncozeb	Rohm & Haas DuPont ELF Atochem	IV	Caution
mancozeb + metalaxyl	Ridomil MZ	Ciba	IV	Caution
maneb	Maneb	ELF Atochem Drexel	III	Caution
mercurous chloride + mercuric chloride	Calo-gran	Grace-Sierra	III	Danger (eyes, skin)
metalaxyl	Ridomil Apron	Ciba	III	Caution
metiram	Polyram	BASF Marshall Thomas	IV	Caution
myclobutanil	Eagle Nova Rally Systhane	Rohm and Haas	I (depends on II formulation) III	Danger Warning Caution
oxycarboxin	Plantvax	Uniroyal	III	Caution
quintozene (PCNB)	Terraclor Turfcide RTU PCNB	Uniroyal Gustafson	III	Caution
piperalin	Pipron	DowElanco	III	Caution
propiconazole	Tilt Orbit	Ciba	III	Caution
streptomycin	Agri-Mycin 17	Merck Ag Vet	IV	Caution
sulfur	Microthiol Special Drexel Sulfur Thiolux Uniflow Golden Dew Signal Alfa Bravo S Super-sul	ELF Atochem Drexel Sandoz Uniroyal Wilbur-Ellis ZENECA Helena Helena Cuproquim	IV	Caution

(continued)

TABLE 3 *(Continued)*

Common name	Trade name	Producer	Toxicity class	Signal word
TCMBT	Busan 30A	Buckman Labs	II	Warning
tebuconazole	Folicur	Bayer	III	Caution
thiabendazole	Mertect	Merck	III	Caution
	Arbotect			
	Storite			
	TBZ			
	Tecto			
	Thibenzole			
thiophanate	Topsin E	ELF Atochem	IV	Caution
		Marshall Thomas		
thiophanate-methyl	Fungo 50	Grace-Sierra	IV	Caution
	Domain			
	Topsin M	ELF Atochem		
thiram	Spotrete	W. A. Cleary	III	Caution
triadimefon	Bayleton	Bayer	III	Caution
triadimenol	Baytan	Bayer	III	Caution
	Bayfidan	Gustafson		
	Baytan 30			
triflumizole	Procure	Uniroyal	I	Danger
triforine	Funginex	Ciba	I	Danger
	Denarin	American		
	Triforine EC	Cyanamid		
		Valent		
triphenyltin hydroxide	Super Tin	Griffin	II	Warning
vinclozolin	Ronilan	BASF	IV	Caution
	Ornalin	Grace-Sierra		
	Vorlan			
zineb	Parzate C	DuPont	IV	Caution
ziram	Ziram 76	ELF Atochem	I	Danger
		UCB		
	Zyban	Grace-Sierra	II	Warning

TABLE 4 TOXICITY CLASSIFICATION AND SIGNAL WORDS
FOR RODENTICIDES AND PREDICIDES

Common name	Trade name	Producer	Toxicity class	Signal word
brodifacoum	Talon Weatherblok Havoc	ZENECA Pitman-Moore	I	Danger
bromadiolone	Bromone Maki Boot Hill	Sanex Cornbelt LiphaTech	III	Caution
chlorophacinone	Rozol Caid	Atomergic LiphaTEch	II (powder) III (bait)	Warning Caution
cholecalciferol	Rampage	Atomergic Motomco	III	Caution
diphacinone	Diphacin Ramik Promar Tomcat	HACCO Motomco	I (depending on II formulation) III	Danger, Warning, Caution
pindone	Pivalyn	Motomco	III	Caution
strychnine sulfate	—	Several	I	Danger–Poison

warfarin	Rodex
	Cov-R-Tox
	Tox-Hid
	Co-Rax
	RAX
		HACCO
		
		Prentiss
			I (depending
			    on concen-
			    tration) or
			III
				Danger
				
				Caution

zinc phosphide	Ridall-Zinc Rodenticide AG Zinc Phosphide	LiphaTech Motomco UAP/HACO	I (depending II on concen- III tration)	Danger–Poison Warning Caution

TABLE 5 TOXICITY CLASSIFICATION AND SIGNAL WORDS FOR NEMATICIDES

Common name	Trade name	Producer	Toxicity class	Signal word
aldicarb	Temik	Rhone-Poulenc	I	Danger
carbofuran	Furadan	FMC	I (flowable)	Danger
			II (granule)	Warning
dazomet	Mylone		III	Caution
dichloropropene	Telone	DowElanco	II	Warning
	Telone II			
	D-D 92			
fenamiphos	Nemacur	Bayer	I	Danger–Poison
metam-sodium	Vapam	ZENECA	I	Danger
	Metam 426	Amvac		
	—	Gilmore		
	Busan 1020	Buckman		
	Metam	UCB Chemicals		
methomyl	Lannate	DuPont	I	Danger–Poison
methyl bromide	Meth-O-Gas	Great Lakes	I	Danger–Poison
methyl bromide + chloropicrin	Brom-O-Gas	Great Lakes	I	Danger–Poison
	Brom-O-Sol			
	Terr-O-Gas			
	TRI-CON	Trical, Inc.		

TABLE 6 TOXICITY CLASSIFICATION AND SIGNAL WORDS FOR COMMODITY OR SPACE FUMIGANTS

Common name	Trade name	Producer	Toxicity class	Signal word
aluminum phosphide (phosphine)	Phostoxin	Degesch America	I	Danger
chloropicrin	Chlor-O-Pic		I	Danger (eyes)
chloropicrin + methyl bromide	Terr-O-Gas	Great Lakes	I	Danger (eyes)
	Brom-O-Gas	Great Lakes		
	Brom-O-Sol			
	TRI-CON			
magnesium phosphide	Fumi-Cel	Degesch America	I	Danger–Poison
	Fumi-Strip			
methyl bromide	Metho-O-Gas	Great Lakes	I	Danger
para dichlorobenzene		Marshall Thomas	II	Warning
sulfuryl fluoride	Vikane	DowElanco	I	Danger

TABLE 7 TOXICITY CLASSIFICATION AND SIGNAL WORDS FOR BIRD REPELLENTS

Common name	Trade name	Producer	Toxicity class	Signal word
aminopyridine	Avitrol	Avitrol	I (powder)	Danger
			IV (bait)	Caution
methiocarb	Mesurol	Bayer	I	Danger

APPENDIX G: STABILITY OF AGRI-CHEMICALS WITH RESPECT TO pH OF CARRIERS/DILUENTS*

Reference/ Source	Common/ Trade name	Chemical/ Technical name	Comments/Rate of hydrolysis time for 50% to decompose ($T_{\frac{1}{2}}$)
I. *Insecticides, Nematicides, Acaricides, Miticides*			
11	Acarben	Chlorobenzilate	Decomposes slowly under alkaline conditions and more rapidly if lime is present
2	Acaramate	Benzomate	Decomposes by strong alkali
16	Acarastop	Clofentezine	pH 9.2 = 4.8 hours
8	Actellic	Pirimiphos Methyl	pH 5 = 7 days, pH 7 = 35 days, pH 8.5 = 12 days, hydrolysis caused by strong acids and alkalies
2	Acrex	Dinobuton	Hydrolyzes by alkalies
2	Afilene	Butocarboxim	Stable at pH 4 to 7
2	Allethrin	Allethrin	Incompatible with alkalies
2	Amaze	Isofenfos	Subject to hydrolysis under alkaline conditions
19	Ambush/Pounce	Permethrin	Optimum stability pH 4
7	Ammo	Cypermethrin	pH 9 = 35 hours, easily hydrolyzes but more stable in acid than in alkaline solutions
11	Aracide	Aramite	Decomposes slowly under alkaline conditions and more rapidly if lime is present
2	Artaban	Benzoximate	Decomposes by strong alkali
2	Azodrin	Monocrotophos	Incompatible with alkaline compounds
2	Bactimos	Bacillus Thuringiensis Var Israelensis	Incompatible with highly alkaline materials
2	Baygon	Propoxur	Stable except under alkaline conditions
2	Baytex	Fenthion	Incompatible with alkaline conditions
1	Bidrin	Dicrotophos	pH 9 = 50 days, pH 1 = 100 days, more stable in acid than in alkaline medium
2	Bioallethrin	D-Trans Allethrin	Avoid high acidic or alkaline conditions
2	Bolstar	Sulprofos	Subject to hydrolysis under alkaline conditions
16	Carbamult	Promecarb	pH 95 = to 7 hours
10	Carzol	Formetanate	pH 5 = 4 days, pH 7 = 14 hours, pH 9 = 3 hours
11	Chlordane	Chlordane	Decomposes slowly under alkaline conditions and more rapidly if lime is present
2	Concord	Fastac	Hydrolyzes under strong alkaline conditions
1	Ciodrin	Crotoxyphos	pH 1 = 87 hours, pH 9 = 35 hours
6	Coral	Coumaphos	Hydrolyzes slowly under alkaline conditions
2	Counter	Terbufos	Hydrolyzes under alkaline conditions
2, 14	Curacron	Profenofos	Unstable under alkaline conditions, water pH of 7 is preferred
5	Cygon	Dimethoate	pH 2 = 21 hours, pH 6 = 12 hours, pH 9 = 48 minutes, presence of iron accelerates decomposition
	Cymbush	Cypermethrin	See *Ammo*

(continued)

Reference/ Source	Common/ Trade name	Chemical/ Technical name	Comments/Rate of hydrolysis time for 50% to decompose ($T\frac{1}{2}$)

I. Insecticides, Nematicides, Acaricides, Miticides (continued)

Reference/ Source	Common/ Trade name	Chemical/ Technical name	Comments
1, 17	Cythion	Malathion	Hydrolyzes rapidly in water above pH 7 and below pH 5. Iron will catalyze decomposition. pH 6 = 7.8 days, pH 7 = 3 days, pH 8 = 19 hours, pH 10 = 2.4 hours
2	Dasanit	Fensulfothion	Incompatible with alkaline
2	DDT	Chlorophenothane	Unstable in the presence of alkalies
	DDVP	Dichlorvos	See *Vapona*
11	Delnav	Dioxathion	Decomposes slowly under alkaline conditions and more rapidly if lime is present
5	Diazinon/ Knoxout	Diazinon	pH 5 = 31 days, pH 7.5 = 185 days, pH 9 = 136 days, most stable near neutral, avoid extreme acid conditions
1	Dibrom	Naled	Over 90% hydrolyzes in 48 hours in aqueous solutions
7, 17	Dicofol	Dicofol	Compatible with all but highly alkaline pesticides. No degradation in 20 days at pH 5. pH 7 = 5 days, pH 105 = 15 minutes
6	Dimecron	Phosphamidon	See *Phosphamidon*
19	Dimetan	Dimethan	Subject to hydrolysis
	Dipel/Vectobac	Bacillus thuringiensis	Stable at pH 4 & 7
2, 5	Dipterex	Trichlorfon	See *Dylox*
7, 17	Di-Syston	Disulfoton	pH 5 = 60 hours, pH 6 = 2 hours, pH 9 = 7.2 hours, subject to hydrolysis under alkaline conditions
	Dursban	Chloropyrifos	pH 4.7 = 63 days, pH 6.9 = 35 days, pH 8.1 = 22 days, pH 10 = 7 days, hydrolyzes by strong alkalies, stable in neutral and in weak acidic solutions
5	Dylox	Trichlorfon	pH 6 = 5340 minutes, pH 7 = 386 minutes, pH 8 = 63 minutes, subject to hydrolysis
7	Elocron	Dioxacarb	pH 5 = 3 days, pH 7 = 60 hours, pH 9 = 20 hours, pH 10 = 2 hours
11	Endrin	Endrin	Decomposes slowly under alkaline conditions and more rapidly if lime is present
2	Entex	Fenthion	See *Baytex*
1, 4	EPN	EPN	Hydrolyzes in alkalies, stable in acidic and neutral solutions. pH 6 = more than 1 year, pH 10 = 8.2 hours
2	Esbiol	Esbiol	Avoid alkaline conditions
1, 17	Ethion	Ethion	Subject to hydrolysis, pH 6 = 37.5 hours, pH 8 = 8.4 weeks, pH 9 = 17.4 days
7, 16	Ficam/Turcam	Bendiocarb	pH 7 = 4 days, pH 9 = 45 minutes
1	Folithion	Fenitrothion	Subject to hydrolysis, slightly more stable than Methyl Parathion
2	Fomothion	Fomothion	Hydrolyzes in alkaline water
14	Fundal/Galecron	Chlordimeform	Alkaline water can adversely affect these compounds. Water pH of 7 is preferred

Reference/ Source	Common/ Trade name	Chemical/ Technical name	Comments/Rate of hydrolysis time for 50% to decompose ($T_\frac{1}{2}$)

I. *Insecticides, Nematicides, Acaricides, Miticides* (continued)

Reference/ Source	Common/ Trade name	Chemical/ Technical name	Comments/Rate of hydrolysis time for 50% to decompose
7	Furadan	Carbofuran	pH 6 = 200 days, pH 7 = 40 days, pH 8 = 5 days, pH 9 = 78 hours, performs best at a pH of 4 to 6
4	Gardona	Tetrachlorvinphos	pH 3 = 54 days, pH 7 = 44 days, pH 10.5 = 3.3 days
4	Guthion	Azinphosmethyl	pH 5 = 17.3 days, pH 7 = 10 days, pH 9 = 12 hours
2	Hostathion	Triazophos	Degrades by alkaline hydrolysis
1, 13, 17	Imidan	Phosmet	pH 4.5 = 13 days, pH 7 = 12 hours, pH 8.34 hours, pH 10 = 1 minute. Activity may be reduced when pH is above 7. Correct pH with buffering or acidifying agent
6	Isolan	Isolan	Subject to hydrolysis
	Kelthane	Dicofol	See *Dicofol*
2	Lance	Cleothocarb	Hydrolyzes under strong alkalies
5, 13, 17	Lannate	Methomyl	pH 9.1 = loss of 5% in 6 hours, stable in slightly acidic solutions. Do not use in highly alkaline mixtures. pH 6 = 54 weeks, pH 7 = 38 weeks, pH 8 = 20 weeks
19	Larvin	Thiodicarb	Stable pH 6, rapidly hydrolyzed at pH 9
2	Lindane	Lindane	Avoid strong alkalies
	Lorsban	Chloropyrifos	See *Dursban*
13	Lorsban 50W	Chloropyrifos	Avoid alkaline materials
	Malathion	Malathion	See *Cythion*
2	Mavrik	Fluvalinate	Do not mix with strong basic products
13	Mavrik Aquaflow	Fluvalinate	Buffer spray water of pH 5 to 7
1	Metasystox	Demeton Metyls	Subject to hydrolysis, pH 6 = 7 hours
1, 2	Metasystox-R	Oxydemeton Methyl	pH 6 = 12.3 hours, unstable in alkaline conditions
19	Mocap	Ethoprophos	Rapid hydrolysis at pH 9
2	Morestan	Oxythioquinox	Subject to hydrolysis under alkaline conditions
	Neguvon	Trichlorfon	See *Dylox*
2	Nemacur	Fenamiphos	Subject to hydrolysis
2	Neo-Pynamin	Tetramethrin	Avoid alkaline conditions
2	Neo-Pynamin Forte	D-Tetramethrin	Avoid alkaline conditions
	Nudrin	Methomyl	See *Lannate*
17	Omite	Propargite	pH 3 = 17 days, pH 6 = 331 days, pH 9 = 1 day
5	Orthene	Acephate	pH 3 = 65 days, pH 9 = 16 days
11	Ovotran	Chlorofenizon	Decomposes slowly under alkaline conditions and more rapidly if lime is present
	Ovex	Chlorofenizon	See *Ovotran*

(continued)

Reference/ Source	Common/ Trade name	Chemical/ Technical name	Comments/Rate of hydrolysis time for 50% to decompose ($T_\frac{1}{2}$)

I. *Insecticides, Nematicides, Acaricides, Miticides* (continued)

Reference/ Source	Common/ Trade name	Chemical/ Technical name	Comments/Rate of hydrolysis
2	Padan	Cartap	Stable in acidic solutions, hydrolyzes slowly at neutral and instantly in alkaline solution
4	Parathion	Parathion Ethyl	pH 5 = 690 days, pH 7 = 120 days, pH 10 = 29 hours, pH 11 = 170 minutes
1	Parathion Methyl	Parathion Methyl	Hydrolyzes 4.3 times faster than Parathion Ethyl
5	Penncap-M	Encapsulated Methylparathion	Turns yellow in alkaline solutions, performs best at pH 4 to 6
7	Phosdrin	Mevinphos	pH 7 = 35 days, pH 11 = 1.4 hours
4, 6	Phosphamidon	Phosphamidon	pH 2 to 5 90% undecomposed after 24 days, pH 4 = 74 days, pH 7 = 13.5 days, pH 10 = 30 hours
1	Phosvel	Leptophos	Hydrolyzes slowly under strong alkaline conditions
19	Pydrin	Fenvalerate	Optimum stability at pH 4
	Pynamin	Allethrin	See *Allethrin*
6	Pyrolan	G22008	Subject to hydrolysis
2	Quinalphos	Quinalphos	Subject to hydrolysis
	Rabon	Tetrachlorvinphos	See *Gardona*
9	Reldan	Chlorophyrifos Methyl	pH 4 = 10 days, pH 6 = 38 days, pH 8 = 3 days, pH 10 = 1.7 days
5	Sevin	Carbaryl	pH 6 = 100 to 150 days, pH 7 = 24 to 30 days, pH 8 = 2 to 3 days, pH 9 = up to 1 day
13	Sevin XLR	Carbaryl	Do not use in water with pH over 8 unless buffer is added
	Sumithion	Fenithrothion	See *Folithion*
5	Supracide	Methidathion	Unstable in alkaline conditions
2	Systox	Demeton	Hydrolyzes under alkaline conditions
7	Tepp	Tepp	pH 6 = 6.8 hours, pH 9 = 3.5 hours, pH 10 = 21 minutes
1, 2	Thimet	Phorate	Subject to hydrolysis under alkaline conditions, pH 8 at 70°C = 2 hours
2	Thiocyclam	Thiocyclam Hydrogen	pH 5 = 181 days, pH 7 = 6 days, pH 9 = 6 Oxalatedays
5, 17	Thiodan	Endosulfan	Undergoes some degree of alkaline, 70% loss after 7 days at pH 7.3 to 8.0
2	Thiometon	Thiometon	Hydrolyzes easily in aqueous solutions
2	Toxaphene	Toxaphene	Unstable in alkaline conditions
1	Vapona	Dichlorvos	pH 7 = 8 hours
17	Vydate	Oxamyl	Stable at pH 4.7, 3% loss in 24 hours at pH 6.9, 45% loss in 24 hours at pH 9.1
2	Zectran	Mexacarbate	Hydrolyzes readily in alkaline conditions
1, 17	Zolone	Phosalone	Hydrolyzes rapidly in alkaline medium, stable at pH 5 to 7, pH 9 = 9 days

Reference/ Source	Common/ Trade name	Chemical/ Technical name	Comments/Rate of hydrolysis time for 50% to decompose ($T_{\frac{1}{2}}$)
II. *Fungicides*			
13	Benlate	Benomyl	Do not mix with alkaline materials
2	Captafol	Captafol	Stable except under alkaline conditions
17	Captan	Captan	pH 4 = 32 hours, pH 10 = 2 minutes, pH 7 = 8.3 hours
2	Cyprex	Dodine	Not compatible with lime or chlorobenzilate
2	Dexon	Fenaminosulf	Decomposition in water accelerated by alkaline conditions
	Difolatan	Captafol	See *Captafol*
	Dodine	Dodine	See *Cyprex*
6	Dyrene	Anilazine	Subject to hydrolysis
2	Fenfuram	Fenfuram	Hydrolyzes under strong alkaline conditions
2	Galben	Benaxyl	Stable in acid and neutral media
	Lesan	Fenaminosulf	See *Dexon*
2	Pancotine	Guazatine	Unstable in alkaline medium
13	Rovral	Iprodione	Chemical breakdown may occur in water with high pH
13	Topsin M	Thiophanate-Methyl	Do not combine with high alkaline materials
III. *Defoliants, Desiccants, Harvest Aids*			
2	DEF	None	Hydrolyzes slowly under alkaline conditions
IV. *Plant Growth Regulators*			
3	Acti-Aid	Cycloheximide	Maximum stability at pH 4 to 5, degrades rapidly at about pH 7
3	Alar 85	Diaminozide	Do not use with alkaline materials
	B-Nine	Diaminozide	See *Alar 85*
3	Cerone	Ethephon	See *Ethrel*
2	Ergostim	Folcisteine	Not compatible with alkaline materials
2	Ethrel	Ethephon	Very stable at pH 3 or less, incompatible with alkaline salt
	Florel	Ethephon	See *Ethrel*
	Kylar	Diaminozide	See *Alar*
2	Pro Gibb	Gibberellic Acid	Hydrolyzes slowly by water, should not be combined with alkaline materials
	Prep	Ethephon	See *Ethrel*
13	Promalin	Gibberellins + Benzyl	A buffered wetting agent should be used. Adenine final spray should not exceed pH 8
V. *Herbicides*			
18	Alanap	Naptalam	Hydrolyzes in solutions with a pH greater than 9.5
6	Atrazine	Atrazine	Decomposes slowly in alkaline solution and more rapidly if lime is present
18	Avenge	Difenzoquat	Stable at a low pH, alkaline conditions cause precipitation
	Betamix	Phenmedipham Desmedipham 50-50 Mixture	See *Betanal*

(continued)

Reference/ Source	Common/ Trade name	Chemical/ Technical name	Comments/Rate of hydrolysis time for 50% to decompose ($T\frac{1}{2}$)
V. *Herbicides* (continued)			
2,16	Betanal	Phenmediphan	Undergoes hydrolysis under alkaline conditions, pH 7 = 5 hours, pH 9 = 10 minutes
2,18	Betanex	Desmedipham	Undergoes hydrolysis under alkaline conditions, $T\frac{1}{2}$: pH 5 = 70 days, pH 7 = 20 hours, pH 9 = 10 minutes
18	Casoron	Dichlobenil	Hydrolyzes rapidly by alkali
18	Carbyne	Barban	Hydrolyzes very rapidly in alkali, $T\frac{1}{2}$ = 58 seconds at 25°C and pH 13
18	Certrol, Mylone	Ioxynil	Hydrolyzes readily by alkali
18	Commando, Suffix	Flamprop Isomer	Stable to hydrolysis with a pH greater than 3, and a pH less than 9
18	Cornox	Benazolin	Stable except to concentrated alkali
2,18	Diquat	Diquat	Stable in neutral or acid solutions, but decomposes in alkaline conditions, unstable with a pH greater than 9, and a pH less than 12
18	Drepamon	Tiocarbazil	Stable to hydrolysis, pH 5.6 to 8.4
18	Faneron	Bromofenoxim	$T\frac{1}{2}$: pH 1 = 41.4 hours, pH 5 = 9.6 hours, pH 9 = 76 hours
18	Goltix	Metamitron	Stable in acid, unstable with pH greater than 10
18	Gramoxone	Paraquat	Stable in acid or neutral conditions
18	Herbit	MCPA-thioethyl	$T\frac{1}{2}$: pH 7 = 22 days, pH 9 = 2 days
18	Lasso	Alachlor	Hydrolyzes under strong acidic or alkaline conditions
18	Mataven, Lancer	Flamprop (Racemate)	Stable to hydrolysis with a pH greater than 2, and a pH less than 7
18	Mesoranil	Aziprotryne	Hydrolyzes slowly in slightly alkaline medium
18	Modown	Bifenox	Stable at pH 5 to 7.3, hydrolyzed rapidly at pH 9
	Paraquat	Paraquat	See *Gramoxone*
6	Princep	Simazine	Decomposes slowly in alkaline solution and more rapidly if lime is present
16	Roundup	Glyphosate	Reported to have an optimum pH of 2.5, alkaline conditions should be avoided
18	Sonar	Fluridone	Stable to hydrolysis with a pH greater than 3, and less than 9
18	Suffix	Benzolyprop-ethyl	Hydrolytically stable at pH 3 to 6
18	Tandex	Karbutilate	Stable in acid medium, $T\frac{1}{2}$: 4.6 days at pH 8
18	Ustilan	Ethidimuron	Unstable to alkali
VI. *Repellents*			
2	Mesurol	Methiocarb	Unstable in highly alkaline medium
2	MGK Repellent 11	None	Avoid alkaline conditions
2	MGK Repellent 326	None	Avoid high acidic or alkaline conditions
2	MGK Repellent 874	None	Avoid extreme acidic or alkaline conditions

Reference/ Source	Common/ Trade name	Chemical/ Technical name	Comments/Rate of hydrolysis time for 50% to decompose ($T\frac{1}{2}$)
VII. *Antibiotics*			
15	Streptomycin 17	Streptomycin Sulfate	Avoid use with alkaline materials. For high alkaline water sources, use of an acceptable acidifying agent may be advisable to bring tank solution to a normal or slightly acid pH.

Source: Modified from Loveland Industries, Inc., *Special Products for Special Needs,* (technical bulletin).

*The information in this technical bulletin has been collected and compiled from several sources. The accuracy of the information has not been verified by Loveland Industries, Inc. Bob Reeves, Technical Services Manager.

Bibliography/References

1. *Organophosphorus Pesticides: Organic & Biological Chemistry,* Eto, M., CRC Press, 1974.
2. *Farm Chemicals Handbook 1989,* Meister Publishing Company.
3. *Plant Growth Regulator Handbook,* 1st Edition, Plant Growth Regulator Working Group, 1977.
4. "pH Effect on Pesticides," article #3004 R, Leffingwell Chemical Company, 1976.
5. *North Dakota Insect Control Guide,* North Dakota State University, Cooperative Extension Service.
6. "pH Effect on Pesticides," Miller Chemical & Fertilizer.
7. "The Effect of pH on Pesticides," Diad Agricultural Services, Ltd.
8. "Actellic Manual 634," Imperial Chemical Industries PLC, 1981.
9. *Chloropyrifos Methyl Technical Information Bulletin,* Dow Chemical USA.
10. *Technical Information,* Carzol, Morton Chemical Company.
11. *Organic Insecticides, Their Chemistry Mode of Action,* Metcalf, R. L., Interscience, 1955.
12. "It Pays To Be Smart About Adjuvants," Whitmore, T. E., Farm Chemicals, February 1985.
13. *Crop Protection Chemicals References,* 5th Edition, Chemical & Pharmaceutical Press, 1989.
14. "How to Prevent Alkaline Hydrolysis," Carver, Lia, The Cotton Farmer, May 1987.
15. *Hopkins Strep 17 label.*
16. "Pest-Asides," No. 3, Western Australia Department of Agriculture, February 1987.
17. "Preventing Decomposition of Agriculture Chemicals by Alkaline Hydrolysis in the Spray Tank," A. J. and H. Riedl, New York Food and Life Sciences Bulletin.
18. *The Pesticide Manual, A World Compendium,* 7th Edition, British Crop Protection Council.
19. *The Pesticide Manual, A World Compendium,* 9th Edition, British Crop Protection Council.

APPENDIX H: COLD WEATHER HANDLING OF LIQUID CHEMICAL PRODUCTS*

The information contained herein is based on information that is believed reliable. However, due to changes that the manufacturer may make in the manufacturing or formulating process, the minimum storage temperature may change. Consult the product label to be sure. If any product is stored below the manufacturer's suggested minimum storage temperature, the most important criterion in determining if the product is usable is the complete absence of crystals. If crystals remain after all effort to redissolve, do not use the product; contact the manufacturer for assistance.

Product	Minimum storage temperature	Comments
Aaccess Penetrator	0	Keep in warm, dry area.
AAtrex 4L	No special handling	Freezes with no damage to product.
Abate 4E	0	If stored below 0°F, do not use product. Contact manufacturer.
Adhere	32	Do not let freeze. Product performance may decrease.
Alanap L	32	If freezing occurs, gradually warm to 60°F to dissolve crystals; roll or shake container to thoroughly mix product before use.
Actellic 5E	32	If frozen, warm to room temperature and agitate.
Alfatox	32	After product warms above 32°F, roll or shake to ensure product is thoroughly mixed.
Altosid SR-10	32	Product may separate; shake well. If not redissolved, contact manufacturer.
Ambush	32	Avoid freezing. Mild agitation will return active ingredient to suspension.
Amiben	32	If exposed to prolonged cold temperatures, place in warm storage (50°F to 80°F) for several hours (several days for drums). Agitate before using by inverting the container several times or by rolling the drum.
Amitrol T	32	Freezes at 32°F, but needs no special handling as temperatures return to normal.
Ammo 2.5 EC	10	Place in warm room (65°F to 70°F). Roll back and forth until redissolved.
Antor	26	Freezes at 25°F, but completely redissolves above that temperature.
Arsenal	10	If product freezes, contact manufacturer.
Atrazine 4L	No special handling	Freezes with no damage to product.
Atrabute II	−30	If stored for long periods of time below 0°F, container should be rolled.
Avadex	32	Place in warm room (72°F) and roll container for several days.
Arosurf-MSF	No special handling	Freezes with no damage to product.
Avenge	40	Freezes at 40°F, but needs no special handling as temperatures return to normal.

Product	Minimum storage temperature	Comments
Bactimos FC	32	Avoid freezing. Container may split; product may separate. Shake before using.
Balan EC	40	Avoid freezing. If frozen, poor weed control may result. Contact manufacturer's representative if product freezes.
Banvel II	15	Freezes at 15°F, but completely redissolves above that temperature. No special handling is required.
Banvel 720	0	Freezes at 0°F, but completely redissolves above 32°F.
Banvel 4E	15	Freezes at 15°F, but completely redissolves above that temperature. No special handling is required.
Banvel 520	20	Freezes at 20°F, but completely redissolves above that temperature.
Banvel CST	0	Freezes at 0°F, but completely redissolves above 32°F.
Basagran	40	Place in warm room and shake can periodically. If crystals completely dissolve, the product is usable. If crystals do not dissolve completely, do not use product and contact the manufacturer's representative.
Basalin	40	Do not allow to freeze. If freezing occurs and crystals are present, contact the manufacturer's representative.
Baytex LC	45	Do not allow to freeze. Product performance could be affected.
Baytex 4	0	If stored below 0°F, do not use product. Contact manufacturer's representative.
Betamac-4	28	Warm to 68°F; agitate thoroughly.
Betamix	15	If product freezes, warm to 50°F; product will redissolve.
Betanal	15	If product freezes, warm to 50°F; product will redissolve.
Betanex	15	If product freezes, warm to 50°F; product will redissolve.
Bexton 4F & Bexton/Atrazine	32	If stored below 32°F, warm to above 50°F and agitate thoroughly before using.
Bicep 4.5L	No special handling	Freezes with no damage to product.
Bladex 4L	No special handling	Freezes with no damage to product.
Blazer 2L	32	Below 32°F, active ingredient will settle out. Warm to 70°F and agitate thoroughly. Product effectiveness is not affected.
Bond	32	Do not let freeze; product performance may decrease.
Bravo 500	32	Do not let freeze. If stored below 32°F, product should be warmed and mildly agitated before using.
Bravo 720	32	Do not let freeze. If stored below 32°F, product should be warmed and mildly agitated before using.
Brominal	5	Do not let product freeze. Contact manufacturer's representative.
Brominal 3+3	0	Prolonged storage below 0°F; agitate thoroughly.
Brominal ME-4	0	Prolonged storage below 0°F; agitate thoroughly.
Brominal Plus	20	Do not let product freeze. Contact manufacturer's representative.
Bronate	4	Warm product to 70°F; agitate until crystals dissolve.

(continued)

Product	Minimum storage temperature	Comments
Bronco	40	After product warms above 40°F, shake to ensure product is thoroughly mixed.
Buctril	4	Warm product to 70°F; agitate until crystals dissolve.
Butoxone Ester	No special handling	Freezes with no damage to product.
Butoxone	32	If allowed to freeze, remix before using.
Butoxone 200	30	If allowed to freeze, warm to 60°F and agitate.
Butoxone SB	32	After product warms above 32°F, roll or shake the can to ensure product is thoroughly mixed.
Butyrac 200	No special handling	Freezes with no damage to product.
Cantrol	28	If product freezes, contact manufacturer's representative.
Carbyne II	40	If product freezes, contact manufacturer's representative.
Cerone	32	Product is not harmed by freezing. However, freezing could rupture container.
Chemhoe 4FL	32	Do not allow product to freeze. If freezing occurs, contact manufacturer.
Chlordane	No special handling	Freezes with no damage to product.
Cide-Kick	No special handling	Freezes with no damage to product.
Citcop 5E	32	Has been stored at −20°F.
Cittowett Plus	40	Freezes at 40°F but completely redissolves above that temperature. No special handling necessary.
Cobex	40	If product freezes, contact manufacturer's representative.
Comite	No special handling	Does not freeze. It will jell but will not separate. Agitation will hasten the product's return to liquid state after it has been warmed.
Command	40	If solid crystals are observed, warm material to above 60°F by placing containers in warm location. Shake or roll container to redissolve crystals.
Compex	No special handling	Product does not freeze.
Conquest 4L	No special handling	Freezes with no damage to product.
Crop Oil Concentrate	32	After product warms above 32°F, roll or shake container to ensure product is thoroughly mixed.
Crop Oil Regular 2%	32	Warm to room temperature. Roll container to ensure mixing.
Crossbow	10	If product is stored below 10°F, agitate before use.
Cygon 2E	32	Place in a warm room (65° to 70°F) and agitate until redissolved.
Cygon 400	45	Do not store below 45°F. If this occurs, contact manufacturer representative.
Cythion E-5	32	Place in warm room (40° to 70°F). Roll or shake container every few hours until redissolved.
Cythion ULV	45	If product freezes, warm to above 45°F to ensure solution.
Cytrol	10	Place in warm room (65° to 70°F). Roll or shake container every few hours until redissolved.
Daconate	No restrictions	Freezes with no damage to product.

Product	Minimum storage temperature	Comments
Daconil 2787	40	Do not let freeze. If stored below 32°F, product should be warmed and mildly agitated before using.
Dasanit 6SC	0	No special handling required.
Defol 6	32	If product freezes, poor weed control could be expected.
Defoamer	No special handling	Freezes with no damage to product.
Devrinol 2E	20	If product reaches below 20°F, warm and agitate immediately.
Diazinon AG500	No restrictions	Freezes with no damage to product.
Diazinon 4E	No restrictions	Freezes with no damage to product.
Dibrom 8	32	Do not let product freeze.
Dibrom 14	32	Do not let product freeze.
Dimethoate 267	32	Place in warm room (65°F to 70°F). Agitate until dissolved.
Dipel 4L	0	Prolonged storage below 0°F. Agitate thoroughly before using.
Diquat	32	Do not let product freeze. If frozen, poor weed control may occur. Contact manufacturer's representative if product freezes.
Direx	No restrictions	Freezes with no damage to product.
DiSyston 8	0	No special handling required.
Dithane FZ	15	If product freezes, contact manufacturer's representative.
Dowfume EB5	−20	If product freezes, contact manufacturer's representative.
Dow General Weed Killer	20	If stored below 30°F, warm to above 50°F and agitate until product is thoroughly mixed.
Dual 8E	−30	Freezes with no damage to product.
Dursban ULV 1.5	0	Do not let product get below 0°F for product performance could be affected.
Dursban 4E	0	If stored below 10°F, warm to 30°F and agitate.
Dursban 2E	0	If product freezes, contact manufacturer.
Dyanap	40	Gentle warming (50°F to 75°F) will redissolve any crystals. Mild agitation, such as rolling or shaking before use, is recommended.
Dyfonate 4E	−40	If stored for long periods below 0°F, the container should be rolled to mix product before use because of possible layering.
Dylox	32	If freezing occurs, contact manufacturer's representative.
Eagle Iron	No temp. listed	If product freezes, gently warm. Then agitate until product is thoroughly mixed.
Easy Spot	40	Do not let freeze; product's performance will drastically decrease.
Embark 2E	40	If product freezes, warm to room temperature to ensure stability.
Envert 171	−15	Freezes with no damage to product.
EPN	65	If product freezes, contact manufacturer's representative. Do not use product.

(continued)

Product	Minimum storage temperature	Comments
Eptam 7E	Does not freeze	If stored for long periods below 0°F, the container should be rolled to mix product before using because of possible layering.
Eradicane Extra	−50	If stored for long periods below 0°F, container should be rolled.
Eradicane 6.7E	Does not freeze	If stored for long periods below 0°F, the container should be rolled to mix product before using because of possible layering.
Esteron BK	20	Warm above 50°F and agitate before using.
Ethion	0	Prolonged storage below 0°F; agitate thoroughly before using.
Exhalt 800	No special handling	Freezes with no damage to product.
Extend	No special handling	Warm product and agitate until product is back into suspension.
Extrazine	No special handling	
Fargo	32	Place in warm room (72°F); then roll container frequently to dissolve crystals.
Farm Bin Spray	32	Place in warm room; mix product to ensure crystals have dissolved.
FC4	32	Freezes with no damage to product.
FC15	32	Freezes with no damage to product.
FC30	32	Freezes with no damage to product.
Fenamine	32	Freezes at 32°F, but needs no special handling as temperature returns to normal.
Fenatrol	No special handling	Freezes with no damage to product.
Fenthion 4E	0	If stored below 0°F, do not use product; contact manufacturer.
Fomark	40	Product can be used after warmed up and agitated; however, it will never perform as well after freezing as it did before.
Furadan Flowable	35	If the product is frozen, warm to room temperature; agitate by shaking or rolling. If substantial sediment remains, contact your manufacturer's representative.
Furloe 4EC	40	Warm product and agitate.
Fusilade 2000	0	Freezing does not affect performance.
Fusilade 4E	0	Freezing does not affect performance.
Garlon 4	20	Warm above 40°F; agitate before use.
Genate Plus 6.7 EC	No special handling	Product will not freeze.
Genep 7 EC	No special handling	Product will not freeze.
Granox Plus Flowable	0	If product freezes, warm to 70°F and agitate.
Goal 1-6E	32	If product freezes, contact manufacturer's representative.
Gramoxone	32	Keep from freezing to avoid container breakage due to expansion of product. Mild agitation is required to return product to normal state.
	32	
Granol Flowable	32	If product freezes, warm to 70°F and agitate.

Product	Minimum storage temperature	Comments
Guthion 2S	32	If product freezes, contact manufacturer's representative.
Herbicide 273	32	If freezing of product occurs, warm product; then shake or roll container to dissolve crystals. If crystals do not dissolve, contact manufacturer's representative.
Hoelon	20	If stored below 20°F, warm and agitate thoroughly before using.
Hyvar XL	0	If freezing of product occurs, warm product; then shake or roll container to dissolve crystals. If crystals do not dissolve, contact manufacturer's representative.
Kelthane EC	10	If product freezes, contact manufacturer's representative.
Kerb 50W	32	To prevent condensation on water-soluble package.
Kocide 606	30	If stored below 30°F, product damage could occur.
Kocide 404S	30	If stored below 30°F, product damage could occur.
Krenite (S)	No special handling	Freezes with no damage to product. Simply warm product.
Kuron	0	Warm to 32°F. Agitate before use.
Laddock	40	Product redissolves when warmed above freezing point.
Landmaster	40	Place in warm room (72°F); then roll or shake container frequently for several days.
Lannate	32	If product is stored below 32°F, contact manufacturer.
Larvacide	No special handling	Product will not freeze.
Lasso/Atrazine	40	If product freezes, place in warm room (72°F). Then roll or shake container frequently for several days.
Lasso 4E	32	Place in warm room (72°F). Then roll or shake the container frequently for several days. For containers larger than 5 gallons, see front panel labeling for storage.
Lasso MT	No special handling	Freezing does not affect performance. If frozen, thaw before use.
Lexone 4L	32	If product freezes, warm and agitate. If product appears to be lumpy, do not use. Contact manufacturer's representative.
Lindane 200	10	If product crystalizes, contact manufacturer.
Linex	No restrictions	Freezes with no damage to product.
Lorox 4L	Avoid freezing	If product freezes, contact manufacturer's representative. Do not use product.
Lodrift	40	If product freezes, warm and agitate.
Lorsban 4E	0	If product freezes, warm to at least 30°F; agitate thoroughly before using. Do not allow to heat above 80°F.
Malathion ULV	45	If product freezes, warm to above 45°F to ensure solution.
Malathion 57%	32	Place in warm room (40°F to 70°F). Roll or shake container every few hours until redissolved.
Marksman	15	Freezes at 15°F, but completely redissolves above that temperature. No special handling required.
Mecomec 4	32	Warm to room temperature and agitate.
Mertect 340F	32	Do not let product freeze.

(continued)

Product	Minimum storage temperature	Comments
Metasystox R	25	If product freezes, warm to 32°F. Agitate to dissolve crystals. If crystals do not dissolve, contact manufacturer.
Methoxychlor EC	32	If product freezes, warm to 68°F. Agitate until redissolved.
Methyl Parathion	32	If stored below 32°F, have product tested by manufacturer before using.
Milocep	No special handling	Freezes with no damage to product.
Milogard	No special handling	Freezes with no damage to product.
Milopro	No special handling	Freezes with no damage to product.
Mineral Oil	32	Warm to room temperature; roll container to ensure mixing.
Mitac	32	If product freezes, warm to above 32°F and agitate before use.
Mobait	32	Do not let freeze; product performance may decrease.
Modown	No special handling	If freezing should occur, place in warm room above 55°F for 24 hours, and agitate thoroughly until crystals completely dissolve. Do not use product and contact manufacturer representative.
Mondak	0	If product freezes, warm above 32°F; agitate before use.
Monitor 4	15	Do not let product get below 15°F.
Nalcotrol I-II	40	If product freezes, warm and agitate mixture.
Nemacur 3	32	If stored below 32°F, contact manufacturer's representative.
Nortron 1.5 EC	18	Freezes with no damage to product.
Nortron 4F	40	Do not let freeze. If this occurs, call manufacturer's representative.
N-Serve 24 & 24E	18	If stored below 18°F, warm to 70°F and agitate to dissolve crystals. If crystals do not dissolve, do not use and contact manufacturer's representative.
Nudrin	32	If stored below 32°F, contact manufacturer.
Oftanol 2E	0	Do not store below 0°F for damage may occur.
Orthotrol	40	If product does freeze, warm and agitate mixture.
Paraquat Plus	32	Do not let freeze. If frozen, poor weed control may occur. Contact manufacturer's representative if product freezes.
Parathion 8#	32	If stored below 32°F, have product tested before using and contact manufacturer's representative.
Pay-Off	No special handling	Freezes with no damage to product.
Penncap M	32 Storage above 32 preferable	If freezing of product occurs, warm and agitate to dissolve crystals. If crystals do not dissolve, do not use and contact manufacturer's representative.
Performance+ Flushing Solution	32	Warm to above 32°F and shake to ensure product is thoroughly mixed.
Phytar 560	−5	If product freezes, warm and agitate so crystals dissolve.
Poast	0	After product warms above 32°F, agitate thoroughly before using.
Pounce	10	Place in warm room (65°F to 70°F). Roll back and forth until redissolved.

Product	Minimum storage temperature	Comments
Pramitol 25E	32	Must be stored above 32°F. If temperature is below this, move to heated storage and contact manufacturer as soon as possible. Product may lose effectiveness when frozen.
Prefar 4E	42	If freezing occurs, warm product and agitate container to dissolve crystals. If crystals do not dissolve, contact manufacturer representative.
Premerge 3	−15	Separation may occur at temperatures below −15°F; warm to 50°F and agitate.
Premerge Plus	35	Separation may occur at temperatures below 35°F; warm to above 60°F and agitate.
Prentox 96C	32	Warm to room temperature and agitate.
Princep 4L	No special handling	Freezes with no damage to product.
Prowl EC	40	If crystals form, place in warm room and agitate. If crystals completely dissolve, product is usable. If crystals do not completely dissolve, do not use product and call manufacturer's representative.
Pydrin	20	Storage at temperatures below 20°F for prolonged periods could result in some crystals forming. If crystals occur, place in warm room (65°F to 70°F) and agitate.
Rad-E-Cate 25	20	If product does freeze, product will dissolve when warmed.
Ramrod Flowable & Ramrod/Atrazine F	20	If frozen, mix well to redissolve when temperatures are above 40°F.
Reldan 4E	10	Do not let product get below 10°F; separation may occur.
Rescue	32	Warm to 60°F and roll the container before using.
Reward	−50	If stored for long periods of time below 0°F, container should be rolled.
Rodeo	10	If product freezes, place in warm room (68°F) for several days to redissolve crystals. Mix well before using.
Roost-No-More	No special handling	Freezes with no damage to product.
Roneet 6E	20	If freezing of product occurs, warm and agitate to dissolve crystals. If crystals do not dissolve, contact manufacturer's representative.
Royal MH-30	40	Do not let product freeze; container may burst.
Roundup	32	If product freezes, place in warm room (68°F) for several days to redissolve crystals. Mix well before using.
Salvo	No special handling	Freezes with no damage to product.
Savit 4F	No special handling	Freezes with no damage to product.
Scourge 18-54	32	If product freezes, warm so crystals may redissolve.
Scourge 4+12	32	If product freezes, warm so crystals may redissolve.
Sencor 4F	No special handling	Freezes with no damage to product.
Serafume	No special handling	Freezes with no damage to product.
Sevimol	No special handling	Freezes with no damage to product.
Sevin 4 Oil	No special handling	As a precaution, store drums upside down.
Sevin XLR Plus	No special handling	Freezes with no damage to product.
Simtrol	No special handling	Freezes with no damage to product.

(continued)

Product	Minimum storage temperature	Comments
Solo	28	If product freezes, contact manufacturer's representative.
Sonalan	40	If product freezes, performance can be affected.
Spectracide	No special handling	Freezes with no damage to product.
Spray Fuse 90	Do not let freeze	Freezing will affect product's performance. Do not freeze.
Spray Oil Additive	32	If stored below 32°F, warm to room temperature and roll or shake container.
Spray Tracer	32	If product freezes, container could burst.
Stampede CM	15	If frozen, warm to 60°F for 24 hours with periodic shaking to redissolve product.
Sta-Put	40	If product does freeze, warm and agitate mixture.
SubDue 2E	32	Do not let freeze; product damage could occur.
Super Cu	No restrictions	Freezes with no damage to product.
Super Six	No special handling	Freezes with no damage to product.
Super Tin	No special handling	Freezes with no damage to product.
Supracide 2E	32	Do not let freeze. If freezing occurs, move product to heated storage and contact manufacturer's representative. Product may lose effectiveness when frozen.
Surfel	0	If freezing occurs, product will redissolve when warmed.
Surflan AS	32	If freezing of product occurs, warm and agitate to dissolve crystals. If crystals do not dissolve, contact manufacturer representative.
Sutan Plus 6.7E	Does not freeze	If stored for long periods below 0°F, container should be rolled to mix product in case layering may have occurred.
Sutazine	−30	If stored for long periods of time below 0°F, container should be rolled.
Systox	0	If freezing occurs, contact manufacturer's representative.
Tackel 2AS	No special handling	Freezes with no damage to product.
Teknar	32	If product freezes, contact manufacturer immediately.
Tenox P	No special handling	Nothing required.
Tenncop 5E	32	Has been stored at −20°F.
Thisul	15	If product freezes, agitate before using.
Telone II	No special handling	Shelf life is 24 months without deterioration.
Telone C-17	No special handling	Shelf life is 24 months without deterioration.
Tordon 22K	15	If stored below 15°F, warm to 30°F and agitate before use.
Tordon 212	0	Warm to 30°F and agitate before using.
Tordon RTU	0	Freezes with no damage to product.
Turbo	No special handling	Freezes with no damage to product.
Toxaphene	32	If freezing of product occurs, warm and agitate to dissolve crystals. If crystals do not dissolve, contact manufacturer's representative.
Treflan 4E	40	Place in warm room and agitate. If crystals completely dissolve, product is usable. If crystals do not completely dissolve, do not use product and call manufacturer's representative.
Treflan MTF	No special handling	May be stored in unheated facilities.

Product	Minimum storage temperature	Comments
Trimec 352	32	Warm to room temperature (68°F) and agitate before using. If crystals are present, contact manufacturer.
Trimec Turf Ester	32	Warm to room temperature (68°F) and agitate before using. If crystals are present, contact manufacturer.
Trimec Bentgrass	32	Warm to room temperature (68°F) and agitate before using. If crystals are present, contact manufacturer.
Trimec Broadleaf	32	Warm to room temperature (68°F) and agitate before using. If crystals are present, contact manufacturer.
Trimec 992	32	Warm to room temperature (68°F) and agitate before using. If crystals are present, contact manufacturer.
Trithion 8E	0	Liquid may separate at lower temperature. Warm to a higher temperature and mix thoroughly to recombine liquid. This will assure uniformity of product.
Unite	20	Warm to room temperature (65°F) and agitate before using.
Urox	40	Warm to room temperature (65°F) and agitate before using.
Vapam	0	Warm to room temperature (65°F) and agitate before using.
Vectobac AS	No special handling	Freezes with no damage to product.
Velpar L	32	Do not allow to freeze. If freezing occurs, contact manufacturer.
Vegomec	Do not let freeze	If freezing occurs, contact manufacturer.
Vernam 7E	Does not freeze	If stored for long periods below 0°F, the container should be rolled before using to mix product because of possible layering.
Vertac MSMA 600	30	If stored below 30°F, warm above 60°F and agitate.
Vertac General	20	Warm to room temperature (65°F) and agitate before using.
Vitavax 200	32	If stored below 32°F, warm above freezing point and agitate thoroughly before using.
Weedhoe 108	30	If product gets below 30°F, warm and agitate.
Weedone CB	−15	Freezes with no damage to product.
Weedone DPC	−15	Freezes with no damage to product.
Weedone 170	−15	Freezes with no damage to product.
Wetter	35	If product freezes, performance can be decreased.
X-77 Spreader	No special handling	Freezes with no damage to product. Simply warm product to use.

2,4-D Products

Product	Minimum storage temperature	Comments
4# Amine		
Cornbelt	10	After product has warmed above 40°F, roll or shake to
Formula 40	10	ensure a homogeneous mixture.
Weedar 64	No special handling	
Weedar 64 A	No special handling	
DMA 4	18	
4# L. V. Ester		
Cornbelt	−32	After product has warmed above 40°F, roll or shake to
Weedone LV4	No special handling	ensure a homogeneous mixture.
Esteron 99C	−40	

(continued)

Product	Minimum storage temperature	Comments
6# Butyl Ester		After product has warmed above 60°F, roll or shake to ensure a homogeneous mixture.
Cornbelt	32	
Esteron 76 BE	40	
6# L. V. Ester		After product has warmed above 40°F, roll or shake to ensure a homogeneous mixture.
Cornbelt	No special handling	
Weedone LV6	No special handling	
2,4-D/2,4,5-T Mixture		After product has warmed above 50°F, roll or shake to ensure a homogeneous mixture.
Esteron Brush Killer	35	
Esteron 245	0	
2# MCP Amine		After product has warmed above 40°F, roll or shake to ensure a homogeneous mixture.
Weedar MCPA	32	
4# MCP Amine		After product has warmed above 40°F, roll or shake to ensure a homogeneous mixture.
Weedar MCPA concentrate	No special handling	
MCP Amine	32	Freezes with no damage to product.
Hi Dep	No special handling	Freezes with no damage to product.
Ultra-Su LV	No special handling	Avoid freezing. If stored below 32°F, product should be warmed and mildly agitated before using.
Dacamine 4D	32	Warm to room temperature. Agitate thoroughly.
Super Brush Killer	32	
D E D Seed Su LV	15	If freezing occurs, gradually warm to 32°F and agitate to ensure a homogeneous mixture.

Cornbelt Chelates

Product	Minimum storage temperature	Comments
3% Calcium	12°F	temperature or product may separate.
7.5% Copper	2°F	Do not let product fall below minimum storage temperature or product may separate.
5% Iron	16°F	Do not let product fall below minimum storage temperature or product may separate.
2.5% Magnesium	12°F	Do not let product fall below minimum storage temperature or product may separate.
5% Manganese	14°F	Do not let product fall below minimum storage temperature or product may separate.
6% Zinc	15°F	Do not let product fall below minimum storage temperature or product may separate.
9% Zinc	−2°F	Do not let product fall below minimum storage
Super iron plus	15°F	temperature or product may separate.
Super blend plus	15°F	Do not let product fall below minimum storage temperature or product may separate.

With all chelates, avoid freezing if at all possible. Separation may be a particular problem with the 3% calcium and the 2.5% magnesium chelates.

*Reproduced courtesy of Dann Watson, Cornbelt Chemical Co., P.O. Box 410, McCook, Nebraska.

Index

B

Back-siphoning
  avoid, 188, 189
  prevention of, 236
Bacteria, plant diseases and, 34–35, 36, 37
Bactericides, 6, 7, 44
Badgers, 63
Bags, dissolvable, 257
Baits, 254
Band applications, 300, 336, 337
Band applicators, calibration of, 338–40
  nozzle method for, 339–40
  refill method for, 339
Band spraying
  described, 338
  rate, 342
BAT, 9
Bats, 51–52
BDAT, 9
Bean leaf beetles, 28
Bears, 55
Beavers, 63
Beekeepers, notification of, pesticide applications and, 227, 238
Bees, 24
  honey. See Honey bees
  metamorphosis of, 18
Beetles, 23
  beneficial
    ground, 31
    lady, 31
  Colorado potato, resistance of, 248
  control of, 1
  harmful
    bean leaf, 28
    black carpet larva, 28
    Japanese, 28
    saw-toothed grain, 28
    striped blister, 28
  metamorphosis of, 18
Best available technology (BAT), 9
Best demonstrated available technology (BDAT), 9
Best management practices (BMP), 9
Biennials, weeds, 78
Biocides, 6
Biological weed control, 83
Biologicals, for insect control, 33
Bird eggs, environmental concerns for, 3–4
Bird repellents
  with insecticides, 7
  toxicity classifications, signal words for, 484
Birds
  environmental concerns for, 3–4
  fish-eating, 3–4
  migration/nesting seasons of, insecticide applications during, 246
  as vertebrate pests, 59–61

Birth defects statement, on pesticide labels, 200, 201
Black carpet larva, 28
Blackbirds, 60–61
Bleach, chlorine. See Chlorine bleach
Blight
  chestnut, 34
  corn, 34
  fire, 37
  trees, 34
Blood tests, cholinesterase, for pesticide applicators, 213–14, 224–25
Blowers, mist, 279
BMP, 9
Bobcats, 56
Bombs, aerosol, 282
Boom sprayers
  calibration of, 325–35
    ground speed and, 327
    nozzle flow rate and, 326–27
      with installed nozzle tips, 333–34
    sprayed width per nozzle and, 327
  covered, 304
  failures of, 334–35
  low-pressure, 275–76
  nozzle tips for, selection of, 328–30
  precalibration checking of, 330–31
  swath marking with, 334
Booms, for aerial applications, 317
Borers, chewing insects as, 16
Botanical pesticides
  modes of action, 215–16
  poisoning from, signs/symptoms of, 216
Botanicals, 32
Brand names, 140
Bristletails, identification of, 22
Broadcast applications, 336, 337
Broadleaf weeds, 81
  defined, 81
Buffering agents, 261
Bugs
  chinch, 27
  hemiptera, 20, 21, 23, 27
    metamorphosis of, 18
Bulk packaging, guidelines for, 372
Burned leaves, sucking insects and, 17
Burning, for weed control, 82–83
Burns, chemical, first aid for, 219
Business people, risks ranking by, 10
Butterflies, 20, 23
  metamorphosis of, 18

C

CAA, 9
Cabbage loopers, 28
Calculations/formulas, 349–70. See also Conversion tables; Formulations

for figuring capacity of sprayer tanks, 360
for figuring volume of ponds, 355
  desired pesticide concentration and, 356
  for mixing, 350–54. See also Mixing
  for pesticide dilution, 360–61
Calibration. See Equipment calibration
Cancellation/suspension, of pesticide registration, in Canada, 117
Cancer/tumor statements, on pesticide labels, 200–201
Canister-type respirators, 209
Cankered trees, plant diseases and, 34
Cankers
  fire blight, 37
  tree, 34
Capacity, of sprayer tanks, figuring, 360
Carbamates, effect on cholinesterase, 215
Carcinogenicity, warning statements for, 200–201
  determining need for, 200
Carnivores, 55–58
Carriers, vegetable oils as, 259, 260
Cartridge respirators, 209
Cattle spray, preparation of, from wettable powders, 352
Caution. See also Signal words
  meaning of, on pesticide labels, 142
  toxicity levels and, 198
CDL, for hazardous material transportation, 372
Centipedes, 21, 25
Centrifugal pumps, 285–86
CEQ, 9
CERCLA, 9, 104
Certification, of pesticide applicators, 97
CFSA, 9
Charcoal, activated, for deactivating residual herbicides, 385–87
Chemical emergency preparedness program, of EPA, hotlines for, 101
Chemical hazards, statements about, on pesticide labels, 144
Chemical Manufacturers Association (CMA), 9
Chemical names. See also Common names
  of pesticides, 141
Chemical Referral Center, telephone numbers for, 446
Chemical resistance, reducing, 248–49
Chemical transfer
  dripless, 308
  volumetric, pressure-free, 308
Chemical Transportation Emergency Center (CHEMTREC), 9, 383–84
  emergency hot line for, 446

Crops
defined, 392
weed control rotation in, 83
Cross contamination, of pesticides,
124–25
liability concerning, 125
Crows, 59, 60
CRP, 11
CSREES, 11
Cubic measures, conversion tables for,
367
CWA, 11, 104

D

Damage, form pesticide applicator
noise, liability and, 124
Danger. *See also* Signal words
meaning on pesticide labels, 142
toxicity levels and, 198
Danger-poison, meaning on pesticide
labels, 142
Days-to-harvest intervals, 191
Dealers, of pesticides, as commercial
vs. private applicators, 110–11
Death of honey bees, liability and, 124
Decontamination, 378–87
of clothing, 381–83
deactivating residual herbicides,
385–87
of pesticides spilled during transportation, 383–85
of protective equipment, 381
of spray equipment, 378–80
of spray personnel, 383
alcohol for, 383
Decontamination sites, under Worker
Protection Standard for
Agricultural Pesticides, 106
Deer, 58–59
Deer mice, 52
Defoliants, 6, 7
nonhazardous to honey bees, 241–42
stability of, 489
Defoliators, chewing insects as, 16
Degradation
controlling, 265, 267
studies of new pesticides, 138
Delphinium, 81
Denaturation, of pesticides, in Canada,
119
Density, of liquids, nozzle flow rate
and, 326–27
Department of Agriculture, U.S. *See*
United States Department of
Agriculture (USDA)
Department of Labor (DOL), 11
Department of the Interior (DOI), 12
pesticides in groundwater and, 102
Department of Transportation (DOT),
11
information, on MSDS, 149

regulation of hazardous materials by,
112–13
transportation of pesticides/
hazardous materials, 371, 372
Dermal exposure, 203–7
prevention of, 205–7
Dermal penetration
factors affecting, 203–4
rates of, 204
Desiccants, 6, 7
nonhazardous to honey bees,
241–42
stability of, 489
Detention, of pesticides, in Canada, 120
Deterioration, of pesticides, signs of,
375
DF. *See* Flowables, dry
Diaphragm pumps, 289, 290
Digestive systems, of insects, 15
Dilution tables, for pesticides, 360–61
Diplopoda, 21, 25
Dipstick gauges, for drums, 364
Diptera, 20, 24
metamorphosis of, 18
Direct injection systems, 306–7
Directed spraying, 300
Dirt, on equipment, 267
Disease transmittal, low rates in bats
of, 51–52
Diseases, plant, 37–49
control of, 44–49
defined, 37–39
development of, 39–41
identification of, 41–43
Dispersal equipment, granular, for aer-
ial applications, 319–20
Displaying, pesticides, in Canada, 119
Disposal
of pesticide waste/containers,
387–91
triple rinse before, 387–89
of pesticides/hazardous materials,
236
telephone numbers on, 446
regulations on, 98
Distribution, of pesticides in Canada,
119
Distributors, name/address of, on pesticide labels, 141
Dogs, 56–57
DOI. *See* Department of the Interior
(DOI)
Dosage, conversion tables for, 366–67
DOT. *See* Department of
Transportation (DOT)
Downy mildew, control in grapes of, 1
Dragonflies, 21
metamorphosis of, 18
Drift, 271–74, 272–74
controlling, 266
during aerial applications, 226–27

factors affecting, 237–38, 272–74
increased, from vegetable oils, 260
liability on, 123
spray
management of, in aerial applications, 323
minimizing, 274
vapor, 272, 273–74
Driver licenses, commercial, 372
Drought, plant diseases and, 41
Drugs, affecting people/animals,
FIFRA and, 100
Drums, dipstick gauges for, 364
Dry formulations, 254–57, 351
baits, 254
conversion tables for, 363
dusts, 254–55
granules, 255
water-dispersible, 256
microencapsulated, 256
pellets, 256
powders
soluble, 256
wettable, 256–57
Ducks, 59–60
Dusters, 280
hand-operated, 282
Dusts, 254–55
mixing, 353
signs of deterioration in, 375
Dutch elm disease, fungus and, 34, 35

E

Eagles, 60, 61
Ecosystems, wildlife and, 51
Efficacy, of pesticides, water pH and,
268
Eggs
environmental concerns and, 3–4
insect development and, 18
Egypt, ancient, agricultural chemicals
in, 1
Elk, 58–59
Elm disease, Dutch, 34, 35
Emergency incidents, chemical, assistance in, 383–84
Emergency Planning and Community
Right to Know Act (EPCRA), 11
Emergency planning notification, for
hazardous substances, 100–101
Emergency release notification, for
hazardous substances, 101
Employees. *See* Workers
Employer duties
on personal protective equipment,
107
on pesticide handlers, 105–7
on respirators, 209
on workers
notification of, 107–8
restricted entry intervals and, 107

House plants, diseases, 34–35, 36
House sparrows, 60–61
Houseflies. *See* Flies
Humans
  drugs affecting, FIFRA and, 100
  health concerns, 4–5
  pesticide accumulation in, 247–48
Humidity, drift control and, 273
Hunger, world, *vs.* world food supply,
    2
Hydrated lime, for reducing hazards,
    from small pesticide spills, 385
Hymenoptera, 20, 24
  metamorphosis of, 18
Hypochlorites, for reducing hazards,
    from small pesticide spills, 385

I
IGRs, 33
Importing, of pesticides, in Canada,
    120–21
Incompatibility, of pesticides, 269
  dealing with, 265
Incubation, plant pathology and, 39
Indemnities, registration of pesticides
    and, 97–98
Indicator plants, plant diseases and, 43
Infections, plant pathology and, 39
Information systems, geographic, 306
Ingestion
  of fungicides, toxicity of, 196
  of herbicides, toxicity of, 195
  of insecticides, toxicity of, 193
  of poisons
    corrosive, signs/symptoms of, 218
    first aid for, 218–19
Ingredients. *See also* Hazardous
    materials
  statements about, on pesticide labels,
    141
Inhalation, toxicity on, 192
Injection, subsurface, 304–5
Injection systems, direct, 306–7
Injectors, soil, 281
Injuries
  to crops, from pesticides, liability
    and, 123
  to persons, from pesticides, liability
    and, 123
Inoculation, plant pathology and, 39
Inorganic chemicals, list of, 32
Inorganic insecticides, 32
Insect control. *See also* Insecticides
  beneficial insects and, 31
  of chewing insects, 16, 17
  identification and, 27, 28
  methods of, 29–30, 31
  outbreaks and, 26
  pests and, 13–14
  steps in, 26, 27, 28, 29
  of sucking insects, 16–17

Insect growth regulators (IGRs), 33
Insecticides. *See also* Federal
    Insecticide, Fungicide, and
    Rodenticide Act (FIFRA);
    Insect control
  applications of, during bird migra-
    tion/nesting seasons, 246
  with bird repellents, 7
  chemicals in, 30
  classification of, 31–33
  defined, 7
  first synthetic organic, 1
  granular applicators for, 280–81
  hazards of, to wildlife, 244
  inorganic, 32
  insect resistance to, 248
  with nematicides, 7
  nonhazardous to honey bees, 241
  organic, 32–33
  poisoning symptoms of, in honey
    bees, 243
  1950s use of, 2
  stability of, 485–88
    water pH and, 268
  for target species, 6, 7
  toxicity classifications and, signal
    words for, 469–72
  toxicity of
    with skin contact, 194
    when ingested, 193
Insectivores
  bats, 51–52
  skunks, 58
Insects, 13–14, 13–33
  basic anatomy, 14–15
  beneficial, 31, 238–39
  centipedes/millipedes, 21, 25
  circulatory system, 15
  classification of, 14
  digestive system, 15
  harmful, 27, 28
  identification of, 19–21, 22–25
  known species of, 13
  metamorphosis in, 18–19, 22, 23, 24
  mouth parts, 15–16
    chewing damage, 16, 17
    sucking damage, 16–17
  nervous system, 15
  outbreaks of, 26
  as pests, 13–14
  pheromone traps for, 396
  relatives of, 21
  resistance of, to insecticides, 248
  respiratory system, 15
  snails/slugs, 21, 25, 26
Inspection
  of pesticide establishments, 97
  of pesticide handling equipment, 106
Instar, 18
Instructions, on pesticide labels,
    144–45

Insurance. *See also* Liability
  for pesticide applicators, 125–26
  accident procedures, 125–26
Integrated pest management (IPM),
    11, 236–37, 392–96
  concept/principle of, 392
  effective, 395
  evolution of, 392–94
  goals of, 394
  implementation of, 395–96
  who benefits from, 394
Interior, U.S. Department of. *See*
    Department of the Interior (DOI)
In-the-row applications, 336, 337
Invert emulsions, 253
IPM. *See* Integrated pest management
    (IPM)
Irish potato famine, 34
Irrigation systems, for pesticide appli-
    cations. *See* Chemigation
Irrigation tasks, during restricted-entry
    intervals, 107
Isolation, plant diseases and, 43
Isoptera, 20, 22
  metamorphosis of, 18

J
Jackrabbits, 54
Japanese beetles, harmful, 28

K
Katydids, 23
  metamorphosis of, 18

L
Label review, for new pesticides, 138,
    139
Labeling, 140
  in Canada, 118
  as restricted use, 111
Labels, 136–84
  contents listed on, 141
  described, 140
  EPA registration numbers on, 141
  establishment numbers on, 141
  flashpoint warnings on, 376
  information on, 136, 140–49
    about adjuvants, 261–62
  instructions on, 144–45
  name/address of manufacturers on,
    141
  pesticide compatibility on, 270
  pesticide type on, 141
  registration of, 138, 139
  respirator information on, 207–9
  signal words on, 142
    precautionary statements follow-
      ing, 142
  toxicity levels and, 198
  specimen, 151–84
  statements on
    cancer/tumor, 200–201

chlorine bleach with liquid
fertilizers, 380
dry formulations, 351
dust, 353
liquid formulations, 350–51
percentage, 352–53
of pesticides, potentiation and, 270
problems during, 265
safety during, 187–89
square feet calculations and, 354
Moles, 52
Molluscicides
for snail control, 25
for target species, 6, 7
Monitoring, of handlers of pesticides,
employer duties and, 106
Monitors, electronic, for sprayers,
305–6
Moose, 58–59
Mosquitoes, 24, 27
metamorphosis of, 18
Moths, 20, 23
armyworm, 28
metamorphosis of, 18
Motor vehicles, commercial, defined,
371
Mountain lions, 57
Mouth-to-mouth/mouth-to-nose artificial
respiration, for adults, 220–21
MSDS. *See* Material safety data sheets
(MSDS)
Muskrats, 63
Mycoplasmas, plant diseases and,
34–35, 36, 38

N

NAPIAP, 11
Narrowleafed weeds, 81
defined, 81
NAS, 11
NASDA, 11
National Academy of Sciences
(NAS), 11
National Agricultural Pesticide
Impact Assessment Program
(NAPIAP), 11
National Association of State
Departments of Agriculture
(NASDA), 11
National Environmental Protection
Agency (NEPA). *See*
Environmental Protection
Agency (EPA)
National Institute for Occupational
Safety and Health
(NIOSH), 11
National Institutes of Health
(NIH), 11
National Permit Discharge
Elimination System
(NPDES), 11

National Pesticide Information
Retrieval System (NPIRS), 11
National Pesticides
Telecommunications Network
(NPTN), 384
telephone number for, 446
National Response Center, telephone
number for, 446
Natural Resources Conservation
Service (NRCS), 11
Needle dropping, in conifers, 42
Nematicides
defined, 44
with insecticides, 7
stability of, 485–88
for target species, 6, 7
toxicity classifications and, signal
words for, 484
Nematodes, plant diseases and, 34–35,
36, 38–39
NEPA, 11
Nervous system, of insects, 15
Nesting seasons, of birds, insecticide
applications during, 246
Neuroptera, 21
defined, 18
NIH, 11
NIOSH, 11
Nitrogen, fixing nodules, 42
Nodules, nitrogen fixing, 42
NOEL, 11
Noise, damage from, by pesticide
applicators, liability and, 124
Nonaccumulative pesticides, 248
Nonbulk packaging, guidelines for,
372
No-observed-effect-level (NOEL), 11
Norway rats, 54
NOS, 372
Not otherwise specified (NOS), 372
Notice of proposed rulemaking
(NPRM), 11
Notification
of beekeepers
about pesticide applications, 238
before pesticide applications,
124, 227
of clients, about restricted entry into
treated fields, 190
emergency planning, for hazardous
substances, 100–101
emergency release, for hazardous
substances, 101
of pesticide spills, 384
of pesticide storage, 377
of physicians, on pesticides used,
224
requirements under Worker
Protection Standard for
Agricultural Pesticides,
105–6

Nozzle tips, 298
installed, spray rate with, determining, 333–34
selection of, for boom sprayers,
328–30
Nozzles, 292–300
for aerial applications, 317–18
broadcast, 296–98
configurations of, for band/directed
spraying, 300, 301
design changes in, 292–93
drift control and, 272, 273
flat-fan
even, 294, 295
flooding, 296, 329–30
off-center, 295
regular, 293–95
flow rate of, 326–27
determination of, 329, 332
for high-pressure sprayers, 277
hollow-cone, 296
disc and core type, 330, 331
importance of, 292
performance of, 292
quick attach, 298, 300
sprayed width of, 327
determining effective, 329
tip numbers on, 300, 301
twin orifice, 296
NPDES, 11
NPIRS, 11
NPRM, 11
NPTN, 384
telephone number for, 446
NRCS, 11
Nuisances, attractive, liability on, 124
Numbers, on nozzle tips, 300
Nurseries, Worker Protection Standard
for Agricultural Pesticides
and, 105–8
Nutria, 63
Nutrient deficiencies, plant diseases
and, 34–35
Nymphs, 18

O

Occupational Safety and Health
Administration (NIOSH),
Hazard Communications
Standard and, 112
Occupational Safety and Health
Administration (OSHA), 11
OCM, 11
Odonata, 21
metamorphosis of, 18
ODW, 11
OES, 11
Office of Compliance Monitoring
(OCM), 11
Office of Drinking Water (ODW), 11
Office of Endangered Species (OES), 11

labeling of. *See* Labeling; Labels
laws, 94–122
leachability of, 235
liability. *See* Liability
marketing of, regulation of, 103
measurement perspectives on, 12
nonaccumulative, 248
notification of physicians on, 224
out-of-date sources, warnings about, 150
pest competition and, 3
poisoning from. *See* Poisoning
products defined as, under Federal Insecticide, Fungicide, and Rodenticide Act, 99–100
public perception of, 8–9, 10
record keeping. *See* Record keeping
registered
classification of, 95, 96
reregistration of, 96
registration of. *See* Registration, of pesticides restricted use. *See* Restricted use pesticides (RUPs)
1960s prescription type, 2
safety. *See* Safety
spills of, emergency release notification for, 101
stability of, 268, 485–91
state authority and, 99
stop sale, use, or removal order for, 97
survey risk ratings of, 10
synthetic, problems with dependence on, 393
terminology of, 5–7
tests for effectiveness/safety of, 136–38
use of
penalties for violations of, 97
record keeping of. *See* Record keeping
users of, 7–8
water concerns and, 3–4
wildlife concerns and, 3–4
world hunger concerns and, 2
Pheromone traps, for insects, 396
Photodeactivation, residual fungicides and, 45
Physical hazards, statements about, on pesticide labels, 144
Phytotoxicity, 243
pesticide compatibility and, 270
Pigeons, 60–61
Pilots, safety of
aerial applications and, 226
exposed to organophosphates, 227
PIMS. *See* Pesticide Incident Monitoring System (PIMS)
Piscicides, 6, 7
Piston pumps, 288–89

Placarding, of hazardous materials, requirements for, 372–73
Plant disease chemicals, toxicity classifications and, signal words for, 480–82
Plant diseases, 34–49
bacterial, 34–35, 36, 37
biological abnormalities and, 35
causes of, 34–35, 36
control of, 44–46
effects of, 34–35
fungal, 34–35, 36, 37
fungicides and, 46–49
inorganic, 46
nonsystemic, 47–48
organic, 46–49
systemic, 48–49
identification of, 41–43
mycoplasmas in, 34–35, 36, 38
nematodes in, 34–35, 36, 38–39
pathology of, 39–41
principles of, 35–37
three conditions of, 40–41
viroids in, 34–35, 36
viruses in, 34–35, 36, 38
weed injury and, 74–75
Plant growth, sucking insects and, 16
Plant growth regulators
stability of, 489
toxicity classifications and, signal words for, 473–79
Plant Quarantine Act (PQA), 11
Plants, danger of pesticides to, 239, 243
Plumbing, for aerial application equipment, 317
Pocket gophers, 52
Poison. *See also* Signal words
classes of, 113
first insect stomach, 1
toxicity levels, 198
vs. remedy quote, 1
Poison Control Centers, 224
Poisoning
accidental, 202
acute, 199
antidotes for, 223–24
from botanical pesticides, symptoms of, 216
first aid for, 217–19
in eyes, 218
general, 217
inducing vomiting in, 219
inhaled, 218
on skin, 218
swallowed, 218–19
from fumigants, symptoms of, 216
gas, first aid for, 223
medical procedures for, on pesticide labels, 143
symptoms of, 210–17
Pollutants, pesticides as, 231

Pollution, air
environmental concerns and, 3–4
pesticide drift and, 237–38
Ponds, volume of
formulas for, 355
pesticide concentration and, 356
POP, 372
Porcupines, 52, 53
Positioning systems, global, 306
in aerial applications, 322
Potato beetles, Colorado, resistance of, 248
Potato leafhoppers, 27
Potatoes
insect control for, 1
Irish famine, 34
yield increases in, 3
Potentially responsible party (PRP), 11
Potentiation, 270
Powders
soluble, 256
wettable. *See* Wettable powders
PPE. *See* Personal protective equipment (PPE)
PQA, 11
Prairie dogs, 53–54
Precautionary statements
following signal words, on pesticide labels, 142
on pesticide labels, 143
Predacides, 6, 7
Predicides, toxicity classifications and, signal words for, 483
Preharvest intervals, statements about, on pesticide labels, 145, 146–47
Pressure, spraying, drift and, 272
Pressure gauges, 291
Pressure regulators, 290
Prevailing winds, plant disease patterns and, 41
Prevention, of weeds, 82
Product development, of new pesticides, 139
Product identification, on MSDS, 145
Pronotum, defined, 14
Protection, plant diseases and, 44
Protective equipment
decontamination of, 381
personal. *See* Personal protective equipment (PPE)
Proventriculus, defined, 15
Provinces, Canadian, pesticide regulations in, 122
PRP, 11
P&TCN, 11
Public perception
environmental concerns and, 3–6
of pesticides, 3–5, 6, 7–8, 10

Subject Index

515

Pumps
  for aerial applications, 316
  for sprayers, 284–90
    centrifugal, 285–86
    comparison of, 286
    diaphragm, 289, 290
    piston, 288–89
    roller, 287–88
    selection of, 284–85
    turbine, 287
Pupae, 18
Puparium, 18
Pyrethrins, synthetic, 217
Pyrethroids, 217

R

Rabbits, 53, 54
Raccoons, 58
Ranunculus, 81
Rates
  of applications, drift and, 237
  conversion tables for, 366–67
  granular, conversion tables for, 365
Rats, 55
  vs. mice, 54
Ravens, 59
RCRA. *See* Resource Conservation and
    Recovery Act (RCRA)
Reactivity data, on MSDS, 148
Recommended maximum containment
    levels (RMCL), 12
Record keeping, 126–35
  of applications, of restricted use pes-
    ticides, 110
  of pesticide applications, 126–28
    methods of, 127–28
    need for, 127
    sample forms for, 128–35
  of pesticide storage, 374
  of pesticides, in Canada, 117–18
Refusal, of pesticide registration, in
    Canada, 117
Registration
  of pesticide-producing establish-
    ments, 96
  of pesticides
    by EPA, 95
    exemption from, in Canada, 115
    indemnities and, 97–98
Registration numbers, for pesticides,
    from EPA, 141
Regulation, of pesticide marketing, 103
Regulators, pressure, 290
Release forms, toxic chemical, 101
Remedy, *vs.* poison quote, 1
Removal, of pesticides, stop order for,
    97
Renewal, of pesticide registration, in
    Canada, 116–17
Repair, of pesticide handling equip-
    ment, 106

Repellents, 6, 7
  bird, toxicity classifications and, sig-
    nal words for, 484
  stability of, 491
Reproductive failure, environmental
    concerns and, 3–4
Reproductive toxicity, warning state-
    ments for, 200, 201
  determining need for, 200
Reregistration, of registered pesticides,
    96
Residue
  on equipment, 267
  excessive, pesticide compatibility
    and, 270
Residue tests, of new pesticides, 138
Residue tolerances, in foods, safe lev-
    els of, 94–95
Resistance
  chemical, reducing, 248–49
  of Colorado potato beetles, 248
  to herbicides, 248
  to insecticides, 248
  pest, 248–49, 393
  plant diseases and, 44
Resource Conservation and Recovery
    Act (RCRA), 11, 389, 391
  Hazardous and Solid Wastes
    Amendments to, 389, 391
  regulation of hazardous wastes
    under, 104
Respiration, artificial. *See* Artificial
    respiration
Respirators
  canister-type, 209
  care/maintenance of, 210, 211
  cartridge, 209
  decontamination of, 381
  employer duties on, 209
  importance of, 207
  information about, on pesticide
    labels, 207–9
  self-contained breathing, 210
  supplied-air, 210
  types of, 209–10
Respiratory exposure, 207–10
  prevention of, 207–10
Respiratory systems, of insects, 15
Restricted entry, into treated fields,
    189–91
  notification of clients about, 190
  warnings to workers about, 190–91
Restricted entry statements, on pesti-
    cide labels, 145
Restricted use pesticides (RUPs), 96,
    111
  applications of
    by noncertified applicators,
      responsibilities of certified
      applicators with, 109–11
    record keeping, 110, 126–28

applicators of, 96, 97
  labeling of, 111, 140
Restricted-entry intervals (REIs)
  determination of, 107
  irrigation tasks during, 107
  limited contact tasks during, 107
  workers and, employer duties on, 107
Resurgence, pest, 393
Rhizomes, of perennial weeds, 80
Rice weevils, 28
Rinsate, collected, 381
Rinsing, before disposal of containers,
    387–89
Risk ratings, of pesticides, 10
RMCL. *See* Recommended maximum
    containment levels (RMCL)
Roaches, 27
Robigus, Ancient Rome and, 34
Rodenticides. *See also* Federal
    Insecticide, Fungicide, and
    Rodenticide Act (FIFRA)
  toxicity classifications and, signal
    words for, 483
Roller pumps, 287–88
Rome
  ancient practice of, 1
  worship of Robigus in, 34
Roof rats, 54
Root feeders, chewing insects as, 16
Roots
  of perennial weeds, 80
  in plant diseases, 42–43
Rope wick applicators, 302
Rotation of crops, for weed control, 83
Rotted fruits, plant diseases and, 34
Row fumigation spraying, rate of, 342
RUPs. *See* Restricted use pesticides
    (RUPs)
Rust, on equipment, 267

S

SAB. *See* Science Advisory Board
    (SAB)
Safe Drinking Water Act (SDWA), 12,
    102, 104
Safety, 185–230. *See also* Material safety
    data sheets (MSDS); Personal
    protective equipment (PPE)
  in aerial applications. *See* Aerial
    applications, safety in
  in applications, 189, 190
  checklist, 229–30
  in chemigation, 235, 236
  in greenhouse operations, 228–29
  in handling/mixing pesticides,
    187–89
  of pesticide handling equipment, 106
  in pesticide selection, 186–87
  restricted entry into treated fields,
    189–91
  toxicity of pesticides, 191–201

Common Pesticide Classifications, Names, and Trade Names

A

Aaccess Penetrator, 493
AASTAR ®, 455
Aaterra ®, 462, 480
AAtrex ®, 80, 88, 195, 234, 241, 447, 462, 473, 493
Abamectin, 462, 468, 469
Abate ®, 241, 462, 472, 493
Acaraben ®, 241, 455, 485
Acaramate ®, 485
Acarastop, 485
Accelerate ®, 462, 476
Accent ®, 379, 462, 478
Access ®, 90, 452, 462, 478
Accord ®, 462, 477
Acephate, 193, 194, 240, 245, 268, 466, 469, 487
Acetamide, 446
Acetanilides, 86
Acetellic ®, 462, 472
Acetic phenoxy compounds, 87
Acetochlor, 446, 465, 467, 473
    with atrazine, 467
Acids, 376
Acifluorfen, 87, 195, 463, 473
    with bentazon, 464
    with imazaquin, 467
Aconite, 1
Acrex ®, 485
Acritet 34-66 ®, 455
Acrobe ®, 462, 469
Acrolein, 446
Acrylonitrile, 455
Actellic ®, 193, 194, 485, 493
Acti-Aid, 489
Acti-Aid Fumigant ®, 455
Actidione ®, 48
Acylalanines, 48
Adhere ®, 493
Adios ®, 462, 469
Aero HCN ®, 455
Afilene ®, 485
AF-SeafloZ-100 ®, 447, 454
Afugan ®, 241
Agri-Mek ®, 447, 462, 469
Agri-Mycin 17 ®, 462, 481
Agritox ®, 241
Alachlor, 80, 88, 195, 200, 234, 241, 446, 463, 465, 466, 473, 490
    with atrazine, 465, 473
    with glyphosate, 463, 473
    with trifluralin, 464, 473
Alanap ®, 88, 242, 489, 493
Alanap-L ®, 462, 478
Alar ®, 489
Ala-Scept ®, 446
Aldicarb, 32, 193, 194, 234, 240, 245, 446, 467, 469, 484

Aldrin, 2, 32, 200
Alfa ®, 462, 481
Alfatox ®, 493
Aliette ®, 48, 196, 462, 481
Aliphatic acids, halogenated, 87
Aliphatic nitrogenous compounds, 47
Allethrin, 241, 485, 488
Ally ®, 379, 462, 478
Allyl alcohol, 455
Allyl Alcohol Weed Seed Killer ®, 455
Alpha-chlorohydrin, 455
Altosid ®, 462, 471
Altosid SR-10 ®, 493
Altratol ®, 447
Aluminum phosphide, 69, 245, 372, 447, 466
Aluminum phosphide (Phosphine), 484
Amaze ®, 485
Amber ®, 462, 479
Ambush ®, 33, 193, 194, 234, 452, 462, 472, 485, 493
Amdro ®, 462, 471
Amercoat Biocide 635 ®, 454
Amercoat ®, 454
Ametryn, 234, 464, 473
Amiben ®, 241, 462, 474, 493
Amino acids, 87
Aminopyridine, 462, 484
Amino-triazole ®, 88
Amitraz, 33, 200, 447, 465, 466, 469
Amitrol T ®, 447, 462, 473, 493
Amitrole, 88, 200, 241, 447, 462, 473
Amizol ®, 447
Ammate ®, 241
Ammo 2.5 EC ®, 493
Ammo ®, 193, 194, 448, 462, 470, 485
Ammonia, 380
Ammonium nitrate, 380
Ammonium salts, 376
Ammonium sulfate, 380
AMS, 86, 241
Ancymidol, 462, 473
Anilazine, 196, 242, 489
Ansar 8100 ®, 462, 476
Ansar ®, 234, 462, 478
Antor ®, 493
Apam ®, 462
Apex ®, 462, 471
Apollo ®, 462, 470
Apollo SC ®, 448
Appex ®, 240
Apron ®, 196, 462, 481
Aqualin ®, 446
Aquathol ®, 88, 462, 476
Aracide, 485
Aramite, 485
Arbotect ®, 462, 482
A-Rest ®, 462, 473

Argold ®, 462, 474
Arosurf-MSF ®, 493
Arrosolo ®, 90, 462, 478
Arsenal ®, 87, 462, 477, 493
Arsenic, 1
Arsenic acid, 200, 447, 464, 465, 473
Arsenic pentoxide, 447
Arsenicals, 87, 240
Arsonate ®, 462, 478
Artaban, 485
Asana ®, 193, 194, 462, 470
Asana XL ®, 450
Assert ®, 87, 462, 477
Assure II ®, 462, 479
as-Triazines (asymmetrical), 88
Asulam, 234, 462, 473
Asulox ®, 234, 462, 473
Atrabute II, 493
Atraol ®, 447
Atrazine, 80, 88, 195, 234, 241, 447, 462, 489, 493
    with acetochlor, 467
    with alachlor, 465, 473
    with bentazon, 465, 466, 473
    with bromoxynil, 473, 474
    with butylate, 467, 473
    with cyanazine, 464, 473
    with dicamba, 465, 473
    with dimethenamid, 465, 473
    with imazethapyr, 463, 473
    with metolachlor, 463, 473
    with propachlor, 479
Atrazine ®, 447
    with Buctril ®, 473, 474
    and Ramrod ®, 479
Atrazine/Bexton, and Bexton 4F, 494
Augus Hot Rod ®, 453
Avadex BW ®, 462, 479
Avadex ®, 88, 455, 493
Avenge ®, 462, 475, 489, 494
Avermectin, 447
Avid ®, 462, 469
Avitrol ®, 447, 462, 484
Azinphos-methyl, 32, 193, 194, 234, 240, 245, 268, 447, 465, 469, 487
Aziprotryne, 490
Azodrin ®, 234, 240, 455, 462, 471, 485
Azofene ®, 462, 472

B

Baam ®, 33
Bacillus thuringiensis, 241, 463, 464, 465, 466, 467, 468, 469, 485, 486
Bacillus thuringiensis berliner, 462, 463, 468, 469
Bacillus thuringiensis spores, 33
Bactec Bernan ®, 462, 469

D-Tetramethrin, 487
D-Trans Allethrin, 485
Dual II ®, 477
Dual ®, 86, 195, 234, 464, 477, 496
 with Broadstrike ®, 90, 476
Duraphos ®, 451
Durathion ®, 451
Dura-Treat ®, 452
Dursban ®, 32, 193, 194, 240, 372,
 373, 486, 496
Du-Ter ®, 48, 454
Dyanap, 496
Dybar ®, 88
Dyfonate ®, 193, 194, 450, 464,
 471, 496
Dylox ®, 193, 194, 241, 464, 472,
 486, 496
Dylox/MSR ®, 451
Dyrene ®, 196, 242, 489
D-z-n Diazinon ®, 464, 470

E

Eagle Iron, 496
Eagle ®, 481
Easy Spot, 496
EBDC, with metalaxyl, 466
Echo 75 ®, 448
Eclipse ®, 464, 471
EDB, 200
Edge ®, 464
ELF ®, 464, 470
Elocron, 486
Embark ®, 464, 477, 496
Encapsulated methylparathion, 488
Endosulfan, 193, 194, 240, 245, 465,
 466, 467, 470, 488
Endothal ®, 88, 195, 242, 476
Endothall, 195, 242, 462, 465, 476
Endrin, 2, 32, 69, 234, 449, 486
Endrin 1.6 EC ®, 449
Endurance ®, 464, 478
Entex ®, 486
Envert 171, 496
Envert 171 ®, 90, 464
Envery 171 ®, 475
Epibloc ®, 455
EPN, 240, 455, 486, 496
Eptam ®, 88, 195, 234, 242, 464,
 476, 497
EPTC, 88, 195, 234, 242, 449, 464, 476
Eradicane ®, 464, 476, 497
Ergostim, 489
Esbiol, 486
Escort ®, 379, 464, 478
Esfenvalerate, 193, 194, 462, 470
Esteron 76 BE, 6# butyl ester, 503
Esteron BK ®, 497
Esteron Brush Killer, 2,4-D/2,4,5-T
 mixture, 503
Esteron 99C, 4# L. V. ester, 502

Esterone 245 ®, 2,4-D/2,4,5-T
 mixture, 503
Ethalfluralin, 195, 467, 476
Ethephon, 463, 464, 476, 489
Ethidimuron, 490
Ethion, 241, 449, 464, 466, 470,
 486, 497
Ethion 8 ®, 449
ETHION-EC ®, 241
Ethofumisate, 466, 476
Ethoprop, 193, 194, 234, 240, 449,
 465, 470
 with disulfoton, 465
Ethoprophos, 487
Ethrel ®, 464, 476, 489
Ethyl parathion, 193, 194
Ethylene dibromide, 455
Etridiazole, 196, 462, 465, 466, 467,
 468, 480
 with disulfoton, with PCNB, 467
 with PCNB, 480
Evik ®, 234, 464, 473
Exhalt 800, 497
Express ®, 379, 464, 479
Extend ®, 497
Extrazine II DF ®, 90
Extrazine II 4L ®, 90
Extrazine II ®, 464, 473
Extrazine ®, 447, 448, 497

F

Facet ®, 464, 479
Fallow Master ®, 90, 464, 475
Famphur, 240
Fanamiral, 196
Faneron, 490
Fargo ®, 88, 497
Far-Go ®, 464, 479
Farm Bin Spray, 497
Fastac, 485
FC4 ®, 497
FC15 ®, 497
FC30 ®, 497
Fenamine, 497
Fenaminosulf, 489
Fenamiphos, 234, 450, 465, 484, 487
Fenarimol, 467, 480
Fenatrol, 497
Fenbutatin, 33
Fenbutatin-oxide, 468
Fenbutatin-oxide (hexakis), 470
Fenfuram, 489
Fenithrothion, 488
Fenitrothion, 240, 450, 470, 486
Fenoxaprop-ethyl, 465, 468, 476
 with fluazifop-P, 464
Fenoxaprop-P, with 2,4-D, with
 MCPA, 467, 475
Fenoxaprop-P-ethyl, 463, 466, 476
 with MCPA, 464

Fenoxycarb, 464, 465, 471
Fenoxyprop-ethyl, 195
Fenpropathrin, 450, 464, 471
Fensulfothion, 240, 455, 486
Fenthion, 193, 194, 240, 245, 450, 462,
 471, 485, 486
Fenthion 4E, 497
Fenuron, 88
Fenvalerate, 33, 245, 450, 462, 471, 488
Ferbam, 463, 481
Fermate ®, 47
Ficam ®, 464, 469
Ficam (R)/Turcam ®, 486
Finesse ®, 90, 379, 464, 474
Fish-Tox ®, 453
Flamprop Isomer, 490
Flamprop (Racemate), 490
Florel ®, 489
Floumeturon, 463
Fluazifop-butyl, 195, 476
Fluazifop-P
 with fenoxaprop-ethyl, 464
 with fomesafen, 467, 468
Fluazifop-P-butyl, 464
 with fomesafen, 476
Flucythrinate, 455, 464, 471
Flumetralin, 466, 476
Flumetsulam, 463, 476
 with clopyralid, 463, 474, 476
 with metolachlor, 476
 with MSMA, 463
 with trifluralin, 476
Flumiclorac pentyl ester, 466, 476
Fluometuron, 242, 465, 476
Fluoroacetamide, 455
Fluoroacetamide/1081 ®, 455
Fluorodifen, 242
Fluridone, 195, 467, 476, 490
Flusilazole, 196, 466, 481
Fluvalinate, 465, 471, 487
Fluxofenim, 463, 476
Folcisteine, 489
Folex ®, 464, 479
Folicur ®, 464, 482
Folithion, 486
Folpet, 196, 242
Fomark, 497
Fomesafen, 466, 476
 with fluazifop-P, 467, 468
 with fluazifop-p-butyl, 476
Fomothion, 486
Fonofos, 194, 450, 464, 471
 with pebulate, 464
Fonophos, 193
Force ®, 453, 464, 472
Fore ®, 242, 464, 481
Forepen-50 ®, 452
Formaldehyde, 200
Formamidines, 33
Formetanate, 240, 485

Classification Index